Biographical Information.

Myron Evans is currently the only scientist on the British Civil List and is the third chemical physicist to be appointed in British history, the other two being John Dalton and Michael Faraday. He is currently the Director of the Alpha Foundation's Institute for Advanced Studies (A.I.A.S.) and the author or editor of some seven hundred scientific papers and forty monographs. These are to be collected by the Library of Congress and a list of these publications has been collected by the Nils Bohr Library of the American Institute of Physics. He was educated at the then University College of Wales Aberystwyth and is sometime Junior Research Fellow of Wolfson College Oxford. Numerous honours include the Harrison Memorial Prize and Meldola Medal of the Royal Society of Chemistry. He has made scientific inferences and discoveries, including the explanation of the far infra red spectra of materials; the first computer simulation of the far infra red; the discovery of the gamma or boson peak of the far infra red; the development of non-equilibrium and non-linear molecular dynamics simulation methods at Aberystwyth, IBM and elsewhere; the development of fundamental methods in statistical mechanics; the application of computer simulation to non-linear optics; the inference of the fundamental Evans spin field $B(3)$ of electromagnetism at Cornell in 1991; and the development of Einstein Cartan Evans unified field theory as described for the first time in this book.

I dedicate this book to my dear wife Larisa.

GENERALLY COVARIANT UNIFIED FIELD THEORY

THE GEOMETRIZATION OF PHYSICS

VOLUME VI

Myron W. Evans

Published 2009 by abramis

www.abramis.com

ISBN 978-1-84549-384-4

© Myron W. Evans 2009

All rights reserved

This book is copyright. Subject to statutory exception and to provisions of relevant collective licensing agreements, no part of this publication may be reproduced, stored in a retrieval system, or transmitted in any form or by any means, without the prior written permission of the author.

Printed and bound in the United Kingdom & USA

This book is sold subject to the conditions that it shall not, by way of trade or otherwise, be lent, re-sold, hired out, or otherwise circulated without the publisher's prior consent in any form of binding or cover other than that which it is published and without a similar condition including this condition being imposed on the subsequent purchaser.

abramis is an imprint of arima publishing

arima publishing
ASK House, Northgate Avenue
Bury St Edmunds, Suffolk IP32 6BB, UK
t: (+44) 01284 700321

www.arimapublishing.com

Contents

1. Spin Connection Resonance in the Bedini Machine
1.1 Introduction
1.2 The Equations of Classical Electrodynamics in General Relativity.
1.3 Systematic Evaluation of Equations for the Bedini Machine.
1.4 Detailed Investigation of the Bedini Machine.
Appendix 1: Reduction of Form Notation to Vector Notation.
Appendix 2: Derivation of the Electric Field in Vector Notation.

2. The Coulomb and Ampere Maxwell Laws in the Friedmann Lemaitre Robertson Walker Metric.
2.1 Introduction.
2.2 The Law of Electrodynamics in the FLRW Metric.
2.3 Discussion.

3. A Critical Evaluation of Standard Model Cosmology with ECE Field Theory.
3.1 Introduction.
3.2 Testing with the Inhomogeneous Field Equation of ECE.
3.3 The Big Bang Cosmological Model.
3.4 The Friedmann Models.
3.5 Recapitulation and Summary.

4. Line Element for a Radiating Electron in a Generally Covariant Unified Field Theory.
4.1 Introduction.
4.2 Change of Polarization of Light due to Gravitation.
4.3 Light Bending in Ricci Flat Spacetime.
4.4 General method for Determining the Line Element of a Radiating Electron.

5. Rank Three Tensors in Gravitation and Electrodynamics.
5.1 Introduction.
5.2 Field Equations in the Base Manifold.
5.3 General Coordinate Transformation and Rotation.
5.4 Canonical Angular Momentum Density.

6. The Fundamental Origin of Curvature and Torsion.

6.1 Introduction.
6.2 Derivation of the Curvature and Torsion Tensors from Commutators of Covariant Derivatives.
6.3 The Jacobi and Bianchi Identities.
6.4 Pure Rotational Limit.

7. A Review of Einstein Cartan Evans (ECE) Field Theory

7.1 Introduction.
7.2 Geometrical Principles
7.3 The Field and Wave Equations of ECE Theory.
7.4 Aharonov Bohm and Phase Effects in ECE Theory.
7.5 Tensor and Vector Laws of Classical Electrodynamics.
7.6 Spin Connection Resonance.
7.7. Effects of Gravitation on Optics and Spectroscopy.
7.8 Radiative Corrections in ECE Theory.
7.9 Summary of Advances made by ECE and Criticism of the Standard Model.
Appendix 1: Homogeneous Maxwell Heaviside Equations.
Appendix 2: The Inhomogeneous Equations.
Appendix 3: Some Examples of Hodge Duals in Minkowski Spacetime.
Appendix 4: Standard Tensorial Formulation of the Homogeneous Maxwell Heaviside Field Equations.
Appendix 5: Illustrating the Meaning of the Connection with Rotation in a Plane.

8. The Incompatibility of the Christoffel Connection with the Bianchi Identity.

8.1 Introduction.
8.2 Development of the Bianchi Identity and its Incompatibility with the Christoffel Connection.
8.3 Interaction of Gravitation and Electromagnetism.

9. The Fundamental Origin of the Bianchi Identity of Cartan Geometry and ECE Theory.

9.1 Introduction.
9.2 Proof of the Bianchi Identity.
9.3 Hodge Dual of the Bianchi Identity.
9.4 Incompatibility of the Christoffel Connection.

10. Development of the Einstein Hilbert (EH) Field Equation into the ECE Field Equation.
10.1 Introduction.
10.2 Self Inconsistency of the Standard Model Riemann Geometry.
10.3 ECE Field Equations.

11. A Rigorous Proof of the Hodge Dual of the Bianchi Identity of Cartan.
11.1 Introduction.
11.2 Proof of the Hodge Dual Bianchi Identitity.

12. A New Theory of Light Deflection due to Gravitation.
12.1 Introduction.
12.2 The Fundamental Geometry.
12.3 The General Geodesic Method.

13. ECE Theory of the Orbit of Binary Pulsars.
13.1 Introduction.
13.2 Orbit of the Hulse Taylor Binary Pulsar.

14. Spin Connection Resonance in the Faraday Disk Generator.
14.1 Introduction.
14.2 Analytical theory.
14.3 Numerical Results.
14.4 Dynamics of the Homopolar Generator.
14.5 Discussion of Design.

15. Orbital ECE Theory and Non Einstein Hilbert orbits in Astronomy and Cosmology.
15.1 Introduction.
15.2 General ECE Orbital Theorem.
15.3 The Equivalence Principle.
15.4 Perturbation Theory of the Orbit of a Binary Pulsar System.
15.5 Graphical Results and Discussion.

16. Generalized Cartan Bianchi Identity and a New Theorem of the Cartan Torsion.
16.1 Introduction.
16.2 Generalized Cartan Bianchi Identity.
16.3 A New Cyclic Identity of the Cartan Torsion.

17. Derivation of the Thomas Precession in Terms of the Irreducible Rotation Generator.
17.1 Introduction.
17.2 Rotation of the Minkowski Spacetime.
17.3 Irreducible Generator of the Cartan Torsion.

18. The Origin of Orbits in Spherically Symmetric Spacetime.
18.1 Introduction.
18.2 Line Element for a Spherically Symmetric Spacetime.
18.3 General Equation for the Metric in ECE Theory.

19. On the Violation of the Bianchi Identity by the Einstein Field Equation and Big Bang Cosmologies.
19.1 Introduction.
19.2 The Tensorial Formats of the Bianchi Identity.

20. The Complete Equation of Classical Dynamics in ECE Theory.
20.1 Introduction
20.2 The Equations of Relativistic Dynamics.

21. Derivation of the Gravitational Red Shift from the Theorem of Orbits
21.1 Introduction
21.2 Line Element and Orbital Equation from the Theorem of Orbits.
21.3 The Gravitational Red Shift.

22. Invariance, Covariance and Duality Properties of the ECE Laws of Dynamics and Electrodynamics.
22.1 Introduction.
22.2 Invariance and Covariance.
22.3 Hodge Duality.

23. The Continuity Equation in ECE Theory.
23.1 Introduction.
23.2 The ECE Equations of Classical Dynamics and Electrodynamics.
23.3 Derivation of the Generally Covariant Continuity Equation.

24. ECE Theory of the Earth's Gravitomagnetic Precession.
24.1 Introduction.
24.2 Calculation of the Gravitomagnetic Angular Frequency.
24.3 Testing Metrics of the Einstein Equation with Eq. (24.1).
24.4 Discussion of the Results of Gravity Probe B.
24.5 Detailed Metrics.
Appendix: Validity of the Dipole Approximation.

25. Explanation of the Cosmological Red Shift.
25.1 Introduction.
25.2 A Summary of Experimental Data that Refute Big Bang.
25.3 ECE Explanation of the Cosmological Red Shift.

26. ECE Theory of the Equinoctial Precession and Galactic Dynamics.
26.1 Introduction
26.2 The Equinoctial Precession of the Earth.
26.3 Gravitomagnetic Explanation of the Equinoctial Precession.
26.4 ECE Equation of Motion of Galaxies.

27. Criticisms of Black Hole Theory.
27.1 Introduction.
27.2 Computer Evaluation of Some Commonly Used Black Hole Metrics.

28. The Conservation Theorem of ECE Field Theory.
28.1 Introduction
28.2 Proof of the Tetrad Postulate.
28.3 Invariance of the Tetrad Postulate.
28.4 Conservation Theorem of ECE Theory.

29. On the Symmetry of the Connection in Relativity and EEC Theory.
29.1 Introduction.
29.2 Antisymmetry of the Connection.

Spin Connection Resonance in the Bedini Machine

by

Myron W. Evans,
Alpha Institute for Advanced Study, Civil List Scientist.
(emyrone@aol.com)

and

H. Eckardt, C. Hubbard, J. Shelburne,
Alpha Institute for Advanced Studies (AIAS).
(www.aias.us, www.atomicprecision.com)

Abstract

Spin connection resonance (SCR) is used to explain theoretically why devices in electrical engineering can use the properties of space-time to induce voltage. Einstein Cartan Evans (ECE) theory has shown why classical electrodynamics is a theory of general relativity in which covariant derivatives are used with the spin connection playing a central role. These concepts are applied to a device known as the Bedini machine.

Keywords: Spin connection resonance, electrodynamics in general relativity, Einstein Cartan Evans theory, Bedini machine.

1.1 Introduction

Recently [1–10] the Einstein Cartan Evans (ECE) field theory has been generally accepted as the first successful unified field theory on the classical and quantum levels. It shows that classical electrodynamics is a theory of general relativity, not of special relativity. In ECE theory the spin connection plays a central role in the structure of the laws of electrodynamics and in the way the electric and magnetic fields are related to the scalar and vector potentials. The ECE equations of classical electrodynamics allow the existence of resonances in potential which can be used to extract electric power from the structure of space-time. This structure is not the vacuum, the latter in relativity theory is a universe devoid of all curvature and torsion. The resonance phenomenon induced by these equations is known as spin connection resonance (SCR). In this paper it is applied to a device known as the Bedini machine [11], which has been patented and which has been shown to be experimentally reproducible and repeatable. In section 1.2 the equations of classical electrodynamics are given in ECE theory. These are given in the vector notation used by engineers, and the reduction of the original differential form equations of ECE theory to the vector equations is given in technical appendices. In section 3 models of the Bedini device are developed, in section 1.4 the occurrence of resonances is identified and graphed using computer algebra to check the derivations.

1.2 The Equations of Classical Electrodynamics in General Relativity

All electromagnetic devices of engineering are governed by these equations, which are the generally covariant form of classical electrodynamics. Each device must be considered separately, and the general equations applied systematically to each device. The electric field in ECE theory is defined in general by the scalar and vector potentials and by the scalar and vector components of the spin connection:

$$\mathbf{E} = -\frac{\partial \mathbf{A}}{\partial t} - c\boldsymbol{\nabla}\phi - c\omega^0 \mathbf{A} + c\phi\boldsymbol{\omega}. \tag{1.1}$$

Here ϕ is the scalar potential, $\mathbf{A}$ is the vector potential, ω^0 is the scalar part of the spin connection and $\boldsymbol{\omega}$ is the vector part of the spin connection (see technical appendices). The Coulomb law in ECE theory [1–10] is

$$\boldsymbol{\nabla} \cdot \mathbf{E} = \frac{\rho}{\epsilon_0} := c\mu_0 J^0 \tag{1.2}$$

where ϵ_0 is the vacuum permittivity and ρ is the scalar part of the inhomogeneous charge current density of ECE theory. The magnetic field in ECE theory is defined by:

$$\mathbf{B} = \boldsymbol{\nabla} \times \mathbf{A} - \boldsymbol{\omega} \times \mathbf{A} \tag{1.3}$$

and the Gauss law of magnetism is:

$$\boldsymbol{\nabla} \cdot \mathbf{B} = \mu_0 j^0 \tag{1.4}$$

where j^0 is the scalar part of the homogeneous charge current density. The Faraday law of induction in ECE theory is:

$$\nabla \times \mathbf{E} + \frac{\partial \mathbf{B}}{\partial t} = c\mu_0 \mathbf{j} \qquad (1.5)$$

where $\mathbf{j}$ is the vector part of the homogeneous charge current density and the Ampère Maxwell law is:

$$\nabla \times \mathbf{B} - \frac{1}{c^2}\frac{\partial \mathbf{E}}{\partial t} = \mu_0 \mathbf{J} \qquad (1.6)$$

where $\mathbf{J}$ is the vector part of the inhomogeneous charge current density.

The explanation of various devices that are reproducible and repeatable depends on the systematic application of these general equations. It has been shown [1–10] that they are resonance equations in general, so that a small driving term can produce a very large amplification of space-time effects through the inter-mediacy of the spin connection. Devices which find no explanation in the standard model can be explained in this way. For example, we consider the Bedini device [11] as one in which an electric pulse produced by the rate of change of a magnetic field is induced in a generator. The electric field pulse produces a pulse of electrons in a battery [11] as controlled by Eqs. (1.1) and (1.2), from which:

$$\nabla \cdot \nabla \phi + \frac{1}{c}\frac{\partial}{\partial t}(\nabla \cdot \mathbf{A}) + \nabla \cdot (\omega^0 \mathbf{A}) - \nabla \cdot (\phi \boldsymbol{\omega}) = -\mu_0 J^0. \qquad (1.7)$$

This equation produces resonances in two ways, each of which gives a resonance equation.

1. If it is assumed that the origin of $\mathbf{E}$ is purely due to ϕ, we obtain the basic resonance equations of paper 63 and 92 of the ECE series [1–10].

2. If it is assumed that the origin of $\mathbf{E}$ is purely magnetic, and that the scalar potential is zero, we have:

$$\frac{1}{c}\frac{\partial}{\partial t}(\nabla \cdot \mathbf{A}) + \nabla \cdot (\omega^0 \mathbf{A}) = -\mu_0 J^0. \qquad (1.8)$$

i.e.

$$\nabla \cdot \left(\frac{1}{c}\frac{\partial \mathbf{A}}{\partial t} + \omega^0 \mathbf{A}\right) = -\mu_0 J^0. \qquad (1.9)$$

which can be integrated to give a resonance equation. It is also possible to produce a time dependent resonance equation from Eqs. (1.1) and (1.6). The Ampère Maxwell law (1.6) is considered to produce a driving term:

$$\frac{\partial \mathbf{E}}{\partial t} = c^2 \left(\boldsymbol{\nabla} \times \mathbf{B} - \mu_0 \mathbf{J} \right)_{driving} = -\boldsymbol{\nabla} \frac{\partial \phi}{\partial t} - \frac{\partial^2 \mathbf{A}}{\partial t^2} - \frac{\partial}{\partial t} \left(c \omega^0 \mathbf{A} \right) + \frac{\partial}{\partial t} \left(c \phi \boldsymbol{\omega} \right) \tag{1.10}$$

so that the most general resonance equation of time-dependent type is:

$$\frac{\partial^2 \mathbf{A}}{\partial t^2} + c \frac{\partial \omega^0}{\partial t} \mathbf{A} + c \omega^0 \frac{\partial \mathbf{A}}{\partial t} = c \frac{\partial \phi}{\partial t} \boldsymbol{\omega} + c \phi \frac{\partial \boldsymbol{\omega}}{\partial t} + c^2 \mu_0 \mathbf{J} - \boldsymbol{\nabla} \frac{\partial \phi}{\partial t} - c^2 \boldsymbol{\nabla} \times \mathbf{B}. \tag{1.11}$$

If there is no charge and current density this equation reduces to:

$$\frac{\partial^2 \mathbf{A}}{\partial t^2} + c \omega^0 \frac{\partial \mathbf{A}}{\partial t} + c \frac{\partial \omega^0}{\partial t} \mathbf{A} = -c^2 \left(\boldsymbol{\nabla} \times \mathbf{B} \right)_{driving}. \tag{1.12}$$

There is resonance in $\mathbf{A}$ under the following conditions:

1. the scalar part, ω^0, of the spin connection is non-zero,
2. the time derivative, $\frac{\partial \omega^0}{\partial t}$, is non-zero,
3. the curl $\boldsymbol{\nabla} \times \mathbf{B}$ is non-zero and also time dependent.

When investigating various claims such as the Bedini machine it is necessary to use equations such as this, which show for example that the magnetic field in the design must be both space and time dependent, and produced by a device that satisfies these requirements. That is an example of a design prediction of ECE theory in engineering.

In addition to Eq. (1.6) there exists the Coulomb law (1.2), which is the resonance equation [1–10]: [1–10]

$$\boldsymbol{\nabla} \cdot \left(c \phi \boldsymbol{\omega} - \boldsymbol{\nabla} \phi - \frac{\partial \mathbf{A}}{\partial t} - c \omega^0 \mathbf{A} \right) = \frac{\rho}{\epsilon_0}. \tag{1.13}$$

In the absence of charge this equation reduces to:

$$\boldsymbol{\nabla} \cdot \left(\frac{\partial \mathbf{A}}{\partial t} + c \omega^0 \mathbf{A} \right) = 0 \tag{1.14}$$

1.2 The Equations of Classical Electrodynamics in General Relativity

so ω^0 may be eliminated between equations (1.12) and (1.14). Eq. (1.14) is:

$$\nabla \cdot \frac{\partial \mathbf{A}}{\partial t} = -c\left(\mathbf{A} \cdot \nabla \omega^0 + \omega^0 \nabla \cdot \mathbf{A}\right). \tag{1.15}$$

Therefore ω^0 is governed by Eqs. (1.12) and (1.15) which must be solved simultaneously. The latter equation can be integrated with the divergence theorem [12]. For any well behaved vector field $\mathbf{V}(\mathbf{r})$ defined with a volume surrounded by a closed surface S:

$$\oint_S \mathbf{V} \cdot \mathbf{n}\, da = \int_V \nabla \cdot \mathbf{V}\, d^3r. \tag{1.16}$$

Thus for the Coulomb law 1.12:

$$\int_V \left(\nabla \cdot \mathbf{E} - \frac{\rho}{\epsilon_0}\right) d^3r = 0 \tag{1.17}$$

i.e.

$$\oint_S \mathbf{E} \cdot \mathbf{n}\, da = \frac{1}{\epsilon_0} \int_V \rho(r)\, d^3r. \tag{1.18}$$

So the integration of Eq. (1.14) is:

$$\oint_S \left(\frac{\partial \mathbf{A}}{\partial t} + c\omega^0 \mathbf{A}\right) \cdot \mathbf{n}\, da = 0 \tag{1.19}$$

i.e.

$$\oint_S \frac{\partial \mathbf{A}}{\partial t} \cdot \mathbf{n}\, da = -c \oint_S \omega^0 \mathbf{A} \cdot \mathbf{n}\, da. \tag{1.20}$$

Eq. (1.20) is a relation between ω^0 and $\mathbf{A}$. The correct way of solving (1.12) is simultaneously with (1.18). This can be carried out numerically for various models of $\nabla \times \mathbf{B}$ produced by various devices. It can be seen that ω^0 can be eliminated and that Eq. (1.12) reduces to an undamped oscillator [1–10] because $\frac{\partial \mathbf{A}}{\partial t}$ is eliminated in favour of $\mathbf{A}$. So in this example $\mathbf{A}$ can be amplified to INFINITY for various models of $\nabla \cdot \mathbf{B}$ acting as a driving force. There is no need to model ω^0 because it can be expressed in terms of $\mathbf{A}$.

1.3 Systematic Evaluation of Equations for the Bedini Machine

If no scalar potential is present, the ECE field equations (1.1–1.6) in the base manifold take the simple form:

$$\nabla \times \mathbf{E} + \dot{\mathbf{B}} = 0 \tag{1.21}$$

$$\nabla \times \mathbf{B} - \frac{1}{c^2}\dot{\mathbf{E}} = 0 \tag{1.22}$$

$$\nabla \cdot \mathbf{B} = 0 \tag{1.23}$$

$$\nabla \cdot \mathbf{E} = 0 \tag{1.24}$$

with the definition equations

$$\mathbf{B} = \nabla \times \mathbf{A} - \boldsymbol{\omega} \times \mathbf{A} \tag{1.25}$$

$$\mathbf{E} = -\dot{\mathbf{A}} - c\,\omega^0 \mathbf{A}. \tag{1.26}$$

Here the dot denotes the time derivative, $\mathbf{A}$ is the vector potential, $\boldsymbol{\omega}$ the vector spin connection and ω^0 the scalar spin connection, both in units of $1/m$. It is more convenient to transform the scalar spin connection to a time frequency:

$$\omega_0 := c\,\omega^0. \tag{1.27}$$

Eqs. (1.21-1.24) represent a system of eight equations and by the right-hand side of Eqs. (1.25-1.26) seven variables are defined. In the most general case the scalar potential Φ is the eights variable so that (1.21)–(1.24) can be solved uniquely. Here we restrict consideration to the case without charges and therefore without a scalar potential.

In classical electrodynamics we have the same equations, but without the spin connection. This leads to an inconsistency for solving the equations. Sometimes solely the fields E and B are considered, then only the equations (1.21)–(1.22) can be used. The Gauss and Coulomb law are tried to be handled as "constraints", but this leads to an over-determined equation system. In other cases (when charges and currents are present) the potentials $\mathbf{A}$ and

Φ are taken as variables. Then only the Eqs. (1.22) and (1.24) can be used, the other two are homogeneous and lead to the trivial solution $\mathbf{A} = 0$. In contrast, ECE theory presents a perfectly well-defined situation with eight equations and eight variables.

There are basically two methods to combine these equations to obtain resonances for particular cases:

1. use (1.21) and (1.22) completely to define driving terms, use (1.25) and (1.26) as basis for resonance solutions,

2. use the terms $\dot{\mathbf{B}}, \dot{\mathbf{E}}$ in (1.21), (1.22) as driving terms, insert curl of (1.25) and (1.26) into (1.21) and (1.22) and use these equations for resonance solutions.

We will see that both methods are not applicable in all possible cases.

In addition to both methods, we have to use one of the equations (1.3), (1.4). The actual choice depends on the case if $\boldsymbol{\omega}$ or ω_0 occurs in the equations (1.21) and (1.22). In the following we work out the distinguished cases 1 and 2 each for Eq. (1.21) (called sub-case a) and Eq. (1.22) (called sub-case b).

1a: Faraday Law as driving term, B field resonance
By definition we have

$$(\boldsymbol{\nabla} \times \mathbf{E})_{driving} = -(\dot{\mathbf{B}})_{driving} \tag{1.28}$$

Inserting the time derivative of (1.25) into (1.28):

$$\boldsymbol{\nabla} \times \dot{\mathbf{A}} - \dot{\boldsymbol{\omega}} \times \mathbf{A} - \boldsymbol{\omega} \times \dot{\mathbf{A}} = (\dot{\mathbf{B}})_{driving} = -(\boldsymbol{\nabla} \times \mathbf{E})_{driving} \tag{1.29}$$

In order to obtain resonance a differential equation of second order in time is required, therefore we take a further time derivative:

$$\boldsymbol{\nabla} \times \ddot{\mathbf{A}} - \ddot{\boldsymbol{\omega}} \times \mathbf{A} - 2\dot{\boldsymbol{\omega}} \times \dot{\mathbf{A}} - \boldsymbol{\omega} \times \ddot{\mathbf{A}} = (\ddot{\mathbf{B}})_{driving} \tag{1.30}$$

This is a resonance equation in $\mathbf{A}$ (for constant $\boldsymbol{\omega}$) as well as in $\boldsymbol{\omega}$ (for constant $\mathbf{A}$). The spin connection can be obtained from simultaneously solving Eq. (1.23). This could be sufficient, if not all components of $\mathbf{A}$ or $\boldsymbol{\omega}$ are different from zero. In the most general case further equations have to be added.

1b: Ampère-Maxwell Law as driving term, E field resonance
In analogy to case 1a we obtain from (1.22):

$$(\boldsymbol{\nabla} \times \mathbf{B})_{driving} = \frac{1}{c^2}(\dot{\mathbf{E}})_{driving} \tag{1.31}$$

and by applying (1.26):

$$\ddot{\mathbf{A}} + \dot{\omega}_0 \mathbf{A} + \omega_0 \dot{\mathbf{A}} = -(\dot{\mathbf{E}})_{driving} \qquad (1.32)$$

This is an equation for a damped resonance for $\omega_0 > 0$. The spin connection can be determined by combining (1.32) with (1.24).

2a: B field definition as driving term, Faraday Law as resonance equation

Taking the magnetic field in (1.21) as driving term gives

$$\boldsymbol{\nabla} \times \mathbf{E} = -(\dot{\mathbf{B}})_{driving}. \qquad (1.33)$$

Inserting (1.26) into (1.33):

$$\boldsymbol{\nabla} \times \dot{\mathbf{A}} + \boldsymbol{\nabla} \times (\omega_0 \mathbf{A}) = (\dot{\mathbf{B}})_{driving} \qquad (1.34)$$

or after taking a further time derivative:

$$\boldsymbol{\nabla} \times \ddot{\mathbf{A}} + \boldsymbol{\nabla} \times (\dot{\omega}_0 \mathbf{A}) + \boldsymbol{\nabla} \times (\omega_0 \dot{\mathbf{A}}) = (\ddot{\mathbf{B}})_{driving} \qquad (1.35)$$

which is the equivalent of (1.30) with the other type of spin connection.

2b: E field definition as driving term, Ampère-Maxwell Law as resonance equation

Starting with Eq. (1.22) we obtain

$$\boldsymbol{\nabla} \times \mathbf{B} = \frac{1}{c^2} (\dot{\mathbf{E}})_{driving} \qquad (1.36)$$

and with (1.25):

$$\boldsymbol{\nabla} \times \boldsymbol{\nabla} \times \mathbf{A} - \boldsymbol{\nabla} \times \boldsymbol{\omega} \times \mathbf{A} = \frac{1}{c^2} (\dot{\mathbf{E}})_{driving} \qquad (1.37)$$

or

$$\begin{aligned}\boldsymbol{\nabla}(\boldsymbol{\nabla} \cdot \mathbf{A}) - \boldsymbol{\nabla}^2 \mathbf{A} - \boldsymbol{\omega}(\boldsymbol{\nabla} \cdot \mathbf{A}) + \mathbf{A}(\boldsymbol{\nabla} \cdot \boldsymbol{\omega}) \\ - (\mathbf{A} \cdot \boldsymbol{\nabla})\boldsymbol{\omega} + (\boldsymbol{\omega} \cdot \boldsymbol{\nabla})\mathbf{A} = \frac{1}{c^2}(\dot{\mathbf{E}})_{driving}.\end{aligned} \qquad (1.38)$$

This is a resonance equation for the space coordinates of $\mathbf{A}$. Investigating time-dependent resonances requires a twofold additional time derivation which makes this equation impractible.

1.4 Detailed Investigation of the Bedini Machine

1.4.1 Description of the Bedini machine

In the book "Free Energy Generation" [13], see also [14], Bedini explains his battery charging device of 1984. He presents some variants of the machine constructed within 20 years. The basic design has remained the same. The patented Bedini machine inventor claims that his machines are able to extract energy from the surrounding space in the form of radiant energy. The authors here attempt to show that the energy produced by these machines is the result of disturbing the local space-time unit volume, creating a resonance effect, which allows energy to flow out of the local unit volume, in the form of asymmetric electromagnetic wave forms into a rectifier circuit, where it can be then sent to a storage device. The mathematical expressions developed are based on ECE field theory. The resulting expressions will allow electrical designers to produce productive circuits, based on this math, since the inventor has not furnished an adequate explanation of the machines' operation. One of the authors has replicated two of the Bedini machines successfully, and others have had success building and operating the machines.

The Bedini machine has several distinct elements (see Fig. 1.1). The input power supply, which can be a battery or a rectified power supply from an external supply, provides the transducer coil and trigger circuitry with energy to pulse the unit volume through the trigger winding. The magnet induces an asymmetric pulse into the transducer core, which induces an e-m pulse in

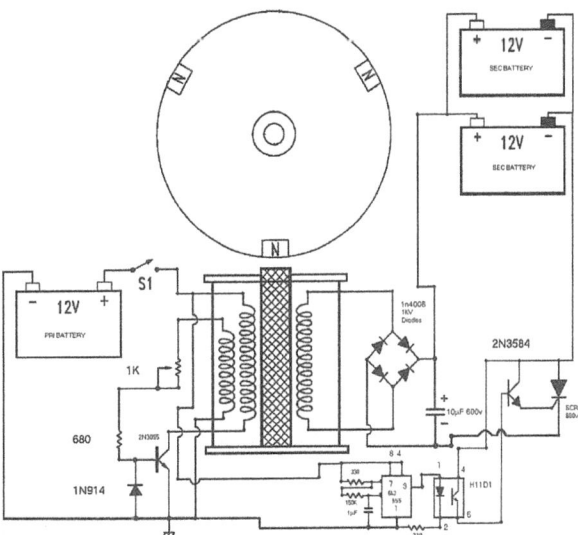

Fig. 1.1. Bedini machine (from [13], p. 47).

the trigger winding, power winding, and generator winding. The trigger pulse causes power to flow into the power winding, giving a boot to the magnet as it goes on by, thereby powering the rotor to the next magnet. The transducer pulse from the coils flows into the unit volume, upsetting the local field, and the resulting return energy is rectified after flowing through the generator winding.

All of these windings of the transducer are separate coils, wound concentrically on a spool, which has a core consisting of mild iron rods, typically 1/16" in diameter. Once the rotor is spun manually, and the power source and storage device are connected, the rotor will accelerate to a select speed determined by a tuning rheostat, and the machine will maintain that speed indefinitely, charging the storage device, using less energy to run than it stores, thereby achieving over unity in its operation. One of the authors has determined that the machine operates more efficiently at 24 Volts DC, than at 12 Volts DC, and the machine operates at almost twice the rpm as compared to 12 Volt operation.

Mr. Bedini has built several demonstrator machines in the kilowatt size, however, one of the authors' machines is only capable of 10-15 watts of output, but this size is adequate to provide meaningful test results. One of the authors is presently building a larger machine to replicate Mr. Bedini's claims of higher power outputs. In addition to the rotor style machines, the inventor has shown solid state designs, which the authors have not replicated yet, but others have, with limited output success. A company using Mr. Bedini's designs is presently marketing a line of battery chargers claiming to use radiant energy to enhance battery life and longevity.

1.4.2 Models of the Bedini machine

Charging of a battery means a flow of ions in the electrolyte in direction reverse to the discharging current. According to the explanations of Bedini, the battery charging process is evoked by high frequency pulses. This type of charging is completely different from the conventional DC charging process where the ion transport is effected by applying a DC voltage. Bedini points out that the high frequency / high voltage oscillations initiate a coupling to spacetime so that the ions resonate and move in the direction opposite to the discharge current. No significant conventional recharging energy has to be expended in this process.

Key of understanding the process is the mechanism of tapping the vacuum background energy, i.e. to evoke a resonant coupling to the spacetime background. As has been shown by ECE field theory [papers 63, 92], a coupling to spacetime background can be achieved by a resonance circuit. Such an original circuit from Bedini is shown in Fig. 1.1.

The key component of the Bedini machine is the trifilar wound coil which acts as a combined transmitter-receiver transducer. In the following we use the working hypothesis that the spacetime coupling takes place by means of this coil. Therefore we need not consider the complex electro-chemical processes in the battery, and an electrical potential Φ can be omittet as already done in Equations (1.21–1.26).

Since we have to model the fields of a cylindrical coil, we choose cylinder coordinates (r, φ, z) for convenience with unit vectors $\mathbf{e}_r, \mathbf{e}_\varphi, \mathbf{e}_Z$ as shown in Fig. 1.2. Inside a conventional coil the magnetic field is parallel to the z direction and the vector potential is tangential to circles around **B**. We assume that the magnetic field maintains its direction in case of resonance. Then the vector spin connection has to lie in the r-φ plane as well as **A**. In the simplest case it is perpendicular to **A**.

Whether type a or b of setcion 3.1 should be chosen for modeling the device, depends on the type of excitation mechanism. Inside the transducer we have

$$(\mathbf{\nabla} \times \mathbf{B})_{driving} \approx 0, \tag{1.39}$$

during the pulsing phase. In the preceding phase when a rotor-mounted magnet approaches the transducer, the moving magnet induces a non-symmetric magnetic field within the iron core of the transducer. Therefore condition (1.39) is not always valid. To obtain a viable model, we make the following additional simplifying assumptions. The B field is in z direction:

$$\mathbf{B} = \begin{pmatrix} 0 \\ 0 \\ B_Z \end{pmatrix}. \tag{1.40}$$

The vector potential in classical electrodynamics then has only a φ and r component:

$$\mathbf{A} = \begin{pmatrix} A_r \\ A_\varphi \\ 0 \end{pmatrix}. \tag{1.41}$$

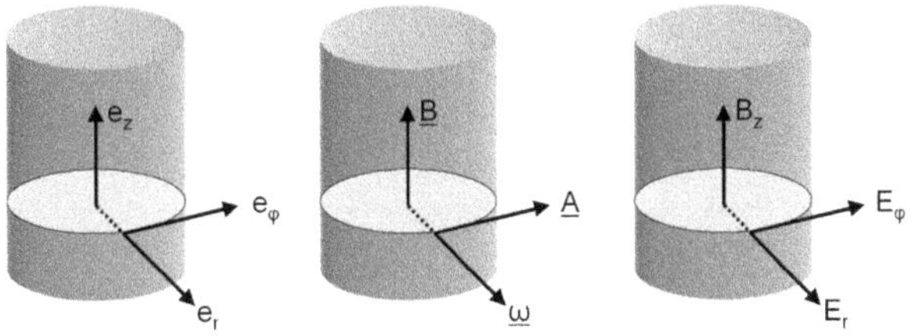

Fig. 1.2. Cylindrical coordinate system and fields in a coil.

Since the spin connection $\boldsymbol{\omega}$ cannot be in parallel to $\mathbf{A}$ and $\mathbf{B}$ according to Eq. (1.25), we choose

$$\boldsymbol{\omega} = \begin{pmatrix} \omega_r \\ \omega_\varphi \\ 0 \end{pmatrix}. \tag{1.42}$$

Due to the rotational symmetry of the device, there cannot be a φ dependence of the fields. In total we have the functional dependencies

$$\begin{aligned} B_Z &= B_Z(r,t) \\ A_r &= A_r(r,t) \\ A_\varphi &= A_\varphi(r,t) \\ \omega_r &= \omega_r(r,t) \\ \omega_\varphi &= \omega_\varphi(r,t) \\ \omega_0 &= \omega_0(r,t) \end{aligned} \tag{1.43}$$

With (1.40–1.42) we have (using the differential operators in cylinder coordinates)

$$\nabla \times \mathbf{A} = \begin{pmatrix} 0 \\ 0 \\ \frac{1}{r}\frac{\partial}{\partial r}(rA_\varphi) - \frac{\partial A_r}{\partial \varphi} \end{pmatrix}, \tag{1.44}$$

$$\boldsymbol{\omega} \times \mathbf{A} = \begin{pmatrix} 0 \\ 0 \\ \omega_r A_\varphi - \omega_\varphi A_r \end{pmatrix}. \tag{1.45}$$

The divergence of a vector $\mathbf{V}$ is in cylindric coordinates:

$$\nabla \cdot \mathbf{V} = \frac{1}{r}\frac{\partial}{\partial r}(rV_r) + \frac{1}{r}\frac{\partial}{\partial \varphi}(V_\varphi) + \frac{\partial}{\partial z}(V_Z). \tag{1.46}$$

We are now ready to apply the methods 1a, 1b, 2a. Starting with 1a, we obtain from Eq. (1.30) with the special form of $\mathbf{A}$ and $\boldsymbol{\omega}$ (1.40-1.45):

$$\frac{1}{r}\frac{\partial}{\partial r}(r\ddot{A}_\varphi) - \frac{\partial \ddot{A}_r}{\partial \varphi} - \ddot{\omega}_r A_\varphi + \ddot{\omega}_\varphi A_r - 2(\dot{\omega}_r \dot{A}_\varphi - \dot{\omega}_\varphi \dot{A}_r) \\ - \omega_r \ddot{A}_\varphi + \omega_\varphi \ddot{A}_r = (\ddot{B}_Z)_{driving} \tag{1.47}$$

From (1.23) follows

$$\nabla \cdot (\nabla \times \mathbf{A} - \boldsymbol{\omega} \times \mathbf{A}) = 0 \tag{1.48}$$

or

$$\frac{\partial}{\partial z}\left(\frac{1}{r}\frac{\partial}{\partial r}(rA_\varphi) - \omega_r A_\varphi + \omega_\varphi A_r\right) = 0. \tag{1.49}$$

This equation is trivially fulfilled. Even if we additionally assume $A_r = \omega_\varphi = 0$ we have one equation with two unknowns A_φ and ω_r so that no unique solution is obtained.

Considering the alternative case 2a we get from Eq. (1.35):

$$\begin{aligned}\frac{1}{r}\frac{\partial}{\partial r}\left(r\ddot{A}_\varphi + r\dot{\omega}_0 A_\varphi + r\omega_0 \dot{A}_\varphi\right) \\ - \frac{\partial}{\partial \varphi}\left(\ddot{A}_r + \dot{\omega}_0 A_r + \omega_0 \dot{A}_r\right) = (\ddot{B}_Z)_{driving}.\end{aligned} \tag{1.50}$$

From Eq. (1.24) follows

$$\nabla \cdot \left(-\dot{\mathbf{A}} - \omega_0 \mathbf{A}\right) = 0 \tag{1.51}$$

or

$$\frac{1}{r}\frac{\partial}{\partial r}\left(r\dot{A}_r + r\omega_0 A_r\right) + \frac{1}{r}\frac{\partial}{\partial \varphi}\left(\dot{A}_\varphi + \omega_0 A_\varphi\right) = 0 \tag{1.52}$$

According to (1.43) Eqs. (1.50) and (1.52) can be simplified to

$$\frac{1}{r}\frac{\partial}{\partial r}\left(r\ddot{A}_\varphi + r\dot{\omega}_0 A_\varphi + r\omega_0 \dot{A}_\varphi\right) = (\ddot{B}_Z)_{driving} \tag{1.53}$$

$$\frac{1}{r}\frac{\partial}{\partial r}\left(r\dot{A}_r + r\omega_0 A_r\right) = 0 \tag{1.54}$$

These are two equations for three unknowns and not unique as before.

Finally we apply case 1b. This is different from the previous ones since the electrical field is considered to be the driving term. From Eq. (1.32) we obtain the two equations

$$\ddot{A}_r + \dot{\omega}_0 A_r + \omega_0 \dot{A}_r = -(\dot{E}_r)_{driving} \tag{1.55}$$

$$\ddot{A}_\varphi + \dot{\omega}_0 A_\varphi + \omega_0 \dot{A}_\varphi = -(\dot{E}_\varphi)_{driving} \tag{1.56}$$

and from Eq. (1.24):

$$\frac{1}{r}\frac{\partial}{\partial r}\left(r\dot{A}_r + r\omega_0 A_r\right) = 0 \tag{1.57}$$

or

$$\dot{A}_r + \left(\omega_0 + r\frac{\partial \omega_0}{\partial r}\right)A_r + r\frac{\partial \dot{A}_r}{\partial r} + r\omega_0\frac{\partial A_r}{\partial r} = 0. \tag{1.58}$$

We see that the spin connection is coupled to the radial part of the vector potential. This indicates that the unit volume interacting with spacetime may be somewhat extended beyond the transducer. The occurrence of $\dot{\mathbf{E}}$ implies a non-vanishing curl of $\mathbf{B}$ according to (1.31).

The result (1.57) can further be simplified by applying the divergence theorem as explained at the end of section 1.2. The surface integral of Eq. (1.19) is to be taken over the cylinder surface of the model. The parts over the circular areas cancel out due to the assumed symmetry in z direction. For the cylindrical part the φ component of the vector potential is perpedicular to the surface normal and does not contribute anything. The only contributing part is the radial component:

$$\int_V \nabla \cdot \mathbf{A}\, d^3r = \int_S (\dot{A}_r + \omega_0 A_r)\, da = 0 \tag{1.59}$$

Since A_r and ω_0 are independent on the individual surface points, the integral can be evaluated trivially and results in

$$\omega_0 = -\frac{\dot{A}_r}{A_r}. \tag{1.60}$$

The equations (1.55, 1.56, 1.60) are three equations for three unknowns A_r, A_φ, ω_0. This set of equations has to be solved numerically to provide guidance to designers in sizing the transducer, designing the trigger and power circuits, and predicting power outputs. Since the unit volume is surrounded by the large number of unit volumes in a spherical configuration(the rest of space), the theoretical power input to the machine transducer is limited by its conductor size and impedance seen looking into the transducer from the space side.

This paper discusses the Bedini machine in particular, but the concept of a transducer acting as a transmitter-receiver for power extraction from the surrounding space should be applicable to other machine designs also.

The inventor has put forth a hypothesis as to how his machines operate, which is non-conventional in its premise. The authors here suggest that the latest ECE theory will provide a rational explanation to the machines' operation, using conventional mathematical notation, and recognized physical theory.

1.4.3 Resonance behaviour of the vector potential

Without doing any numerical calculations, we can demonstrate that resonance solutions for Eqs. (1.55, 1.56) exist. We assume a harmonic time dependence

$$A_r = A_1(r)\sin(\omega t) \qquad (1.61)$$
$$\omega_0 = \omega_1(r)\sin(\omega t) \qquad (1.62)$$

with a frequency ω (not to be confused with the spin connection ω_0) and radius dependent functions A_1 and ω_1. Let's further denote the right-hand side of (1.55) by f_1, then this equation can be written:

$$2A_1\omega_1\omega \cos(\omega t)\sin(\omega t) - A_1\omega^2 \sin(\omega t) = -f_1. \qquad (1.63)$$

For $\omega t = \pi/4$ we have

$$\sin(\omega t) = \cos(\omega t) = \frac{1}{\sqrt{2}} \qquad (1.64)$$

and (1.63) simplifies to

$$A_1\left(\omega_1\omega - \frac{\omega^2}{\sqrt{2}}\right) = -f_1 \qquad (1.65)$$

which gives the solution for A_r:

$$A_1 = \frac{f_1}{\frac{\omega^2}{\sqrt{2}} - \omega_1\omega}. \qquad (1.66)$$

There is resonance when the denominator approaches zero, i.e.

$$\omega_1 = \frac{\omega}{\sqrt{2}}. \qquad (1.67)$$

If we had defined (1.61, 1.62) by the cosine function, we had got the same value for ω_0 with a negative sign. From this simple model we learn that the spin connection can assume both signs (in contrast to a real frequency) and show up sharp resonances for certain phases of the time period. This is in accordance with the experimental findings. From the original Eqs. (1.55, 1.56) we would expect a damped oscillation, but these equations are non-linear and therefore some unexpected results can occur, in this case an undamped oscillation.

1.4.4 Computation of the energy balance

The theory should provide a method to estimate the energy balance of the Bedini machine. According to the previous section it is assumed that the excess energy comes from the spacetime processes in the extended unit volume, where they are evoked by the transducer. So a calculation has to compare the energy density of the input fields $(\mathbf{E})_{driving}$ or $(\mathbf{B})_{driving}$ to the energy of the total fields being present in the resonance case. The result may depend on whether we consider the energy of the force fields only or whether we include the effects on the spacetime potential $\mathbf{A}$. In the first case we can define the energy densities for input and output:

$$u_{in} = \frac{\epsilon_0}{2}(\mathbf{E}^2)_{driving} + \frac{1}{2\mu_0}(\mathbf{B}^2)_{driving}, \tag{1.68}$$

$$u_{out} = \frac{\epsilon_0}{2}\mathbf{E}^2 + \frac{1}{2\mu_0}\mathbf{B}^2. \tag{1.69}$$

The resulting total energies then are obtained by integrating over the unit volume and time:

$$E_{in} = \int u_{in} \, d^3r \, dt \tag{1.70}$$

$$E_{out} = \int u_{out} \, d^3r \, dt \tag{1.71}$$

and the "coefficient of performance" is

$$COP = \frac{E_{out}}{E_{in}}. \tag{1.72}$$

Alternatively, the output energy can be related to the spacetime potential. From the minimal prescription of momentum density p

$$p \to p + eA \tag{1.73}$$

we can define the kinetic energy density of the field by

$$u = \frac{e^2 A^2}{2m} \tag{1.74}$$

where m is the "mass" of the field volume. According to the de Broglie equation

$$m = \frac{\hbar \omega}{c^2} \tag{1.75}$$

the mass corresponds to a frequency w. This leads to the expression

$$u = u_{out} = \frac{e^2 c^2}{2\hbar\omega} A^2. \tag{1.76}$$

1.4.5 Analytical and numerical solutions

The equations to be solved for the model we have developed (Eqs. 1.55, 1.56, 1.60) read

$$\ddot{A}_r + \dot{\omega}_0 A_r + \omega_0 \dot{A}_r = -f_1 \tag{1.77}$$

$$\ddot{A}_\varphi + \dot{\omega}_0 A_\varphi + \omega_0 \dot{A}_\varphi = -f_2 \tag{1.78}$$

$$\omega_0 = -\frac{\dot{A}_r}{A_r}. \tag{1.79}$$

with driving terms $f_1(r)$ and $f_2(r)$. Instead of Eq. (1.79) we can alternatively use its original form (1.58) without application of the divergence theorem:

$$\dot{A}_r + \left(\omega_0 + r\frac{\partial \omega_0}{\partial r}\right) A_r + r\frac{\partial \dot{A}_r}{\partial r} + r\omega_0 \frac{\partial A_r}{\partial r} = 0. \tag{1.80}$$

The difference is that the original form represents a differential equation in r while the r differentiation has vanished in the other form. Thus Eqs. (1.77–1.79) are only to be solved in the time domain which is a great alleviation. In this case Eq. (1.79) can be inserted into (1.77). Then all terms on the left cancel out, leading to the condition

$$f_1 = 0. \tag{1.81}$$

Obviously this is a compatibility condition, indicating that a driving force f_1 cannot be applied. The second Equation (1.78) can be solved analytically by computer algebra and gives the particular solution

$$A_\varphi = \omega_0 f_2 \left(e^{-\omega_0 t} \int \frac{e^{\omega_0 t}}{\omega_0 \dot{\omega}_0 t - \dot{\omega}_0 + \omega_0^2} dt - \int \frac{1}{\omega_0 \dot{\omega}_0 t - \dot{\omega}_0 + \omega_0^2} dt \right) \tag{1.82}$$

As already made plausible in section (1.4.3), this is a resonance equation if the denominator goes to zero. This means that resonances occur at solutions of the differential equation

$$\omega_0 \dot{\omega}_0 t - \dot{\omega}_0 + \omega_0^2 = 0. \tag{1.83}$$

Computer algebra gives for this equation the general solution

$$\omega_0 \, t - log\,(\omega_0) = c \qquad (1.84)$$

with a constant c. This is a transcendent equation for ω_0. Since c is arbitrary, there is an infinite number of resonances in the whole interval of real numbers for ω_0.

All further investigations are made by a numerical model. As we have seen by analysing Eq. (1.77) the vector potential A_r can be chosen freely. Considering the Bedini machine, such a radial component can only be created by an asymmetric disturbance of the field potential of the transducer coil. This is achieved by the magnets of the wheel passing the transducer. We model these pulses by a sinoidal function:

$$A_r(t) = A_1 \, sin^6(\omega t) \qquad (1.85)$$

with an arbitrary amplitude A_1 and a time frequency ω. This function and its time derivative are shown in Figs. 1.3 and 1.4 for three frequencies. With this ansatz, Eq. (1.79) takes the form

$$\omega_0(t) = -6\,\omega\,cot(\omega t). \qquad (1.86)$$

This function has vertical tangents where the values approach infinity, see Fig. 1.5 for a plot of $|\omega_0|$ for three frequencies in a logarithmic scale. Consequently, the derivative shows also this behaviour (Fig. 1.6).

Eq. (1.78) has been solved numerically for A_φ. The driving force f_2 was assumed to be in proportion to the "symmetry breaking" potential A_r. With ω_0 having the singular behaviour, the solution spans a remarkable order of magnitude and is vulnerable to numerical instabilities. Therefore the solution was checked by inserting it back into Eq. (1.78) and checking for equality with f_2. In all cases the equality was maintained within sufficient precision. The result (Fig. 1.7) shows giant resonance peaks over 15 orders of magnitude which occur in coincidence with the structure of ω_0. Obviously these peaks correspond to the peak signals in the Bedini machine. The time frequency is to be identified with the passing rate of the magnets over the transducer. To make comparison even more appropriate, in Fig. (1.8) the derivative of A_φ is shown which should correspond to the induced voltage

$$U_{ind} = -\dot{A}_\varphi. \qquad (1.87)$$

The structure is very similar to that of A_φ itself.

Next we have tested the dependence of the solution on the driving force f_2. It results that A_φ is practically insensitive to the form of f_2, provided the value is different from zero where ω_0 has its poles. It is even sufficient to take

1.4 Detailed Investigation of the Bedini Machine   19

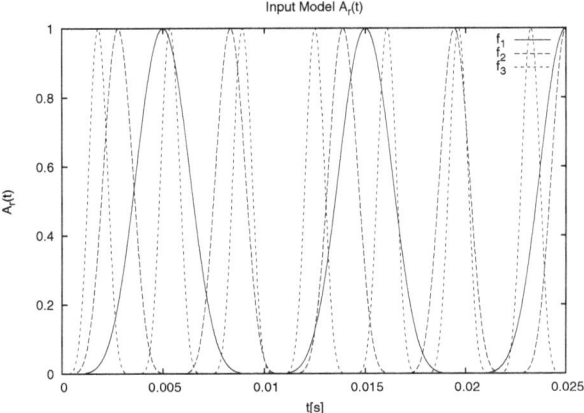

Fig. 1.3. Radial component of vector potential A_r for three frequencies f=50 Hz, 90 Hz, 140 Hz.

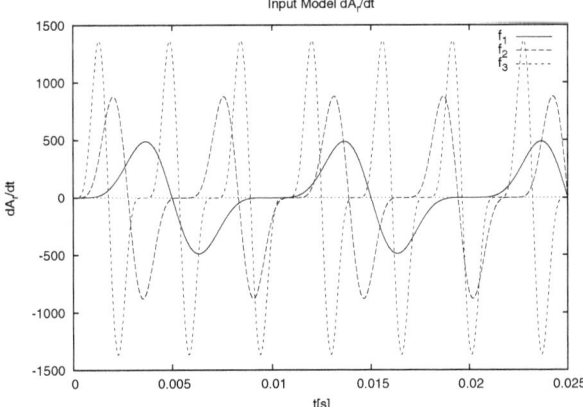

Fig. 1.4. Time derivative $\dot{A}_r$ for three frequencies f=50 Hz, 90 Hz, 140 Hz.

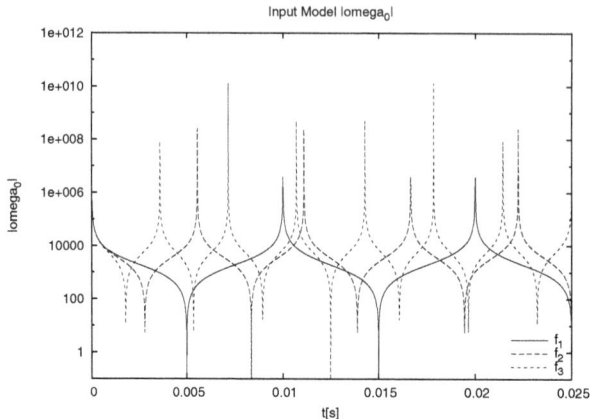

Fig. 1.5. $|\omega_0|$ for three frequencies f=50 Hz, 90 Hz, 140 Hz.

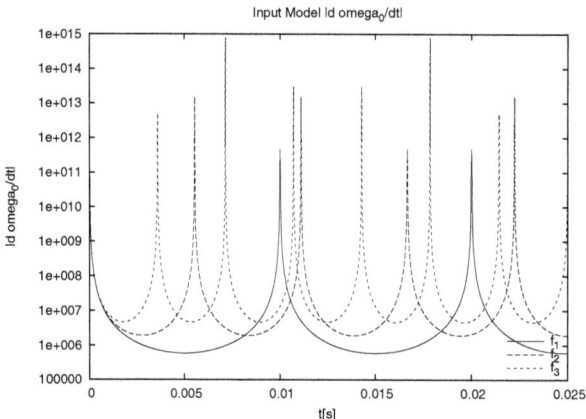

Fig. 1.6. Time derivative $|\dot{\omega}_0|$ for three frequencies f=50 Hz, 90 Hz, 140 Hz.

1.4 Detailed Investigation of the Bedini Machine

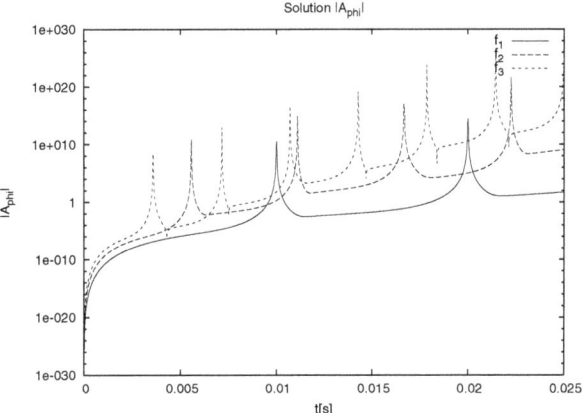

Fig. 1.7. Solution $|A_\varphi(t)|$ for three frequencies f=50 Hz, 90 Hz, 140 Hz.

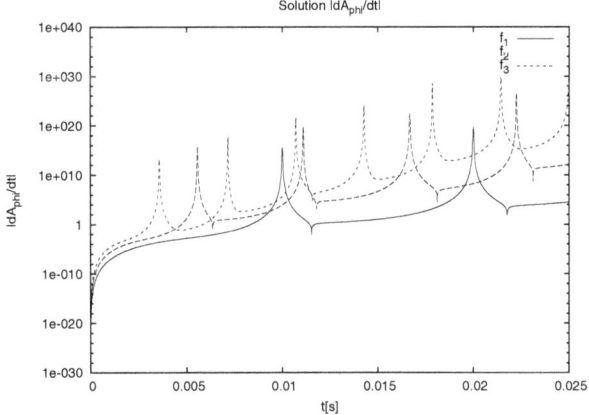

Fig. 1.8. Time derivative $|\dot{A}_\varphi(t)|$ for three frequencies f=50 Hz, 90 Hz, 140 Hz.

a spike pulse of one percent of the time period. Fig. 1.9 shows the result for a constant value of f_2.

Since the zero crossing of A_r is essential for the resonances, we have modified Eq. (1.85) by adding a constant value of 0.001, thus displacing the curve of Fig. 1.3 by this value from zero. The result (Fig. 1.10) shows a far smaller resonance structure indicating that resonances are very sensitive to the form of A_r via ω_0.

Next we inspect the development of the maximum amplitude. In Fig. 1.11 the maximum difference over the first six time periods is plotted in dependence of the time frequency. Obviously the resonance is most dramatic for low frequencies. In the next figure (Fig. 1.12) the maximum amplitude difference was recorded over a constant simulated time of 0.1 sec. To avoid numerical instabilities inferred by the calculation we used a modified A_r input value as discussed for Fig. 1.10 (shifted by 0.1 upwards, no zero crossing). Solutions are stable in the low frequency range but there are windows of instability for higher frequencies. We argue that the differential equation (1.78) can show chaotic behaviour and must be carefully evaluated.

Finally we present the amount of transferred energy integrated over time. According to Eq. (1.76) this is proportional to

$$u(t) = \int_0^t \frac{A_\varphi^2(t')}{\omega} dt'. \tag{1.88}$$

This term is represented in Fig. 1.13. Since A_φ crosses zero at the resonances (remember that the modulus is shown in the figures), a considerable amount of energy is pushed back to the vacuum after having been transferred to the system, but there is enough energy left after each resonance peak so that the energy in the system rises considerably.

As a last item in this section let us consider the radius dependence of the fields which can not be determined from Eqs. (1.77–1.79) as discussed above. Therefore let's start from Eqs. (1.77) and (1.80):

$$\ddot{A}_r + \dot{\omega}_0 A_r + \omega_0 \dot{A}_r = -f_1 \tag{1.89}$$

$$\dot{A}_r + \left(\omega_0 + r\frac{\partial \omega_0}{\partial r}\right) A_r + r\frac{\partial \dot{A}_r}{\partial r} + r\omega_0 \frac{\partial A_r}{\partial r} = 0 \tag{1.90}$$

We will make an ansatz for A_r and compute the solution for ω_0 which is compatible with this. We choose

$$A_r = C e^{-\alpha r - i\beta t} \tag{1.91}$$

1.4 Detailed Investigation of the Bedini Machine

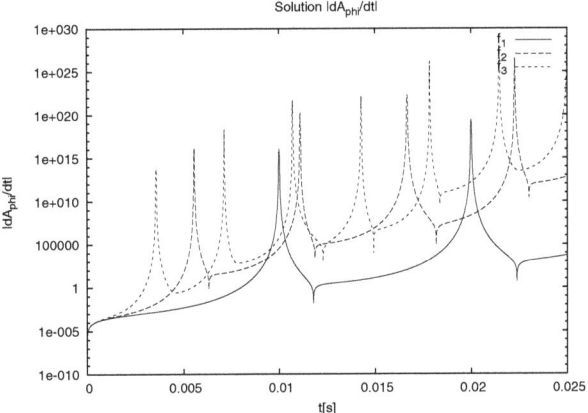

Fig. 1.9. Time derivative $|\dot{A}_\varphi(t)|$ with $f_2 = 1$ for three frequencies f=50 Hz, 90 Hz, 140 Hz.

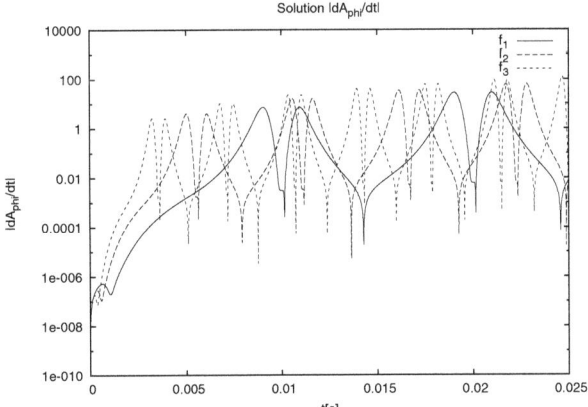

Fig. 1.10. Time derivative $|\dot{A}_\varphi(t)|$ with A_r shifted by 0.001 for three frequencies f=50 Hz, 90 Hz, 140 Hz.

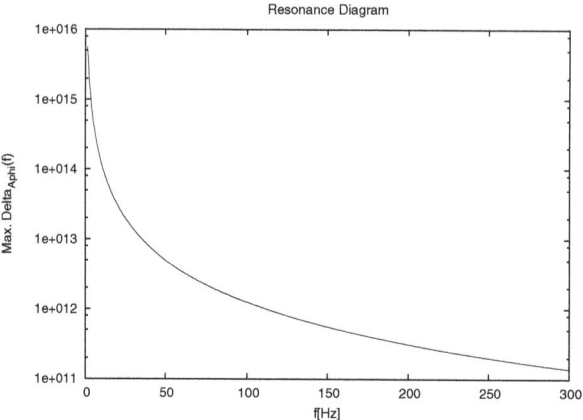

Fig. 1.11. Resonance behaviour: amplitude of A_φ after 6 periods of frequency f.

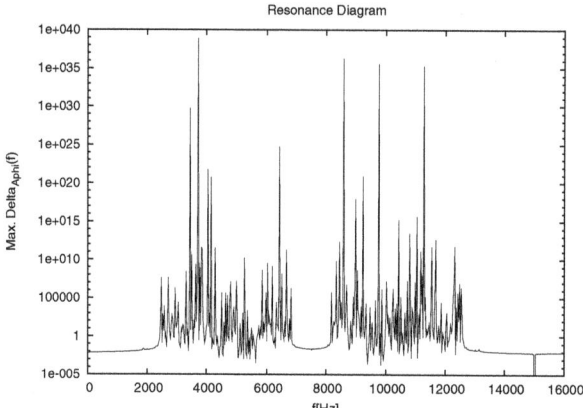

Fig. 1.12. Resonance behaviour: maximum amplitude of A_φ within 0.1 sec runtime, A_r shifted by 0.1.

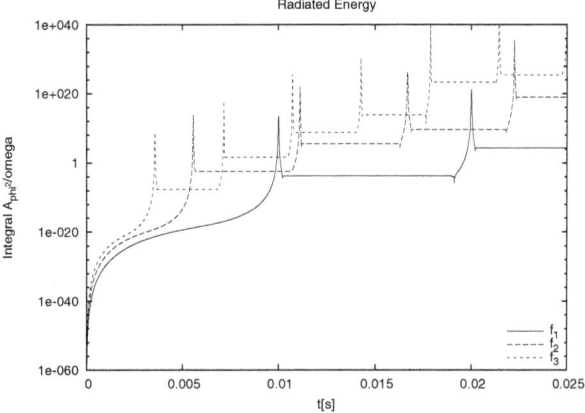

Fig. 1.13. Integral over radiated energy for three frequencies f=50 Hz, 90 Hz, 140 Hz.

which is a conventional approach for a radially decreasing vector potential which oscillates in time with frequency β. Inserting this into Eq. (1.89) results in a differential equation for ω_0 in the variable t:

$$-\left(\beta^2 + i\,\omega_0\,\beta - \dot{\omega}_0\right)\,e^{-\alpha\,r-i\,\beta}\,C = -f_1 \tag{1.92}$$

The solution of this equation is

$$\omega_0(t) = c(r)\,e^{i\beta t} + i\,\beta - \frac{f_1 t}{C}e^{\alpha r+i\,\beta t} \tag{1.93}$$

with a constant $c(r)$ which is dependent on r in general. Inserting (1.93) into (1.90) yields

$$(i\,\alpha\,\beta\,r - \omega_0\,\alpha\,r + \dot{\omega}_0\,r - i\,\beta + \omega_0)\,e^{-\alpha\,r-i\,\beta\,t}\,C = 0 \tag{1.94}$$

which has the solution

$$\omega_0(r) = \frac{c(t)}{r}\,e^{\alpha\,r} + i\,\beta. \tag{1.95}$$

Since both solutions (1.93) and (1.95) must be compatible, we have to assume

$$c(r) = 0. \tag{1.96}$$

By comparison of both equations for ω_0 (1.93 and 1.95) we find

$$c(t) = -\frac{f_1 t}{C} e^{i\beta t} \tag{1.97}$$

and finally

$$\omega_0(r) = -\frac{f_1 t}{C} e^{\alpha r + i\beta t} + i\beta. \tag{1.98}$$

We see that the spin connection has a diverging behaviour in space as well as in time which is consistent with the results of the numerical model.

1.4.6 Summary and discussion

The Bedini device has been analysed by analytical and numerical methods. Based on a model of cylindrical symmetry of the transducer, which is considered to be the essential part for spacetime coupling, the following mechanism of spacetime interaction could be identified:

Under undisturbed conditions, the magnetic field in the transducer is cylindrically symmetric. The radial part of the vector potential must vanish due to the Gauss law. The passing magnets of the wheel distort the symmetry of the magnetic field in the transducer by inducing an asymmetric signal. This leads to a radial component of the vector potential which was not present before. The vector potential changes in time and therefore induces an electric field. Consequently, the Coulomb law has to be fulfilled as an additional condition, in this case for a vanishing charge density (the electric field is completely a radiated field). The ECE field equations show that for the Coulomb law the radial component of the vector potential has either to be zero, or the scalar spin connection must exist to compensate a non-vanishing radial component of the vector potential. The latter case is fulfilled in the Bedini device and leads to the observed resonant behaviour.

A model has been developed which takes a timely varying radial vector potential A_r as a given input. By means of the Coulomb law, a spin connection ω_0 is produced. The zero crossings of A_r lead to a singular value of the spin connection, leading in turn to very high values of the φ component of the vector potential. These are the giant resonances which are strong enough in practice to transfer significant amounts of energy from spacetime to the machine.

For some spacetime resonance experiments it is reported that there is a glowing or fluorescent light effect around the apparatus when it is at resonance. At the same time the measured current of the driving mechanism takes a minimum. This can qualitatively be explained by analyzing the contributions of the ECE electric charge current density. It is given in general by

$$J = (\tilde{R} \wedge A - \omega \wedge \tilde{F}) \tag{1.99}$$

(in short hand notation) for the Hodge duals of curvature $\tilde{R}$ and the electromagnetic field $\tilde{F}$ as well as the potential A. At spin connection resonance, it may happen that the term $\omega \wedge \tilde{F}$ outweights the curvature term. Then the charge current can become significantly smaller while the region of space has a high energy density due to the large spin connection term. Obviously no "negative energy" is required to explain the effect.

The only experimental feature which cannot be directly related to our model is the required behaviour of the driving current. According to Bedini, the motor pulse acts as driving force for the spacetime resonance and must be very short and sharp without oscillations. The model calculations showed that the form of the driving force is not important as long as it is different from zero at the diverging time positions of the spin connection.

Based on the results of this paper we can give some recommendations for further investigations and improvements of the Bedini design, under the prerequisite that our model is correct:

1. The vector potential A_r has to be provided in a way to have zero crossings. This could be enforced by positioning magnets with alternating polarity on the wheel.

2. Since mechanical parts limit the lifetime of a device, a design without moving parts is desirable. The principles of the design can be retained by replacing the wheel by a rotating electromagnetic field (for example based on a three-phase AC voltage). Then arbitrary rotation frequencies can be applied without mechanical restrictions.

3. Effects of the asymmetry of the signal inducing the resonance should be investigated. For example it could be tested if a linear motion of a magnet perpendicular to the transducer would evoke resonance effects too.

A

Appendix 1: Reduction of Form Notation to Vector Notation

In differential form notation the electromagnetic field in ECE theory is:

$$F^a = d \wedge A^a + \omega^a_b \wedge A^b \tag{A.1}$$

which in tensor notation is 1–10:

$$F^a_{\mu\nu} = \partial_\mu A^a_\nu - \partial_\nu A^a_\mu + \omega^a_{\mu b} A^b_\nu - \omega^a_{\nu b} A^b_\mu. \tag{A.2}$$

The electromagnetic potential is:

$$A^a_\mu = A^{(0)} q^a_\mu \tag{A.3}$$

where q^a_μ is a rank two mixed index tensor defined by:

$$V^a = q^a_\mu V^\mu. \tag{A.4}$$

Here V^a and V^μ are four vectors in different frames of reference labeled a and μ in four dimensional space-time. Consider a particular example of Eq. (A.2):

$$F^1_{23} = \partial_2 A^1_3 - \partial_3 A^1_2 + \omega^1_{2b} A^b_3 - \omega^1_{3b} A^b_2. \tag{A.5}$$

Either side of the equation there are rank three tensors whose components must correspond to each other on both sides. Thus:

$$F^1_{23} = \left(\partial_2 A_3 - \partial_3 A_2\right)^1 + \left(\omega_{2b} A^b_3 - \omega_{3b} A^b_2\right)^1. \tag{A.6}$$

Inside the brackets on the right hand side are anti-symmetric tensor components which correspond to the components of an axial vector (magnetic field) or polar vector (electric field). The magnetic vector components are defied by:

$$B_i^1 = \frac{1}{2}\epsilon_{ijk}F_{jk}^1 \qquad (A.7)$$

thus:

$$B_1^1 = \frac{1}{2}\left(\epsilon_{123}F_{23} + \epsilon_{132}F_{32}\right)^1 = F_{23}^1. \qquad (A.8)$$

This is recognized as the X component:

$$B_X = B_1^1 \qquad (A.9)$$

of the magnetic field:

$$\mathbf{B} = B_X\mathbf{i} + B_Y\mathbf{j} + B_Z\mathbf{k}. \qquad (A.10)$$

Similarly:

$$B_Y = B_2^2 = F_{31}^2, \qquad (A.11)$$

$$B_Z = B_3^3 = F_{12}^3. \qquad (A.12)$$

These results were checked by computer in paper 93 of the ECE series [1–10]. So Eq. (A.6) becomes:

$$\mathbf{B} = \nabla \times \mathbf{A} - \boldsymbol{\omega}_b \times \mathbf{A}^b. \qquad (A.13)$$

In this notation:

$$\left(\boldsymbol{\omega}_b \times \mathbf{A}^b\right)_X = \left(\omega_{3b}A_2^b - \omega_{2b}A_3^b\right)^1 \qquad (A.14)$$

where the minus sign has been introduced following the usage of previous papers.

These results are obtained in the special case:

$$a = \mu \qquad (A.15)$$

in Eq. (A.4). This means that the vectors V^a and V^μ are written in the same frame of reference. Thus q_μ^a is diagonal in this special case:

$$V^0 = q_0^0 V^0, \; V^1 = q_1^1 V^1, \; V^2 = q_2^2 V^2, \; V^3 = q_3^3 V^3, \qquad (A.16)$$

and from Eq. (A.3), A^a_μ must be diagonal also. So in Eq. (A.14)

$$(\boldsymbol{\omega}_b \cdot \mathbf{A}^b)_X = (\omega_{32} A_2^2 - \omega_{23} A_3^3)^1 = \omega^1_{32} A_2^2 - \omega^1_{23} A_3^3. \qquad (A.17)$$

Similarly:

$$(\boldsymbol{\omega}_b \times \mathbf{A}^b)_Y = \omega^2_{13} A_3^3 - \omega^2_{31} A_1^1, \qquad (A.18)$$

$$(\boldsymbol{\omega}_b \times \mathbf{A}^b)_Z = \omega^3_{21} A_1^1 - \omega^3_{12} A_2^2. \qquad (A.19)$$

Therefore the meaning of the b index is given by Eqs. (A.17) to (A.19). The final result is:

$$\mathbf{B} = \boldsymbol{\nabla} \times \mathbf{A} - \boldsymbol{\omega} \times \mathbf{A} \qquad (A.20)$$

as used in previous papers on SCR [1]- [10]. The spin connection has been reduced here to a vector $\boldsymbol{\omega}$. The components of this vector in analogy with Eqs. (A.9) to (A.12) are:

$$\omega_X = \omega^1_1 = \omega^1_{32} = -\omega^1_{23}, \qquad (A.21)$$
$$\omega_Y = \omega^2_2 = \omega^2_{31} = -\omega^2_{13}, \qquad (A.22)$$
$$\omega_Z = \omega^3_3 = \omega^3_{12} = -\omega^3_{21}. \qquad (A.23)$$

So:

$$\boldsymbol{\omega} = \omega_X \mathbf{i} + \omega_Y \mathbf{j} + \omega_Z \mathbf{k}. \qquad (A.24)$$

Finally if we adopt the complex circular basis [1]- [10]:

$$\mathbf{B}^{(3)*} = \boldsymbol{\nabla} \times \mathbf{A}^{(3)*} - i\boldsymbol{\omega}^{(1)} \times \mathbf{A}^{(2)} \qquad (A.25)$$

and if:

$$\boldsymbol{\omega}^{(1)} = g\mathbf{A}^{(1)} \qquad (A.26)$$

we obtain the $\mathbf{B}^{(3)*}$ spin field:

$$\mathbf{B}^{(3)*} = -ig\mathbf{A}^{(1)} \times \mathbf{A}^{(2)}. \qquad (A.27)$$

B

Appendix 2: Derivation of the Electric Field in Vector Notation

For the electric field we consider:

$$F^i_{0i} = (\partial_0 A_i - \partial_i A_0)^1 + \omega^i_{0i} A^i_i - \omega^i_{i0} A^0_0, \ i = 1, 2, 3 \tag{B.1}$$

which is equivalent in vector notation to:

$$\mathbf{E} = -\boldsymbol{\nabla}\phi - \frac{\partial \mathbf{A}}{\partial t} - c\omega^0 \mathbf{A} + c\phi\boldsymbol{\omega}. \tag{B.2}$$

Therefore

$$-\left(\boldsymbol{\nabla}\phi + \frac{\partial \mathbf{A}}{\partial t}\right)_X = (\partial_0 A_1 - \partial_1 A_0)^1, \tag{B.3}$$

$$-\left(\boldsymbol{\nabla}\phi + \frac{\partial \mathbf{A}}{\partial t}\right)_Y = (\partial_0 A_2 - \partial_2 A_0)^2, \tag{B.4}$$

$$-\left(\boldsymbol{\nabla}\phi + \frac{\partial \mathbf{A}}{\partial t}\right)_Z = (\partial_0 A_3 - \partial_3 A_0)^3, \tag{B.5}$$

and

$$-\left(c\omega^0 \mathbf{A} - c\phi\boldsymbol{\omega}\right)_X = \omega^1_{01} A^1_1 - \omega^1_{10} A^0_0, \tag{B.6}$$

$$-\left(c\omega^0 \mathbf{A} - c\phi\boldsymbol{\omega}\right)_Y = \omega^2_{02} A^2_2 - \omega^2_{20} A^0_0, \tag{B.7}$$

$$-\left(c\omega^0 \mathbf{A} - c\phi\boldsymbol{\omega}\right)_Z = \omega^3_{03} A^3_3 - \omega^3_{30} A^0_0. \tag{B.8}$$

Thus:

$$\mathbf{A} = A_X \mathbf{i} + A_Y \mathbf{j} + A_Z \mathbf{k}. \tag{B.9}$$

Appendix 2: Derivation of the Electric Field in Vector Notation

where

$$A_X = A_1^1, \ A_Y = A_2^2, \ A_Z = A_3^3. \tag{B.10}$$

and

$$\boldsymbol{\omega} = \omega_X \mathbf{i} + \omega_Y \mathbf{j} + \omega_Z \mathbf{k}. \tag{B.11}$$

where

$$\omega_X = \omega_{10}^1, \ \omega_Y = \omega_{20}^2, \ \omega_Z = \omega_{30}^3. \tag{B.12}$$

The scalar part of the spin connection is defined by:

$$c\omega^0 = -\omega_{01}^1 = -\omega_{02}^2 = -\omega_{03}^3 \tag{B.13}$$

and the scalar potential is defined by:

$$c\phi = -A_0^0. \tag{B.14}$$

So the electric and magnetic field in general relativity (ECE theory) are:

$$\mathbf{E} = -\boldsymbol{\nabla}\phi - \frac{\partial \mathbf{A}}{\partial t} - c\omega^0 \mathbf{A} + c\boldsymbol{\omega}\phi. \tag{B.15}$$

$$\mathbf{B} = \boldsymbol{\nabla} \times \mathbf{A} - \boldsymbol{\omega} \times \mathbf{A}. \tag{B.16}$$

References

[1] M. W. Evans, "Generally Covariant Unified Field Theory" (Abramis, Suffolk, 2005 onwards), volumes one to four, volume five in prep. (Papers 71 to 93 on www.aias.us).

[2] L. Felker. "The Evans Equations of Unified Field Theory" (Abramis, Suffolk, 2007).

[3] K. Pendergast, "Crystal Spheres" (Abramis to be published, preprint on www.aias.us, 2008).

[4] M. W. Evans, Acta Phys. Polonica, **38**, 2211 (2007).

[5] M. W. Evans and H. Eckardt, Physica B, 400, 175 (2007).

[6] M. W. Evans, Physica B, 403, 517 (2008).

[7] M. W. Evans, Omnia Opera of www.aias.us, 1992 to present.

[8] M. W. Evans and H. Eckardt, Paper 63 of www.aias.us, published in volume four of ref. (1).

[9] M. W. Evans, reviews in Adv. Chem. Phys., vols. 119(2) and 119(3) (2001)

[10] M. W. Evans and L. B. Crowell, "Classical and Quantum Electrodynamics and the B(3) Field" (World Scientific, 2001); M. W. Evans and J.-P. Vigier, "The Enigmatic Photon" (Kluwer, Dordrect, 1994 to 2002, hardback and softback), in five volumes.

[11] J. Bedini, US patents on the Bedini devices: US Patent No. 6,392,370 (2002), 6,545,444 (2003), 20020097013 (2002), 20020130633 (2002); industry certification test report (German TUV, 2002) under http://www.icehouse.net/john34/bedinibearden.html.

[12] E. G. Milewski, ed., "Vector Analysis Problem Solver" (Education Association, New York, 1984).

[13] J. Bedini and T. E. Bearden, "Free Energy Generation", Cheniere Press, 2006.

[14] http://tech.groups.yahoo.com/group/Bedini_Monopole.

The Coulomb and Ampère Maxwell Laws in the Friedmann Lemaître Robertson Walker Metric

by

Myron W. Evans,
Alpha Institute for Advanced Study, Civil List Scientist.
(emyrone@aol.com and www.aias.us)

Abstract

The Coulomb and Ampère Maxwell laws are derived in the Friedmann Lemaître Robertson Walker metric and it is shown that these laws of classical electrodynamics depend on the evolution of the universe. This is the result of a generally covariant unified field theory in which all sectors are rigorously objective. The charge density and current density of these laws depend on the age of the universe, a new result in physics. The experimental implications of this result are discussed in terms of the change in the optical properties of light deflected by gravitation.

Keywords: Coulomb law, Ampère Maxwell law, Friedmann Lemaître Robertson Walker metric, generally covariant unified field theory.

2.1 Introduction

Recently a generally covariant unified field theory has been developed on the basics of general relativity, that physics is geometry, and that objectivity in physics is measured by geometry. The latter in Einstein Cartan Evans (ECE) field theory is represented by standard Cartan geometry [1–10] in which space-time torsion is present in general as well as space-time curvature. Within this geometrical structure the laws of classical electrodynamics have been unified with the laws of gravitation [11] for several representative metrics.

In Section 2.2 the implications of this result are developed for the most used metric in cosmology, the Friedmann Lemaître Robertson Walker (FLRW) metric [12] that is an exact solution of the Einstein Hilbert field equation. Many such exact solutions are now known [13] but few of these are likely to give scientifically acceptable charge and current densities of the Coulomb and Ampère Maxwell laws. This procedure has been made possible by the development [1–10] of ECE theory (www.aias.us and www.atomicprecision.com). The charge and current densities of the FLRW metric are given in Section 2 for the usual cases developed in cosmology, a flat, closed and open universe [12]. In Section 2.3 the results are analyzed, and the FLRW metric criticized on the geometrical ground given by Crothers [11]. Some elementary errors [12] in units used in cosmology are corrected. The main result is that these laws of classical electrodynamics depend on the evolution and age of the universe. They are not immutable laws as in the Maxwell Heaviside field theory of special relativity. As Crothers has argued [11] the singularity in FLRW at $t = 0$ does not signify Big Bang, it is due to a geometrical misconception carried through uncritically in the standard model literature. The correct treatment of metrics of this type has been given by Eddington [14]. This paper is intended to develop the laws of electrodynamics in the FLRW metric as usually used in cosmology, in a more rigorous treatment in forthcoming work, the rigorously correct Crothers metric [11] will be used.

2.2 The Laws of Electrodynamics in the FLRW Metric

The FLRW metric is [12]:

$$ds^2 = -c^2 dt^2 + a^2(t) \left(\frac{dr^2}{1 - kr^2} + r^2(d\theta^2 + \sin^2\theta d\phi^2) \right) \qquad (2.1)$$

and is an exact solution of the Einstein Hilbert (EH) field equation. Here a is dimensionless, k is in inverse square meters and r is in meters squared. The usual habit in cosmology is to regard k as "dimensionless". If this is done then $k = -1$ is an open universe, $k = 0$ is a flat universe and $k = 1$ is a closed universe. We replace this habit as follows in rigorous S.I. units:

$$k = 0, \quad \pm 1 \ m^{-2}. \qquad (2.2)$$

Matter and energy in this metric may be modeled by a perfect fluid [12], whose canonical energy momentum density is:

$$T_{\mu\nu} = (\boldsymbol{p} + \rho) U_\mu U_\nu + \rho g_{\mu\nu}. \qquad (2.3)$$

Here ρ and p are respectively the energy density and pressure as measured in the rest frame, and U^μ is the four velocity of the fluid. If both the fluid

2.2 The Laws of Electrodynamics in the FLRW Metric

and metric are isotropic then their frames coincide and they are known as "co-moving" [12]. The fluid is at rest in co-moving coordinates. In this case:

$$U^\mu = (1,0,0,0) \tag{2.4}$$

and

$$T_{\mu\nu} = \begin{bmatrix} \rho & 0 & 0 & 0 \\ 0 & & & \\ 0 & & g_{(ij)}p & \\ 0 & & & \end{bmatrix}. \tag{2.5}$$

Lowering an index it is found that [12]:

$$T^\mu{}_\nu = \text{diag}\,(-\rho, p, p, p) \tag{2.6}$$

whose trace is:

$$T = T^\mu{}_\mu = -\rho + 3p. \tag{2.7}$$

An equation of state [12] must now be defined. The simplest choice is:

$$p = w\rho \tag{2.8}$$

where w is a constant independent of time. The law of conservation of energy is then given by [12]:

$$\frac{\dot\rho}{\rho} = -3(1+w)\frac{\dot a}{a} \tag{2.9}$$

i.e.

$$\rho \propto a^{-3(1+w)}. \tag{2.10}$$

Dust and radiation are examples of cosmological fluids. The former is defined as collision-less and non-relativistic with no pressure (w is zero). Examples are stars and galaxies. The energy density of dust is dominated by the rest energy, which is proportional to the number density. In this paper we will restrict consideration to dust defined in this way. The EH equation is written [12] in the form:

$$R_{\mu\nu} = 8\pi G \left(T_{\mu\nu} - \frac{1}{2} g_{\mu\nu} T \right) \tag{2.11}$$

where $R_{\mu\nu}$ is the Ricci tensor, G is Newton's constant, $g_{\mu\nu}$ is the symmetric metric and T the trace of $T_{\mu\nu}$. For:

$$\mu = 0, \nu = 0 \tag{2.12}$$

Eq. (2.11) gives, for the FLRW metric (2.1):

$$\frac{-3\ddot{a}}{a} = 4\pi G \left(\rho + 3p\right) \tag{2.13}$$

where:

$$\dot{a} = \frac{da}{dt}. \tag{2.14}$$

For

$$\mu = i, \quad \nu = j \tag{2.15}$$

we obtain:

$$\frac{\ddot{a}}{a} + 2\left(\frac{\dot{a}}{a}\right)^2 + \frac{2k}{a^2} = 4\pi G(\rho - p). \tag{2.16}$$

Re-arrangement of terms gives the Friedmann equations [12]:

$$\frac{\ddot{a}}{a} = -\frac{4}{3}\pi G(\rho + 3p) \tag{2.17}$$

and

$$\left(\frac{\dot{a}}{a}\right)^2 = \frac{8\pi G}{3}\rho - \frac{k}{a^2}. \tag{2.18}$$

In ref. [11] it was found that the charge density of the Coulomb law given by this metric is:

$$J^0 = -3\phi\frac{\ddot{a}}{a} \tag{2.19}$$

in Vm^{-2} (S.I. units). Here ϕ is the primordial scalar potential in volts (J/C) of the ECE theory. Using the first Friedmann equation (2.17):

$$J^0 = 4\pi\phi G(\rho + 3p) \tag{2.20}$$

2.2 The Laws of Electrodynamics in the FLRW Metric

For dust:

$$p = 0 \tag{2.21}$$

and we obtain:

$$J^0 = 4\pi\phi G\rho. \tag{2.22}$$

The Coulomb law in this case is therefore:

$$\nabla \cdot \mathbf{E} = \rho_e/\epsilon_0 \tag{2.23}$$

where the electric charge density ρ_e is defined by:

$$\rho_e = 4\pi\epsilon_0 \phi G \rho \tag{2.24}$$

The S.I. units of this equation are as follows:

$$\rho_e = Cm^{-3}; \epsilon_0 = J^{-1}C^2m^{-1}; \phi = \text{volt} = JC^{-1};$$
$$G = mkgm^{-1}; \rho = kgm\,m^{-3}; J^0 = \text{volt}\,m^{-2}. \tag{2.25}$$

The precise self consistency of ECE theory is demonstrated by the fact that Eq. (2.24) is the same exactly in structure as Eq. (5.102) vol. 1 of ref. (1). Eq (2.24) shows that charge density cannot exist without mass density, and the implications of this result were discussed in chapter five vol. 1 of ref. (1). For example the electron's charge is always found with the electron's mass. This finding has now been explained from first principles using the required generally covariant unified field theory [1–11]. In Maxwell Heaviside (MH) theory there is no explanation for it, because MH is special relativity whose flat or Minkowski metric is:

$$ds^2 = -c^2 dt^2 + dr^2 + r^2(d\theta^2 + \sin^2\theta d\phi^2). \tag{2.26}$$

In ECE theory [11] the Minkowski metric produces zero charge density in the Coulomb law because curvature and torsion are zero in the Minkowski metric.

The radial and angular components of the current density of the FLRW metric [11] were found to be as follows:

$$J_r = \frac{A^{(0)}}{\mu_0}\left(\frac{1-kr^2}{a^2}\right)\left(\frac{2}{a^2}(k+\dot{a}^2)+\frac{\ddot{a}}{a}\right) \tag{2.27}$$

and

$$J_\theta = J_\phi \sin^2\theta = \frac{A^{(0)}}{\mu_0 r^2 a^2}\left(\frac{2}{a^2}(k+\dot{a}^2)+\frac{\ddot{a}}{a}\right), \quad (2.28)$$

where $A^{(0)}$ is the magnitude of the primordial vector potential of ECE theory and μ_0 is the vacuum permeability. The S.I. units of the ratio of these two factors are as follows:

$$\frac{A^{(0)}}{\mu_0} = \frac{Js\,C^{-1}m^{-1}}{Js^2\,C^{-2}m^{-1}} = Cs^{-1} = \text{ampere}, \quad (2.29)$$

so (A/μ_0) has the units of Ampères (C/s). The results for current density depend on the parameter k, i.e. on whether the universe is closed, flat or open. The results for charge density however are independent of k.
a) For a flat universe:

$$k = 0 \quad (2.30)$$

and

$$J_r = r^2 J_\theta = r^2 \sin^2\theta J_\phi = \frac{A^{(0)}}{\mu_0}\left(\frac{4\pi G\rho}{a^2}\right). \quad (2.31)$$

b) For a closed universe:

$$k = 1m^{-2} \quad (2.32)$$

and

$$J_r = \frac{A^{(0)}}{\mu_0}\left(4\pi G\rho\left(\frac{1-kr^2}{a^2}\right)\right) \quad (2.33)$$

with:

$$r^2 J_\theta = r^2 \sin^2\theta J_\phi = \frac{A^{(0)}}{\mu_0}\left(\frac{4\pi G\rho}{a^2}\right) \quad (2.34)$$

c) For an open universe:

$$k = -1m^{-2} \quad (2.35)$$

2.2 The Laws of Electrodynamics in the FLRW Metric

and:

$$J_r = \frac{A^{(0)}}{\mu_0} \left(4\pi G\rho \left(\frac{1 - kr^2}{a^2} \right) \right) \tag{2.36}$$

with:

$$r^2 J_\theta = r^2 \sin^2\theta J_\phi = \frac{A^{(0)}}{\mu_0} \left(\frac{4\pi G\rho}{a^2} \right). \tag{2.37}$$

In each case the factor a^2 appears in the denominator. This is defined for a flat universe by:

$$a \propto t^{2/3} \tag{2.38}$$

for a closed universe by:

$$a \propto \left(\frac{1 - \cos\phi}{\phi - \sin\phi} \right) t \tag{2.39}$$

and for an open universe by:

$$a \propto \left(\frac{1 - \cosh\phi}{\phi - \sinh\phi} \right) t. \tag{2.40}$$

Unfortunately the S.I. units of a^2 given in ref. (12) are incorrect, so we adopt a proportionality sign for a. In ref. (12) a constant C is defined by:

$$C := \frac{8\pi G}{3} \rho a^3 \tag{2.41}$$

which must have the units of inverse square meters because a is unitless. However, in ref. (12) the following equations are used respectively for a flat, closed and open universe:

$$\begin{aligned} a &= \frac{C}{2}(1 - \cos\phi), t = \frac{C}{2}(\phi - \sin\phi), k = 1, \\ a &= \left(\frac{9C}{4} \right)^{1/3} t^{2/3}, k = 0, \\ a &= \frac{C}{2}(\cosh\phi - 1), t = \frac{C}{2}(\sinh\phi - \phi), k = -1. \end{aligned} \tag{2.42}$$

It is seen that the units in these equations are not correct. This is a problem of usage in standard model cosmology. The units in the results of this paper are rigorously correct S.I. units of charge density (C/m^3) and current density$(C/s/m^2)$.

2.3 Discussion

The main result is that the Coulomb and Ampère Maxwell laws of classical electrodynamics depend on the evolution of the universe from an initial:

$$t = 0. \tag{2.43}$$

This is usually interpreted as Big Bang [12]. At $t = 0$ the charge density of Eq. (2.24) is finite but the current densities are infinite. If the volume at this initial instant is zero, then the mass density and charge density of Eq. (2.24) become infinite also. However, Crothers [11] has shown that there is an irretrievable error in this theory because of confusion in geometry. The Universe may only go to a finite radius at the initial instant. We accept this criticism by Crothers [11] and in future work will replace the FLRW metric by the rigorously correct Crothers metric evaluated in ref. (11). The purpose of this paper is to show that in a generally covariant unified field theory the laws of electrodynamics evolve, they are not immutable as in the Maxwell Heaviside field theory. The FLRW metric can only be used as a guide to this overall result. As shown in ref. (11), the correct treatment of the metric, following Eddington [14] leads to a universe that is infinite in extent. The standard model FLRW metric leads to the flawed idea of Big Bang. In ref. (11), several other metrics were used to evaluate the charge density of the Coulomb law and the current density of the Ampère Maxwell laws. The FLRW metric is simply the one most used in cosmology [12].

In a generally covariant unified field theory the charge density of the Coulomb law and the current density of the Ampère Maxwell law must be well defined and well behaved. Few metrics of the many now known [13] will satisfy this requirement. For example the charge and current densities must be free of mathematical singularities. In the FLRW metric a singularity occurs at the initial point in time, so at that point the metric is not well behaved in general relativity. As argued already the source of this is incorrect [11] evaluation of geometry. However, for finite t the charge density of FLRW appears to be well behaved and finite. Unfortunately it is not possible to fully evaluate the behavior of the current densities for finite t because of elementary errors in units in ref. (12). It may be concluded that the current densities of the Ampère Maxwell law from the standard model FLRW metric (which is flawed [11] geometrically) are different for a flat, closed and open universe, and provided that t is not zero, they have no mathematical singularities.

In order to test these conclusions experimentally the properties of light deflected by gravitation can be analyzed for changes in electrodynamic and kinematic properties, for example changes in polarization [1–10] which are known to occur from objects such as a white dwarf. The reason for this method is that the current density of the Ampère Maxwell law is defined in ECE by curvature [11], so the existence of the current density changes the vacuum Ampère Maxwell law from:

$$\nabla \times \boldsymbol{B} = \frac{1}{c^2} \frac{\partial \boldsymbol{E}}{\partial t} \tag{2.44}$$

to

$$\nabla \times \boldsymbol{B} = \frac{1}{c^2} \frac{\partial \boldsymbol{E}}{\partial t} + \mu_0 \boldsymbol{J}. \tag{2.45}$$

Similarly the vacuum Coulomb law is changed from

$$\nabla \cdot \boldsymbol{E} = 0 \tag{2.46}$$

to

$$\nabla \cdot \boldsymbol{E} = \rho_e / \epsilon_0 \tag{2.47}$$

In other words charge and current density exist in ECE where there appears to be no mass present, for example at the edges of a very heavy object such as a white dwarf. The existence of $\boldsymbol{J}$ in eq. (2.45) changes the polarization from for example circular to elliptical [1–10].

Acknowledgments

The British Government is thanked for a Civil List Pension and the staff of AIAS and others for many interesting discussions.

References

[1] M. W. Evans, "Generally Covariant Unified Field Theory" (Abramis Academic, Suffolk, 2005 onwards), volumes one to four, volume five in prep. (papers 71 to 93 on www.aias.us).

[2] L. Felker, "The Evans Equations of Unified Field Theory" (Abramis Academic, Suffolk, 2007).

[3] K. Pendergast, "Mercury as Crystal Spheres" (Abramis Academic, in prep., and on www.aias.us, a short history of ECE theory).

[4] The AIAS websites www.aias.us and www.atomicprecision.com.

[5] M. W. Evans, Acta Phys. Polon., **38**, 2211 (2007).

[6] M. W. Evans and H. Eckardt, Physica B, in press (2007).

[7] M. W. Evans, Physica B, in press (2007).

[8] M. W. Evans et al., precursor theories to ECE (1992 to 2003 on the Omnia Opera section of www.aias.us).

[9] M. W. Evans and L. B. Crowell, "Classical and Quantum Electrodynamics and the $\bm{B}^{(3)}$ Field" (World Scientific, 2001); M. W. Evans and J.-P. Vigier, "The Enigmatic Photon" (Kluwer, Dordrecht 1994 to 2002 in hardback and softback), in five volumes.

[10] M. W. Evans (ed.), "Modern Non-linear Optics", a special topical issue of I. Prigogine and S. A. Rice (series eds.), "Advances in Chemical Physics" (Wiley, New York, 2001, second edition) vols. 119(1) to 119(3); ibid.., M. W. Evans and S. Kielich (eds.), (Wiley, New York, 1992, 1993 and 1997, first edition), vols. 85(1) to 85(3).

[11] M. W. Evans, H. Eckardt and S. Crothers, paper 93 on www.aias.us.

[12] S. P. Carroll, "Space-time and Geometry: an Introduction to General Relativity" (Addison Wesley, New York, 2004).

[13] H. Stephani, D. Kramer, M. MacCallum, C. Hoenselaers and E. Hertl, "Exact Solutions of Einstein's Field Equation" (Cambridge Univ. Press., second edition, 2003).

[14] A. S. Eddington, "The Mathematical Theory of Relativity" (Cambridge Univ. Press, second edition, 1960).

A Critical Evaluation of Standard Model Cosmology with Einstein Cartan Evans (ECE) Field Theory

by

Myron W. Evans,
Alpha Institute for Advanced Study, Civil List Scientist.
(emyrone@aol.com and www.aias.us)

and

H. Eckardt and S. J. Crothers,
Alpha Institute for Advanced Study.
(www.aias.us)

Abstract

Some claims of standard model cosmology are tested with the inhomogeneous field equation of ECE field theory. The only rigorously correct line elements available at present are those given by Crothers, because they are the only line elements that are geometrically correct as well as being exact solutions of the Einstein Hilbert (EH) field equation. A small sample of rigorously correct line elements is used to produce charge/current densities of the inhomogeneous ECE field equation and it is found that at present there is no rigorously correct metric available that produces electromagnetic radiation in ECE theory. The reason is that the standard model of cosmology is based on an incorrect appreciation of differential geometry and must be disregarded for this reason.

Keywords: ECE theory, Einstein Hilbert field equation, charge/current density, Crothers lien elements, criticisms of standard model cosmology.

3.1 Introduction

Recently a generally covariant unified field theory has been developed [1–10] that is based rigorously on the philosophy of general relativity [11]. This is known as Einstein Cartan Evans (ECE) field theory because it is based on the well known differential geometry of Cartan. The latter geometry extends Riemann geometry by use of the Cartan torsion. ECE theory has been tested extensively (www.aias.us) against experimental data and is based directly on Cartan geometry. The mathematical correctness of ECE theory is obvious, because Cartan geometry is a valid geometry, but nevertheless ECE theory has also been tested exhaustively and the results accepted by the international community of scientists (feedback to www.aias.us over three and a half years). Recently the theory has been applied using line elements [12] which are exact solutions of the Einstein Hilbert (EH) field equation. The basic equation used for this test of standard model cosmology is the inhomogeneous ECE field equation given in Section 3.2. This has a simple structure when written in differential geometry. It becomes a little more complicated in other notations, but at the same time can be reduced to the familiar vector notation of the Maxwell Heaviside (MH) field theory. The familiar laws of electrodynamics in ECE theory take the same form as in MH theory, but are written in ECE in a space-time that has both curvature and torsion. In MH theory the equations are written in a space-time that has no curvature and no torsion - the Minkowski space-time of special relativity. Thus ECE unifies electrodynamics and gravitation in a natural way - based directly on geometry. MH cannot unify the two fundamental fields because it is developed in a space-time that is flat and not generally covariant. The Minkowski space-time supports only Lorentz covariance as is well known.

In order to apply ECE theory, line elements must be found that are suitable for the gravitational sector of ECE. Charge/current densities are calculated from these line elements [1–10] in various approximations. In the first approximation used in this paper and previous papers on this topic, the gravitational torsion is assumed to be negligible compared with the gravitational curvature. This is a situation that exists for example in the solar system. In this approximation line elements can be used which are solutions to the EH field equation, in which gravitational torsion is absent. Crothers [13–15] has shown that such line elements must also be well behaved geometrically as well as being exact solutions to the EH field equation. At present the Crothers metrics are the only ones that are acceptable, because they are the only ones that are rigorously correct geometrically. In Section two, a small sample of Crothers metrics is used to compute charge and current densities of the ECE Coulomb and Ampère Maxwell laws. The well known line elements of the standard model are incorrect fundamentally because they violate differential geometry at a fundamental level. This incorrectness has led to the crude fallacies of Big Bang, Black Holes, and dark matter. All inferences based on this pseudo-science are false because they are based on incorrect mathematics. In

Sections 2 and 3 the obvious errors in the standard model are illustrated with a few examples. Because of these errors it is concluded that at present there exists no rigorously correct line element that is able to produce electromagnetic radiation in a generally covariant unified field theory, i.e. a theory that is demanded by the philosophy of relativity.

3.2 Testing with the Inhomogeneous Field Equation of ECE

In the simplest type of notation [1–10] the inhomogeneous ECE field equation in the approximation of vanishing gravitational torsion is:

$$d \wedge \widetilde{F} = A^{(0)} (\widetilde{R} \wedge q)_{\text{grav}} \tag{3.1}$$

where $\widetilde{F}$ is the Hodge dual of the electromagnetic field form F and $\widetilde{R}$ is the Hodge dual of the curvature or Riemann form of ECE theory. The subscript in Eq. (3.1) means that the gravitational sector is described by the wedge product $\widetilde{R} \wedge q$, where q is the tetrad form. In vector notation Eq. (3.1) becomes two laws of ECE theory, the Coulomb and Ampère Maxwell laws:

$$\boldsymbol{\nabla} \cdot \boldsymbol{E} = \frac{\rho}{\epsilon_0} \tag{3.2}$$

and

$$\boldsymbol{\nabla} \times \boldsymbol{B} - \frac{1}{c^2} \frac{\partial \boldsymbol{E}}{\partial t} = \mu_0 \boldsymbol{J}. \tag{3.3}$$

Here $\boldsymbol{E}$ is the electric field strength in volts per meter, ρ is the charge density in Cm^{-3}, ϵ_0 is the S.I. vacuum permittivity, $\boldsymbol{B}$ is the magnetic flux density in tesla, $\boldsymbol{J}$ is the current density in Cs^{-1} meter^{-2}, and μ_0 is the S.I. vacuum permeability. These are the same laws as in MH theory, but ECE derives them from geometric first principles, and is able to compute the charge density and components of the current density from line elements used in the theory of gravitation. The choice of line elements is important, because not only must they be exact solutions of the EH equation, but must also be geometrically correct [13–15]. These requirements are described in more detail by Crothers in Section 23.3 of Chapter 23 of www.aias.us. Here we base our discussion on that Section 23.3, using the same notation. The results are further discussed in Section 23.3 of this paper.

The first example discussed in this paper is the Schwarzschild class of static vacuum solutions. As shown by Eddington [16] and Crothers [13–15], there is an infinite number of possible solutions of the EH equation for this class of metrics. The most general form of the line element for this class of metrics has been given by Crothers [13–15] and is:

$$ds^2 = A(C(r))^{1/2}dt^2c^2 - B(C(r))^{1/2}d(C(r))^{1/2}$$
$$- C(r)(d\theta^2 + \sin^2\theta d\phi^2) \tag{3.4}$$

where:

$$C(r) := C(|r - r_0|). \tag{3.5}$$

Here $A(C(r))^{1/2}$, $B(C(r))^{1/2}$ and $C(r)$ are a priori unknown positive valued analytic functions that must be determined by the intrinsic geometry of the line element and associated boundary conditions. In the class of vacuum solutions the Einstein tensor vanishes:

$$G_{\mu\nu} = 0. \tag{3.6}$$

The radius of curvature [13–15] is defined by:

$$R_c(r) = (C(r))^{1/2}. \tag{3.7}$$

Using (3.4) in the EH field equation gives:

$$ds^2 = \left(1 - \frac{\alpha}{(C(r))^{1/2}}\right)c^2t^2 - \left(1 - \frac{\alpha}{(C(r))^{1/2}}\right)^{-1}d(C(r))^{1/2}$$
$$- C(r)(d\theta^2 + \sin^2\theta d\phi^2). \tag{3.8}$$

Crothers has shown furthermore [17] that the admissible form of $C(r)$ that satisfies the intrinsic geometry of the line element and also the required boundary conditions must be

$$(C(r))^{1/2} = R_c(r) = (|r - r_0|^n + \alpha^n)^{1/n}, \quad r \in R, n \in R^+, r \neq r_0, \tag{3.9}$$

where r_0 and n are entirely arbitrary constants and α is a constant that depends on the mass of the gravitational field, but which cannot be identified with a point mass M. The line element (3.8) is well defined on

$$-\infty < r < r_0 < r < \infty \tag{3.10}$$

and has a singularity if and only if:

$$r = r_0. \tag{3.11}$$

3.2 Testing with the Inhomogeneous Field Equation of ECE

Since r is never equal to r_0 in Eq. (3.9), no such singularity occurs. There is no black hole singularity. Numerous other errors of the standard model have also been pointed out by Crothers [13–15]. These are irretrievable errors and so standard model cosmology must be discarded and replaced by Crothers metrics. The solution of the EH equation obtained originally by Karl Schwarzschild [18] is the special case:

$$n = 3, r_0 = 0, r > r_0. \tag{3.12}$$

Using this line element it was found by computer algebra that the charge and current densities vanish. The reason for this is that the line element (8) is Ricci flat, i.e. all components of the Ricci tensor vanish, and consequently the Ricci scalar curvature. All the line elements of the spherically symmetric and static Schwarzschild class will give this result, because they are vacuum line elements. The vacuum is defined as Ricci flat. There is an infinite number of such line elements that are exact solutions of the Einstein Hilbert equation, but only the Crothers class is acceptable as also being rigorously correct in differential geometry. In this case the inhomogeneous ECE field equations have the same vector form precisely as the Maxwell Heaviside inhomogeneous field equations. They are the vacuum Coulomb law:

$$\nabla \cdot \boldsymbol{E} = 0 \tag{3.13}$$

and the vacuum Ampère Maxwell law:

$$\nabla \times \boldsymbol{B} - \frac{1}{c^2}\frac{\partial \boldsymbol{E}}{\partial t} = \boldsymbol{0}. \tag{3.14}$$

For any line element that is not Ricci flat a finite charge/current density is obtained, as in Chapter 93 of www.aias.us. The correct way of computing the ECE charge current density from the original line element of Schwarzschild is from eq. (3.8) of this paper. In the generalization (3.4) of the Minkowski element, the a priori unknown A and B must be functions of $C^{1/2}$. The line element (8) is obtained from the line element (4) using the Einstein Hilbert field equation, so the line element (8) must be used to compute ECE charge and current densities, as in this paper. A Lehnert type [19] vacuum current charge density may conceivably be obtained from a line element that is not Ricci flat in the absence of canonical energy-momentum density. In that case, the "vacuum" is a different one from that of the Schwarzschild class, all of whose members are Ricci flat. The usual definition of "vacuum" in general relativity is therefore that the Ricci tensor vanishes and the canonical energy momentum density tensor vanishes. In the Schwarzschild vacuum the ECE charge current density vanishes for all $A^{(0)}$ as we have seen. So ECE is rigorously self-consistent conceptually and mathematically. It is the first successful generally covariant unified field theory that can be used in electrical engineering with the inclusion of the spin connection.

The second example given in this section is Crothers' generalization of the line element for the incompressible sphere of fluid obtained in 1916 by Schwarzschild [20]. In the notation used by Crothers in Section 23.3 of paper 23 his generalization is:

$$ds^2 = \left(\frac{3}{2}\left(\cos|\chi_a - \chi_0| - \cos|\chi - \chi_0|\right)^2 c^2 dt^2 \right.$$
$$\left. -\frac{3}{\kappa\rho_0}d\chi^2 - \frac{3}{\kappa\rho_0}\sin^2|\chi - \chi_0|\left(d\theta^2 + \sin^2\theta d\phi^2\right)\right). \tag{3.15}$$

It was found by computer algebra (using a program written by HE) that this line element obeys the fundamental equation:

$$R \wedge q = 0 \tag{3.16}$$

usually known as the first Bianchi identity. (In fact it was discovered by Ricci and Levi-Civita.) Having checked the line element in this way the computer algebra was used to find that the charge density is proportional to:

$$J^0 = \phi\left(\frac{4\cos(|\chi - \chi_0|)\kappa\rho_0}{(\cos(|\chi - \chi_0|) - 3\cos(|\chi_a - \chi_0|))^3}\right) \tag{3.17}$$

and the current densities to:

$$J_r = \frac{A^{(0)}}{\mu_0}\left(\frac{\cos(|\chi - \chi_0|)\kappa^2\rho_0^2}{9(\cos(|\chi - \chi_0|) - 3\cos(|\chi_a - \chi_0|))} + \frac{2}{9}\kappa^2\rho_0^2\right) \tag{3.18}$$

$$J_\theta = J_\phi \sin^2\theta = \frac{A^{(0)}}{\mu_0}\left(\frac{\cos(|\chi - \chi_0|)\kappa^2\rho_0^2}{9(\cos(|\chi - \chi_0|) - 3\cos(|\chi_a - \chi_0|))}\right.$$
$$\left. + \frac{\kappa^2\rho_0^2}{9\sin(|\chi - \chi_0|)^2} - \frac{(\cos(|\chi - \chi_0|) - 1)(\cos(|\chi - \chi_0|) + 1)\kappa^2\rho_0^2}{9\sin(|\chi - \chi_0|)^4}\right). \tag{3.19}$$

These are graphed in Figs. (3.1) to (3.3). These results pertain to the interior of the sphere only and depend on a non-zero primordial voltage $cA^{(0)}$ being present, proportional to the electronic charge -e regarded as a fundamental constant. Outside the sphere of incompressible fluid the charge and current densities vanish even for non-zero $cA^{(0)}$. As pointed out by Crothers, two line elements are needed for a source of the gravitational field, one for the interior of the source, another for the exterior, where the gravitational

3.2 Testing with the Inhomogeneous Field Equation of ECE

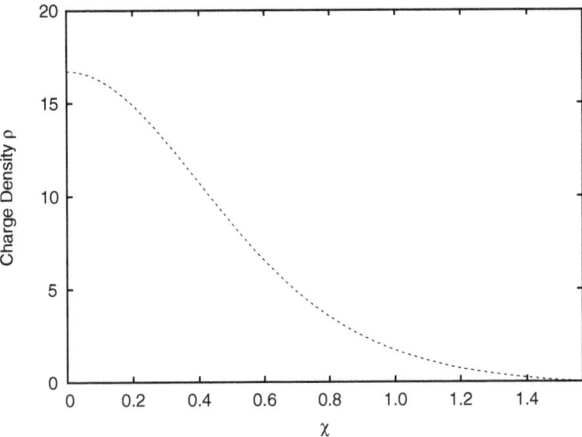

Fig. 3.1. Homogeneous Fluid Sphere, charge density ρ for $\rho_0 = 1, \kappa = 1$, $\chi_a = 1, \chi_0 = 0$.

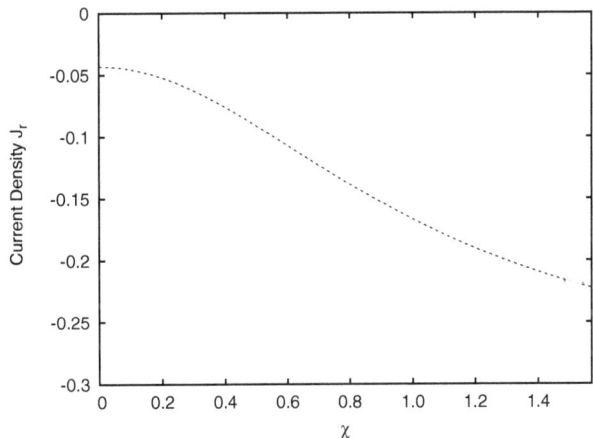

Fig. 3.2. Homogeneous Fluid Sphere, current density J_r for $\rho_0 = 1, \kappa = 1$, $\chi_a = 1, \chi_0 = 0$.

field is modeled mathematically by the center of mass, a purely mathematical concept. This is explained further in Section 3 of this paper. So if a classical electron is modeled like this, it has charge and current density in its interior, but not around it. Obviously this conflicts with the laws of classical electrodynamics, which show that an electron is a source for an electric field, and if it moves with time, radiates. To describe this correctly, a rigorous Crothers type line element is needed that gives a charge current density both in the

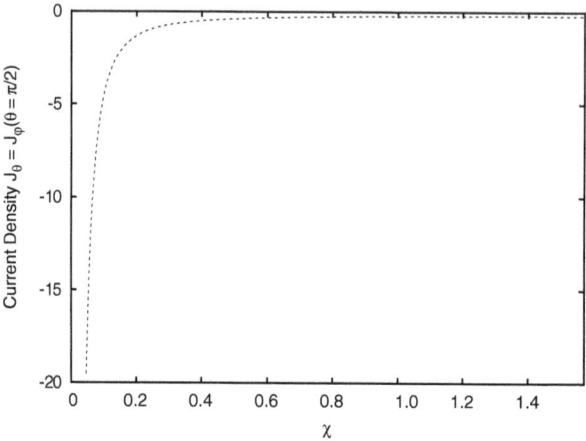

Fig. 3.3. Homogeneous Fluid Sphere, current density J_ϑ, J_φ for $\rho_0 = 1$, $\kappa = 1, \chi_a = 1, \chi_0 = 0$.

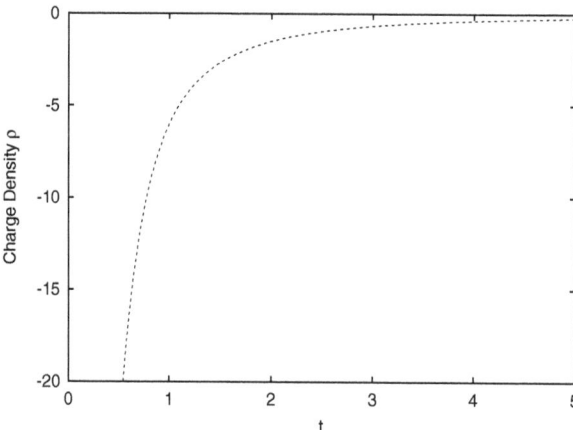

Fig. 3.4. FLRW metric, charge density ρ for $R(t) = t^2, k = .5, r = 1$.

interior and exterior of a source. None of the line elements of the standard model can be accepted because they are geometrically incorrect.

In anticipation of the next section we present the results of the cosmological charge and current densities of the FLRW metric, Eq. (3.20). Figs. (3.4) to (3.6) show the time dependence of these quantities for a fixed radius where the time-dependent curvature radius increases quadratically. All quantities tend to zero over time for an expanding universe. If the universe is contracting (Figs. (3.7) to (3.8)), the cosmological quantities tend to explode. This would only be meaningful if it would appear in a restricted volume. However

3.2 Testing with the Inhomogeneous Field Equation of ECE   53

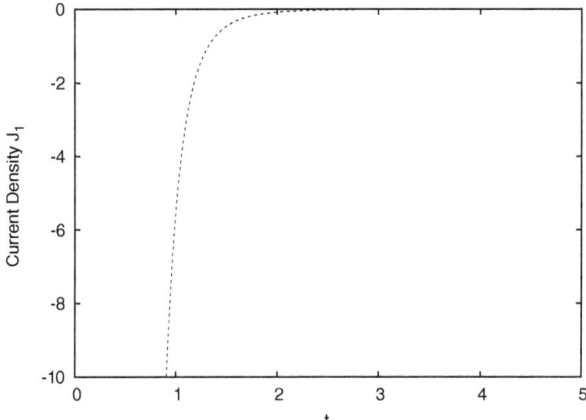

Fig. 3.5. FLRW metric, current density J_r for $R(t) = t^2, k = .5, r = 1$.

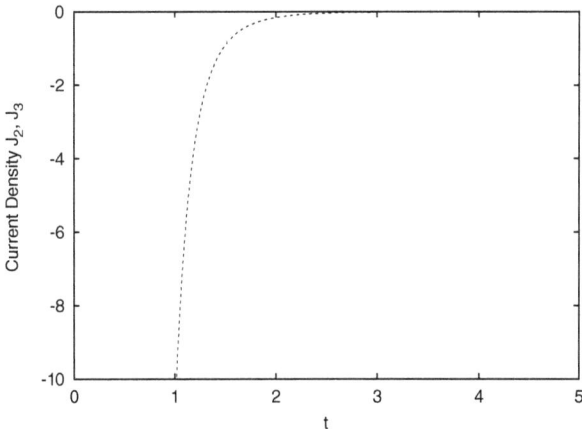

Fig. 3.6. FLRW metric, current density J_ϑ, J_φ for $R(t) = t^2, k = .5, r = 1$.

54   3 A Critical Evaluation of Standard Model Cosmology

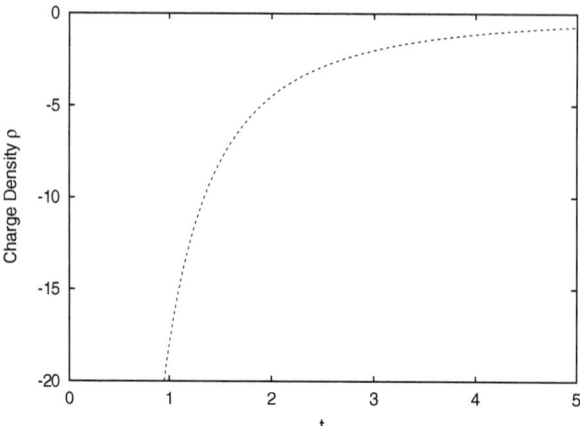

Fig. 3.7. FLRW metric, charge density ρ for $R(t) = t^{-2}, k = .5, r = 1$.

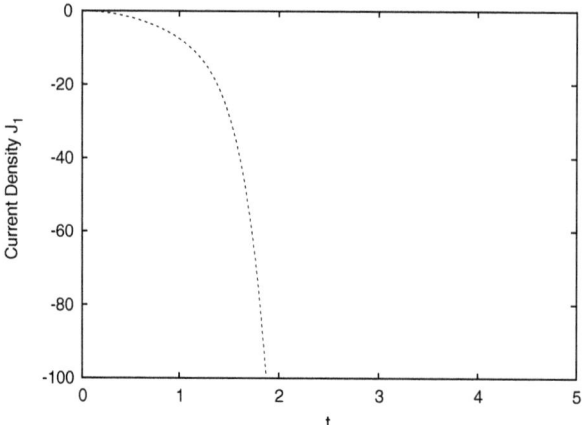

Fig. 3.8. FLRW metric, current density J_r for $R(t) = t^{-2}, k = .5, r = 1$.

3.2 Testing with the Inhomogeneous Field Equation of ECE   55

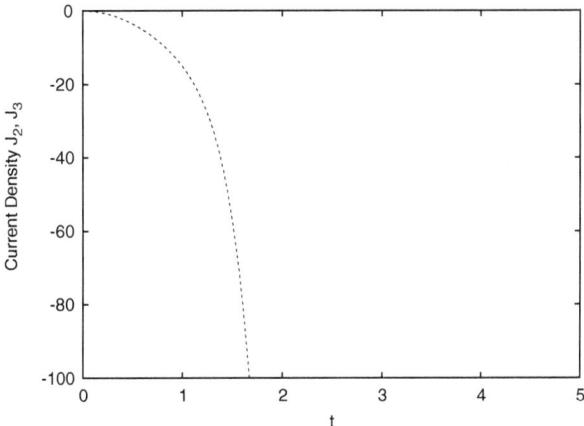

Fig. 3.9. FLRW metric, current density J_ϑ, J_φ for $R(t) = t^{-2}, k = .5, r = 1$.

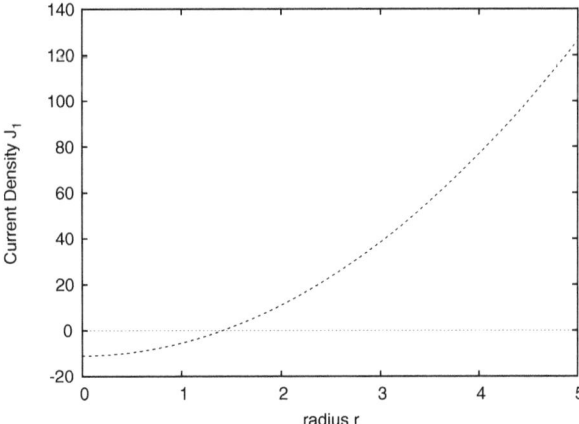

Fig. 3.10. FLRW metric, current density, r dependence of J_r for $R(t) = t^2, t = 1, k = .5$.

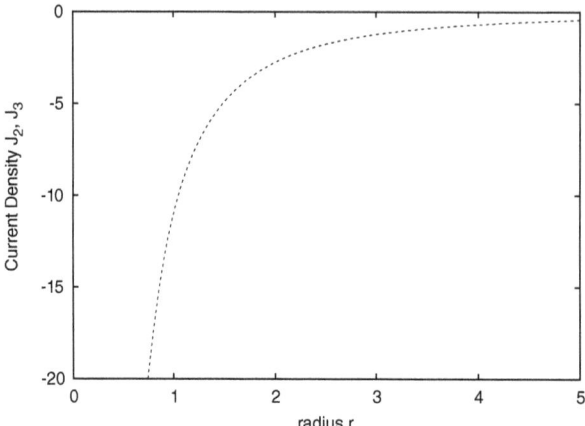

Fig. 3.11. FLRW metric, current density, r dependence of J_ϑ, J_φ for $R(t) = t^2, t = 1, k = .5$.

Fig. (3.10) shows that the radial part of the current density grows indefinitely with radius while the angular dependence (Fig. (3.11)) disappears. This is no meaningful physical behaviour and the significance of the FLRW metric is indeed strongly relativated in the next section.

3.3 Concerning the Standard (Big Bang) Cosmological Model

3.3.1 Non-Static Spherically Symmetric Metric Manifolds

It has been frequently claimed by the proponents of the Standard (Big Bang) Cosmological Model that cosmology truly became a science with the advent of Einstein's General Theory of Relativity and the subsequent works of Friedmann, Lemaître, Robertson, and Walker. The essential theoretical elements underlying the alleged Big Bang cosmology are codified in what has become known as the Friedmann-Lemaître-Robertson-Walker (FLRW) line-element. This line-element has three standard forms:

$$ds^2 = dt^2 - \frac{R^2(t)}{\left(1 + \frac{k}{4}r^2\right)^2}[dr^2 + r^2(d\theta^2 + \sin^2\theta d\varphi^2)], \tag{3.20}$$

$$ds^2 = dt^2 - R^2(t)\left[\frac{d\bar{r}^2}{1 - k\bar{r}^2} + \bar{r}^2(d\theta^2 + \sin^2\theta d\varphi^2)\right], \tag{3.21}$$

$$ds^2 = dt^2 - \frac{R^2(t)}{k}[d\chi^2 + \sin^2\chi(d\theta^2 + \sin^2\theta d\varphi^2)]. \tag{3.22}$$

3.3 Concerning the Standard (Big Bang) Cosmological Model

The path to these line-elements is through a tortuous series of transformations of coordinates and assumptions. The starting point is to assume that:

(a) the desired line-element can be written in the form $ds^2 = dt^2 + g_{ij}dx^i dx^j$, $(i, j = 1, 2, 3)$;

(b) spacetime is spatially homogeneous and isotropic for any observer, located anywhere in the Universe and at rest with respect to the mean motion of matter in the observer's neighbourhood.

In accordance with these assumptions it is next supposed that the sought for line-element can be expressed in the spherically symmetric general form

$$ds^2 = e^\nu dt^2 - e^\lambda dr^2 - e^\mu(r^2 d\theta^2 + r^2\sin^2\theta d\varphi^2) + 2adrdt, \quad (3.23)$$

here ν, λ, μ are functions of r and t. By a series of coordinate transformations this line-element is reduced to the the form,

$$ds^2 = dt^2 - e^\mu(dr^2 + r^2 d\theta^2 + r^2 \sin^2\theta d\varphi^2), \quad (3.24)$$

wherein $\mu = \mu(r, t) = f(r) + h(t)$.

Now comes a crucial step: to find, in terms of (3.24), a solution to Einstein's field equations $R_{\rho\sigma} - \frac{1}{2}g_{\rho\sigma}R = \kappa T_{\rho\sigma} \neq 0$, without specific knowledge of $T_{\rho\sigma}$. By spatial isotropy it is asserted that $T_{11} = T_{22} = T_{33}$, by which a 2nd-order differential equation is obtained for $f(r)$. Note however that one cannot obtain an expression leading to an explicit expression for $h(t)$, which remains a priori unknown. Solving the differential equation for $f(r)$, the line-element (3.20) is obtained, and by implicit coordinate transformations, line-elements (3.21) and (3.22) are obtained. To obtain (3.21) from (3.20), set

$$\bar{r} = \frac{r}{1 + \frac{k}{4}r^2}, \quad (3.25)$$

and to obtain (3.22) from (3.21), set

$$\bar{r} = \frac{1}{\sqrt{k}}\sin\chi. \quad (3.26)$$

It must be noted that in the line-elements (3.20), (3.21) and (3.22), the quantities r, $\bar{r}$, and χ do not denote distances, radial or otherwise, in the spacetime they equivalently describe. Line-elements (3.20), (3.21) and (3.22) share the same intrinsic geometrical structure as the usual line-element, in spherical coordinates, for Minkowski space, upon which they are fundamentally based. This is so because a geometry is completely determined by the form of its line-element [21, 22].

Since line-elements (3.20), (3.21) and (3.22) share the same basic intrinsic geometry, that same intrinsic geometry must be applied to all these line-elements, to determine, in each case, quantities associated with the spacetime they describe. The radius of curvature and the proper radius are therefore obtained for each line-element from the components of the metric tensor thereof and the fixed geometrical relations between them [23–26]. In the case of (3.20), the radius of curvature R_c, is

$$R_c = \frac{rR(t)}{1 + \frac{k}{4}r^2}, \tag{3.27}$$

and the proper radius R_p is

$$R_p = R(t) \int \frac{dr}{1 + \frac{k}{4}r^2}. \tag{3.28}$$

In the case of (3.21),

$$R_c = \bar{r}R(t),$$
$$R_p = R(t) \int \frac{d\bar{r}}{\sqrt{1 - k\bar{r}^2}}. \tag{3.29}$$

In case of (3.23),

$$R_c = \frac{R(t)}{\sqrt{k}} \sin \chi,$$
$$R_p = \frac{R(t)}{\sqrt{k}} \int d\chi. \tag{3.30}$$

Note that in each case, $R_c \neq R_p$ in general.

Now it is also assumed by the Standard Model cosmologists that in (3.20), $0 \leq r < \infty$, in (3.21), $0 \leq \bar{r} < \infty$, and in (3.22), $0 \leq \chi \leq \pi$ [22, 27, 28]. However, no Big Bang cosmologist has ever proved that these domains on the respective variables are valid. They have all only ever assumed that they are valid. That the assumptions are false is rather easily demonstrated [25, 26]. The correct intervals are $0 \leq r < \frac{2}{\sqrt{k}}$, $0 \leq \bar{r} < \frac{1}{\sqrt{k}}$ and $2n\pi \leq \frac{\pi}{2} + 2n\pi$ ($n = 0, 1, 2, ...$).

Notwithstanding the fact that $R(t)$ is a priori unknown, it is also assumed by the Big Bang cosmologists that $R(t)$ is well-defined in line-elements (3.20) to (3.22), simply because those line-elements satisfy the field equations. However, that satisfaction of the field equations is a necessary but insufficient condition for a solution to Einstein's gravitational field has not been realisrd by the Standard Model cosmologists. In addition to the field equations, the intrinsic geometry of the line-element and boundary conditions must also be satisfied. Application of the intrinsic geometry of line-elements (3.20), (3.21) and (3.22) shows that a well-defined $R(t)$ therein does not exist: $R(t)$ is necessarily infinite for all values of the time t. This means that Einstein's Universe, insofar as the FLRW configuration is concerned, is infinite and unbounded in both space and time, and is therefore actually independent of time [25, 26].

3.4 The Friedmann Models

Friedmann's equation is dependent upon the assumption of a well-defined $R(t)$ [27, 29, 30]:

$$\dot{R}^2 + \bar{k} = \frac{8\pi G}{3}\rho R^2, \qquad (3.31)$$

where $\rho = \rho(t)$ is the proper density of a Universe modelled by the tensor for a perfect fluid,

$$T_{\mu\nu} = (\rho + p)u_\mu u_v - pg_{\mu\nu}, \qquad (3.32)$$

where $p = p(t)$ is pressure and u_μ is the covariant world-velocity of fluid particles. Both ρ and p are functions of t alone owing to the assumption of homogeneity. The derivation of (3.31) completely ignores the issue of satisfaction of the intrinsic geometry of the line-element (3.20), and its equivalents. Also, setting to zero the covariant derivative of (3.32), as $T^{\mu\nu}{}_{,\mu} - 0$, the Big Bang cosmologists obtain the equation of continuity, as

$$\dot{\rho} + (\rho + p)\frac{3\dot{R}}{R} = 0, \qquad (3.33)$$

where the validity of this expression is also contingent upon the validity of the assumption of a well-defined $R(t)$ in metric (3.20) and its equivalents.

Now by setting $p = 0$ (i.e. for a matter-dominated Universe), the Standard model proponents reduce (3.33) to,

$$\rho R^3 = \text{const.} \qquad (3.34)$$

and Friedmann's equation to,

$$\dot{R}^2 + \bar{k} = \frac{A^2}{R}, \qquad (3.35)$$

where $A > 0$ is a constant that subsumes $R(t_0)$, where t_0 is alleged to be the current age of the Universe.

The Hubble relation is defined by the Big Bang cosmologists as:

$$H(t) = \frac{\dot{R}(t)}{R(t)}. \tag{3.36}$$

The "present day" value of Hubble's "constant" is claimed to be $H_0 = H(t_0)$. Using this in Friedmann's equation (3.31), the Big Bang cosmologists obtain their so-called "critical density",

$$\dot{\rho}_c = \frac{3H_0^3}{8\pi G}, \tag{3.37}$$

and their so-called "deceleration parameter", with $p = 0$,

$$q_0 = \frac{4\phi G \rho_0}{3H_0^2} = \frac{\rho_0}{2\rho_c}, \tag{3.38}$$

where $p_0 = p(t_0)$. Again, all these results completely ignore the intrinsic geometry of the line-element and rest upon the unproven assumption that $R(t)$ is a priori well-defined in the line-element.

The so-called "Friedmann" models involve solving (3.35) for the three cases $\kappa = 0, \pm 1$, and are therefore based upon the unproven assumption of a well-defined $R(t)$ in the line-element (3.20) and its equivalents. Yet no relativist has ever proved that there exists an a priori well-defined $R(t)$, because none of them have ever realised that $R(t)$ must satisfy the intrinsic geometry of the line-element (3.20), and its equivalents.

Since it is easily proved [25] that $R(t)$ is necessarily infinite for all values of time t, the whole Standard (Big Bang) Cosmological Model is fallacious.

3.5 Recapitulation and Summary

The Big Bang Cosmology makes the assumptions (a) and (b) given above, and obtains the line-element (3.20), and by implicit coordinate transformations, metrics (3.21) and (3.22), satisfying the field equations.

The range on the variables $r, \bar{r}$ and χ in line-elements (3.20), (3.21) and (3.22) respectively are never deduced by the Big Bang proponents by an application of the intrinsic geometry of the line-elements. Instead, they have merely assumed, erroneously, that $0 \leq r < \infty, 0 \leq \bar{r} < \infty$ and $0 \leq \chi \leq \pi$.

The intrinsic geometry of the line-element has been completely ignored by the relativists, or more accurately, has gone thoroughly unrecognised so that it has never been applied by them.

3.5 Recapitulation and Summary

Ignorance of the intrinsic geometry of the line-element manifests in the unproven (and demonstrably false) assumption that $R(t)$ is a priori well-defined.

Using the false assumption that $R(t)$ is well-defined, the Friedmann models and all the other Big Bang paraphernalia are constructed.

The Standard Cosmology and black hole proponents have never realised that there is a clear geometrical distinction between the radius of curvature (from the Gaussian curvature) and the proper radius (the radial geodesic) in the non-Efcleethean[1] spherically symmetric pseudo-Riemannain metric manifold of Einstein's gravitational field. They have therefore never realised that on the usual Minkowski line-element in spherical coordinates, the radius of curvature and the proper radius are identical (owing to the fact that Minkowski space is pseudo-Efcleethean). They have failed to understand that a geometry is entirely determined by its line-element. Furthermore, they have never realised that, in general, the quantity r appearing in their usual line-elements is neither a radius nor a distance in the spacetime described by those line-elements, but is in fact only a parameter for the radius of curvature and the proper radius of those line-elements.

In short, Big Bang cosmology and black holes are based upon fatal errors in the elementary differential geometry of a spherically symmetric metric manifold, and so they are entirely false.

Contrary to the now almost daily claims by the astronomers and astrophysicists, nobody has ever found a black hole. The alleged signatures of a black hole are an infinitely dense singularity and an event horizon. Hundreds of black holes have alleged to have been discovered, yet not one instance of an infinitely dense singularity or one instance of an event horizon has been identified. This amplifies that fact that the black hole did not come from observations. The notion of the black hole did not exist before it was conjured up from General Relativity. It is an entirely theoretical object that has not been found in Nature. But the theoretical derivation of the black hole is a gross violation of differential geometry. therefore, it is fallacious.

Similarly, before it was conjured up from General Relativity, the Big Bang concept did not exist. Hubble's relation has been reformulated as a red-shift/cosmological recessional velocity relation, from a red-shiftldistance relation, tenuous to begin with, to give some facade of physical validity. Being

---

[1] For the geometry due to Efcleethees, usually and abominably rendered as Euclid.

also a creature of pure theory, allegedly derived from General Relativity, Big Bang fails completely since it too is due to fatal errors in differential geometry. One cannot interpret observations in terms of concepts derived from General Relativity by erroneous mathematics. Yet that is precisely what the astronomers and astrophysical relativists have always done.

As for the CMB, its most likely source is the oceans of the Earth [31–38]. In any event it cannot be the afterglow of a Big Bang, as so commonly claimed.

References

[1] M. W. Evans, "Generally Covariant Unified Field Theory" (Abramis Academic, Suffolk, 2005 onwards), vols. 1–4, volume 5 in prep. (papers 71 to 93 on www.aias.us).

[2] M. W. Evans et al., papers 1–95 to date on www.aias.us and www.atomicprecision.com, also Omnia Opera Section on www.aias.us, 1992 to present for precursor theories.

[3] L. Felker, "The Evans Equations of Unified Field Theory" (Abramis Academic, Suffolk, 2007).

[4] K. Pendergast, "Mercury as Crystal Spheres" (Abramis Academic, Suffolk, in prep., preprint on www.aias.us).

[5] M. W. Evans, Acta Phys. Polon., **38**, 2211 (2007).

[6] M. W. Evans and H. Eckardt, Physica B, in press (2007).

[7] M. W. Evans, Physica B, in press (2007).

[8] M. W. Evans and L. B. Crowell, "Classical and Quantum Electrodynamics and the $\mathbf{B}^{(3)}$ Field" (World Scientific, 2001).

[9] M. W. Evans and J.-P. Vigier, "The Enigmatic Photon" (Kluwer, Dordrecht, 1994 to 2002 hardback and softback), in five volumes.

[10] M. W. Evans (ed.), "Modern Non-Linear Optics" in I. Prigogine and S. A. Rice (series eds.), "Advances in Chemical Physics" (Wiley, New York, 2001, second edition), vols. 119(1) to 119(3); ibid M. W. Evans and S. Kielich (eds.), (Wiley, New York, 1992, 1993 and 1997, first edition) vols. 85(1) to 85(3).

[11] S. P. Carroll, "Space-time and Geometry: an Introduction to General Relativity" (Addison Wesley, New York, 2005, online 1997 notes).

[12] Papers 93 and 95 of www.aias.us .

[13] S. J. Crothers, "A Brief History of Black Holes", Progress in Physics, 2, 54–57 (2005).

[14] S. J. Crothers, " Symmetric Metric Manifolds and the Black Hole Catastrophe" (www.aias.us, 2007).

[15] S. J. Crothers, "On the "Size" of Einstein's Spherically Symmetric Universe", progress in Physics, vol. 3, in press, (2007).

[16] A. S. Eddington, "The Mathematical Theory of Relativity" (Cambridge University Press, 2nd ed., 1960).

[17] S. J. Crothers, "On the Geometry of the General Solution for the Vacuum Field of the Point Mass", Progress in Physics, **2**, 3–14 (2005).

[18] K. Schwarzschild, "On the Gravitational Field of a Mass Point According to Einstein's Theory", Sitz. Preuss. Akad., Phys. Math., K1, 189 (1916).

[19] B.Lehnert, a review in ref. (10), vol. 119(2), pp. 1 ff.

[20] K. Schwarzschild, "On the Gravitational Field of a Sphere of Incompressible Fluid According to Einstein's Theory", Sitz. Preuss. Akad., Phys. Math., K1, 424 (1916).

[21] Levi-Civita T. The Absolute Differential Calculus, Dover Publications Inc., New York, 1977.

[22] Tolman R. C. Relativity Thermodynamics and Cosmology, Dover Publications Inc., New York, 1987.

[23] Crothers S. J. On the geometry of the general solution for the vacuum field of the point-mass. *Progress in Physics*, v. 2, 3–14, 2005, (www.ptep-online.com/index_files/2005/PP-02-01.PDF)

[24] Crothers S. J. Spherically Symmetric Metric Manifolds and the Black Hole Catastrophe, (www.aias.us).

[25] Crothers S. J. On the 'Size' of Einstein's Spherically Symmetric Universe. *Progress in Physics*, v. 3, p–pp, 2007, (www.ptep-online.com/index_files/2007/PP-11-10.PDF).

[26] Evans M. W., Eckardt H., Crothers S. J. The Coulomb and Ampere Maxwell Laws in Generally Covariant Unified Field Theory, paper 93, (www.aias.us).

[27] Misner C. W., Thorne K. S., Wheeler J. A. Gravitation. W. H. Freeman and Company, New York, 1973.

[28] Landau, L., Lifshitz, E. The Classical Theory of Fields, Addison-Wesley Publishing Company, Inc., Reading, Massachusettes, 1951.

[29] Foster, J., Nighingale, J. D. A Short Course in General Relativity, 2nd Ed., Springer-Verlag, New York, Inc., 1995.

[30] d'Inverno, R. Introducing Einstein's Relativity, Clarendon Press, Oxford 1992.

[31] Robitaille P-M. L. WMAP: a radiological analysis II. Spring Meeting of the American Physical Society Northwest Section, G1.0005, May 19–20, 2006.

[32] Robitaille P.-M. L. WMAP: a radiological analysis. *Progress in Physics*, 2007, v. 1, 3–18, (www.ptep-online.com/index_files/2007/PP-08-01.PDF).

[33] Robitaille P.-M. L. On the origins of the CMB: insight from the COBE, WMAP, and Relikt-1 Satellites. *Progress in Physics*, 2007, v. 1, 19–3, (www.ptep-online.com/index_files/2007/PP-08-02.PDF).

[34] Robitaille P.-M. L. On the Earth Microwave Background: absorption and scattering by the atmosphere. *Progress in Physics*, 2007, v. 3, 3–4, (www.ptep-online.com/index_files/2007/PP-11-11.PDF).

[35] Robitaille P.-M. L., Rabounski D. COBE and the absolute assignment of the CMB to the Earth. American Physical Society March Meeting, L20.00007, March 5-9, 2007.

[36] Robitaille P.-M. L., On the Nature of the Microwave Background at the Lagrange 2 Point. Part 1. *Progress in Physics*, 2007, v. 4., 74–83, (www.ptep-online.com/index_files/2007/PP-11-11.PDF).

[37] Rabounski D. The relativistic effect of the deviation between the CMB temperatures obtained by the COBE satellite. *Progress in Physics*, 2007, v. 1, 24–26, (www.ptep-online.com/index_files/2007/PP-08-03.PDF).

[38] Borissova L., Rabounski D. On the nature of the Microwave Background at the Lagrange 2 Point. Part II. *Progress in Physics*, 2007, v. 4., 84–95, (www.ptep-online.com/index_files/2007/PP-11-12.PDF).

Line Element for a Radiating Electron in a Generally Covariant Unified Field Theory

by

Myron W. Evans,
Alpha Institute for Advanced Study, Civil List Scientist.
(emyrone@aol.com and www.aias.us)

Abstract

A method is given to determine the four-dimensional line element for a radiating electron in a generally covariant unified field theory. The radiating electron is described classically as an object of finite volume and mass, carrying the electron charge −e. Starting from the most general line element in four dimensions and curvilinear coordinates, the Einstein Hilbert field equation is solved with a canonical energy momentum density tensor modeled in various ways, one example being a rotating electron. It is shown that the charge current density is in general directly proportional to the Ricci tensor in Einstein Cartan Evans (ECE) unified field theory, so disappears in Ricci flat space-times. However light bending due to gravity and changes of its polarization, occur in Ricci flat space-times. The complete system of equations is given for the line element of a radiating electron in ECE theory.

Keywords: ECE theory, radiating electron, line element, light bending and changes of polarization due to gravity.

4.1 Introduction

Recently a generally covariant unified field theory has been developed which expresses the equations of classical gravitation and electromagnetism in terms of geometry, as required by the philosophy of general relativity [1]. Standard Cartan geometry [2] is used for this purpose, and a unified view of classical and quantum physics physics has emerged through Einstein Cartan Evans (ECE) unified field theory [3–12]. It has been shown that the equations of classical electrodynamics in a generally covariant unified field theory have the same vector form as in the Maxwell Heaviside (MH) field theory, but they are expressed in a space-time with curvature and torsion rather than in the flat Minkowski space-time without curvature or torsion. In ECE theory furthermore the electric and magnetic fields are related to the scalar and vector potentials with the scalar and vector parts of the spin connection of Cartan geometry, the object that indicates that space-time is curving and spinning. The charge-current density in ECE theory is built up from elements of the Ricci tensor, so in Ricci flat space-times the charge-current density vanishes. This result proves the conceptual self-consistency of ECE theory, because the charge-current density vanishes in a Ricci flat vacuum.

In Section 4.2, the dielectric formulation of ECE theory [3–12] is used to show that light grazing a massive object is deflected and also changes its polarization. This is a straightforward result of the ECE Ampère Maxwell law with finite current density in general. When the current density is zero the deflection is due to a change in phase velocity from c to v, as in the theory of refraction. This results in a phase change and change of polarization from circular to elliptical. Such a process is shown to occur in a Ricci flat space-time where the light travels along a null geodesic.

In Section 4.3 the general equations due to Crothers [13–15] are given for line elements in the class of Ricci flat space-times and it is emphasized that in general the geodesic proper radius is not the same as the radius of curvature. This is the basic flaw at the root of theories that give a singularity at initial event, such as Big Bang or black hole.

Finally in Section 4.4 a method is developed for defining the line element by starting from the most general type of metric in curvilinear coordinates in four dimensions and using models for the canonical energy momentum density in the Einstein Hilbert (EH) field equation.

4.2 Change of Polarization of Light Due to Gravitation

The basis of this calculation is the ECE Ampère Maxwell law [3–12]:

$$\boldsymbol{\nabla} \times \boldsymbol{B} - \frac{1}{c^2}\frac{\partial \boldsymbol{E}}{\partial t} = \mu_0 \boldsymbol{J}_{\text{grav}} \qquad (4.1)$$

where the current density $\boldsymbol{J}_{\text{grav}}$ is calculated from the space-time around a mass M. In order to generate $\boldsymbol{J}_{\text{grav}}$ there must be present a primordial voltage $cA^{(0)}$ and electric charge $-e$. There is no satisfactory metric currently in existence for the gravitational field external to a point mass, as argued rigorously by Crothers [13–15] the available metrics [16] confuse the radius of curvature of a line element with its geodesic proper radius. There are also

4.2 Change of Polarization of Light Due to Gravitation

other serious errors in the standard model literature [13–15] so in this paper a method is suggested of deducing the line element of a radiating electron that is an exact solution of EH and which also obeys the fundamental rules [13–15] of differential geometry. In this section the dielectric theory of ECE [3–12] is used to illustrate how it is possible to proceed without knowledge of the precise line element.

In this theory the space-time around M is considered to be a dielectric:

$$\boldsymbol{E} = \frac{1}{\epsilon_0}(\boldsymbol{D} - \boldsymbol{P}), \quad \boldsymbol{B} = \mu_0(\boldsymbol{H} + \boldsymbol{M}) \tag{4.2}$$

where $\boldsymbol{P}$ is the polarization, $\boldsymbol{M}$ is the magnetization, $\boldsymbol{D}$ is the electric displacement, $\boldsymbol{H}$ is the magnetic field strength, $\boldsymbol{B}$ is the magnetic flux density, $\boldsymbol{E}$ is the electric field strength and ϵ_0 and μ_0 are the permittivity and permeability in vacuo respectively. The vacuum is defined as flat or Minkowski space-time with no matter present, so the connection vanishes and the Christoffel symbols and Riemann tensor elements are all zero. From Eq. (4.2) in Eq. (4.1)

$$\nabla \times \boldsymbol{H} - \frac{\partial \boldsymbol{D}}{\partial t} = \boldsymbol{J} - \left(\nabla \times \boldsymbol{M} + \frac{\partial \boldsymbol{P}}{\partial t}\right). \tag{4.3}$$

The current density is defined in terms of $\boldsymbol{M}$ and $\boldsymbol{P}$ as follows:

$$\boldsymbol{J} := \nabla \times \boldsymbol{M} + \frac{\partial \boldsymbol{P}}{\partial t} \tag{4.4}$$

so Eq. (4.1) becomes:

$$\nabla \times \boldsymbol{H} - \frac{\partial \boldsymbol{D}}{\partial t} = 0 \tag{4.5}$$

where

$$\boldsymbol{H} = \frac{1}{\mu}\boldsymbol{B}, \quad \boldsymbol{D} = \epsilon \boldsymbol{E}. \tag{4.6}$$

Here μ and ϵ are the permeability and permittivity respectively of the space-time around the mass M. The solution of Eq. (4.5) is

$$\boldsymbol{H} = \frac{H^{(0)}}{\sqrt{2}}(i\boldsymbol{i} + \boldsymbol{j})\exp(i(\omega t - \kappa Z)) \tag{4.7}$$

$$\boldsymbol{D} = \frac{D^{(0)}}{\sqrt{2}}(\boldsymbol{i} - i\boldsymbol{j})\exp(i(\omega t - \kappa Z)) \tag{4.8}$$

provided that:
$$\kappa H^{(0)} = \omega D^{(0)} \tag{4.9}$$

i.e. provided that
$$\frac{B^{(0)}}{\mu} = \frac{\omega}{\kappa} \epsilon E^{(0)}. \tag{4.10}$$

If:
$$\epsilon\mu = \epsilon_0 \mu_0 = \frac{1}{c^2} \tag{4.11}$$

then Eq. (4.10) becomes
$$E^{(0)} = cB^{(0)} \tag{4.12}$$

which is the vacuum result of Minkowski spacetime, i.e. the vacuum result in MH theory. Eq. (4.5) is:
$$\nabla \times \left(\frac{B}{\mu}\right) - \frac{\partial}{\partial t}(\epsilon E) = 0. \tag{4.13}$$

In the first approximation it is assumed that ϵ and μ are r and t independent so:
$$\nabla \times B - \epsilon\mu \frac{\partial E}{\partial t} = 0 \tag{4.14}$$

where the phase velocity of the wave is defined by:
$$\epsilon\mu = \frac{1}{v^2}. \tag{4.15}$$

The solution of eq. (4.14) is the plane wave:
$$B = \frac{B^{(0)}}{\sqrt{2}} (i\boldsymbol{i} + \boldsymbol{j}) e^{i\phi'} \tag{4.16}$$

$$E = \frac{E^{(0)}}{\sqrt{2}} (\boldsymbol{i} - i\boldsymbol{j}) e^{i\phi'} \tag{4.17}$$

where
$$v = \frac{\omega}{\kappa} \tag{4.18}$$

4.2 Change of Polarization of Light Due to Gravitation

and

$$\phi' = \omega t - \kappa Z. \tag{4.19}$$

The vacuum plane wave has the properties:

$$c = \frac{\omega}{\kappa}, \tag{4.20}$$

$$\phi = \omega t - \kappa Z. \tag{4.21}$$

So the effect of the current density $\boldsymbol{J}_{\text{grav}}$ is to slow the phase velocity of the plane wave form c to v. This is a refractive index effect similar to light at an interface [16]. It is known from the EH theory that light in a null geodesic is bent by gravitation. This result is usually obtained from a purely kinematic consideration - the photon of mass m interacting with the gravitating mass M. The light goes into orbit and so its phase velocity is slowed from c to v.

The real part of Eq. (4.17) is obtained from the de Moivre theorem:

$$e^{i\phi'} = \cos\phi' + i\sin\phi' \tag{4.22}$$

giving:

$$Re(\boldsymbol{E}) = \frac{E^{(0)}}{\sqrt{2}}(\boldsymbol{i}\cos\phi' + \boldsymbol{j}\sin\phi'). \tag{4.23}$$

The equivalent vacuum result is:

$$Re(\boldsymbol{E}) = \frac{E^{(0)}}{\sqrt{2}}(\boldsymbol{i}\cos\phi + \boldsymbol{j}\sin\phi). \tag{4.24}$$

Using:

$$\cos\phi' = a\cos\phi \tag{4.25}$$
$$\sin\phi' = b\sin\phi' \tag{4.26}$$

Eq. (4.23) becomes:

$$Re(\boldsymbol{E}) = \frac{E^{(0)}}{\sqrt{2}}(a\boldsymbol{i}\cos\phi + b\boldsymbol{j}\sin\phi) \tag{4.27}$$

so the circularly polarized plane wave of Eq. (4.24) is changed to the elliptically polarized plane wave (4.27).

4 Line Element for a Radiating Electron

In a Ricci flat space-time the charge/current density vanishes, so:

$$R_{\mu\nu} = 0, R = 0, \rho = 0, \boldsymbol{J} = \boldsymbol{0}. \tag{4.28}$$

The line element that is usually used to produce the kinematic theory of light bending due to gravity is:

$$ds^2 = \left(1 - \frac{2mG}{rc^2}\right) c^2 dt^2 - \left(1 - \frac{2mG}{rc^2}\right)^{-1} dr^2 - r^2 \left(d\theta^2 + \sin^2\theta d\phi^2\right) \tag{4.29}$$

which has serious flaws in it [13–15] and is wrongly attributed to Schwarzschild. The line element (4.29) is Ricci flat by construction [2] but its Christoffel symbols and Riemann tensor elements may be non-zero. In the kinematic theory of light bending by gravitation a null geodesic is constructed from (4.29), but the Ricci tensor elements are all zero. The phase velocity of the plane wave in this case is given by Eq. (4.18) with:

$$\boldsymbol{J}_{\text{grav}} = 0 \tag{4.30}$$

and this results in a change of polarization. So whenever light is bent by gravity, its polarization also changes in ECE theory. This is as observed experimentally [3–12].

4.3 Light Bending in Ricci Flat Space-Times

Crothers has shown [13–15] that Ricci flat space-times have the form:

$$ds^2 = \left(1 - \frac{\alpha}{C^{1/2}}\right) c^2 dt^2 - \left(1 - \frac{\alpha}{C^{1/2}}\right)^{-1} d(C^{1/2}) - C(r)(d\theta^2 + \sin^2\theta d\phi^2) \tag{4.31}$$

where the radius of curvature is defined as:

$$R_c = C^{1/2} = (|r - r_0|^n + \alpha^n)^{1/n}. \tag{4.32}$$

The important points in the analysis by Crothers [13–15] are as follows.

1. C is not determined by the EH field equation.

2. Any C can be used in Eq. (4.31) without changing the spherical symmetry or violating the field equation.

3. C must be asymptotically Minkowskian.

4. There is a difference between the radius of curvature and the geodesic proper radius R_p

   The latter is defined for the Crothers line element [13–15]:

   $$ds^2 = AC^{1/2}c^2 dt^2 - BC^{1/2}d(C^{1/2}) - C(r)(d\theta^2 + \sin^2\theta d\phi^2) \quad (4.33)$$

   by:

   $$\begin{aligned} R_p &= \int_0^{R_p} dR_p = \int_{R_c(r_0)}^{R_c(r)} \left(B(R_c(r))^{1/2} dR_c(r) \right) \\ &= \int_{r_0}^{r} (B(R_c(r)))^{1/2} \left(\frac{dR_c(r)}{dr} \right) dr. \end{aligned} \quad (4.34)$$

5. It is not possible to assume that

   $$0 \leqslant R_c(r) < \infty \quad (4.35)$$

   if

   $$0 \leqslant r < \infty. \quad (4.36)$$

6. The Ricci flat space-times include the following classes: Schwarzschild, Kerr-Newman, Kerr, charged Kerr, and the exterior of an incompressible spherical fluid [13–15]. All these describe the gravitational field in terms of a center of mass, a pure mathematical construct. In physics a center of mass is contained within an identically non-zero volume, and the line elements inside and outside the volume are in general different [13–15].

Light bending occurs in a Ricci flat space-time because the Christoffel symbols and Riemann tensor elements may be non-zero while by construction [2, 13–15]:

$$R_{\mu\nu} = 0, R = 0. \quad (4.37)$$

The usual line element used to describe light bending is Eq. (4.29), which is wrongly attributed to Schwarzschild and which is a special case of Eq. (4.31) [13–15]. In the kinematic theory of light bending a photon of mass m is attracted by the mass M along a null geodesic [3–12] constructed from Eq. (4.29). The line elements (4.29) and (4.31) both give in ECE theory:

$$\rho = 0, \boldsymbol{J} = \boldsymbol{0} \quad (4.38)$$

so the ECE laws for light bending in a Ricci flat space-time are the four equations:

$$\nabla \cdot \mathbf{B} = 0 \tag{4.39}$$

$$\nabla \times \mathbf{E} + \frac{\partial \mathbf{B}}{\partial t} = 0 \tag{4.40}$$

$$\nabla \cdot \mathbf{E} = 0 \tag{4.41}$$

$$\nabla \times \mathbf{B} - \frac{1}{v^2}\frac{\partial \mathbf{E}}{\partial t} = 0 \tag{4.42}$$

where the phase velocity of the plane wave is defined by Eq. (4.18) of Section 4.2. These considerations are also true for the Ricci flat line element outside a sphere of incompressible fluid [13–15]:

$$ds^2 = \left(1 - \frac{\alpha}{R_c}\right)c^2 dt^2 - \left(1 - \frac{\alpha}{R_c}\right)^{-1} dR_c^2 - R_c^2(d\theta^2 + \sin^2\theta d\phi^2). \tag{4.43}$$

Outside the sphere:

$$\rho = 0, \ \mathbf{J} = \mathbf{0} \tag{4.44}$$

but inside the sphere:

$$\rho \neq 0, \ \mathbf{J} \neq \mathbf{0}. \tag{4.45}$$

The irretrievable problems with line element (4.29) are well known and discussed in Chapter 23 of www.aias.us. In addition, the standard model does not correctly distinguish between the passive mass:

$$M = \rho_0 V \tag{4.46}$$

and the active mass:

$$m = \frac{\alpha}{2}. \tag{4.47}$$

In Eq. (4.43), in the Crothers notation of Chapter 23:

$$R_c = (|r - r_0|^n + \epsilon^n)^{1/3}, \tag{4.48}$$

$$\alpha = \left(\frac{3}{\kappa \rho_0}\right)^{1/2} \sin^3|\chi_a - \chi_0|, \tag{4.49}$$

$$\epsilon = \left(\frac{3}{\kappa\rho_0}\right)^{1/2} \left(\frac{3}{2}\sin^3|\chi_a - \chi_0| - \frac{9}{4}\cos|\chi_a - \chi_0|\right.$$
$$\left. \cdot \left(|\chi_a - \chi_0| - \frac{1}{2}\sin^2|\chi_a - \chi_0|\right)\right)^{1/3}. \tag{4.50}$$

The original Schwarzschild result of 1916 is obtained for:

$$n = 3, r_0 = 0, \chi_0 = 0, r > 0, \chi_a > 0. \tag{4.51}$$

The geodesic proper radius is determined by the line element of the interior of the sphere and in general:

$$\alpha \neq M. \tag{4.52}$$

These fundamental properties of geometry irretrievably invalidate the theory of Big Bang and of black holes and dark matter. None exist in natural philosophy.

4.4 General Method for Determining The Line Element of a Radiating Electron

The most general line element in four dimensions and in curvilinear coordinates is

$$ds^2 = A(dx_0)^2 - B(dx_1)^2 - C(dx_2)^2 - D(dx_3)^2. \tag{4.53}$$

where A, B, C and D are a priori unknown. Computer algebra was used to evaluate all the Christoffel symbols, Riemann tensor elements, Ricci tensor elements, scalar curvature and Einstein tensor elements of the line element (4.53). It was checked that it obeys the usually named "first Bianchi identity", which in differential form notation is:

$$R \wedge q = 0 \tag{4.54}$$

and in tensor notation is:

$$R_{\sigma\mu\nu\rho} + R_{\sigma\rho\mu\nu} + R_{\sigma\nu\rho\mu} = 0. \tag{4.55}$$

Therefore the left hand side of the Einstein Hilbert equation:

$$R_{\mu\nu} - \frac{1}{2}Rg_{\mu\nu} = kT_{\mu\nu} \tag{4.56}$$

can be evaluated in terms of A, B, C, and D. The complexity of the algebra is no longer a factor because computational packages such as Maxima can be used, in this case code written by Dr. Horst Eckardt. In Eq. (4.56), $R_{\mu\nu}$ is the Ricci tensor, R is the scalar curvature, $g_{\mu\nu}$ is the symmetric metric, k is the Einstein constant and $T_{\mu\nu}$ is the canonical energy momentum density tensor of Noether. Thus $T_{\mu\nu}$ can be expressed in terms of A, B, C and D in the curvilinear coordinate system (x_0, x_1, x_2, x_3) or any other coordinate system. To determine A, B, C, and D a system of simultaneous equations is solved numerically for given initial and boundary conditions.

Particular models are then introduced for $T_{\mu\nu}$ in order to simplify the problem and to find what A, B, C, and D are needed for the model. The tensor $T_{\mu\nu}$ determines the curvature of space-time in Eq. (4.56) and its properties are described for example by Ryder [17]. The angular momentum density for example is a rank three tensor defined by:

$$M^{0\mu\nu} = T^{0\mu}x^\nu - T^{0\nu}x^\mu \tag{4.57}$$

and the angular momentum tensor is:

$$M^{\mu\nu} = \int M^{0\mu\nu} dx^3. \tag{4.58}$$

The Noether Theorem gives the conservation of energy momentum density as follows:

$$D_\mu T^{\mu\nu} = 0. \tag{4.59}$$

The three components [17] of the spin angular momentum are M^{12}, M^{23}, and M^{31} and the three components of the orbital angular momentum are M^{01}, M^{02} and M^{03}. The conservation of angular momentum is given from the Noether Theorem as:

$$D_0 M^{\mu\nu} = 0 \tag{4.60}$$

and in curved space-time:

$$D_\rho M^{\rho\mu\nu} = 0 \tag{4.61}$$

where:

$$M^{\rho\mu\nu} = T^{\rho\mu}x^\nu - T\rho^\nu x^\mu. \tag{4.62}$$

4.4 General Method for Determining The Line Element of a Radiating Electron

Thus:

$$(D_\rho T^{\rho\mu})x^\nu + T^{\rho\mu}D_\rho x^\nu - (D_\rho T^{\rho\nu})x^\mu - T^{\rho\nu}D_\rho x^\mu = 0. \tag{4.63}$$

Using Eq. (4.59):

$$T^{\rho\mu}D_\rho x^\nu = T^{\rho\nu}D_\rho x^\mu. \tag{4.64}$$

The left hand side is true when

$$\rho = \nu \tag{4.65}$$

and the right hand side is true when:

$$\rho = \mu. \tag{4.66}$$

Thus, self consistently:

$$T^{\nu\mu} = T^{\mu\nu}. \tag{4.67}$$

The orbital angular momentum in classical dynamics is:

$$\mathbf{J} = \mathbf{p} \times \mathbf{r} \tag{4.68}$$

where p is linear momentum and r is distance, so it is seen by comparison of Eqs. (4.68) and (4.57) that:

$$M^{012} = T^{01}x^2 - T^{02}x^1, \quad M^{12} = \int M^{012} d^3x \tag{4.69}$$

and so on are tensor representations of Eq. (4.68). By definition:

$$\begin{aligned} x^\mu &= (ct, X, Y, Z) \\ &= (x^0, x^1, x^2, x^3), \end{aligned} \tag{4.70}$$

$$\begin{aligned} p^\mu &= \left(\frac{En}{c}, p_X, p_Y, p_Z\right) \\ &= (p^0, p^1, p^2, p^3), \end{aligned} \tag{4.71}$$

Therefore

$$T^{00} = p^0/V = E_n/(cV) \tag{4.72}$$
$$T^{01} = p^1/V = p_X/(V) \tag{4.73}$$
$$T^{02} = p^2/V = p_Y/(V) \tag{4.74}$$
$$T^{03} = p^3/V = p_Z/(V). \tag{4.75}$$

These considerations can now be applied to the radiating electron, which in ECE theory [3–12] is described by the vector equations:

$$\nabla \cdot \boldsymbol{B} = 0 \tag{4.76}$$
$$\nabla \times \boldsymbol{E} + \frac{\partial \boldsymbol{B}}{\partial t} = 0 \tag{4.77}$$
$$\nabla \cdot \boldsymbol{E} = \rho/\epsilon_0 \tag{4.78}$$
$$\nabla \times \boldsymbol{B} - \frac{1}{c^2}\frac{\partial \boldsymbol{E}}{\partial t} = \mu_0 \boldsymbol{J}. \tag{4.79}$$

The charge density and current density are found from the components of the Ricci tensor of Eq. (4.56). The electron is considered as possessing an orbital angular momentum, a finite volume V, and a charge $-e$. So the system of equations (4.56) reduces to:

$$R_{00} - \frac{1}{2}Rg_{00} = kp_0(\boldsymbol{r},t)/V \tag{4.80}$$
$$R_{01} - \frac{1}{2}Rg_{01} = kp_1(\boldsymbol{r},t)/V \tag{4.81}$$
$$R_{02} - \frac{1}{2}Rg_{02} = kp_2(\boldsymbol{r},t)/V \tag{4.82}$$
$$R_{03} - \frac{1}{2}Rg_{03} = kp_3(\boldsymbol{r},t)/V \tag{4.83}$$

with:

$$T_{00}(\boldsymbol{r},t) = p_0(\boldsymbol{r},t)/V \tag{4.84}$$
$$T_{01}(\boldsymbol{r},t) = p_1(\boldsymbol{r},t)/V \tag{4.85}$$
$$T_{02}(\boldsymbol{r},t) = p_2(\boldsymbol{r},t)/V \tag{4.86}$$
$$T_{03}(\boldsymbol{r},t) = p_3(\boldsymbol{r},t)/V. \tag{4.87}$$

Therefore Eqs. (4.80) to (4.83) defined A, B, C, and D in Eq. (4.52) in a particular coordinate system in terms of

$$p_0(\boldsymbol{r},t), p_1(\boldsymbol{r},t), p_2(\boldsymbol{r},t), \text{ and } p_3(\boldsymbol{r},t). \tag{4.88}$$

4.4 General Method for Determining The Line Element of a Radiating Electron

Using the minimal prescription:

$$p_\mu = eA_\mu \tag{4.89}$$

we have defined A, B, C, and D in terms of the four potential A_μ of ECE theory, and thus in terms of the electric and magnetic fields of a radiating electron and the spin connection [3–12]:

$$\boldsymbol{B} = \boldsymbol{\nabla} \times \boldsymbol{A} - \boldsymbol{\omega} \times \boldsymbol{A} \tag{4.90}$$

$$\boldsymbol{E} = -\boldsymbol{\nabla}\phi - \frac{\partial \boldsymbol{A}}{\partial t} - c\omega^0 \boldsymbol{A} + c\boldsymbol{\omega}, \tag{4.91}$$

where

$$A^\mu = (A^0, \boldsymbol{A}) = (\phi, \boldsymbol{A}) \tag{4.92}$$

The system of equations (4.80) to (4.83) thus becomes:

$$R_{00} - \frac{1}{2}Rg_{00} = ekA_0(\boldsymbol{r},t)/V \tag{4.93}$$

$$R_{01} - \frac{1}{2}Rg_{01} = ekA_1(\boldsymbol{r},t)/V \tag{4.94}$$

$$R_{02} - \frac{1}{2}Rg_{02} = ekA_2(\boldsymbol{r},t)/V \tag{4.95}$$

$$R_{03} - \frac{1}{2}Rg_{03} = ekA_3(\boldsymbol{r},t)/V \tag{4.96}$$

where:

$$A_\mu = (A_0, -\boldsymbol{A}) \tag{4.97}$$

and where A^μ must obey Eqs. (4.76)–(4.79), (4.90) and (4.91). Using computer algebra Eqs. (4.76)–(4.79) can be expressed in terms of A^μ and ω^μ, where:

$$\omega^\mu = (\omega^0, \boldsymbol{\omega}). \tag{4.98}$$

The charge current density can be expressed [3–12] as

$$J^{\kappa\nu} = -A^{(0)}R^{\kappa\nu} = -A^{(0)}R^\kappa{}_{\mu\mu\nu}. \tag{4.99}$$

Therefore this system of equations defines a radiating electron in ECE theory. They must be solved self-consistently using computer algebra and packages for any given method.

4 Line Element for a Radiating Electron

The Ricci flat solution is:

$$R_{00} - \frac{1}{2}Rg_{00} = 0 \tag{4.100}$$

$$R_{01} - \frac{1}{2}Rg_{01} = 0 \tag{4.101}$$

$$R_{02} - \frac{1}{2}Rg_{02} = 0 \tag{4.102}$$

$$R_{03} - \frac{1}{2}Rg_{03} = 0 \tag{4.103}$$

and in this case A, B, C, and D of Eq. (4.52) are constrained by Eqs. (4.100) to (4.103). It is seen that this system of equations goes further than the standard model by considering the following contradiction of the standard model. In the latter it is possible to have a plane wave such as:

$$\mathbf{A} = \frac{A^{(0)}}{\sqrt{2}}(\mathbf{i} - i\mathbf{j})\exp(i(\omega t - \kappa Z)) \tag{4.104}$$

propagating in free space without a source:

$$\rho = 0, \mathbf{J} = \mathbf{0}. \tag{4.105}$$

This is a well known philosophical contradiction of the standard model, one of many, because the field of force exists without a source for the field. In ECE the equations of this type of plane wave are:

$$R^{\kappa\nu} = 0 \tag{4.106}$$

and

$$R_{01} - \frac{1}{2}Rg_{01} = ek\frac{A^{(0)}}{\sqrt{2}}e^{i(\omega t - \kappa Z)} \tag{4.107}$$

$$R_{02} - \frac{1}{2}Rg_{02} = -iek\frac{A^{(0)}}{\sqrt{2}}e^{i(\omega t - \kappa Z)} \tag{4.108}$$

$$R_{03} - \frac{1}{2}Rg_{03} = 0. \tag{4.109}$$

Eq. (4.106) contradicts Eqs. (4.107) and (4.108) because the Ricci tesnor is zero in Eq. (4.106) and non-zero in Eqs. (4.107) and (4.108). If the scalar potential is considered to be:

$$A^0 = \phi = \frac{\phi^{(0)}}{\sqrt{2}}\exp(i(\omega t - \kappa Z)) \tag{4.110}$$

there is another contradiction:

$$R_{00} - \frac{1}{2}Rg_{00} = ekc\frac{\phi^{(0)}}{\sqrt{2}}e^{i(\omega t - \kappa Z)}, \qquad (4.111)$$

so it is seen that ECE is both rigorously self consistent and also a generally covariant unified field theory which describes the radiating electron with the equations of classical gravitation and classical electrodynamics in one self consistent geometrical framework.

Acknowledgments

The British Government is thanked for a Civil List Pension and the staff of AIAS and many other colleagues for interesting discussions.

References

[1] A. Einstein, "The Meaning of Relativity" (Princeton Univ. Press, 1921–1954).
[2] S. P. Carroll, "Space-time and Geometry–an Introduction to General Relativity" (Addison Wesley, New York, 2004, notes of 1997 online).
[3] M. W. Evans, "Generally Covariant Unified Field Theory" (Abramis Academic, Suffolk, 2005 onwards), volumes 1–4 to date, volume five in prep. (papers 71 to 93 of www.aias.us).
[4] L. Felker, "The Evans Equations of Unified Field Theory" (Abramis, 2007).
[5] K. Pendergast, "Mercury as Crystal Spheres" (www.aias.us, to be published by Abramis).
[6] M. W. Evans, Acta Phys. Polonica, **38**, 2211 (2007).
[7] M. W. Evans and H. Eckardt, Physica B, in press (2007); M. W. Evans, Physica B, in press (2007).
[8] M. W. Evans at al., Omnia Opera section of www.aias.us, 1992 to present.
[9] M. W. Evans and L. B. Crowell, "Classical and Quantum Electrodynamics and the B(3) Field" (World Scientific, 2001).
[10] M. W. Evans (ed.), "Modern Non-linear Optics" in I. Prigogine and S. A. Rice, "Advances in Chemical Physics" (Wiley, 2001, second edition), vols. 119(1) to 119(3); ibid., M. W. Evans and S. Kielich (eds.), first edition (1992, 1993 and 1997), vols. 85(1) to 85(3).
[11] M. W. Evans and J.-P. Vigier, "The Enigmatic Photon" (Kluwer, Dordrecht, 1994 to 2002, hardback and softback) in five volumes.
[12] M. W. Evans and A. A. Hasanein, "The Photomagneton in Quantum Field Theory" (World Scientific, 1994).
[13] S. J. Crothers, "A Brief History of Black Holes", Progress in Physics, **2**, 54 (2005).
[14] S. J. Crothers, "Symmetric Metric Manifolds and the Black Hole Catastrophe" (www.aias.us, 2007).
[15] S. J. Crothers, "On the Geometry of the General Solution for the Vacuum Field of the Point Mass", Progress in Physics, 2, 3 (2005).
[16] J. D. Jackson, "Classical Electrodynamics" (Wiley, 1999, third edition).
[17] L. H. Ryder, "Quantum Field Theory" (Cambridge Univ. Press, 1996, 2nd ed.).

5

Rank Three Tensors in Unified Gravitation and Electrodynamics

by

Myron W. Evans,
Alpha Institute for Advanced Study, Civil List Scientist.
(emyrone@aol.com and www.aias.us)

Abstract

The role of base manifold rank three tensors is discussed in generally covariant unified field theory. It is shown that the rank three tensor in a space-time with torsion and curvature is in general dual to a rank one four-vector, a duality that follows from fundamental considerations of general coordinate transformation within the Einstein group. When considering the rotation group, a sub-group of the Einstein group, the rank three tensor is anti-symmetric in its lower two indices. Examples include the canonical angular energy-momentum density tensor, the Cartan torsion tensor, and the electromagnetic field tensor. The field equations of classical electrodynamics are expressed in terms of such a rank three tensor and reduced to the familiar Maxwell Heaviside vector format with the inclusion of the spin connection.

Keywords: Generally covariant unified field theory, rank three tensors, duality, general coordinate transformation, field equations of classical electrodynamics.

5.1 Introduction

In the Maxwell Heaviside field theory of classical electrodynamics [1] it is well known that the electromagnetic field tensor is a rank two tensor in the Minkowski space-time. This is a flat space-time with no torsion and no curvature and is covariant under the Lorentz transformation. As such it cannot be unified in a meaningful manner with the generally covariant Riemannian space-time of the classical theory of gravitation [2]. Recently [3–12] a generally covariant unified field theory has been proposed based on standard differential geometry in a space-time with torsion and curvature present simultaneously in general. In this theory, known as Einstein Cartan Evans (ECE) field theory, all sectors are generally covariant, i.e. covariant under the general coordinate transformation as required by the philosophy of relativity. The electromagnetic field is within a primordial voltage the Cartan torsion form [2]. This is a vector valued differential two-form, an anti-symmetric tensor with an additional label a that comes from a tangent space-time at point P to a base manifold. Electromagnetic field equations have been deduced in ECE theory from the Bianchi identity and Cartan structure equations. It is shown in Section 5.2 that these field equations can be reduced to equations in the base manifold. The latter constitute a particular solution of the more general differential form equations and involve the use of a rank three electromagnetic field tensor in the base manifold, a space-time with non-zero torsion and curvature in general. In Section 5.3 it is shown that such a tensor is in general dual to a rank one four-vector in the base manifold, such a duality arises from considerations of the general coordinate transform in the Einstein group, of which the rotation group is a sub-group. For the rotation group the rank three tensor is anti-symmetric in its lower two indices, and components of such a rank three tensor can be identified with the components of a vector. In this way the ECE field equations can be reduced to vector equations which have the same structure as the familiar Maxwell Heaviside field equations but with the key addition of the spin connection. The ECE vector equations are written not in a Minkowski space-time, but in one with torsion and curvature, and the charge-current density is proportional to the Ricci tensor of Riemannian geometry. In Section 5.4 finally it is argued that the rank three Cartan torsion tensor in the base manifold is proportional to the canonical angular energy-momentum density from the Noether Theorem as well as being directly proportional to the rank three electromagnetic field tensor in ECE theory. ECE is therefore a fully consistent and generally covariant unified field theory whose origins are rigorously geometrical as required by the philosophy of relativity [2].

5.2 Field Equations in the Base Manifold

The homogeneous ECE equation [3–12] in differential form notation is:

$$D \wedge F^a = R^a{}_b \wedge A^b = A^b \wedge R^a{}_b \tag{5.1}$$

where F^a denotes the electromagnetic field form, $R^a{}_b$ the curvature form, A^b the electromagnetic potential form. The electromagnetic form is related to

5.2 Field Equations in the Base Manifold

the electromagnetic tensor $F^\kappa_{\mu\nu}$ as follows:

$$F^a_{\mu\nu} = q^a_\kappa F^\kappa_{\mu\nu} \tag{5.2}$$

where the base manifold indices have been restored in Eq. (5.2). The left hand side of Eq. (5.1) is:

$$D \wedge F^a = D_\mu F^a_{\nu\sigma} + D_\sigma F^a_{\mu\nu} + D_\nu F^a_{\sigma\mu} \tag{5.3}$$

and the tetrad postulate is:

$$D_\mu q^a{}_\kappa = 0. \tag{5.4}$$

Therefore by Leibnitz's Theorem:

$$\begin{aligned} D_\mu(q^a_\kappa F^\kappa{}_{\nu\sigma}) + D_\sigma(q^a{}_\kappa F^\kappa{}_{\mu\nu}) + D_\nu(q^a{}_\kappa F^\kappa{}_{\sigma\nu}) \\ = q^a_\kappa(D_\mu F^\kappa{}_{\nu\sigma} + D_\sigma F^\kappa{}_{\mu\nu} + D_\nu F^\kappa{}_{\sigma\mu}). \end{aligned} \tag{5.5}$$

Similarly the right hand side of Eq. (5.1) is:

$$R^a{}_b \wedge A^b = A^{(0)} q^a{}_\kappa (R^\kappa{}_{\sigma\mu\nu} + R^\kappa{}_{\nu\sigma\mu} + R^\kappa{}_{\mu\nu\sigma}) \tag{5.6}$$

so:

$$q^a_\kappa(D_\mu F^\kappa{}_{\sigma\nu} + D_\sigma F^\kappa{}_{\mu\nu} + D_\nu F^\kappa{}_{\sigma\mu}) = q^a_\kappa A^{(0)}(R^\kappa{}_{\sigma\mu\nu} + R^\kappa{}_{\nu\sigma\mu} + R^\kappa{}_{\mu\nu\sigma}). \tag{5.7}$$

A particular solution of Eq. (5.7) is:

$$D_\mu F^\kappa{}_{\nu\sigma} + D_\sigma F^\kappa{}_{\mu\nu} + D_\nu F^\kappa{}_{\sigma\mu} = A^{(0)}(R^\kappa{}_{\sigma\mu\nu} + R^\kappa{}_{\nu\sigma\mu} + R^\kappa{}_{\mu\nu\sigma}). \tag{5.8}$$

By definition:

$$D_\mu F^\kappa{}_{\nu\sigma} = \partial_\mu F^\kappa{}_{\nu\sigma} + \omega^\kappa{}_{\mu b} F^b{}_{\nu\sigma} \tag{5.9}$$

where $\omega^\kappa{}_{\mu b}$ is the appropriate form of the spin connection. The structure of Eq. (5.8) is therefore:

$$\partial_\mu F^\kappa{}_{\nu\sigma} + \partial_\sigma F^\kappa{}_{\mu\nu} + \partial_\nu F^\kappa{}_{\sigma\mu} = \frac{A^{(0)}}{\mu_0}(j^\kappa{}_{\mu\nu\sigma} + j^\kappa{}_{\sigma\mu\nu} + j^\kappa{}_{\nu\sigma\mu}). \tag{5.10}$$

For all practical purposes [3–12] the homogeneous current is zero, so Eq. (5.10) is:

$$\partial_\mu \widetilde{F}^{\kappa\mu\nu} = 0 \qquad (5.11)$$

Its Hodge dual [3–12] is:

$$\partial_\mu F^{\kappa\mu\nu} = -\frac{A^{(0)}}{\mu_0} R^\kappa{}_\mu{}^{\mu\nu}. \qquad (5.12)$$

Here μ_0 is the vacuum permeability in S.I. units. It is seen that rank three tensors enter into the field equations (5.11) and (5.12). These are shown later in this paper to be in general dual to a four vector as follows:

$$V^\mu = V^\mu{}_{\rho\sigma}\,\epsilon^{\rho\sigma} \qquad (5.13)$$

where for rotation [13]:

$$\epsilon^{\rho\sigma} = -\epsilon^{\sigma\rho}. \qquad (5.14)$$

The rotation group is a sub group of the Lorentz group, which is in turn a sub group of the Poincaré group and of the Einstein group. In general therefore the electromagnetic field tensor is the anti-symmetric rank three tensor:

$$F^\mu{}_{\rho\sigma} = -F^\mu{}_{\sigma\rho}. \qquad (5.15)$$

The Cartan torsion tensor is well known [2–12] to be:

$$\begin{aligned} T^\mu{}_{\rho\sigma} &= -T^\mu{}_{\sigma\rho} \\ &= \Gamma^\mu{}_{\rho\sigma} - \Gamma^\mu{}_{\sigma\rho} \end{aligned} \qquad (5.16)$$

where $\Gamma^\mu{}_{\rho\sigma}$ is the general gamma connection. Another well known example of a rank three anti-symmetric tensor is the canonical angular momentum-energy density tensor [13]. These three tensors have the same fundamental anti-symmetry in their lower two indices and are shown later in this paper to be proportional to each other. Indeed, the fundamental ECE hypothesis in tensor notation is [3–12]:

$$F^\mu{}_{\rho\sigma} = A^{(0)} T^\mu{}_{\rho\sigma}. \qquad (5.17)$$

Eq. (5.12) therefore has a clear interpretation in terms of a rank three tensor whose elements are electric and magnetic field three-vector components, and in terms of a Ricci type tensor $R^\kappa{}_\mu{}^{\mu\nu}$. The technical correctness of Eq. (5.12) has been demonstrated by computer algebra [3–12]. It has also been checked by computer algebra that in a Ricci flat space-time:

$$R^{\kappa}{}_{\mu}{}^{\mu\nu} = 0 \tag{5.18}$$

meaning that:

$$\partial_\mu F^{\kappa\mu\nu} = 0 \tag{5.19}$$

self-consistently. Eq. (5.19) translates into the following two vector equations:

$$\nabla \cdot \boldsymbol{E} = 0 \tag{5.20}$$

$$\nabla \times \boldsymbol{B} - \frac{1}{c^2}\frac{\partial \boldsymbol{E}}{\partial t} = \boldsymbol{0} \tag{5.21}$$

which have the same vector structure as the vacuum Coulomb law and vacuum Ampère Maxwell law of Maxwell Heaviside (MH) field theory. However, Eqs. (5.20) and (5.21) of ECE theory are written in a space-time with torsion and curvature, not in a flat Minkowski space-time as in MH theory. In ECE theory the charge-current density is recognized as being

$$J^{\kappa\nu} = -\frac{A^{(0)}}{\mu_0} R^{\kappa}{}_{\mu}{}^{\mu\nu} \tag{5.22}$$

and to be proportional directly to a Ricci type tensor $R^{\kappa}{}_{\mu}{}^{\mu\nu}$. The vacuum in ECE theory is defined as being Ricci flat in a generally covariant unified field theory. This is self-consistently the vacuum solution of the Einstein Hilbert (EH) field equation (5.14).

5.3 General Coordinate Transformation and Rotation

The general coordinate transformation of a four-vector V^μ is defined [2] as:

$$V^{\mu'} = \frac{\partial x^{\mu'}}{\partial x^\mu} V^\mu. \tag{5.23}$$

The Lorentz transform is a special case of Eq. (5.23):

$$V^{\mu'} = \Lambda^{\mu'}{}_\mu V^\mu \tag{5.24}$$

and is a special kind of coordinate transformation [2]:

$$x^{\mu'} = \Lambda^{\mu'}{}_\mu x^\mu. \tag{5.25}$$

Eq (5.23) describes the behavior of vectors under arbitrary changes of coordinates and basis elements. The complete vector field is constant under the change of coordinates:

$$V = V^\mu \partial_\mu = V^{\mu'} \partial_{\mu'}. \tag{5.26}$$

A rotation is an example of a coordinate transformation, a rotation can be the rotation of a vector with fixed coordinates or the rotation of coordinates with fixed vector. Rotation in an arbitrary manifold is represented by the rotation group, which is a sub group of the Lorentz group. The latter is itself a subgroup of the Poincaré group [13], which is a subgroup of the Einstein group. Rotation generators are proportional to angular momentum generators [13] and are anti-symmetric tensors.

We now denote:

$$\nu = \mu' \tag{5.27}$$

and denote:

$$\epsilon^\mu{}_\nu = \partial x^\mu / \partial x^\nu. \tag{5.28}$$

Therefore:

$$\epsilon^{\rho\sigma} = -\epsilon^{\sigma\rho} = g^{\rho\kappa}\epsilon^\sigma{}_\kappa \tag{5.29}$$

where $g^{\rho\kappa}$ is the inverse metric tensor in an arbitrary manifold [2]. For rotations:

$$\epsilon^{\rho\sigma} = -\epsilon^{\sigma\rho} \tag{5.30}$$

and the Kronecker delta [2] is defined by:

$$\delta^\mu{}_\sigma = g^{\mu\nu} g_{\nu\sigma}. \tag{5.31}$$

It is seen that the coordinate four-vector can be represented by:

$$x^\mu = \epsilon^\mu{}_\nu x^\nu = X^\mu{}_{\rho\sigma} \epsilon^{\rho\sigma}. \tag{5.32}$$

This means that there exists a rank three tensor $X^\mu{}_{\rho\sigma}$ from considerations of coordinate transformation alone. For rotations:

$$X^\mu{}_{\rho\sigma} = -X^\mu{}_{\sigma\rho} \tag{5.33}$$

5.3 General Coordinate Transformation and Rotation

because:

$$\epsilon^{\rho\sigma} = -\epsilon^{\sigma\rho}. \tag{5.34}$$

Therefore there exists the following duality between V^μ and $V^\mu{}_{\rho\sigma}$:

$$V^\mu = V^\mu{}_{\rho\sigma}\,\epsilon^{\rho\sigma} \tag{5.35}$$

meaning that any four-vector V^μ can be expressed as a rank three tensor $V^\mu{}_{\rho\sigma}$ in the arbitrary base manifold. In 3-D Euclidean space there is a similar duality between an axial three-vector V_i and a rank two anti-symmetric tensor V_{jk}:

$$V_i = \frac{1}{2}\epsilon_{ijk} V_{jk} \tag{5.36}$$

where ϵ_{ijk} is the rank three totally anti-symmetric unit tensor in three dimensional Euclidean space.

In general, Eq. (5.35) is:

$$\frac{1}{2}V^\mu = V^\mu{}_{01}\,\epsilon^{01} + V^\mu{}_{02}\,\epsilon^{02} + V^\mu{}_{03}\,\epsilon^{03} + V^\mu{}_{12}\,\epsilon^{12} + V^\mu{}_{13}\,\epsilon^{13} + V^\mu{}_{23}\,\epsilon^{23}. \tag{5.37}$$

This can be expressed as a four dimensional matrix with a structure similar to the electromagnetic field matrix of MH theory [1] but with an additional upper index:

$$V^\mu{}_{\rho\sigma} = \begin{bmatrix} 0 & V^\mu{}_{01} & V^\mu{}_{02} & V^\mu{}_{03} \\ V^\mu{}_{10} & 0 & V^\mu{}_{12} & V^\mu{}_{13} \\ V^\mu{}_{20} & V^\mu{}_{21} & 0 & V^\mu{}_{23} \\ V^\mu{}_{30} & V^\mu{}_{31} & V^\mu{}_{32} & 0 \end{bmatrix}. \tag{5.38}$$

If for example rotation about the $Z = 3$ axis is considered, then:

$$\mu = 3,\,\epsilon^{12} = -\epsilon^{21} = 1 \tag{5.39}$$

all other $\epsilon^{\mu\nu}$ being zero, so:

$$V_Z = V^3 = 2V^3{}_{12} \tag{5.40}$$

Here V_Z is the Z component of the three-vector:

$$\boldsymbol{V} = V_X \boldsymbol{i} + V_Y \boldsymbol{j} + V_Z \boldsymbol{k}. \tag{5.41}$$

Eq. (5.40) is the magnetic field relation used in the ECE theory of classical electrodynamics to reduce it to the same vector format as the MH theory, but with the key addition of the spin connection [3–12]. The three axial vector relations are:

$$V_X = V^1 = 2V^1{}_{23} \tag{5.42}$$
$$V_Y = V^2 = 2V^2{}_{13} \tag{5.43}$$
$$V_Z = V^3 = 2V^3{}_{12}. \tag{5.44}$$

The three polar vector relations are:

$$V_X = V^1 = 2V^0{}_{01} \tag{5.45}$$
$$V_Y = V^2 = 2V^0{}_{02} \tag{5.46}$$
$$V_Z = V^3 = 2V^0{}_{03} \tag{5.47}$$

and define the electric field relations used in the ECE theory of electrodynamics [3–12]. The electromagnetic field tensor in ECE theory is directly proportional as follows to the Cartan torsion tensor:

$$F^\mu{}_{\sigma\rho} = A^{(0)} T^\mu{}_{\sigma\rho} \tag{5.48}$$

and both arise from the rotation generator in the arbitrary manifold. The Cartan torsion form is defined by the addition of a Minkowski space-time at a point P in the arbitrary base manifold. The Cartan torsion form is defined by the first Cartan structure equation [2]:

$$T^a = d \wedge q^a + \omega^a{}_b \wedge q^b \tag{5.49}$$

and this definition is equivalent to:

$$T^\mu{}_{\sigma\rho} = \Gamma^\mu{}_{\sigma\rho} - \Gamma^\mu{}_{\rho\sigma} \tag{5.50}$$

using the tetrad postulate [2–12]:

$$D_\mu q^a{}_\nu = 0. \tag{5.51}$$

5.4 Canonical Angular Momentum-Energy Density

This kind of rank three tensor arises from considerations of the Noether Theorem [13], which is derived from a lagrangian method. It is also anti-symmetric in its lower two indices and is defined as:

$$J^\mu{}_{\rho\sigma} = -\frac{1}{2}(T^\mu{}_\rho x_\sigma - T^\mu{}_\rho x_\rho). \tag{5.52}$$

where:

$$T^{\mu\nu} = T^{\nu\mu} = g^{\nu\kappa}T^\mu{}_\kappa \tag{5.53}$$

is the symmetric canonical energy-momentum density tensor used in the EH field equation:

$$G_{\mu\nu} = kT_{\mu\nu} \tag{5.54}$$

where $G_{\mu\nu}$ is the Einstein tensor and k is the Einstein constant. From the Noether Theorem the action is invariant under spatial rotations:

$$\delta x^i = \epsilon^{ij}x^j, \epsilon^{ij} = -\epsilon^{ji} \qquad (i,j = 1,2,3) \tag{5.55}$$

as generalized in Section 5.3 of this paper and the canonical angular energy-momentum density:

$$J^{\mu\rho\sigma} = -T^\mu{}_\kappa X^{\kappa\rho\sigma} \tag{5.56}$$

is a conserved Noether current:

$$D_\mu J^\mu{}_{\rho\sigma} = 0. \tag{5.57}$$

This relation is analogous to the conservation of canonical energy-momentum density:

$$D^\nu T_{\mu\nu} = 0 \tag{5.58}$$

and the EH field equation was derived from:

$$D^\mu G_{\mu\nu} = kD^\nu T_{\mu\nu}$$
$$= 0 \tag{5.59}$$

where the left hand side is the second Bianchi identity [2–12].

It is well known experimentally [15] that the electromagnetic field has angular momentum, and as we have argued the canonical form of this is $J^\mu{}_{\rho\sigma}$. So the latter is proportional to $F^\mu{}_{\rho\sigma}$, and the ECE hypothesis makes $F^\mu{}_{\rho\sigma}$ proportional to the Cartan torsion tensor $T^\mu{}_{\rho\sigma}$. All are rank three tensors anti-symmetric in their lower two indices and all derive from the rotation generator. These considerations give the results:

$$F^\mu{}_{\rho\sigma} = \frac{c}{e\omega} J^\mu{}_{\rho\sigma} = \frac{E^{(0)}}{\omega} T^\mu{}_{\rho\sigma} \tag{5.60}$$

Using the quantum relation [3–12]:

$$\left.\begin{array}{l} eA^{(0)} = \hbar\kappa, \\ E^{(0)} = cB^{(0)} = \kappa C A^{(0)} = \omega A^{(0)} \end{array}\right\} \tag{5.61}$$

we obtain:

$$J^\mu{}_{\rho\sigma} = \hbar\kappa^2 (\Gamma^\mu{}_{\rho\sigma} - \Gamma^\mu{}_{\sigma\rho}) \tag{5.62}$$

so that the anti-symmetric connection can be defined as:

$$\Gamma^\mu{}_{\rho\sigma} = -\frac{1}{2\hbar\kappa^2} g^{\mu\kappa} T_{\kappa\rho} x_\sigma. \tag{5.63}$$

This is a self consistent result because in EH theory there is no angular momentum and no Cartan torsion, so that the connection in EH theory is the symmetric Christoffel connection [2]:

$$\Gamma^\mu{}_{\rho\sigma} = \Gamma^\mu{}_{\sigma\rho}. \tag{5.64}$$

Therefore individual vector components of magnetic flux density and electric field strength may be defined as follows:

$$E_X = E^0{}_{01} = (c^2/(e\omega)) J^0{}_{01} \tag{5.65}$$
$$E_Y = E^0{}_{02} = (c^2/(e\omega)) J^0{}_{02} \tag{5.66}$$
$$E_Z = E^0{}_{03} = (c^2/(e\omega)) J^0{}_{03} \tag{5.67}$$
$$B_X = B^1{}_{23} = (c^2/(e\omega)) J^1{}_{23} \tag{5.68}$$
$$B_Y = B^3{}_{12} = (c^2/(e\omega)) J^3{}_{12} \tag{5.69}$$
$$B_Z = B^2{}_{31} = (c^2/(e\omega)) J^2{}_{31} \tag{5.70}$$

where:

$$E^0{}_{01} = \partial_0 A^0_1 - \partial_1 A^0_0 + \omega^0{}_{0b} A^b{}_1 - \omega^0{}_{1b} A^b{}_0 \qquad (5.71)$$

$$B^1{}_{23} = \partial_2 A^1_3 - \partial_3 A^1_2 + \omega^1{}_{2b} A^b{}_3 - \omega^1{}_{3b} A^b{}_2 \qquad (5.72)$$

and where:

$$J^0{}_{01} = -\frac{1}{2}(T^0{}_0 x_1 - T^0{}_1 x_0) \qquad (5.73)$$

etc.

and:

$$J^1{}_{23} = -\frac{1}{2}(T^1{}_2 x_3 - T^1{}_3 x_2), \qquad (5.74)$$

etc.

Acknowledgments

The British Government is thanked for the award of a Civil List Pension and the Telesio-Galilei Association for funding of this work.

References

[1] J. D. Jackson, "Classical Electrodynamics" (Wiley, New York, 1999, 3rd Ed).

[2] S. P. Carroll, "Space-time and Geometry: an Introduction to General Relativity" (Addison-Wesley, New York, 2004, 1997 notes available from web).

[3] M. W. Evans, "Generally Covariant Unified Field Theory" (Abramis Academic, Suffolk, 2005 onwards), in four volumes to date (ibid. vol. 5, papers 71 to 93 on www.aias.us).

[4] L. Felker, "The Evans Equations of Unified Field Theory" (Abramis Academic, Suffolk, 2007).

[5] The papers and educational articles on www.aias.us and www.atomicprecision.com.

[6] K. Pendergast, "Mercury as Crystal Spheres" (an introduction to ECE theory on www.aias.us, Abramis Academic in preparation).

[7] The Omnia Opera articles on www.aias.us from 1992 to present giving the precursor theories of ECE and other details.

[8] M. W. Evans and L. B. Crowell, "Classical and Quantum Electrodynamics and the B(3) Field" (World Scientific, 2001).

[9] M. W. Evans (ed.), "Modern Nonlinear Optics", a special topical issue in three parts of I. Prigogine and S. A. Rice (series eds.), "Advances in Chemical Physics" (Wiley Interscience, new York, 2001, 2nd Ed.), vols 119(1) to 119(3); ibid., M. W. Evans and S. Kielich (eds.), first editions, vols, 85(1) to 85(3), (Wiley Interscience, New York, 1992, 1993 and 1997).

[10] M. W. Evans and J.-P. Vigier, "The Enigmatic Photon" (Kluwer, Dordrecht, 1994 to 2002, hardback and softback), in five volumes.

[11] M. W. Evans and A. A. Hasanein, "The Photomagneton in Quantum Field Theory" (World Scientific, 1994).

[12] M. W. Evans, Physica B, **182**, 227 and 237 (1992).

[13] L. H. Ryder, "Quantum Field Theory" (Cambridge Univ. Press, 1996, 2nd ed.).

[14] M. W. Evans, H. Eckardt and S. P. Crothers, paper 93 on www.aias.us.

[15] P. W. Atkins, "Molecular Quantum Mechanics" (Oxford Univ. Press, 2nd and subsequent editions from 1983).

6

The Fundamental Origin of Curvature and Torsion

by

Myron W. Evans,
Alpha Institute for Advanced Study, Civil List Scientist.
(emyrone@aol.com and www.aias.us)

Abstract

The fundamental origin of curvature and torsion is discussed in terms of commutators of covariant derivatives in the general manifold. Detailed proofs are given of the origin of the curvature and torsion tensors. A proof of the Jacobi operator identity is given and this identity is used to prove that the conventional second Bianchi identity is true if and only if torsion is zero, and if and only if accompanied by a novel operator identity neglected in the literature. Finally group theoretical considerations are used to prove that in the case of rotation, the Riemann and torsion tensors can be interpreted as group structure constants. This proof in turn leads to the conclusion that the equations of classical electrodynamics take the same vectorial form in Einstein Cartan Evans (ECE) and Maxwell Heaviside field theory, but in different base manifolds.

Keywords: Curvature, torsion, covariant derivatives, Jacobi identity, Bianchi identity, rotational origin of Cartan torsion, Einstein Cartan Evans (ECE) field theory.

6.1 Introduction

Recently a generally covariant unified field theory has been developed [1–8] in which the electromagnetic sector is represented by Cartan geometry in which appears curvature and torsion. This theory is known as Einstein Cartan

Evans (ECE) theory because it is based on an extension of the Riemann geometry used by Einstein to include Cartan's torsion [9]. In ECE theory the electromagnetic field is directly proportional to the Cartan torsion. The latter is represented in Cartan differential geometry by a vector valued differential two-form defined by:

$$T^a = d \wedge q^a + \omega^a{}_b \wedge q^b \tag{6.1}$$

where q^a is the Cartan tetrad, a vector valued differential one-form [9], and where $\omega^a{}_b$ is the spin connection. The index a is defined in a tangential Minkowski space-time at a point P in the base manifold, a manifold which represents a four dimensional space-time with torsion and curvature. Using the tetrad postulate [1–9]:

$$D_\mu q^a_\nu = 0 \tag{6.2}$$

the definition (6.1) becomes equivalent to the definition of the Cartan torsion tensor in the base manifold:

$$T^\kappa_{\mu\nu} = \Gamma^\kappa_{\mu\nu} - \Gamma^\kappa_{\nu\mu} \tag{6.3}$$

where $\Gamma^\kappa_{\mu\nu}$ is the connection of the base manifold. In the Riemann geometry used by Einstein to develop general relativity the connection is the Christoffel connection:

$$\Gamma^\kappa_{\mu\nu} = \Gamma^\kappa_{\nu\mu} \tag{6.4}$$

so in Einsteinian general relativity and cosmology, torsion is zero. In ECE theory the electromagnetic potential and field are directly proportional respectively to the tetrad and torsion forms:

$$A^a_\mu = A^{(0)} q^a_\mu \tag{6.5}$$
$$F^a_{\mu\nu} = A^{(0)} T^a_{\mu\nu}. \tag{6.6}$$

Therefore in the base manifold, the electromagnetic field becomes a rank three tensor proportional to the torsion tensor:

$$F^\kappa_{\mu\nu} = A^{(0)} T^\kappa_{\mu\nu}. \tag{6.7}$$

It has been shown [1–8] that the definition (6.7) leads to the equations of classical electrodynamics in the same vectorial notation as the Maxwell Heaviside vector theory but written in a base manifold with torsion and curvature, not in the Minkowski space-time. The Cartesian components of the electric and magnetic field in ECE theory are defined by elements of the rank three torsion tensor as follows:

$$\left.\begin{array}{ll} E_X = E^1{}_{01}, & B_X = B^1{}_{23}, \\ E_Y = E^2{}_{02}, & B_Y = B^2{}_{31}, \\ E_Z = E^3{}_{03}, & B_Z = B^3{}_{12}. \end{array}\right\} \qquad (6.8)$$

In this paper a further fundamental proof of equation (6.8) is given using the fundamental definition [9] of the Riemann tensor and torsion tensor in terms of commutators of covariant derivatives (or round trip in the base manifold). The derivation of the Riemann tensor and torsion tensor (6.3) using this method is given in detail in Section 6.2. In Section 6.3 the proof of the Jacobi identity is given. The Jacobi identity [9–10] is exact and is valid both for covariant derivatives and group generators [10]. In this section the Jacobi identity is used to show under what circumstances the second Bianchi identity [9] is valid. This is important because the second Bianchi identity is used directly in the derivation of the Einstein Hilbert (EH) field equation. It is found that the second Bianchi identity and EH field equation are valid if and only if the Cartan torsion tensor (6.3) is zero, and if and only if:

$$\begin{aligned} R^\rho{}_{\sigma\mu\nu} D_\kappa + R^\rho{}_{\sigma\kappa\mu} D_\nu + R^\rho{}_{\sigma\nu\kappa} D_\mu &= 0, \\ R^a{}_b \wedge D &= 0, \end{aligned} \qquad (6.9)$$

which is a new differential operator relation which appears to have been hitherto neglected in the literature. In Eq. (6.9), $R^a{}_b$ is the curvature or Riemann differential form [1–9] and D represents the covariant derivative in differential geometry. The conventional second Bianchi identity is usually written as the reverse of Eq. (6.9):

$$D \wedge R^a{}_b = 0. \qquad (6.10)$$

In the presence of the torsion form however the rigorously correct Bianchi identity is [1–9]:

$$D \wedge T^a := R^a{}_b \wedge q^b \qquad (6.11)$$

and there is only one Bianchi identity (see paper 88 of www.aias.us). The second one can be derived from Eq. (6.11).

In Section 6.4 finally, the rotational limit of Cartan geometry is considered using the round trip method and it is shown that in this limit the Cartan torsion tensor (6.3) can be considered to be a group structure constant. These considerations lead directly to the interpretation (6.8) of the electric and magnetic field components in ECE theory (see papers 93 and following on www.aias.us).

6.2 Derivation of the Curvature and Torsion Tensors from Commutators of Covariant Derivatives

Although this proof is well known it is usually given in textbooks [9] without sufficient detail for understanding by non-specialists. It gives a fundamental interpretation for both curvature and torsion in terms of a round trip in the general base manifold. This method is also used in field theory [10] and so the curvature, torsion and field tensors have the same fundamental origin. This is therefore a fundamental justification for the basic ECE hypothesis, that the electromagnetic field tensor is directly proportional to the Cartan torsion, they are both commutators of covariant derivatives. The origin of both the Riemann and Cartan torsion tensors is parallel transport around a closed loop:

$$\delta V^\rho = (\delta a)(\delta b) A^\nu B^\mu R^\rho{}_{\sigma\mu\nu} V^\sigma \tag{6.12}$$

which can be represented by a commutator of covariant derivatives [9–10]. The covariant derivative of a tensor in a given direction measures [9] how much the tensor changes relative to what it would have been if it had been parallel transported. The commutator of covariant derivatives measures the difference between parallel transporting the tensor one way and then the other, versus the opposite ordering. In flat or Minkowski space-time the result is zero, it makes no difference which sense the process takes place. In flat space-time the covariant derivatives become ordinary derivatives so the following operator is zero:

$$[\partial_\mu, \partial_\nu] = \partial_\mu \partial_\nu - \partial_\nu \partial_\mu = 0. \tag{6.13}$$

The commutator of covariant derivatives is however an operator which is not zero:

$$[D_\mu, D_\nu] \neq 0. \tag{6.14}$$

Such commutators are well known in the theory of rotation generators, angular momentum, group theory and quantum mechanics [10]. They also appear in differential geometry as the wedge product [1–9] of two one-forms, which is defined by:

$$A^a \wedge B^b = [A^a_\mu, A^b_\nu]. \tag{6.15}$$

The Riemann and torsion tensors are defined [9] by:

$$[D_\mu, D_\nu] V^\rho = D_\mu(D_\nu V^\rho) - D_\nu(D_\mu V^\rho) \tag{6.16}$$

6.2 Derivation of the Curvature and Torsion Tensors

where V^ρ is a four vector in a base manifold with curvature and torsion. On the right hand side of Eq. (6.16) the covariant derivatives act on rank two tensors contained within the brackets. The rule [1–9] for the covariant derivative of a rank two tensor then gives:

$$[D_\mu, D_\nu]V^\rho = \partial_\mu(D_\nu V^\rho) - \Gamma^\lambda_{\mu\nu} D_\lambda V^\rho + \Gamma^\rho_{\mu\sigma} D_\nu V^\sigma \\ - \partial_\nu(D_\mu V^\rho) + \Gamma^\lambda_{\nu\mu} D_\lambda V^\rho - \Gamma^\rho_{\nu\sigma} D_\mu V^\sigma \quad (6.17)$$

within which are defined:

$$\left.\begin{aligned} D_\nu V^\rho &= \partial_\nu V^\rho + \Gamma^\rho{}_{\nu\lambda} V^\lambda, \\ D_\lambda V^\rho &= \partial_\lambda V^\rho + \Gamma^\rho{}_{\lambda\sigma} V^\sigma, \\ D_\nu V^\sigma &= \partial_\nu V^\sigma + \Gamma^\sigma{}_{\nu\lambda} V^\lambda. \end{aligned}\right\} \quad (6.18)$$

Therefore there are equations such as:

$$\partial_\mu(D_\nu V^\rho) = \partial_\mu \partial_\nu V^\rho + (\partial_\mu \Gamma^\rho_{\nu\lambda}) V^\lambda + \Gamma^\rho_{\nu\lambda} \partial_\mu V^\lambda \\ = \partial_\mu \partial_\nu V^\rho + (\partial_\mu \Gamma^\rho_{\nu\sigma}) V^\sigma + \Gamma^\rho_{\nu\sigma} \partial_\mu V^\sigma. \quad (6.19)$$

The dummy or summation indices are now re-arranged as follows:

$$\lambda \to \sigma \quad (6.20)$$

This gives:

$$[D_\mu, D_\nu]V^\rho = \partial_\mu \partial_\nu V^\rho + (\partial_\mu \Gamma^\rho_{\nu\sigma}) V^\sigma + \Gamma^\rho_{\nu\sigma} \partial_\mu V^\sigma \\ - \Gamma^\lambda_{\mu\nu} \partial_\lambda V^\rho - \Gamma^\lambda_{\mu\nu} \Gamma^\rho_{\lambda\sigma} V^\sigma \\ + \Gamma^\rho_{\mu\sigma} \partial_\nu V^\sigma + \Gamma^\rho_{\mu\sigma} \Gamma^\sigma_{\nu\lambda} V^\lambda \\ - \partial_\nu \partial_\mu V^\rho - (\partial_\nu \Gamma^\rho_{\mu\sigma}) V^\sigma - \Gamma^\rho_{\mu\sigma} \partial_\nu V^\sigma \\ + \Gamma^\lambda_{\nu\mu} \partial_\lambda V^\rho + \Gamma^\lambda_{\nu\mu} \Gamma^\rho_{\lambda\sigma} V^\sigma \\ - \Gamma^\rho_{\mu\sigma} \partial_\nu V^\sigma - \Gamma^\rho_{\mu\sigma} \Gamma^\sigma_{\nu\lambda} V^\lambda \quad (6.21)$$

which can be re-arranged to give:

$$[D_\mu, D_\nu]V^\rho = (\partial_\mu \Gamma^\rho_{\nu\sigma} - \partial_\nu \Gamma^\rho_{\mu\sigma} + \Gamma^\rho_{\mu\lambda} \Gamma^\lambda_{\nu\sigma} - \Gamma^\rho_{\nu\lambda} \Gamma^\lambda_{\mu\sigma}) V^\sigma \\ - (\Gamma^\lambda_{\mu\nu} - \Gamma^\lambda_{\nu\mu})(\partial_\lambda V^\rho + \Gamma^\rho_{\lambda\sigma} V^\sigma). \quad (6.22)$$

Finally this is expressed as:

$$[D_\mu, D_\nu]V^\rho = R^\rho{}_{\sigma\mu\nu} V^\sigma - T^\lambda_{\mu\nu} D_\lambda V^\rho \quad (6.23)$$

where the Riemann tensor is defined [9] by:

$$R^\rho{}_{\sigma\mu\nu} = \partial_\mu \Gamma^\rho_{\nu\sigma} - \partial_\nu \Gamma^\rho_{\mu\sigma} + \Gamma^\rho_{\mu\lambda}\Gamma^\lambda_{\nu\sigma} - \Gamma^\rho_{\nu\lambda}\Gamma^\lambda_{\mu\sigma} \quad (6.24)$$

and the torsion tensor is defined by:

$$T^\lambda_{\mu\nu} = \Gamma^\lambda_{\mu\nu} - \Gamma^\lambda_{\nu\mu}. \quad (6.25)$$

The overall result:

$$[D_\mu, D_\nu]V^\rho = R^\rho{}_{\sigma\mu\nu}V^\sigma - T^\lambda_{\mu\nu}D_\lambda V^\rho \quad (6.26)$$

is true irrespective of the symmetry of the metric and connection, and irrespective of the metric compatibility condition [9]. The use of the commutator of covariant derivatives means the Riemann and torsion tensors are always anti-symmetric in their last two indices:

$$R^\rho{}_{\sigma\mu\nu} = -R^\rho{}_{\sigma\nu\mu}, \quad T^\lambda_{\mu\nu} = -T^\lambda_{\nu\mu} \quad (6.27)$$

indicating their rotational or commutative or anti-symmetric origin. Both curvature and torsion are kinds of rotation, or bending and twisting. However, it is important to note that there is no symmetry restriction on the first two indices of the Riemann tensor in general. The Riemann tensor is anti-symmetric in its first two indices if and only if the metric compatibility condition is used [9]. If it is assumed that the metric is symmetric:

$$g_{\mu\nu} = g_{\nu\mu}. \quad (6.28)$$

It follows form metric compatibility [9] that the connection is symmetric:

$$\Gamma^\lambda_{\mu\nu} = \Gamma^\lambda_{\nu\mu} \quad (6.29)$$

and that torsion vanishes. In Cartan geometry [1–9], the torsion is not zero in general, so the metric and connection are not symmetric in general. The conventional first Bianchi identity is true if and only if the metric and connection are symmetric, and if and only if the torsion is zero. In differential form notation the first Bianchi identity in the absence of torsion is:

$$R^a{}_b \wedge q^b = 0 \quad (6.30)$$

and in tensor notation it is:

$$R_{\sigma\mu\nu\rho} + R_{\sigma\rho\mu\nu} + R_{\sigma\nu\rho\mu} = 0. \quad (6.31)$$

However it is important to note that Eqs. (6.30) and (6.31) are special cases. The rigorously correct first Bianchi identity is [1–9]:

$$D \wedge T^a := R^a{}_b \wedge q^b \qquad (6.32)$$

not zero in general, and the rigorously correct second Bianchi identity is a re-expression of Eq. (6.32), and not an independent identity (paper 88 of www.aias.us). Historically, the first Bianchi identity was given by Ricci and Levi-Civita and not by Bianchi.

In the Minkowski space-time:

$$[D_\mu, D_\nu] = [\partial_\mu, \partial_\nu] = 0, \qquad (6.33)$$

$$R^\rho{}_{\sigma\mu\nu} = T^\lambda{}_{\mu\nu} = 0, \qquad (6.34)$$

and this is the space-time of Maxwell Heaviside field theory.

The Riemann and torsion tensors are constructed from the connection and are true for any connection, whether metric compatible or not. Although the connection is non-tensorial, the Riemann and torsion tensors are true tensors by construction [1–9] and all the equations of Riemann and Cartan geometry are generally covariant, i.e. tensorial under the general coordinate transformation. This means that equations of physics based on these geometries, such as ECE theory [1–8] are rigorously objective equations of physics, they are the same in form to an observer moving arbitrarily with respect to another, in a frame of reference moving arbitrarily with respect to another frame of reference. This is the essence of the essentially geometrical philosophy of general relativity as is well known [9]. The essence of the matter is that physics is geometry, all physics is geometry, not just gravitation. Otherwise we do not have a self consistent basic philosophy of physics. Maxwell Heaviside (MH) field theory does not obey this philosophy because the MH field theory is defined in Minkowski space-time, and MH theory is Lorentz covariant by construction. It contains no connection and is not generally covariant.

For a tensor of any rank [9]:

$$[D_\rho, D_\sigma]\chi^{\mu_1...\mu_k}_{\nu_1...\nu_l} = -T^\lambda{}_{\rho\sigma}D_\lambda\chi^{\mu_1...\mu_k}_{\nu_1...\nu_l} + R^{\mu_1}{}_{\lambda\rho\sigma}\chi^{\lambda\mu_2...\mu_k}_{\nu_1...\nu_l} + R^{\mu_2}{}_{\lambda\rho\sigma}\chi^{\mu_1\lambda...\mu_k}_{\nu_1...\nu_l} + \cdots$$
$$- R^\lambda{}_{\nu_1\rho\sigma}\chi^{\mu_1...\mu_k}_{\lambda\nu_2...\nu_l} + R^\lambda{}_{\nu_2\rho\sigma}\chi^{\mu_1...\mu_k}_{\nu_1\lambda...\nu_l} - \cdots \qquad (6.35)$$

The commutator of two vector fields X and Y is a third vector field [9] with components:

$$[X, Y]^\mu = X^\lambda \partial_\lambda Y^\mu - Y^\lambda \partial_\lambda X^\mu. \qquad (6.36)$$

The curvature and torsion tensors can be though of as multi-linear maps [9], the torsion being a map from two vector fields to a third:

$$T(X,Y) = D_X Y - D_Y X - [X,Y] \qquad (6.37)$$

and the curvature as a map from three vector fields to a fourth [9]:

$$R(X,Y)Z = D_X D_Y Z - D_Y D_X Z - D_{[X,Y]} Z \qquad (6.38)$$

where:

$$D_X = X^\mu D_\mu. \qquad (6.39)$$

Cartan's geometry [1–9] expresses these results in an elegant and concise way through his two well known structure equations [9]:

$$T^a = D \wedge q^a, \qquad (6.40)$$
$$R^a{}_b = D \wedge \omega^a{}_b. \qquad (6.41)$$

6.3 The Jacobi and Bianchi Identities

The Jacobi identity [1–10] is an exact identity used in field theory and general relativity. It is an operator identity that applies to covariant derivatives [9] and group generators [10] alike. It is very rarely proven in all detail however and so the following is a detailed proof. It is necessary to prove that:

$$[[D_\lambda, D_\rho], D_\sigma] + [[D_\rho, D_\sigma], D_\lambda] + [[D_\sigma, D_\lambda], D_\rho] := 0 \qquad (6.42)$$

which is the Jacobi identity, an exact identity. The proof expands the commutators as follows

$$\begin{aligned} \text{L.H.S} =\ & (D_\lambda D_\rho - D_\rho D_\lambda) D_\sigma - D_\sigma (D_\lambda D_\rho - D_\rho D_\lambda) \\ & + (D_\rho D_\sigma - D_\sigma D_\rho) D_\lambda - D_\lambda (D_\rho D_\sigma - D_\sigma D_\rho) \\ & + (D_\sigma D_\lambda - D_\lambda D_\sigma) D_\rho - D_\rho (D_\sigma D_\lambda - D_\lambda D_\sigma) \end{aligned} \qquad (6.43)$$

and this expansion is regarded as an expansion of algebra:

$$\begin{aligned} \text{L.H.S} =\ & D_\lambda D_\rho D_\sigma - D_\rho D_\lambda D_\sigma - D_\sigma D_\lambda D_\rho + D_\sigma D_\rho D_\lambda \\ & + D_\rho D_\sigma D_\lambda - D_\sigma D_\rho D_\lambda - D_\lambda D_\rho D_\sigma + D_\lambda D_\sigma D_\rho \\ & + D_\sigma D_\lambda D_\rho - D_\lambda D_\sigma D_\rho - D_\rho D_\sigma D_\lambda + D_\rho D_\lambda D_\sigma := 0 \end{aligned} \qquad (6.44)$$

Q.E.D.

6.3 The Jacobi and Bianchi Identities

In field theory [10] the Jacobi identity is used to define field equations, and the commutator of covariant derivatives defines the field $G_{\mu\nu}$ through a constant g:

$$[D_\mu, D_\nu] = -igG_{\mu\nu}. \tag{6.45}$$

The idea of covariant derivative in field theory is borrowed from general relativity [10] and in condensed notation in field theory there exist commutators such as:

$$[D_\mu, D_\nu] = [\partial_\mu - igA_\mu, \partial_\nu - igA_\nu]$$
$$= -ig(\partial_\mu A_\nu - \partial_\nu A_\mu - ig[A_\mu, A_\nu]) \tag{6.46}$$

which have been extensively developed in to precursor theories of ECE such as O(3) electrodynamics (see Omnia Opera section of www.aias.us from 1992 to 2003). In Ryder's [10] eq. (3.173) for example there appears a field equation:

$$D_\rho G_{\mu\nu} + D_\mu G_{\nu\rho} + D_\nu G_{\rho\mu} = 0 \tag{6.47}$$

which in the notation of differential geometry [1–9] is:

$$D \wedge G = 0. \tag{6.48}$$

Eq. (6.48) is similar to the Bianchi identity of differential geometry, which becomes the ECE homogeneous field equation [1–8]:

$$D \wedge F^a = A^{(0)}(R^a{}_b \wedge q^b) \tag{6.49}$$

or

$$d \wedge F^a = A^{(0)}(R^a{}_b \wedge q^b - \omega^a{}_b \wedge T^b). \tag{6.50}$$

It is clear that both Ryder's field equation (6.48) and the ECE field equation (6.49) share a common origin in the commutator of covariant derivatives, but ECE theory is developed in a more general manifold than the type of field theory used by Ryder [10]. The latter is restricted to the Minkowski manifold only.

Restricting consideration of Eq. (6.26) to the torsion free case it becomes:

$$[D_\mu, D_\nu]V^\rho = R^\rho{}_{\sigma\mu\nu}V^\sigma \tag{6.51}$$

which can be expanded as:

$$[D_\mu, D_\nu]V^0 = R^0{}_{0\mu\nu}V^0 + R^0{}_{1\mu\nu}V^1 + R^0{}_{2\mu\nu}V^2 + R^0{}_{3\mu\nu}V^3$$
$$[D_\mu, D_\nu]V^1 = R^1{}_{0\mu\nu}V^0 + R^1{}_{1\mu\nu}V^1 + R^1{}_{2\mu\nu}V^2 + R^1{}_{3\mu\nu}V^3$$
$$[D_\mu, D_\nu]V^2 = R^2{}_{0\mu\nu}V^0 + R^2{}_{1\mu\nu}V^1 + R^2{}_{2\mu\nu}V^2 + R^2{}_{3\mu\nu}V^3$$
$$[D_\mu, D_\nu]V^3 = R^3{}_{0\mu\nu}V^0 + R^3{}_{1\mu\nu}V^1 + R^3{}_{2\mu\nu}V^2 + R^3{}_{3\mu\nu}V^3. \qquad (6.52)$$

In the torsion free case the following Riemann tensor elements vanish:

$$R^0{}_{0\mu\nu} = R^1{}_{1\mu\nu} = R^2{}_{2\mu\nu} = R^3{}_{3\mu\nu} = 0 \qquad (6.53)$$

because [9] in this case:

$$R^\rho{}_{\sigma\mu\nu} = -R^\sigma{}_{\rho\mu\nu}. \qquad (6.54)$$

Therefore

$$[D_\mu, D_\nu]\begin{bmatrix} V^0 \\ V^1 \\ V^2 \\ V^3 \end{bmatrix} = \begin{bmatrix} 0 & R^0{}_1 & R^0{}_2 & R^0{}_3 \\ -R^0{}_1 & 0 & R^1{}_2 & R^1{}_3 \\ -R^0{}_2 & -R^1{}_2 & 0 & R^2{}_3 \\ -R^0{}_3 & -R^1{}_3 & -R^2{}_3 & 0 \end{bmatrix}_{\mu\nu} \begin{bmatrix} V^0 \\ V^1 \\ V^2 \\ V^3 \end{bmatrix} \qquad (6.55)$$

and it is possible to define an operator equation similar to Eq. (6.45) of field theory:

$$[D_\mu, D_\nu] = R_{\mu\nu} := \begin{bmatrix} 0 & R^0{}_{1\mu\nu} & R^0{}_{2\mu\nu} & R^0{}_{3\mu\nu} \\ -R^0{}_{1\mu\nu} & 0 & R^1{}_{2\mu\nu} & R^1{}_{3\mu\nu} \\ -R^0{}_{2\mu\nu} & -R^1{}_{2\mu\nu} & 0 & R^2{}_{3\mu\nu} \\ -R^0{}_{3\mu\nu} & -R^1{}_{3\mu\nu} & -R^2{}_{3\mu\nu} & 0 \end{bmatrix} \qquad (6.56)$$

illustrating the relation between field theory and general relativity.

The conventionally named second Bianchi identity [9] may be derived from Eq. (6.51) as follows:

$$([D_\kappa, [D_\mu, D_\nu]])V^\rho = (D_\kappa[D_\mu, D_\nu] - [D_\mu, D_\nu]D_\kappa)V^\rho$$
$$= D_\kappa(D_\mu(D_\nu V^\rho) - D_\nu(D_\mu V^\rho)) - D_\mu D_\nu(D_\kappa V^\rho)$$
$$+ D_\nu D_\mu(D_\kappa V^\rho)$$
$$= D_\kappa([D_\mu, D_\nu]V^\rho) - [D_\mu, D_\nu](D_\kappa V^\rho)$$
$$= (D_\kappa[D_\mu, D_\nu])V^\rho + [D_\mu, D_\nu]D_\kappa V^\rho - [D_\mu, D_\nu]D_\kappa V^\rho \qquad (6.57)$$

6.3 The Jacobi and Bianchi Identities

using the Leibnitz Theorem. Therefore:

$$(D_\kappa, [D_\mu, D_\nu])V^\rho = (D_\kappa[D_\mu, D_\nu])V^\rho. \tag{6.58}$$

Therefore the Jacobi identity in this case becomes:

$$(D_\kappa[D_\mu, D_\nu] + D_\nu[D_\kappa, D_\mu] + D_\mu[D_\nu, D_\kappa])V^\rho = 0 \tag{6.59}$$

i.e.:

$$D_\kappa(R^\rho{}_{\sigma\mu\nu}V^\sigma - T^\lambda{}_{\mu\nu}D_\lambda V^\rho) + D_\nu(R^\rho{}_{\sigma\kappa\mu}V^\sigma - T^\lambda{}_{\kappa\mu}D_\lambda V^\rho)$$
$$+ D_\mu(R^\rho{}_{\sigma\nu\kappa}V^\sigma - T^\lambda{}_{\nu\kappa}D_\lambda V^\rho) = 0. \tag{6.60}$$

The conventional second Bianchi identity is a special case of this equation when the torsion vanishes, so:

$$D_\kappa(R^\rho{}_{\sigma\mu\nu}V^\sigma) + D_\nu(R^\rho{}_{\sigma\kappa\mu}V^\sigma) + D_\mu(R^\rho{}_{\sigma\nu\kappa}V^\sigma) = 0 \tag{6.61}$$

Using the Leibnitz Theorem:

$$D_\kappa(R^\rho{}_{\sigma\mu\nu}V^\sigma) = (D_\kappa R^\rho{}_{\sigma\mu\nu})V^\sigma + R^\rho{}_{\sigma\mu\nu}D_\kappa V^\sigma \tag{6.62}$$

and it is seen that the second Bianchi identity:

$$D_\kappa R^\rho{}_{\sigma\mu\nu} + D_\nu R^\rho{}_{\sigma\kappa\mu} + D_\mu R^\rho{}_{\sigma\nu\kappa} = 0 \tag{6.63}$$

is true if and only if the following operator identity is also true:

$$R^\rho{}_{\sigma\mu\nu}D_\kappa + R^\rho{}_{\sigma\kappa\mu}D_\nu + R^\rho{}_{\sigma\nu\kappa}D_\mu = 0 \tag{6.64}$$

In differential form notation Eq. (6.64) is [1–9]:

$$R^a{}_b \wedge D = 0 \tag{6.65}$$

and Eq. (6.63) is:

$$D \wedge R^a{}_b = 0. \tag{6.66}$$

Therefore both Eqs. (6.65) and (6.4) severely restrict the validity of EH general relativity and cosmology.

6.4 Pure Rotational Limit

In previous work [1–8] the pure rotational limit of ECE theory has been considered to be defined by the duality of $R^a{}_b$ and T^d in the Minkowski tangent space-time:

$$R^a{}_b = -\frac{\kappa}{2}\epsilon^a{}_{bd}T^d. \tag{6.67}$$

In this section the pure rotational limit is considered to be the special case from Eq. (6.26) where:

$$R^\rho{}_{\sigma\mu\nu}V^\rho = -T^\lambda{}_{\mu\nu}D_\lambda V^\rho \tag{6.68}$$

which is similar to Eq. (6.67) but written in the base manifold. With Eq. (6.26), Eq. (6.68) becomes a rotation generator type equation:

$$[D_\mu, D_\nu]V^\rho = -2T^\lambda{}_{\mu\nu}D_\lambda V^\rho \tag{6.69}$$

giving the operator equation:

$$[D_\mu, D_\nu] = -2T^\lambda{}_{\mu\nu}D_\lambda \tag{6.70}$$

in which the covariant derivatives obey the Jacobi identity (6.42). The covariant derivatives appearing in Eq. (6.70) can also be considered as group generators [9–10]. For example the group generators of SO(3) obey:

$$[I_i, I_j] = i\epsilon_{ijk}I_k \tag{6.71}$$

where the group structure constant [10] is:

$$C_{ijk} = i\epsilon_{ijk} \tag{6.72}$$

and:

$$C_{lim}C_{mjk} + C_{ljm}C_{mki} + C_{lkm}C_{mij} = 0. \tag{6.73}$$

The group structure constant is defined by the adjoint representation [10]

$$C_{imn} = (I_i)_{mn} \tag{6.74}$$

where:

$$I_1 = \begin{bmatrix} 0 & 0 & 0 \\ 0 & 0 & -i \\ 0 & i & 0 \end{bmatrix}, I_2 = \begin{bmatrix} 0 & 0 & i \\ 0 & 0 & 0 \\ -i & 0 & 0 \end{bmatrix}, I_3 = \begin{bmatrix} 0 & -i & 0 \\ i & 0 & 0 \\ 0 & 0 & 0 \end{bmatrix}. \qquad (6.75)$$

There is a clear similarity between Eqs. (6.70) and (6.71), except that Eq. (6.70) is written in the general manifold and Eq. (6.71) is Euclidean. Therefore it is possible to think of the torsion tensor in Eq. (6.70) as a group structure constant in the general manifold. The generators of the group are the covariant derivatives in the general manifold. Taking the analogy further the SU(3) group [10] is defined by Gell-Mann matrices which obey the commutator relation:

$$\left[\frac{\lambda a}{2}, \frac{\lambda b}{2}\right] = if_{abc}\frac{\lambda c}{2}. \qquad (6.76)$$

The group structure constant in this case is defined by [10] if_{abc}.

It therefore follows that if Eq. (6.70) is considered to be rotational in nature, analogous to Eqs. (6.71) or (6.76), the possible values of $T^\lambda_{\mu\nu}$ are the totally anti-symmetric $T^1{}_{23}$, $T^2{}_{31}$ and $T^3{}_{12}$. These are space-like and play a role analogous to ϵ_{123}, ϵ_{231}, and ϵ_{312} in Euclidean space-time. If the upper indices are held constant and Hodge duals are performed on the lower two indices we obtain $T^1{}_{01}$, $T^2{}_{02}$, and $T^3{}_{03}$. In ECE theory [1–8] they define the components of the electric and magnetic fields of the generally covariant electromagnetic sector (Eqs. (6.8) and (6.50)). In the case of pure rotation the electro-dynamical equations of ECE in vector notation are the free space values:

$$\left.\begin{aligned} \boldsymbol{\nabla} \cdot \boldsymbol{B} &= 0 \\ \boldsymbol{\nabla} \times \boldsymbol{E} + \partial \boldsymbol{B}/\partial t &= \boldsymbol{0} \\ \boldsymbol{\nabla} \cdot \boldsymbol{E} &= 0 \\ \boldsymbol{\nabla} \times \boldsymbol{B} - \frac{1}{c^2}\partial \boldsymbol{E}/\partial t &= \boldsymbol{0}. \end{aligned}\right\} \qquad (6.77)$$

More generally in the laboratory, and for all practical purposes, they become [1–8] the familiar vectorial laws:

$$\left.\begin{aligned} \boldsymbol{\nabla} \cdot \boldsymbol{B} &= 0 \\ \boldsymbol{\nabla} \times \boldsymbol{E} + \partial \boldsymbol{B}/\partial t &= \boldsymbol{0} \\ \boldsymbol{\nabla} \cdot \boldsymbol{E} &= \rho/\epsilon_0 \\ \boldsymbol{\nabla} \times \boldsymbol{B} - \frac{1}{c^2}\partial \boldsymbol{E}/\partial t &= \mu_0 \boldsymbol{J} \end{aligned}\right\} \qquad (6.78)$$

but now written in the general manifold.

Acknowledgments

The British Government is thanked for a Civil List Pension and the staff of AIAS and many others for interesting discussions. The Telesio-Galilei Association is thanked for funding and one of its 2008 Gold Medals, four of which were awarded to AIAS staff members.

References

[1] M. W. Evans, "Generally Covariant Unified Field Theory" (Abramis Academic, Suffolk, 2005 onwards), volumes one to four. Volume Five in prep. (Papers 71–93 on www.aias.us).

[2] L .Felker, "The Evans Equations of Unified Field Theory" (Abramis Academic, Suffolk, 2007).

[3] K. Pendergast, "Mercury as Crystal Spheres" (preprint on www.aias.us, an introduction to ECE theory and historical review, Abramis Academic, in prep).

[4] M. W. Evans et al. Omnia opera Section of www.aias.us (1992 to 2003).

[5] M. W. Evans and L. B. Crowell, "Classical and Quantum Electrodynamics and the B(3) Field" (World Scientific, 2001).

[6] M. W. Evans, (ed.), "Modern Non-linear Optics", a special topical issue in three parts of I. Prigogine and S. A. Rice (series eds.), "Advances in Chemical Physics" (Wiley Interscience, New York, 2001, second edition), vols. 119(1) to 119(3), ibid. first edition, cd. M. W. Evans and S. Kielich, (Wiley Interscience, 1992, 1993, and 1997), vols. 85(1) to 85(3).

[7] M. W. Evans and J.-P. Vigier, "The Enigmatic Photon" (Kluwer, 1994 to 2002, hardback and softback), in five volumes.

[8] M. W. Evans, Physica B, **182**, 227, 237 (1992).

[9] S. P. Carroll, "Space-time and Geometry: an Introduction to Geneal Relativity" (Addison wesley, New York, 2004).

[10] L. H. Ryder, "Quantum Field Theory" (Cambridge Univ. Press, 1996, 2nd. Ed.).

7

A Review of Einstein Cartan Evans (ECE) Field Theory

by

Myron W. Evans,
Alpha Institute for Advanced Study, Civil List Scientist.
(emyrone@aol.com and www.aias.us)

Abstract

The development of ECE theory from Spring 2003 to present is reviewed in major themes, which include: geometrical principles, field and wave equations, phase theory and experimental effects, the unified laws of classical dynamics and electrodynamics, spin connection resonance and applications to new energy, experiments to detect the effects of gravitation in optics and electrodynamics, the theory of radiative corrections, the development of the fundamentals of general relativity, and technical appendices and equation flow charts.

Keywords: Review of ECE theory, major themes.

7.1 Introduction

The well accepted Einstein Cartan Evans (ECE) field theory [1–12] is reviewed in major themes of development from Spring 2003 to present in approximately 103 papers and volumes summarized on www.aias.us and www.atomicprecision.com. Recently a third website, www.telesio-galilei.com, has been associated with these two main websites of the theory. Additionally, these websites contain educational articles by members of the Alpha Institute for Advanced Study (AIAS) and the Telesio-Galilei Association, and also contain an Omnia Opera listing most of the collected works of the present author, including precursor theories to ECE theory from 1992 to present. Most original papers are available by hyperlink for scholarly study. It is seen in detail from the feedback activity sites of the three main sites that ECE theory is fully accepted. All the 103 papers to date are read by someone, somewhere every month, and detailed summaries of the feedback are available on www.aias.us. Additionally ECE theory has been published in the traditional manner: in four journals with anonymous reviewers, (three of them standard model journals), and is constantly internally refereed by AIAS staff. The latter are like minded professionals who have worked voluntarily on ECE theory and in the development of AIAS. Computer algebra (Maxima program) has been developed to check hand calculations of ECE theory and to perform calculations that are too complicated to carry out by hand. Therefore a review of the main themes of development and main discoveries of ECE theory is timely.

The ECE theory is a suggestion for the development of a generally covariant unified field theory based on the principles of general relativity, essentially that natural philosophy is geometry. This principle has been proposed since ancient times in many ways, but its most well known manifestation is probably the work of Albert Einstein from about 1906 to 1915, culminating in the proposal of the well known Einstein Hilbert (EH) field equation of gravitation. This work by Einstein and contemporaries is very well known, but a brief summary is given here. After several false starts Einstein proposed in 1915 that the so called "second Bianchi identity" of Riemann geometry be proportional to a form of the Noether Theorem in which the covariant derivative vanishes of the canonical energy-momentum tensor. It is much less well known that in so doing, Einstein used the only type of geometry then available to him: Riemann geometry without torsion. The EH field equation follows from this proposal by Einstein as a special case:

$$G_{\mu\nu} = kT_{\mu\nu} \qquad (7.1)$$

where $G_{\mu\nu}$ is the Einstein tensor, k is the Einstein constant, and $T_{\mu\nu}$ is the canonical energy - momentum tensor. Eq. (7.1) is a special case of the Einstein proposal of 1915:

$$D^\mu G_{\mu\nu} = kD^\mu T_{\mu\nu} = 0 \qquad (7.2)$$

where on the left hand side appears geometry, and on the right hand side appears natural philosophy. David Hilbert proposed the same equation at about the same time using Lagrangian principles, but Hilbert's work was motivated by Einstein's ideas, so the EH equation is usually attributed to Einstein. The EH equation applies however only to gravitation, whereas ECE has unified general relativity with the other fields of nature besides gravitation. The other fundamental fields are thought to be the electromagnetic, weak and strong fields. ECE has also unified general relativity with quantum mechanics by discarding the acausality and subjectivity of the Copenhagen School, and by deriving objective and causal wave equations from geometrical first principles. The two major and well accepted achievements of ECE theory are therefore the unification of fields using geometry, and the unification of relativity and quantum mechanics. This review is organized in sections outlining the main themes and discoveries of ECE theory, and into detailed technical appendices dealing with basics. These appendices include flow charts of the inter-relation of the main equations.

In Section 7.2 the geometrical first principles of ECE theory are summarized briefly, the theory is based on a form of geometry developed [13] by Cartan and first published in 1922. This geometry is fully self-consistent and well known - it can be regarded as the standard differential geometry taught in good universities. The dialogue between Einstein and Cartan on this geometry is perhaps not as well known as the dialogue between Einstein and Bohr, but is the basis for the development of ECE theory. It is named "Einstein Cartan Evans" field theory because the present author set out to suggest a completion of the Einstein Cartan dialogue. This dialogue was part of the attempt by Einstein and many others to complete general relativity by developing a generally covariant unified field theory on the principles of a given geometry. For many reasons this unification did not come about until Spring of 2003, when ECE theory was proposed. The main obstacles to unification were adherence in the standard model to a U(1) sector for electromagnetism, the neglect of the ECE spin field B(3), inferred in 1992, and adherence to the philosophy of the Copenhagen School. Standard model proponents adhere to these principles at the time of writing, but ECE proponents now adopt a different natural philosophy, since it may be claimed objectively from feedback data that ECE is a new school of thought.

In Section 7.3 the main field and wave equations of ECE are discussed in summary. They are derived from the well known principles of Cartan's geometry. The gravitational, electromagnetic, weak and strong fields are unified by Cartan's geometry, each is an aspect of the same geometry. The field equations are based on the one true Bianchi identity given by Cartan, using different representation spaces. The wave equations are derived from the tetrad postulate, the very fundamental requirement in natural philosophy and relativity theory that the complete vector field be invariant under the general transformation of coordinates. To translate Cartan to Riemann geometry requires use of the tetrad postulate. Therefore both the Bianchi identity and tetrad postulate are fundamentals of standard differential geometry and their use in ECE theory is entirely standard mathematics [13].

In Section 7.4 the unification of phase theory made possible by ECE is summarized in terms of the main discoveries and points of development. The main point of development in this context is the unification of apparently disparate phases such as the electromagnetic phase, the Dirac and Wu Yang phases, and the topological phases. ECE theory presents a unified geometrical approach to each phase, and this approach also gives a straightforward geometrical explanation of the Aharonov Bohm effects and "non-locality". The

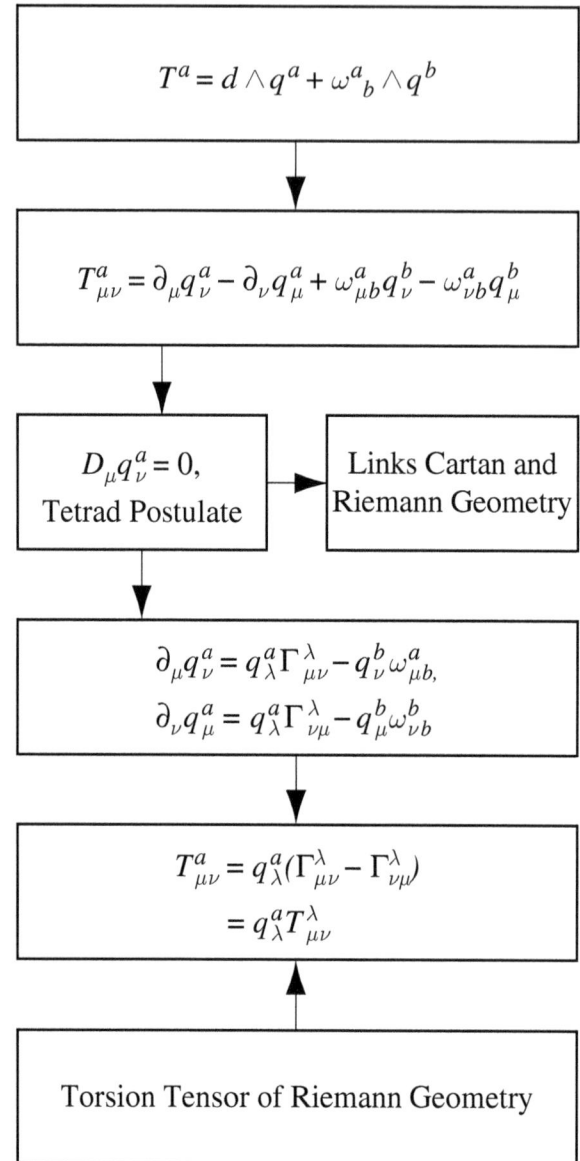

Flowchart 7.1. First Cartan Structure Equation.

electromagnetic phase for example is developed in terms of the B(3) spin field [14] and some glaring shortcomings of the standard model are corrected. Thus, apparently simple and well known effects such as reflection are developed self-consistently with ECE, while in the standard model they are at

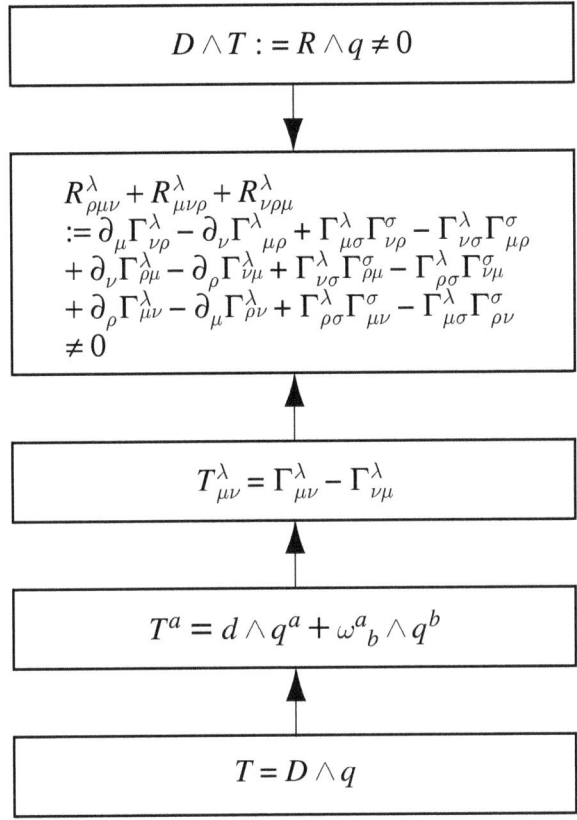

Flowchart 7.2. The Bianchi Identity.

odds with fundamental symmetry [1-12]. The standard model development of the Aharonov Bohm effects is also incorrect mathematically, obscure, controversial and convoluted, while in ECE theory it is straightforward.

In Section 7.5 the ECE laws of classical dynamics and electrodynamics are summarized in the language of vectors, the language used in electrical engineering. The equations of electrodynamics in ECE theory reduce to the four laws: Gauss law of magnetism, Faraday law of induction, Coulomb law and Ampère Maxwell law. In ECE theory they are the same in vector notation as in the familiar Maxwell Heaviside (MH) field theory, but in ECE are written in a different space-time. In ECE the electromagnetic field is the spinning of space-time, represented by the Cartan torsion, while in MH the field is a nineteenth century concept still used uncritically in the contemporary standard model of natural philosophy. The space-time of MH is the flat and static Minkowski space-time, while in ECE the space-time is dynamic with non-zero curvature and torsion. This difference manifests itself in the relation

116 7 A Review of Einstein Cartan Evans (ECE) Field Theory

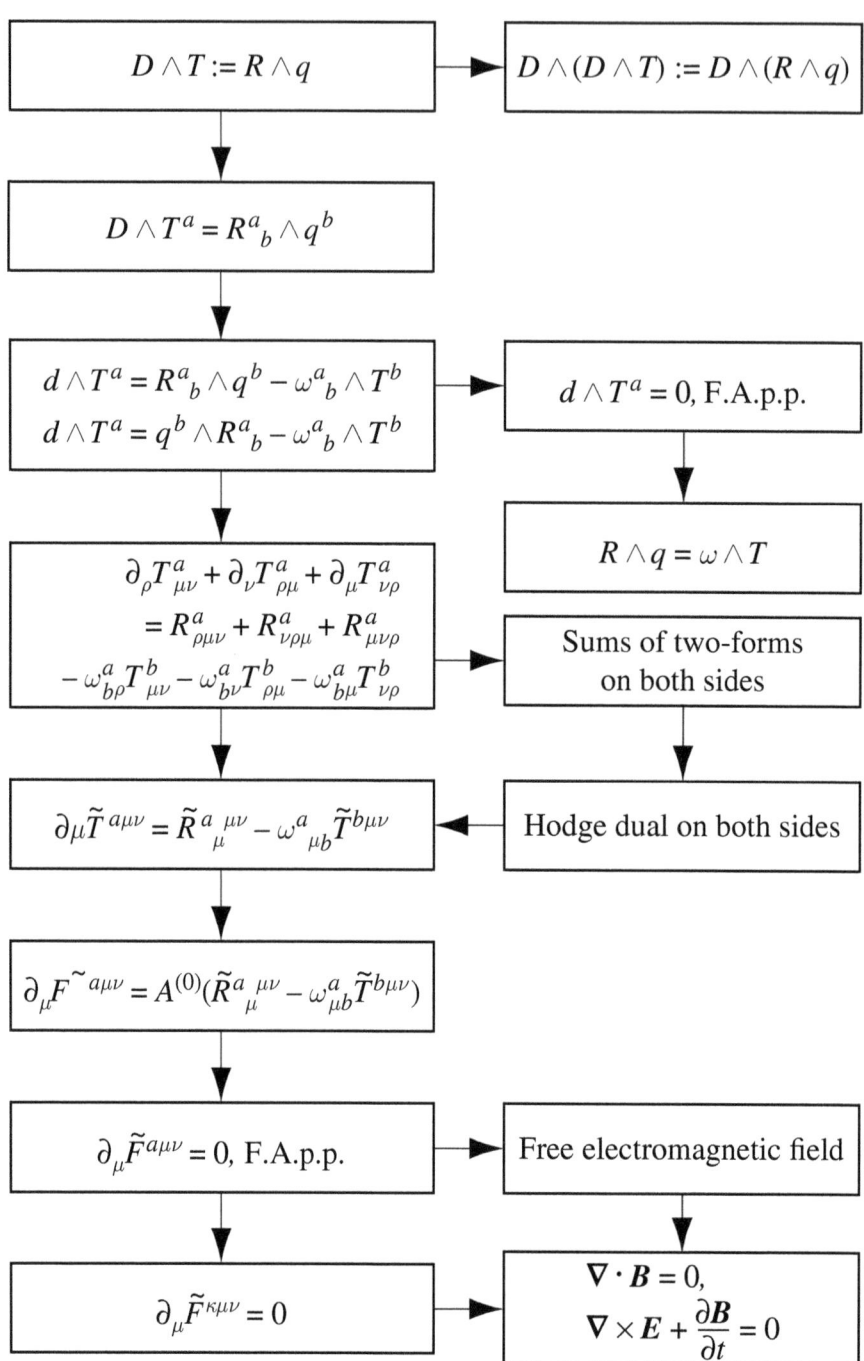

Flowchart 7.3. Homogeneous ECE Field Equation.

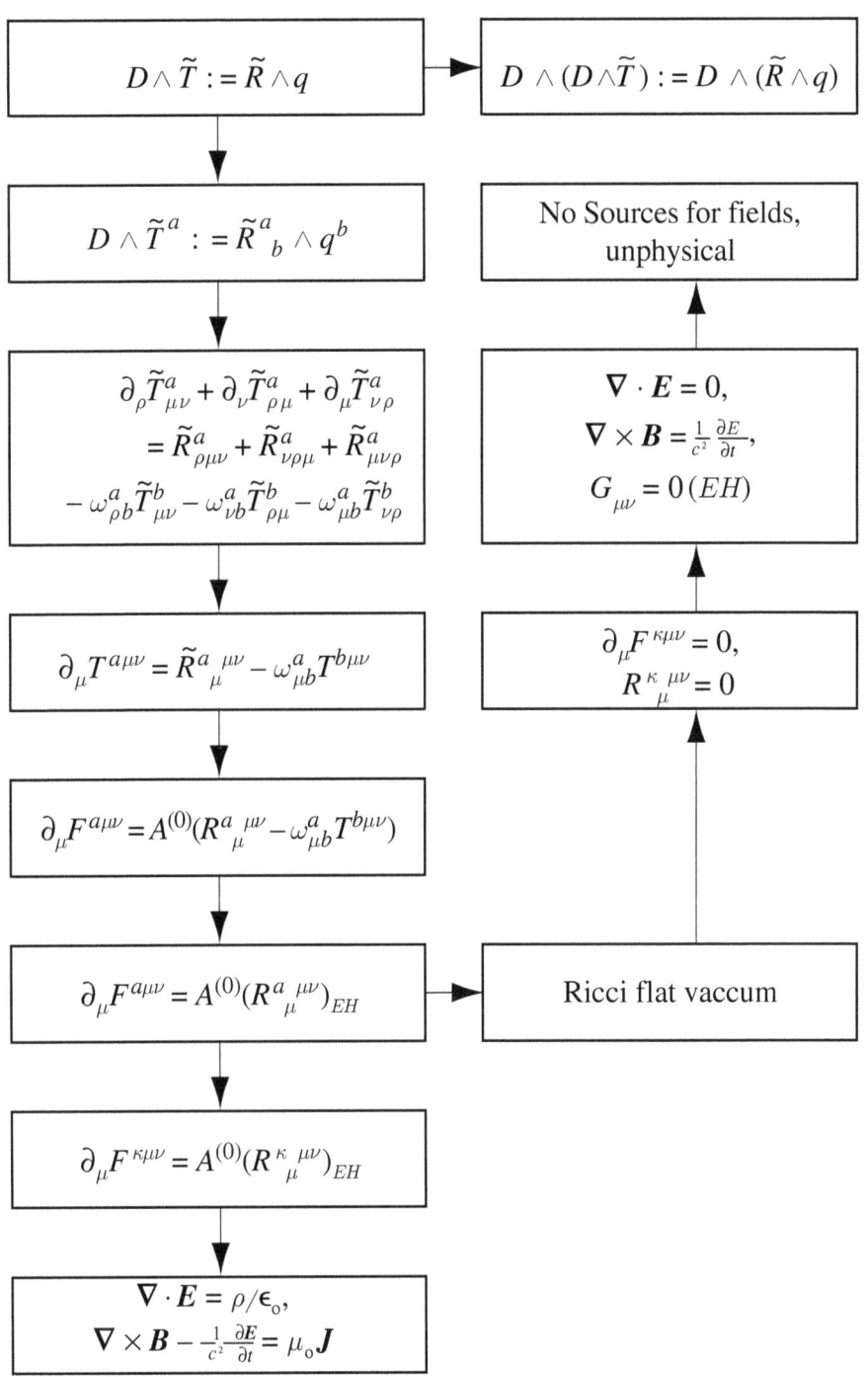

Flowchart 7.4. In homogeneous ECE Field Equation.

118   7 A Review of Einstein Cartan Evans (ECE) Field Theory

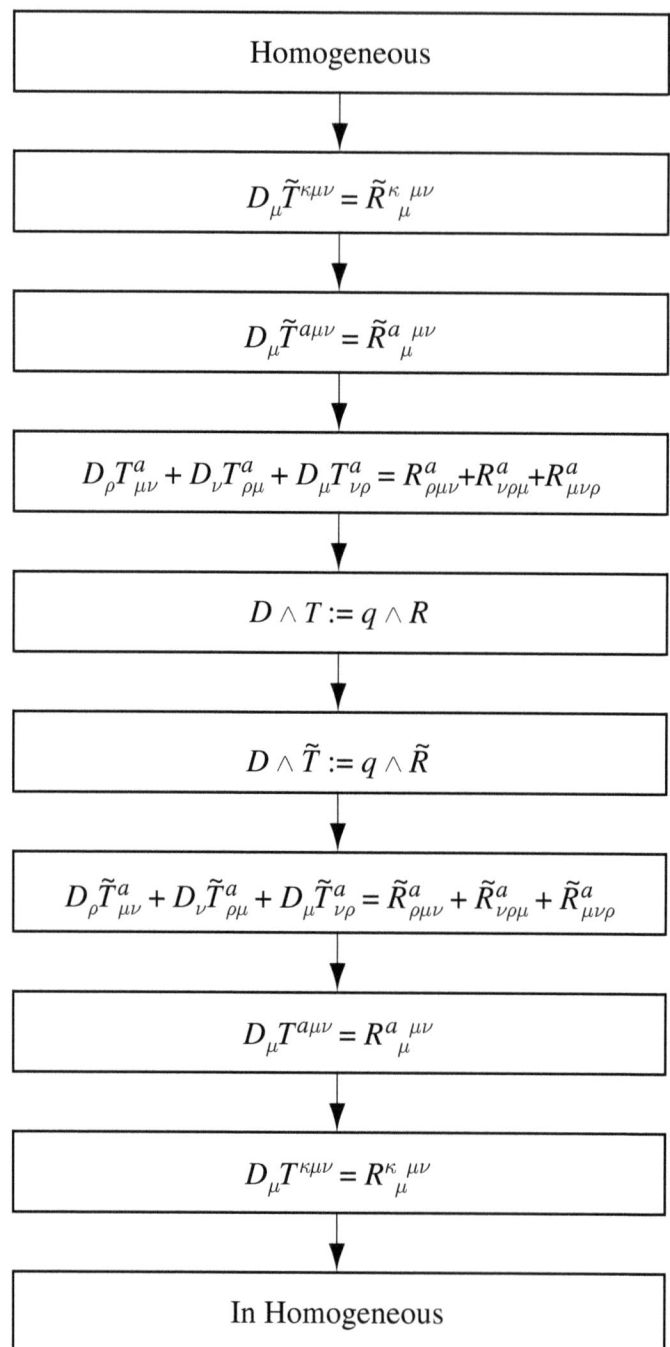

Flowchart 7.5. The Basic Field Equations.

7.1 Introduction

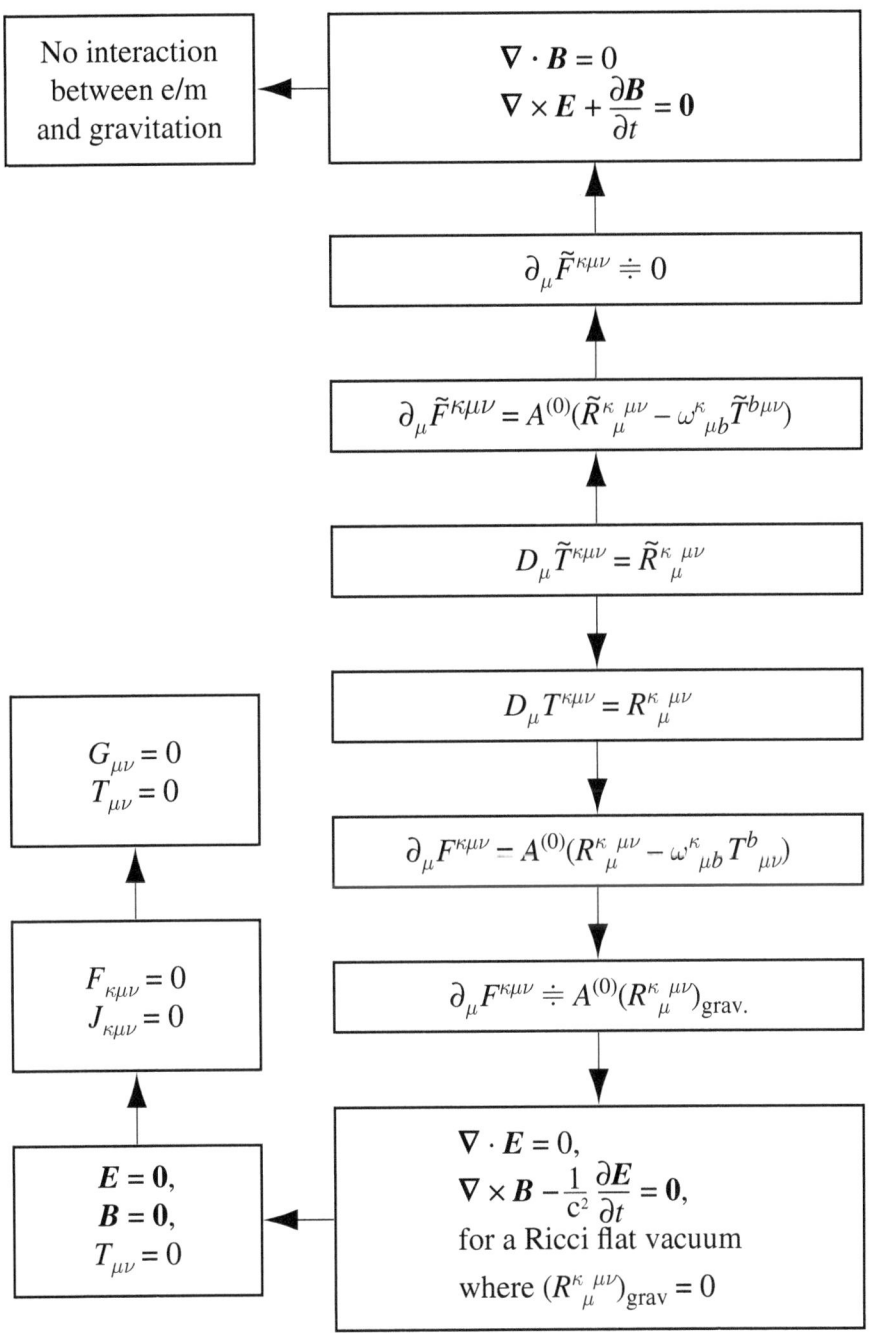

Flowchart 7.6. Approximations to the Basic Field Equations.

120   7 A Review of Einstein Cartan Evans (ECE) Field Theory

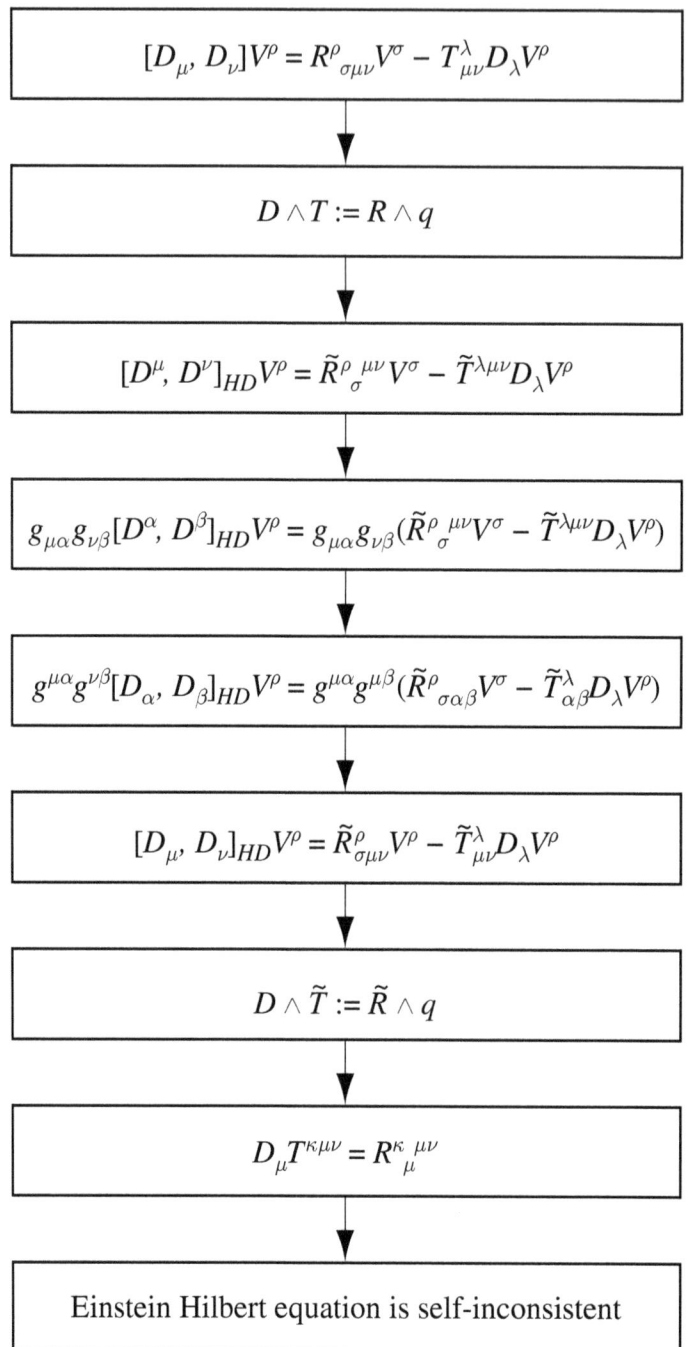

Flowchart 7.7. Hodge Dual of the Bianchi Identity.

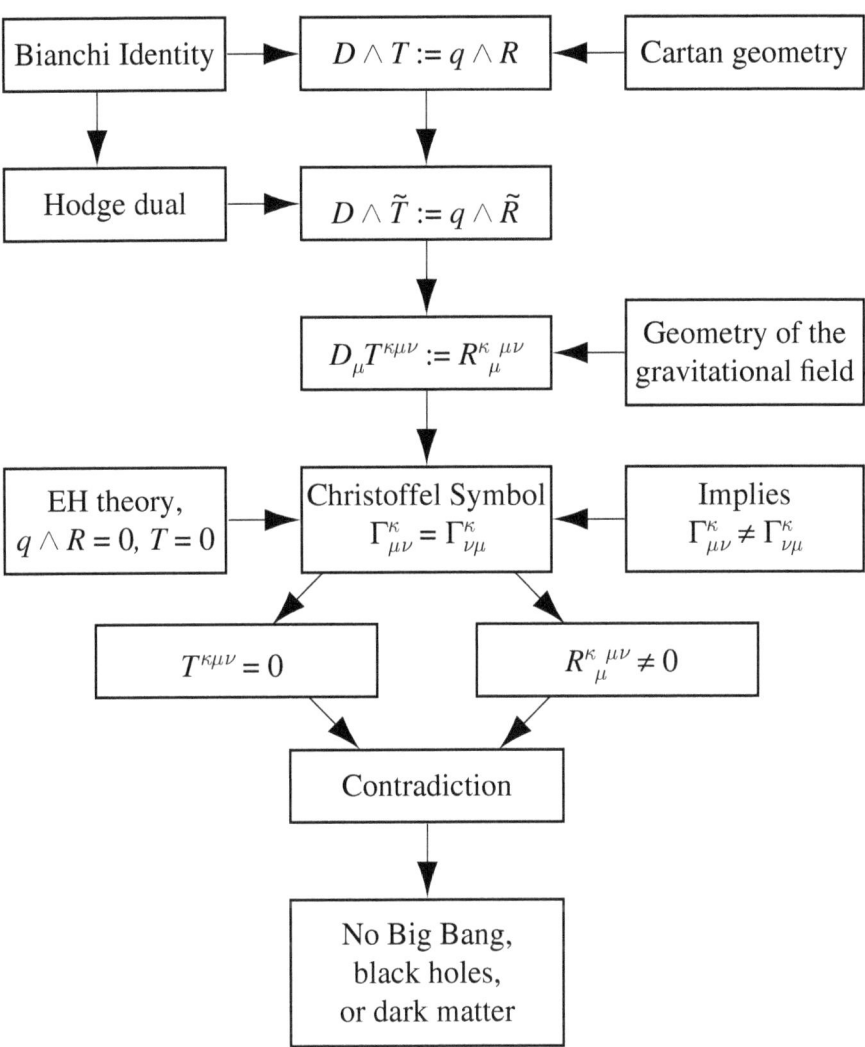

Flowchart 7.8. Self Inconsistency of General Relativity.

between the fields and potentials in ECE, a relation which includes the spin connection.

In Section 7.6, spin connection resonance (SCR) is discussed, concentrating as usual on the main discoveries and points of development of the ECE theory. In theory, SCR is of great practical utility because the equations of classical electrodynamics become resonance equations of the type first inferred by the Bernoulli's and Euler. Therefore a new source of electric power has been discovered in ECE theory - this source is the Cartan torsion of space-

122 7 A Review of Einstein Cartan Evans (ECE) Field Theory

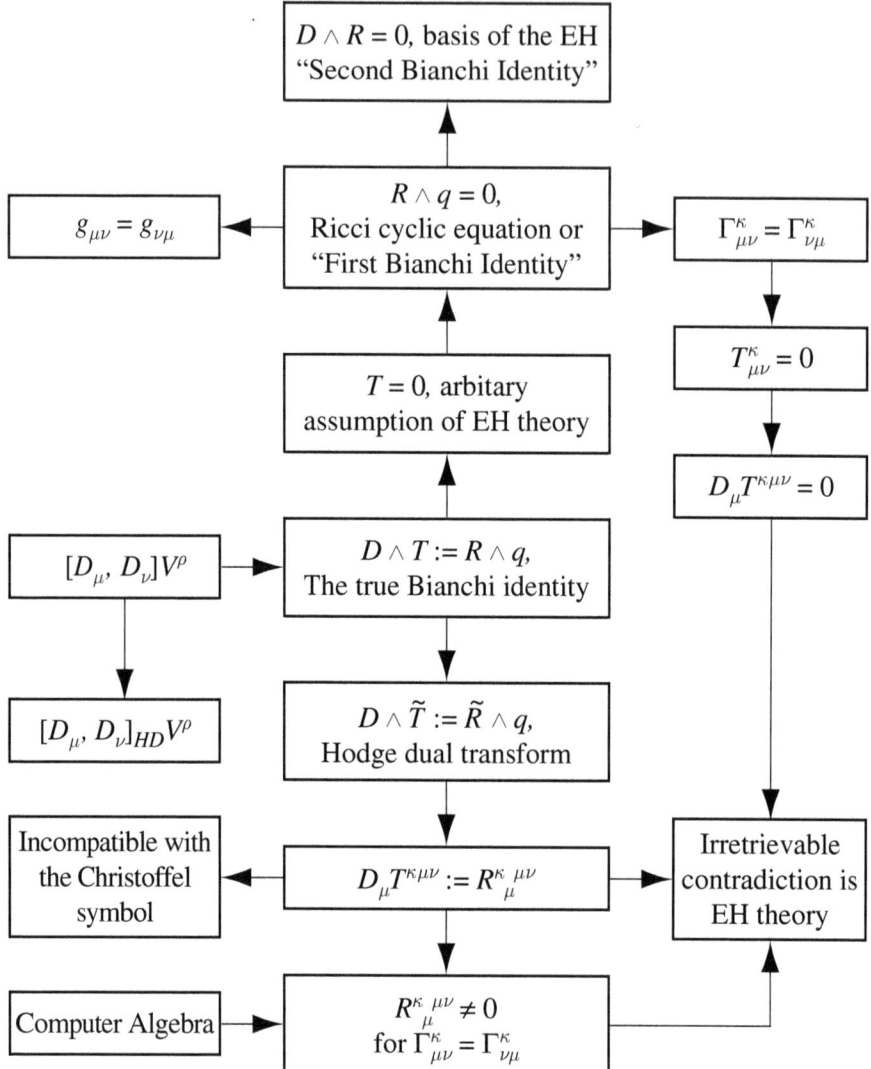

Flowchart 7.9. IrretrievableFlaws in the Geometry of the Einstein Hilbert Field Theory.

time. Amplification occurs in principle through SCR, the spin connection itself being the property of the four-dimensional space-time with curvature and torsion which is the base manifold of ECE theory. It is well known [15] that these resonance equations are equivalent to circuits that can be used to amplify electric power. In all probability these circuits were the ones designed by Tesla empirically.

7.1 Introduction

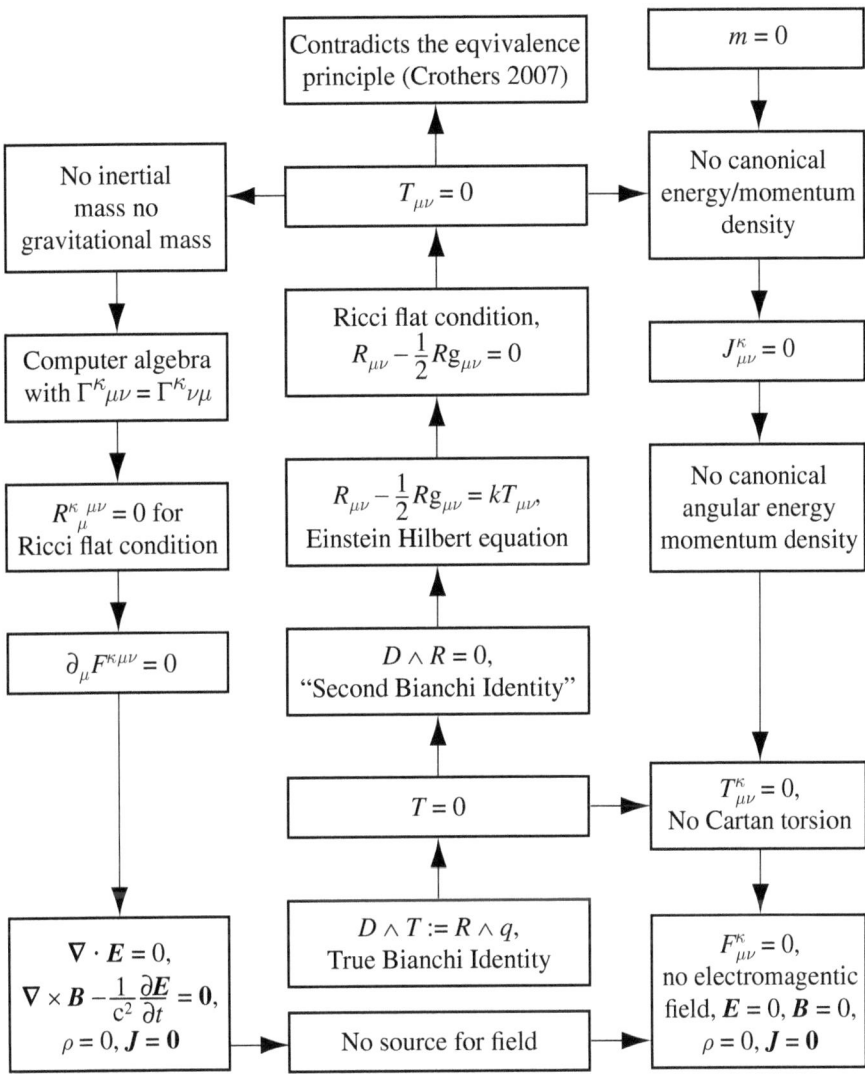

Flowchart 7.10. Irretrievable Contradiction in the Ricci Flat Condition.

In Section 7.7 the utility of ECE as a unified field theory is illustrated through the effects of gravitation in optics and spectroscopy. These are exemplified by the effect of gravitation on the ring laser gyro (Sagnac effect) and on radiatively induced fermion resonance (RFR). RFR itself is of great potential utility because it is a form of electron and proton spin resonance induced not by a permanent magnet, but by a circularly polarized electromagnetic field. This is known as the inverse Faraday effect (IFE) [16] from which the

ECE spin field B(3) was inferred in 1992 [17]. The spin field signals the fact that in a self consistent philosophy, classical electrodynamics must be part of a generally covariant field theory. This is incompatible with the U(1) sector of special relativity still used to describe electrodynamics in the standard model. Any proposal for a unified field theory based on U(1) cannot be generally covariant in all sectors, leaving ECE as the only satisfactory unified field theory at the time of writing.

In Section 7.8 the well known radiative corrections [18] are developed with ECE theory, and a summary of the main points of progress illustrated with the anomalous g factor of the electron and the Lamb shift. It is shown that claims to accuracy of standard model quantum electrodynamics (QED) are greatly exaggerated. The accuracy is limited by that of the Planck constant, the least accurately known fundamental constant appearing in the fine structure constant. There are glaring internal inconsistencies in standards laboratories tables of data on the fundamental constants, and QED is based on a number of what are effectively adjustable parameters introduced by ad hoc procedures such as dimensional renormalization The concepts used in QED are vastly complicated and are not used in the ECE theory of the experimentally known radiative corrections. The Feynman perturbation method is not used in ECE: it cannot be proven to converge as is well known, i.e. needs many terms of increasing complexity which must be evaluated by computer. So ECE is a fundamental theory of quantized electrodynamics from the first principles of general relativity, while QED is a theory of special relativity needing adjustable parameters, acausal and subjective concepts, and therefore of dubious validity.

In Section 7.9, finally, it is shown that EH theory has several fundamental shortcomings. As described on ww.telesio-galilei.com EH has been quite severely criticized down the years by several leading physicists. Notably, Crothers [19] has criticized the methods of solution of EH, and has shown that uncritically accepted concepts are in fact incompatible with general relativity. These include Big Bang, dark hole and dark matter theory and the concept of a Ricci flat space-time. He has also shown that the use of the familiar but mis-named "Schwarzschild metric" is due to lack of scholarship and understanding of Schwarzschild's original papers of 1916. ECE has revealed that the use of the familiar Christoffel symbol is incompatible with the one true Bianchi identity of Cartan. This section suggests a development of the EH equation into one which is self consistent.

Several technical appendices give basic details which are not usually given in standard textbooks, but which are nevertheless important to the student. These appendices also contain flow charts inter-relating the main concepts and equations of ECE.

7.2 Geometrical Principles

The ECE theory is based on the two structure equations of Cartan, and the Bianchi identity of Cartan geometry. During the course of development of the theory a useful short-hand notation has been used in which the indices are removed in order to reveal the basic structure of the equations. In this notation the two Cartan structure equations are:

$$T = D \wedge q = d \wedge q + \omega \wedge q \tag{7.3}$$

and

$$R = D \wedge \omega = d \wedge \omega + \omega \wedge \omega \tag{7.4}$$

and the Bianchi identity is:

$$D \wedge T = d \wedge T + \omega \wedge T := R \wedge q. \tag{7.5}$$

In this notation T is the Cartan torsion form, ω is the spin connection symbol, q is the Cartan tetrad form, and R is the Cartan curvature form. The meaning of this symbolism is defined in all detail in the ECE literature [1–12], and the differential form is defined in the standard literature [13]. The purpose of this section is to summarize the main advances in basic geometry made during the development of ECE theory.

The Bianchi identity (7.5) is basic to the field equations of ECE, and its structure has been developed considerably [1–12]. It has been shown to be equivalent to the tensor equation:

$$\begin{aligned}
R^\lambda_{\rho\mu\nu} &+ R^\lambda_{\mu\nu\rho} + R^\lambda_{\nu\rho\mu} \\
&:= \partial_\nu \Gamma^\lambda_{\rho\mu} - \partial_\rho \Gamma^\lambda_{\nu\mu} + \Gamma^\lambda_{\nu\sigma}\Gamma^\sigma_{\rho\mu} - \Gamma^\lambda_{\rho\sigma}\Gamma^\sigma_{\nu\mu} \\
&+ \partial_\rho \Gamma^\lambda_{\mu\nu} - \partial_\mu \Gamma^\lambda_{\rho\nu} + \Gamma^\lambda_{\rho\sigma}\Gamma^\sigma_{\mu\nu} - \Gamma^\lambda_{\mu\sigma}\Gamma^\sigma_{\rho\nu} \\
&+ \partial_\mu \Gamma^\lambda_{\nu\rho} - \partial_\nu \Gamma^\lambda_{\mu\rho} + \Gamma^\lambda_{\mu\sigma}\Gamma^\sigma_{\nu\rho} - \Gamma^\lambda_{\nu\sigma}\Gamma^\sigma_{\mu\rho}
\end{aligned} \tag{7.6}$$

in which a cyclic sum of three Riemann tensors is identically equal to the sum of three fundamental definitions of the same Riemann tensors. These fundamental definitions originate in the commutator of covariant derivatives acting on a four-vector in the base manifold. The latter is four dimensional space-time with BOTH curvature and torsion [1–13]. This operation produces:

$$[D_\mu, D_\nu]V^\rho = R^\rho{}_{\sigma\mu\nu}V^\sigma - T^\lambda_{\mu\nu}D_\lambda V^\rho \tag{7.7}$$

where the torsion tensor is:

$$T^\lambda_{\mu\nu} = \Gamma^\lambda_{\mu\nu} - \Gamma^\lambda_{\nu\mu}. \tag{7.8}$$

The curvature or Riemann tensor cannot exist without the torsion tensor, and the definition (7.7) has been shown to be equivalent to the Bianchi identity (7.6).

The second advance in basic geometry is the inference [1–12] of the Hodge dual of the Bianchi identity. In short-hand notation this is:

$$D \wedge \tilde{T} := \tilde{R} \wedge q \tag{7.9}$$

and is equivalent to:

$$[D_\mu, D_\nu]_{HD} V^\rho = \tilde{R}^\rho{}_{\sigma\mu\nu} V^\sigma - \tilde{T}^\lambda_{\mu\nu} D_\lambda V^\rho \tag{7.10}$$

where the subscript HD denotes Hodge dual. From these considerations it may be inferred that the Bianchi identity and its Hodge dual are the tensor equations:

$$D_\mu \tilde{T}^{\kappa\mu\nu} = \tilde{R}^\kappa{}_\mu{}^{\mu\nu} \tag{7.11}$$

and

$$D_\mu T^{\kappa\mu\nu} = R^\kappa{}_\mu{}^{\mu\nu} \tag{7.12}$$

in which the connection is NOT the Christoffel connection. Computer algebra [1–12] has shown that the tensor $R^\kappa{}_\mu{}^{\mu\nu}$ is not zero in general for line elements that use the Chrstoffel symbol, while $T^{\kappa\mu\nu}$ is always zero for the Christoffel symbol. So the use of the latter is inconsistent with the tensor equation (7.12). Therefore the neglect of torsion makes EH theory internally inconsistent, so standard model general relativity and cosmology are also internally inconsistent at a basic level. In short-hand notation the geometry used in EH is:

$$R \wedge q = 0 \tag{7.13}$$

which in tensor notation is known as "the first Bianchi identity":

$$R^\kappa{}_{\mu\nu\rho} + R^\kappa{}_{\rho\mu\nu} + R^\kappa{}_{\nu\rho\mu} = 0 \tag{7.14}$$

in the standard model literature. However, this is not an identity, because it conflicts with equation (7.5), and is true if and only if the Christoffel symbol and symmetric metric are used [1–13]. Eq. (7.14) was actually inferred by Ricci and Levi-Civita, not by Bianchi. So it is referred to in the ECE literature as the Ricci cyclic equation.

In the course of development of ECE theory a similar problem was found with what is referred to in the standard model literature as "the second Bianchi identity". In shorthand notation this is given [13] as:

$$D \wedge R = 0 \tag{7.15}$$

but again this neglects torsion. In tensor notation Eq. (7.15) is:

$$D_\rho R^\kappa{}_{\sigma\mu\rho} + D_\mu R^\kappa{}_{\sigma\nu\rho} + D_\nu R^\kappa{}_{\sigma\rho\mu} = 0. \tag{7.16}$$

It has been shown [1–12] that Eq. (7.15) should be:

$$D \wedge (D \wedge T) := D \wedge (R \wedge q) \tag{7.17}$$

which is found by taking $D\wedge$ on both sides of Eq. (7.15). Eq. (7.17) has been given in tensor notation [1–12], and reduces to Eq. (7.16) when:

$$T^\lambda_{\mu\nu} = 0. \tag{7.18}$$

However, Eq. (7.18) is inconsistent with the fundamental operation of the commutator of covariant derivatives on the four vector, Eq. (7.7). So in the ECE literature the torsion is always considered self-consistently. From the fundamentals [13] of Eq. (7.7) there is no a priori reason for neglecting torsion, and in fact the torsion tensor is always non-zero if the curvature tensor is non-zero. This fact precludes the use of the Christoffel symbol, making EH theory self-inconsistent.

These are the main geometrical advances made during the course of the development of ECE theory, which is the only self-consistent theory of general relativity. It has also been pointed out by Crothers [19] that methods of solution of the EH equation are geometrically incorrect, and must be discarded. It is thought that these errors have been repeated uncritically for ninety years because few have the necessary technical ability to understand the geometry of general relativity in sufficient depth, and that the prestige of Einstein has precluded or inhibited due criticism.

7.3 The Field and Wave Equations of ECE Theory

The wave equation of ECE was the first to be developed historically [1–12], and methods of derivation of the wave equation were subsequently simplified and clarified. The field equations were subsequently developed from the

Bianchi identity discussed in Section 7.2. This section summarizes the main equations and methods of derivation. More detail of the equations is given in technical appendices. The field equations are relevant to classical gravitation and electrodynamics, and the wave equation to causal and objective quantum mechanics. Full details of derivations are available in the literature [1–12], the aim of this section is to summarize the main inferences of ECE theory to date.

The Bianchi identity (7.5) and its Hodge dual (7.9) become the homogeneous and inhomogeneous field equations of ECE respectively. These field equations apply to the four fundamental fields of force: gravitational, electromagnetic, weak and strong and can be used to describe the interaction of the fundamental fields on the classical level. For example the electromagnetic field is described by making the fundamental hypothesis:

$$A = A^{(0)} q \tag{7.19}$$

where the shorthand (index-less) notation has been used. Here A represents the electromagnetic potential form and $cA^{(0)}$ is a primordial quantity with the units of volts, a quantity which is proportional to the charge, $-e$, on the electron. The hypothesis (7.19) implies that:

$$F = A^{(0)} T \tag{7.20}$$

where F is shorthand notation for the electromagnetic field form. The homogeneous ECE field equation of electrodynamics follows from the Bianchi identity (7.5):

$$d \wedge F + \omega \wedge F = A^{(0)} R \wedge q \tag{7.21}$$

and the inhomogeneous ECE field equation follows from the Hodge dual (7.9) of the Bianchi identity:

$$d \wedge \widetilde{F} + \omega \wedge \widetilde{F} = A^{(0)} \widetilde{R} \wedge q. \tag{7.22}$$

Therefore the ECE field equations are duality invariant, a basic symmetry which means that they transform into each other by means of the Hodge dual [1–12]. The Maxwell Heaviside (MH) field equations of the standard model do not have this fundamental symmetry and in differential form notation the MH equations are:

$$d \wedge F = 0 \tag{7.23}$$

and
$$d \wedge \widetilde{F} = \widetilde{J}/\epsilon_0 \qquad (7.24)$$

where $\widetilde{J}$ denotes the inhomogeneous charge/current density and ϵ_0 is the S. I. vacuum permittivity. Duality symmetry is broken by the fact that there is no homogeneous charge current density (J) in MH theory (the right hand side of Eq. (7.23) is zero). The absence of J in the standard model is made the basis for gauge theory as is well known, and also made the basis for the absence of a magnetic monopole.

The ECE field equations (7.21) and (7.22) are re-arranged as follows in order to define the homogeneous (J) and inhomogeneous ($\widetilde{J}$) charge current densities of ECE theory:

$$d \wedge F = J/\epsilon_0 = A^{(0)}(R \wedge q - \omega \wedge T) \qquad (7.25)$$

and

$$d \wedge \widetilde{F} = \widetilde{J}/\epsilon_0 = A^{(0)}(\widetilde{R} \wedge q - \omega \wedge \widetilde{T}). \qquad (7.26)$$

Both equations are generally covariant because they originate in the Bianchi identity. The interaction of electromagnetism with gravitation occurs whenever J is non-zero. In MH theory such an interaction cannot be described, because MH theory is developed in Minkowski space-time. The latter has no curvature and in general relativity cannot describe gravitation at all. For all practical purposes in the laboratory there is no interaction of electromagnetism and gravitation, so Eq. (7.25) reduces to:

$$d \wedge F = 0. \qquad (7.27)$$

Therefore ECE theory explains in this way why there is no magnetic monopole observable in the laboratory. The standard model has no physical explanation for this, and indeed asserts that gauge theory is mathematical in nature. ECE theory does not use gauge theory, and adopts Faraday's original point of view that the potential A is a physically effective entity. There are therefore important philosophical differences between ECE and the standard model of classical electrodynamics, in which the potential is mathematical in nature.

Therefore the structure of the ECE field equations is a simple one based directly on the Bianchi identity. The structure is seen the most clearly using the shorthand notation of Eqs. (7.25) and (7.26) where all indices are omitted. The notation of classical electrodynamics varies from subject to subject. In advanced field theory the elegant but concise differential form notation is used, and also the tensor notation. In electrical engineering the vector notation is used. In ECE theory all three notations have been developed [1–12] in all detail, and the ECE field equations developed into a vector form that is identical to the MH equations. The main differences between ECE and MH is firstly that the former is written in a four dimensional space-time with curvature and torsion both present. This is a dynamic space-time whose connection must be more general than the Christoffel connection. The MH equations, although having the same vector form as ECE, are written in the Minkowski space-time of special relativity. This is often referred to as "flat space-time", whose metric is time and space independent. Secondly the relation between the field and potential in ECE includes the connection, whereas in MH the connection is not present. The inclusion of the connection has the all important effect of making the equations of classical electrodynamics resonance equations of the Bernoulli/Euler type. This property means that it is possible to describe well known phenomena such as those first observed by Tesla, and to produce circuits that take electric power from a new source, the Cartan torsion.

The concise tensorial expression of the equations (7.25) and (7.26) is in general [1–12]

$$D_\mu \widetilde{F}^{a\mu\nu} = A^{(0)} \widetilde{R}^a{}_\mu{}^{\mu\nu} \qquad (7.28)$$

and

$$D_\mu F^{a\mu\nu} = A^{(0)} R^a{}_\mu{}^{\mu\nu} \qquad (7.29)$$

where the covariant derivative appears on one side and a Ricci type curvature tensor on the other. It has been shown [1–12] that these reduce in the laboratory, and for all practical purposes, to:

$$\partial_\mu \widetilde{F}^{a\mu\nu} = 0 \qquad (7.30)$$

and

$$\partial_\mu F^{a\mu\nu} = A^{(0)} R^a{}_\mu{}^{\mu\nu}. \qquad (7.31)$$

The index a in these equations comes from the well known [13] tangent space-time of Cartan geometry. However, it has been shown [1–12] that Eqs. (7.30) and (7.31) can be written in the base manifold as a special case of Eqs. (7.28) and (7.29), whereupon we arrive at:

$$\partial_\mu \widetilde{F}^{\kappa\mu\nu} = 0 \qquad (7.32)$$

and

$$\partial_\mu F^{\kappa\mu\nu} = A^{(0)} R^\kappa{}_\mu{}^{\mu\nu}. \qquad (7.33)$$

7.3 The Field and Wave Equations of ECE Theory

Therefore the electromagnetic field tensor in general relativity (ECE theory) develops into a three index tensor. In special relativity (MH theory) it is a two-index tensor as is well known. The equivalents of (7.32) and (7.33) in MH theory are the tensor equations:

$$\partial_\mu \widetilde{F}^{\mu\nu} = 0 \tag{7.34}$$

and

$$\partial_\mu F^{\mu\nu} = J^\nu/\epsilon_0. \tag{7.35}$$

The meaning of the three-index field tensor has been developed [1–12] in detail. It originates in the well known [18] three index angular energy/momentum tensor density, $J^{\kappa\mu\nu}$ which is proportional to the three index Cartan torsion tensor. It is well known that the electromagnetic field carries angular momentum which in the Beth effect [20] is experimentally observable Therefore the Cartan torsion tensor is the expression of this well known angular energy/momentum density tensor of Minkowski space-time [18] in a more general manifold with curvature and torsion. The meaning of the vector form of the ECE field equations is further developed in Section 7.5.

The classical field equations of gravitation in ECE are also based directly on the Bianchi identity and its Hodge dual. The EH equation, as argued already, is incompatible with the Bianchi identity in its rigorously correct form, Eq. (7.5), so during the course of development of ECE theory the well known EH equation has been developed with the proportionalities:

$$T^{\kappa\mu\nu} = kJ^{\kappa\mu\nu} \tag{7.36}$$

and

$$R^\kappa{}_\mu{}^{\mu\nu} = kT^\kappa{}_\mu{}^{\mu\nu} \tag{7.37}$$

which give:

$$D_\mu J^{\kappa\mu\nu} = T^\kappa{}_\mu{}^{\mu\nu}. \tag{7.38}$$

This novel field equation of classical gravitation is based directly on the tensorial formulations (7.11) and (7.12) of the Bianchi identity. The Newton inverse square law for example has been derived straightforwardly from Eq. (7.38) in the limit where the connection goes to zero:

$$\partial_\mu J^{\kappa\mu\nu} \doteq T^\kappa{}_\mu{}^{\mu\nu} \tag{7.39}$$

whereupon we obtain:

$$\nabla \cdot \boldsymbol{g} = kc^2 \rho_m \tag{7.40}$$

an equation which is equivalent to the Newton inverse square law. Here $\boldsymbol{g}$ is the acceleration due to gravity, k is Einstein's constant, ρ_m and is the mass density in kilograms per cubic meter. Similarly the Coulomb inverse square law can be obtained straightforwardly [1–12] by considering the same type of limit of the inhomogeneous ECE field equation:

$$D_\mu F^{\kappa\mu\nu} = A^{(0)} R^\kappa{}_\mu{}^{\mu\nu}. \tag{7.41}$$

The appropriate limit in this case is:

$$\partial_\mu F^{\kappa\mu\nu} \doteq A^{(0)} R^\kappa{}_\mu{}^{\mu\nu} \tag{7.42}$$

and leads to the Coulomb inverse square law:

$$\nabla \cdot \boldsymbol{E} = \rho_e / \epsilon_0 \tag{7.43}$$

where ρ_e is the charge density in coulombs per cubic meter. These procedures illustrate one aspect of the unified nature of ECE, because both laws are obtained from the Bianchi identity. Many other examples of the unification properties of ECE have been discussed [1–12].

In order to unify the electromagnetic and weak fields in a field equation, the representation space is chosen to be SU(2) instead of O(3) and the parity violating nature of the weak field carefully considered. Similarly the electromagnetic and strong fields are unified with an SU(3) representation space, and we have already discussed the unification of the electromagnetic and gravitational fields. Any permutation or combination of fields may be unified, and several examples have been given [1–12] in various contexts. These are discussed further in Section 7.7.

The ECE wave equation was developed [1–12] from the tetrad postulate [13]:

$$D_\mu q^a_\nu = 0 \tag{7.44}$$

via the identity:

$$D^\mu (D_\mu q^a_\nu) := 0. \tag{7.45}$$

This was re-expressed as the ECE Lemma:

$$\Box q^a_\lambda = R q^a_\lambda \tag{7.46}$$

in which appears the scalar curvature:

$$R = q_a^\lambda \partial^\mu (\Gamma^\nu_{\mu\lambda} q_\nu^a - \omega^a_{\mu b} q_\lambda^b). \tag{7.47}$$

Here tensor notation is used, $\omega^a_{\mu b}$ being the spin connection and $\Gamma^\nu_{\mu\lambda}$ the general connection. The Lemma becomes the ECE wave equation using a generalization to all fields of the Einstein gravitational equation [1–13]:

$$R = -kT. \tag{7.48}$$

Here T is an index contracted energy momentum tensor. The main wave equations of physics were all obtained [1–12] as limits of Eq. (7.46), notably the Proca and Dirac wave equations. In so doing however the causal realist philosophy of Einstein and de Broglie was adhered to. This is the original philosophy of wave mechanics. The Schrödinger and Heisenberg equations were also obtained as non-relativistic quantum limits of the ECE wave equation, but the Heisenberg indeterminacy principle was not used in accord with the basic philosophy of relativity and with recent experimental data [21] which refute the uncertainty principle by as much as nine orders of magnitude.

7.4 Aharonov Bohm and Phase Effects in ECE Theory

The well known Aharonov Bohm (AB) effects have been observed using magnetic, electric and gravitational fields [1–12] and as shown by ECE theory are ubiquitous for ALL electromagnetic and optical effects, including phase effects: the subject of this section. These must all be explained by general relativity, and not by the obsolete special relativistic methods of the standard model. Therefore it is important to define the various AB conditions in ECE theory. In so doing a unified description of phase effects such as the electromagnetic, Dirac, Wu Yang and Berry phases may also be developed.

In general, the AB condition is defined in ECE theory by the first Cartan structure equation (adopting the index-less short-hand notation [1–12]):

$$T = D \wedge q := d \wedge q + \omega \wedge q. \tag{7.49}$$

Using the ECE hypothesis:

$$A = A^{(0)} q \tag{7.50}$$

this becomes:

$$F = D \wedge A := d \wedge A + \omega \wedge A \tag{7.51}$$

where F is short-hand for the electromagnetic field form and where A is short-hand for the electromagnetic potential form. The AB effects in ECE theory [1–12] were developed with the spin connection term $\omega \wedge A$ in Eq. (7.51). The accepted notation [13] of Cartan geometry uses the tangent space-time indices without the base manifold indices, because the latter are always the same on both sides of an equation of Cartan geometry. So in the standard notation Eq. (7.51) is:

$$F^a = d \wedge A^a + \omega^a{}_b \wedge A^b \tag{7.52}$$

This denotes that the electromagnetic field is a vector-valued two-form and the potential is a vector-valued one-form. In the standard model the spin connection is zero and the standard relation between field and potential is:

$$F = d \wedge A. \tag{7.53}$$

In Eq. (7.53), F is a scalar-valued two-form, and A is a scalar valued one-form [13] The spin connection is zero in Eq. (7.53) because the latter is written in a Minkowski space-time. In the standard model, classical electrodynamics is still represented by the MH equations, which are Lorentz covariant, but not generally covariant. In other words the MH equations are those of special relativity and not general relativity as required by the philosophy of relativity and objectivity. The latter demands that every equation of physics should be an equation of a generally covariant unified field theory. It is well known that the standard model complies with this only in its gravitational sector: the electro-weak and strong fields of the standard model are sectors of special relativity only. The standard model does not comply with general relativity, notably standard model quantum mechanics is philosophically different from relativity (Einstein Bohr dialogue). ECE complies rigorously with the philosophy of general relativity in all its sectors, and unifies all sectors with geometry as required. In ECE the spin connection is ALWAYS non-zero because the fundamental space-time being used is not a flat space-time, it always contains both torsion and curvature in all sectors of the generally covariant unified field theory [1–12]. Torsion and curvature are ineluctably inter-related in the Bianchi identity (Section 7.2), and during the course of development of ECE theory it was shown that there is only one true Bianchi identity, which always links torsion to curvature and vice versa. This is an important mathematical advance of ECE theory, another (Section 7.2) being the development of the Hodge dual of the Bianchi identity.

It has been shown [1–12] that there is a fundamental error in the standard model explanation of the magnetic AB effect [22]. In differential form notation the standard explanation is based on the two equations:

$$F = d \wedge A, d \wedge F = 0 \tag{7.54}$$

7.4 Aharonov Bohm and Phase Effects in ECE Theory

whose mathematical structure implies:

$$d \wedge (d \wedge A) = 0. \tag{7.55}$$

It is well known that this structure is invariant under the archetypical gauge transformation:

$$A \to A + d\chi \tag{7.56}$$

because of the Poincaré Lemma:

$$d \wedge d\chi := 0. \tag{7.57}$$

As explained in paper 56 of the ECE series (www.aias.us), the standard model uses the mathematical result (7.57) to claim that:

$$\oint d\chi = \int_s d \wedge d\chi \neq 0. \tag{7.58}$$

This claim is incorrect because it does not agree with the Stokes Theorem. The latter applies [23] in non simply connected spaces. The Poincaré Lemma (7.57) implies therefore that:

$$\oint d\chi = \int_s d \wedge d\chi := 0 \tag{7.59}$$

in all types of spaces, including non simply connected spaces and there cannot be an Aharonov Bohm effect due to the contour integral of $d\chi$. The incorrect claim of the standard model [22] is that non simply connected spaces allow $\oint d\chi$ to be non-zero. A counter example to this claim was given in paper 56 of www.aias.us. in full detail.

The explanation of the Aharonov Bohm (AB) effects in ECE theory is not based on the mathematical abstractions of gauge theory but on Einstein's philosophy of relativity and Faraday's philosophy of the potential as a physically effective entity (the electrotonic state). This philosophy of Faraday was also accepted by Maxwell and his followers. The idea that the potential is a mathematical abstraction is based on the perceived redundancy exemplified by Eq. (7.57), and this idea has been made into the basis of the mathematical gauge theory of the standard model, developed in the late twentieth century. It appears in standard model textbooks such as that of Jackson for example [1–24]. The idea of a mathematical potential and a physical field in classical electrodynamics is contradicted by the well known minimal prescription of field theory and quantum electrodynamics, where the PHYSICAL momentum eA is added to the momentum p. The idea of an abstract potential ran into trouble following the demonstration by Chambers of the first AB effect to be observed, the magnetic AB effect. It is well known that Chambers placed a magnetic iron whisker between the apertures of a Young interferometer and isolated the magnetic field from interfering electron beams. Therefore, if the potential is mathematical as claimed in gauge theory, it should have no effect on the electronic interference pattern. The experimental result showed a shift in the interference pattern, and so contradicts the standard model, meaning that Faraday was correct: the potential is a physically effective entity. The same results were later obtained experimentally in the electric and gravitational AB effects. As argued in this section, various phase effects also indicate the existence of an electromagnetic AB effect if interpreted by general relativity, of which ECE theory is an example.

The AB effect in ECE theory is summarized as follows:

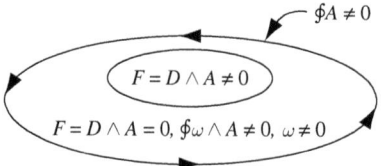

Fig. 7.1. ECE Explanation of the Aharonov Bohm Effect.

It has been shown [1–12] that the observable phase shift of the Chambers experiment in ECE theory is:

$$\Delta\phi = \frac{e}{\hbar}\Phi \tag{7.60}$$

where

$$\Phi = \oint A := -\int_s \omega \wedge A \tag{7.61}$$

in short-hand or index free notation. In the area between the inner and outer rings in Fig. (7.1):

$$F = D \wedge A = 0, A \neq 0, \omega \neq 0. \tag{7.62}$$

The electromagnetic field (F) is zero by experimental arrangement. However, the potential (A) and the spin connection (ω) are not zero in general in this same region between the inner and outer rings. The phase shift is due therefore to the contour integral around A in Eq. (7.61), as indicated in Fig. (7.1). Therefore ECE theory gives a simple explanation of the AB effects as being due to a physical A and a physical ω. The latter indicates that the

7.4 Aharonov Bohm and Phase Effects in ECE Theory

ECE space-time is not a Minkowski space-time as in the attempted standard model explanation of the AB effect. In the standard model the equivalent of Fig. (7.1) is:

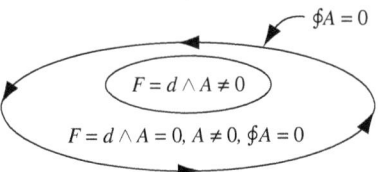

Fig. 7.2. Standard Attempt at Explaining the Aharonov Bohm Effect.

and the contour integral of A is zero. In the standard model the contour integral of the potential is zero in the area between the inner and outer rings of Fig. (7.2) because:

$$F = d \wedge A = 0, A \neq 0, \tag{7.63}$$

$$\int_s d \wedge A = \oint A = 0. \tag{7.64}$$

So when F is zero in the standard model, so is $d \wedge A$. It is possible therefore for A to be non-zero in the standard model while F is zero, but the incorrect twentieth century idea of a non-physical A means that in the standard model A must have no physical effect. In the end analysis this is pure obscurity and has caused great confusion. Such ideas are bad physics and must be discarded sooner or later. The only clear thing about the attempted standard model explanation of the magnetic AB effect is that in the area between the two rings of Fig. (7.2):

$$\int_s F = \int_s d \wedge A = \oint A = 0. \tag{7.65}$$

So the contour integral of A is zero by the Stokes Theorem and there is no AB effect contrary to experiment. Therefore in the standard model, when F is zero the contour integral of A is always zero even though A itself may be non-zero. In other words Stokes' Theorem implies that when F or $d \wedge A$ is zero in the standard model, the contour integral of A must vanish even though A itself may be non-zero. As we have seen, adding a $d\chi$ in an assumed non simply connected space-time does not solve this problem.

In ECE theory the presence of the spin connection ensures that when F is zero, $d \wedge A$ is not zero in general and the contour integral of A is not zero, meaning a phase shift as observed, Eq. (7.61). The way that such an ECE contour integral must be evaluated has been explained carefully [1–12]. Therefore the AB effects show that ECE is preferred experimentally over the standard model. This is one out of many experimental advantages of ECE theory over the standard model. A table of about thirty such advantages is available on the www.aias.us website and in the fourth volume of ref. (7.1). As argued already, the standard model has attempted to obfuscate its way out of the AB paradox by adding $d\chi$ to A and claiming that the AB effect is due to a non-zero contour integral of $d\chi$ when the contour integral of A is zero. Paper 56 of ECE (www.aias.us) shows that this claim is incorrect mathematically, and even if it were correct just leads to obscure ideas, notably that [22] space-time itself must be non-simply connected. This is typical of bad physics - the obscurantism of the twentieth century in natural philosophy with its plethora of nigh incomprehensible and unprovable ideas. In contrast, the twenty first century ECE theory explains the AB effect using the older but experimentally provable philosophy of Faraday, Maxwell and Einstein. Therefore one of the key philosophical advances of ECE theory is to discard standard model gauge theory as being obscurantist and meaningless. In so doing, ECE adheres to Baconian philosophy: the theory is fundamentally changed to successfully and simply explain data that clearly refute the old theory (in this case the old theory is gauge theory).

For self-consistency there should be an AB effect whenever there is present a field and its concomitant potential. So an electromagnetic AB effect should be ubiquitous throughout electrodynamics and optics. This is indeed the case, as manifested for example [1–12] in various well known phase effects interpreted according to general relativity (exemplified in turn by ECE theory). Therefore and in general the electromagnetic AB condition is:

$$\left.\begin{array}{r}F = d \wedge A + \omega \wedge A = 0, \\ A \neq 0, \omega \neq 0,\end{array}\right\} \tag{7.66}$$

and for the gravitational field the AB condition is:

$$\left.\begin{array}{r}T = d \wedge q + \omega \wedge q = 0, \\ q \neq 0, \omega \neq 0.\end{array}\right\} \tag{7.67}$$

This short-hand notation has been translated in full detail [1–12] into three other notations: differential form, tensor and vector because notation is not standardized and different subjects use different notations. In the vector notation of classical electrodynamics [24] and electrical engineering, Eq. (7.66) splits into two equations. The first defines the magnetic field in terms of the vector potential and the spin connection vector. This was developed further in paper 74 of ECE theory (www.aias.us) and published in a standard model journal, Physica B [25]. In paper 74 the context was a balance condition for magnetic motors, but the same equation is also an AB condition. It is:

$$\boldsymbol{B} = \boldsymbol{\nabla} \times \boldsymbol{A} - \boldsymbol{\omega} \times \boldsymbol{A} = \boldsymbol{0}. \tag{7.68}$$

7.4 Aharonov Bohm and Phase Effects in ECE Theory

For spin torsion [1–12] in gravitation the equivalent equation is:

$$\boldsymbol{T} = \boldsymbol{\nabla} \times \boldsymbol{q} - \boldsymbol{\omega} \times \boldsymbol{q} = \boldsymbol{0}. \tag{7.69}$$

In ECE every kind of magnetic field is defined by:

$$\boldsymbol{B} = \boldsymbol{\nabla} \times \boldsymbol{A} - \boldsymbol{\omega} \times \boldsymbol{A} \tag{7.70}$$

for self consistency. The spin connection vector is ubiquitous because it is a property of space-time itself. This is pure relativity of Einstein, but is still missing from the standard model of electrodynamics. The latter is still based on the well known vector development due to Heaviside of the original quaternionic Maxwell equations, and predates the philosophy of relativity.

If an electromagnetic AB effect is being considered the potential in Eq. (7.68) may be modeled by a plane wave as in paper 74 (www.aias.us). In that case the AB condition becomes a Beltrami condition:

$$\boldsymbol{\nabla} \times \boldsymbol{A}^{(1)} = -\kappa \boldsymbol{A}^{(1)} \tag{7.71}$$

$$\boldsymbol{\nabla} \times \boldsymbol{A}^{(2)} = \kappa \boldsymbol{A}^{(2)} \tag{7.72}$$

$$\boldsymbol{\nabla} \times \boldsymbol{A}^{(3)} = 0 \boldsymbol{A}^{(3)} \tag{7.73}$$

which can be developed in turn into a Helmholtz wave equation:

$$(\nabla^2 + \kappa^2)\boldsymbol{A}^{(1)} = \boldsymbol{0}. \tag{7.74}$$

Considering the X component for example:

$$\frac{\partial^2 A_X^{(1)}}{\partial Z^2} + \kappa^2 A_X^{(1)} = 0 \tag{7.75}$$

which is an undamped Bernoulli/Euler resonance equation without a driving force on the right hand side [1–12]. It is also a free space wave equation without a source. It is however a wave equation in the potential ONLY, there being no magnetic field present by Eq. (7.68). In other words there is no radiated electromagnetic field but there is a radiated potential field. This is an example of an electromagnetic AB effect. In ECE theory the radiated potential without field may have a physical effect, in this case an electrodynamic or optical effect.

These arguments of ECE theory go to the root of what is meant by a photon and what is meant by the electromagnetic field. In the standard model there are two approaches to electromagnetic phenomena. As argued already in this Section, the electromagnetic field F is physical but the electromagnetic potential A is unphysical in the standard model on the classical level, whereas in standard model quantum electrodynamics the minimal prescription is used with a physical potential. Also in the standard model there are other concepts such as virtual photons which occur in Feynman's version of quantum electrodynamics. During the course of ECE development however [1-12] the claimed accuracy of the Feynman type QED has been shown to be an exaggeration by several orders of magnitude. It is possible to see this through the fact that accuracy of the fine structure constant is limited by the accuracy of the Planck constant (paper 85 on www.aias.us). The standards laboratory data on fundamental constants were shown in this paper to be self-inconsistent. Finally, Feynman's QED method is based on what are essentially adjustable parameters, in other words it is based on obscurantist concepts such as dimensional renormalization, concepts which cannot be proven experimentally and so distill down to parameters that are adjusted to give a good fit of theory to experiment. It is also well known that the series summation used in the Feynman calculus cannot be proven a priori to converge, and thousands of terms have to be evaluated by computer even for the simplest of problems such as one electron interacting with one photon. The situation in quantum chromodynamics is much more complicated and much worse. In QCD it takes Nobel Prizes to prove renormalization, which is just an adjustable parameter. In a subject such as chemistry, such methods are impractical and are never used. They are therefore confined to ultra-specialist physics and even then are of dubious validity. This is typical of bad science, to claim that a theory is fundamental when it is not. It is well known [1-12] that there are many weaknesses in the standard model of electrodynamics, for example it is still not able to describe the Faraday disk generator of Dec. 26th, 1831 whereas ECE has offered a straightforward explanation.

In ECE the field and potential are both physical [1-12] on both the classical and quantum levels, and in ECE there is no distinction between relativity and wave mechanics. These ideas of natural philosophy all become aspects of the same geometry, and in ECE this is the standard differential geometry of Cartan routinely taught in mathematics. The field, potential and photon are defined by this geometry. In the standard model there is also a distinction between locality and non-locality, a distinction which enters into areas such as quantum entanglement and one photon Young interferometry, in which one photon appears to self-interfere. In ECE [1-12] there is no distinction between locality and non-locality because of the ubiquitous spin connection of general relativity. Thus, in ECE theory, the AB effects are effects of general relativity, and the labels "local" and "non-local" becomes meaningless - all is geometry in four-dimensional space-time.

Having described the essentials of the AB effects, the various phase effects developed during the course of the development of ECE theory [1-12] have been understood by a similar application of the Stokes theorem:

$$\int_s F = \int_s D \wedge A = \oint A + \int_s \omega \wedge A \qquad (7.76)$$

in which the covariant exterior derivative $D\wedge$ appears. The use of this type of Stokes Theorem has been exemplified in volume 1 of ref. (1) by integrating around a helix and by closing the contour in a well defined way. This type of integration was used in the development in ECE theory of the well known Dirac and Wu Yang phases, and in a generalization of the well known Berry phase as for example in the well studied paper 6 of the ECE series (www.aias.us). in which the origin of the Planck constant was discussed. (The extent to which the 103 or so individual ECE papers are studied is measured accurately through the feedback software of www.aias.us, and there can be no doubt that they are all well studied by a high quality of readership.) In the development of the electromagnetic phase with ECE theory [1–12] it has been demonstrated that the phase is due to the well known B(3) spin field of ECE theory, first inferred in 1992 from the inverse Faraday effect. This generally relativistic development of the electromagnetic phase is closely related to the AB effects and resolves basic problems in the standard model electromagnetic phase [1–12]. It has therefore been shown that the B(3) field is ubiquitous in optics and electrodynamics, because it derives from the ubiquitous spin connection of space-time itself.

These considerations have also been developed for the topological phases, such as that of Berry, using for self consistency the same methodology as for the electromagnetic, Dirac and Wu Yang phases [1–12]. These well known phases are again understood in ECE theory in terms of Cartan geometry by use of the Stokes Theorem with $D\wedge$ in place of $d\wedge$. All phase theory in physics becomes part of general relativity, and this methodology has been linked to traditional Lagrangian methods based on the minimization of action.

7.5 Tensor and Vector Laws of Classical Dynamics and Electrodynamics

The tensor law for the homogeneous field equation has been shown [1–12] to be:

$$\partial_\mu \widetilde{F}^{\kappa\mu\nu} = 0. \qquad (7.77)$$

For each κ index the structure of the matrix is:

$$\widetilde{F}^{\mu\nu} = \begin{bmatrix} 0 & cB_X & cB_Y & cB_Z \\ -cB_X & 0 & -E_Z & E_Y \\ -cB_Y & E_Z & 0 & -E_X \\ -cB_Z & -E_Y & E_X & 0 \end{bmatrix} = \begin{bmatrix} 0 & \widetilde{F}^{01} & \widetilde{F}^{02} & \widetilde{F}^{03} \\ \widetilde{F}^{10} & 0 & \widetilde{F}^{12} & \widetilde{F}^{13} \\ \widetilde{F}^{20} & \widetilde{F}^{21} & 0 & \widetilde{F}^{23} \\ \widetilde{F}^{30} & \widetilde{F}^{31} & \widetilde{F}^{32} & 0 \end{bmatrix}. \tag{7.78}$$

The Gauss law of magnetism in ECE theory has been shown to be obtained from:

$$\kappa = \nu = 0 \tag{7.79}$$

and so:

$$\partial_1 \widetilde{F}^{010} + \partial_2 \widetilde{F}^{020} + \partial_3 \widetilde{F}^{030} = 0 \tag{7.80}$$

i.e.:

$$\nabla \cdot \boldsymbol{B} = 0 \tag{7.81}$$

with:

$$\boldsymbol{B} = \boldsymbol{B}_X \boldsymbol{i} + \boldsymbol{B}_Y \boldsymbol{j} + B_Z \boldsymbol{k} \tag{7.82}$$

and:

$$B_X = B^{001}, B_Y = B^{002}, B_Z = B^{003}. \tag{7.83}$$

These are orbital magnetic field components of the Gauss law of magnetism.

In ECE theory the Faraday law of induction is a spin law of electrodynamics defined by:

$$\left. \begin{array}{l} \partial_0 \widetilde{F}^{\kappa 01} + \partial_2 \widetilde{F}^{\kappa 21} + \partial_3 \widetilde{F}^{\kappa 31} = 0 \\ \partial_0 \widetilde{F}^{\kappa 02} + \partial_1 \widetilde{F}^{\kappa 12} + \partial_3 \widetilde{F}^{\kappa 32} = 0 \\ \partial_0 \widetilde{F}^{\kappa 03} + \partial_1 \widetilde{F}^{\kappa 13} + \partial_2 \widetilde{F}^{\kappa 23} = 0 \end{array} \right\}. \tag{7.84}$$

The ECE Faraday law of induction for all practical purposes is [1–12]:

$$\nabla \times \boldsymbol{E} + \frac{\partial \boldsymbol{B}}{\partial t} = \boldsymbol{0} \tag{7.85}$$

7.5 Tensor and Vector Laws of Classical Dynamics and Electrodynamics

where the spin electric and magnetic components are:

$$\left.\begin{array}{l} E_X = E^{332} = -E^{323}, \ B_X = -B^{110} = B^{101}, \\ E_Y = E^{113} = -E^{131}, \ B_Y = -B^{220} = B^{202}, \\ E_Z = E^{221} = -E^{112}, \ B_Z = -B^{330} = B^{303}. \end{array}\right\} \quad (7.86)$$

The ECE Ampère Maxwell law is another spin law [1–12]:

$$\nabla \times \boldsymbol{B} - \frac{1}{c^2} \frac{\partial \boldsymbol{E}}{\partial t} = \mu_0 \boldsymbol{J} \quad (7.87)$$

where the components have been identified as:

$$\left.\begin{array}{l} E_X = E^{101}, \ B_X = B^{332}, \\ E_Y = E^{202}, \ B_Y = B^{113}, \\ E_Z = E^{303}, \ B_Z = B^{221}. \end{array}\right\} \quad (7.88)$$

Therefore in these two spin laws different components appear in ECE theory of the electric and magnetic fields. In the MH theory of special relativity these components are not distinguishable.

Finally the Coulomb law has been shown to be [1–12]:

$$\nabla \cdot \boldsymbol{E} = \rho/\epsilon_0 \quad (7.89)$$

and is an orbital law of electromagnetism as is the Gauss law of magnetism. In ECE theory these individual spin and orbital components are proportional to individual components of the three index Cartan torsion tensor and three index angular energy/momentum density tensor. Therefore ECE theory comes to the important conclusion that there are orbital and spin components of the electric field, and orbital and spin components of the magnetic field. The orbital components occur in the Gauss law of magnetism and Coulomb law and the spin components in the Faraday law of induction and the Ampère Maxwell law. This information, given by a generally covariant unified field theory, is not available in Maxwell Heaviside (MH) theory of the un-unified, special relativistic, field.

Therefore each law develops an internal structure which is summarized in Table 7.1. There are two orbital laws (Gauss and Coulomb) and two spin laws (Faraday law of induction and Ampère Maxwell law). In each law the components of the electric and magnetic fields are proportional to components of the well known [18] angular energy/momentum density tensor. Therefore for example the static electric field is distinguished form the radiated electric field. This is correct experimentally, it is well known that the static electric field exists between two static or unmoving charges, while the radiated

Table 7.1 Components of the Laws of Classical Electrodynamics

Law	Electric Field	Magnetic Field	Type
Gauss		$B_X = B^{001}$, $B_Y = B^{002}$, $B_Z = B^{003}$	orbital
Coulomb	$E_X = -E^{001}$, $E_Y = -E^{002}$, $E_Z = -E^{003}$		orbital
Faraday	$E_X = E^{332}$, $E_Y = E^{113}$, $E_Z = E^{221}$	$B_X = B^{101}$, $B_Y = B^{202}$, $B_Z = B^{303}$	spin
Ampère Maxwell	$E_X = -E^{101}$, $E_Y = -E^{202}$, $E_Z = -E^{303}$	$B_X = B^{332}$, $B_Y = B^{113}$, $B_Z = B^{221}$	spin

electric field requires accelerated charges for its existence. By postulate the components of the electric and magnetic fields are also proportional to components of the Cartan tensor, a rank three tensor in the base manifold (4-D space-time with torsion and curvature).

In tensor notation the inhomogeneous ECE field equation in the base manifold has been shown to be, for all practical purposes [1–12]:

$$\partial_\mu F^{\kappa\mu\nu} = \frac{1}{\epsilon_0} J^{\kappa\nu} = cA^{(0)} R^\kappa{}_\mu{}^{\mu\nu}. \qquad (7.90)$$

The vacuum is defined by:

$$R^\kappa{}_\mu{}^{\mu\nu} = 0 \qquad (7.91)$$

and is Ricci flat by construction. This result is consistent with the fact that the vacuum solutions of the EH equation are Ricci flat by construction. In a Ricci flat space-time there is no canonical energy momentum density [1–12] and so there are no electric and magnetic fields because there is no angular energy/momentum density. However, as in the theory of the Aharonov Bohm effects developed in Section 7.4, there may be non-zero potential and spin connection in a Ricci flat vacuum. Similarly, in the latter type of vacuum the Ricci tensor vanishes but the Christoffel connection and metric of EH

7.5 Tensor and Vector Laws of Classical Dynamics and Electrodynamics

theory do not vanish. Crothers has recently criticized the concept of the Ricci flat vacuum [19] as contradicting the Einstein equivalence principle. He has also shown that the mis-named Schwarzschild metric is inconsistent with the concept of a Ricci flat vacuum and with the geometry of the EH equation. Crothers has also argued that ideas such as Big Bang, black holes and dark matter are inconsistent with the EH equation.

The Coulomb law is the case:

$$\nu = 0 \tag{7.92}$$

of Eq. (7.90). During the course of development of ECE theory it has been shown by computer algebra that for all Ricci flat solutions of the EH equation:

$$R^\kappa{}_\mu{}^{\mu\nu} = 0 \tag{7.93}$$

but for all other solutions of the EH equation the right hand sides of Eq. (7.90) are non zero for the Christoffel connection. This result introduces a basic paradox in the EH equation as discussed already in this review paper.

The Ampère Maxwell law is the case:

$$\nu = 1, 2, 3 \tag{7.94}$$

in Eq. (7.90) and in tensor notation the ECE Ampère Maxwell law is:

$$\begin{aligned}
\partial_0 F^{\kappa 01} + \partial_2 F^{\kappa 21} + \partial_3 F^{\kappa 31} &= cA^{(0)}(R^\kappa{}_0{}^{01} + R^\kappa{}_2{}^{21} + R^\kappa{}_3{}^{31}) \\
\partial_0 F^{\kappa 02} + \partial_1 F^{\kappa 12} + \partial_3 F^{\kappa 32} &= cA^{(0)}(R^\kappa{}_0{}^{02} + R^\kappa{}_1{}^{12} + R^\kappa{}_3{}^{32}) \\
\partial_0 F^{\kappa 03} + \partial_1 F^{\kappa 13} + \partial_2 F^{\kappa 23} &= cA^{(0)}(R^\kappa{}_0{}^{03} + R^\kappa{}_1{}^{13} + R^\kappa{}_2{}^{23})
\end{aligned} \tag{7.95}$$

Therefore it is inferred that the time-like index is 0 and the space-like indices are 1, 2 and 3. The left hand side of Eq. (7.89) is a scalar and so

$$\kappa = 0 \tag{7.96}$$

is identified with a scalar index. So Eq. (7.89) of the Coulomb law is:

$$\partial_1 F^{010} + \partial_2 F^{020} + \partial_3 F^{030} = cA^{(0)}(R^0{}_1{}^{10} + R^0{}_2{}^{20} + R^0{}_3{}^{30}) \tag{7.97}$$

and is the orbital ECE Coulomb law. In vector notation this law is:

$$\nabla \cdot E = \frac{\rho}{\epsilon_0} \tag{7.98}$$

where:

$$E_X = E^{010}, E_Y = E^{020}, E_Z = E^{030},$$
$$\rho = \epsilon_0 c A^{(0)}(R^0{}_1{}^{10} + R^0{}_2{}^{20} + R^0{}_3{}^{30}). \qquad (7.99)$$

The S.I. units of this law are:

$$A^{(0)} = JsC^{-1}m^{-1}, R = m^{-2}, \epsilon_0 = J^{-1}c^2m^{-1}, \rho = Cm^{-3}. \qquad (7.100)$$

In Eq. (7.90):

$$\left.\begin{array}{r} cA^{(0)} = JC^{-1} = \text{volts}, \\ \boldsymbol{E} = \text{volt } m^{-1}, \boldsymbol{\nabla}\cdot\boldsymbol{E} = \text{volt } m^{-2}, \\ cA^{(0)}R = \text{volt } m^{-2}, \\ \rho/\epsilon_0 = JC^{-1}m^{-2} = \text{volt } m^{-2}, \end{array}\right\} \qquad (7.101)$$

thus checking the S. I. units for self consistency. In the Ricci flat vacuum:

$$\boldsymbol{\nabla}\cdot\boldsymbol{E} = 0 \qquad (7.102)$$

which is consistent with:

$$R^0{}_1{}^{10} + R^0{}_2{}^{20} + R^0{}_3{}^{30} = 0 \qquad (7.103)$$

for vacuum solutions of the EH equation as argued already. However, for complete internal consistency the Christoffel symbol cannot be used, because it is not internally consistent with the Bianchi identity as argued already in this review paper.

It is possible to define a curvature scalar of the Coulomb law as:

$$R_{(0)} := R^0{}_1{}^{10} + R^0{}_2{}^{20} + R^0{}_3{}^{30} \qquad (7.104)$$

so that:

$$\boldsymbol{\nabla}\cdot\boldsymbol{E} = \frac{\rho}{\epsilon_0} = cA^{(0)}R_{(0)} \qquad (7.105)$$

and that the charge density of the Coulomb law becomes:

$$\rho = cA^{(0)}\epsilon_0 R_{(0)} \qquad (7.106)$$

7.5 Tensor and Vector Laws of Classical Dynamics and Electrodynamics

in coulombs per cubic meter. In the Cartesian system of coordinates the electric field components of the Coulomb law are:

$$E_X = E^{010}, E_Y = E^{020}, E_Z = E^{030} \tag{7.107}$$

and are proportional to these same components of the three index angular energy momentum density tensor. They are anti-symmetric in their last two indices:

$$E^{010} = -E^{001} \text{etc.} \tag{7.108}$$

In tensor notation the ECE Ampère Maxwell law is given by Eq. (7.95), i.e.:

$$\left.\begin{aligned} \partial_\mu F^{\kappa\mu\nu} = cA^{(0)} R^\kappa{}_\mu{}^{\mu\nu}, \\ \nu = 1, 2, 3 \end{aligned}\right\} \tag{7.109}$$

and in vector notation by:

$$\nabla \times \boldsymbol{B} - \frac{1}{c^2} \frac{\partial \boldsymbol{E}}{\partial t} = \mu_0 \boldsymbol{J}. \tag{7.110}$$

In the Cartesian system:

$$\boldsymbol{J} = J_X \boldsymbol{i} + J_Y \boldsymbol{j} + J_Z \boldsymbol{k} \tag{7.111}$$

where:

$$\left.\begin{aligned} J_X &= \frac{A^{(0)}}{\mu_0} (R^1{}_0{}^{01} + R^1{}_2{}^{21} + R^1{}_3{}^{31}), \\ J_Y &= \frac{A^{(0)}}{\mu_0} (R^2{}_0{}^{02} + R^2{}_1{}^{12} + R^3{}_2{}^{32}), \\ J_Z &= \frac{A^{(0)}}{\mu_0} (R^3{}_0{}^{03} + R^3{}_1{}^{13} + R^3{}_2{}^{23}), \end{aligned}\right\} \tag{7.112}$$

and self consistently in the vacuum:

$$J_X = J_Y = J_Z = 0 \tag{7.113}$$

for Ricci flat space-times. As argued this result has been demonstrated by computer algebra [1–12]. In the Ampère Maxwell law the electric and magnetic field components are proportional to spin angular energy momentum density tensor components of the electromagnetic field as follows:

$$\left. \begin{array}{l} E^{\kappa\mu\nu} = \dfrac{c^2}{e\omega} J^{\kappa\mu\nu}, \\ B^{\kappa\mu\nu} = \dfrac{c}{e\omega} J^{\kappa\mu\nu}. \end{array} \right\} \qquad (7.114)$$

The electric field components of the Coulomb law and the magnetic field components of the Gauss law are all orbital angular energy density tensor components of the electromagnetic field. The angular energy momentum density tensor may be defined as [18]:

$$J^{\kappa\mu\nu} = -\frac{1}{2}(T^{\kappa\mu}x^\nu - T^{\kappa\nu}x^\mu) \qquad (7.115)$$

using the symmetric canonical energy momentum density tensor:

$$T^{\kappa\mu} = T^{\mu\kappa} \qquad (7.116)$$

and the components of the electric and magnetic fields are components of $J^{\kappa\mu\nu}$ as follows:

$$E^{00i} = \frac{c^2}{e\omega} J^{00i}, i = 1, 2, 3, \text{(orbital)},$$
$$E^{ii0} = \frac{c^2}{e\omega} J^{ii0}, i = 1, 2, 3, \text{(spin)}, \qquad (7.117)$$
$$B^{112} = \frac{c}{e\omega} J^{112}, B^{221} = \frac{c}{e\omega} J^{221}, B^{331} = \frac{c}{e\omega} J^{331}.$$

The two index angular energy/momentum tensor of the electromagnetic field is an integral over the three index density tensor. Ryder gives one example of such an integral in Minkowski space-time [18]:

$$M^{\mu\nu} = \int M^{0\mu\nu} d^3x. \qquad (7.118)$$

Therefore the four laws of electrodynamics in ECE theory are:

$$\nabla \cdot \mathbf{B} = 0, \qquad (7.119)$$

$$\nabla \times \mathbf{E} + \frac{\partial \mathbf{B}}{\partial t} = \mathbf{0}, \qquad (7.120)$$

7.5 Tensor and Vector Laws of Classical Dynamics and Electrodynamics

$$\nabla \cdot \boldsymbol{E} = \rho/\epsilon_0, \tag{7.121}$$

$$\nabla \times \boldsymbol{B} - \frac{1}{c^2}\frac{\partial \boldsymbol{E}}{\partial t} = \mu_0 \boldsymbol{J}, \tag{7.122}$$

and therefore have the same vector structure as the familiar MH equations. However, as argued in this section, the ECE theory gives additional information. In the four laws the components of the magnetic and electric fields are as follows. The Gauss law of magnetism in ECE theory is, for all practical purposes (FAPP):

$$\nabla \cdot \boldsymbol{B} = 0 \tag{7.123}$$

which is an orbital law in which the components of the magnetic field are proportional to orbital components of the angular momentum/energy density tensor and are:

$$\boldsymbol{B} = B^{001}\boldsymbol{i} + B^{002}\boldsymbol{j} + B^{003}\boldsymbol{k}. \tag{7.124}$$

The Faraday law of induction in ECE is a spin law with electric and magnetic field components as follows:

$$\boldsymbol{E} = E^{332}\boldsymbol{i} + E^{113}\boldsymbol{j} + E^{221}\boldsymbol{k}, \tag{7.125}$$

$$\boldsymbol{B} = B^{101}\boldsymbol{i} + B^{202}\boldsymbol{j} + B^{303}\boldsymbol{k}. \tag{7.126}$$

The Coulomb law in ECE is an orbital law with electric field components as follows:

$$\boldsymbol{E} = E^{010}\boldsymbol{i} + E^{020}\boldsymbol{j} + E^{030}\boldsymbol{k}, \tag{7.127}$$

Finally the Ampère Maxwell law in ECE is a spin law with electric and magnetic field components as follows:

$$\boldsymbol{E} = E^{110}\boldsymbol{i} + E^{220}\boldsymbol{j} + E^{330}\boldsymbol{k}, \tag{7.128}$$

$$\boldsymbol{B} = B^{332}\boldsymbol{i} + B^{113}\boldsymbol{j} + B^{221}\boldsymbol{k}. \tag{7.129}$$

As argued in Section 7.4 of this review paper, the relation between field and potential is different in ECE theory and contains the spin connection [1–12]. The various notations for the relation between field and potential in ECE theory are collected here for convenience. In the index-less notation:

$$F = d \wedge A + \omega \wedge A \tag{7.130}$$

which is based on the first Cartan structure equation:

$$T = d \wedge q + \omega \wedge q. \qquad (7.131)$$

In the standard notation of differential geometry:

$$F^a = d \wedge A^a + \omega^a{}_b \wedge A^b. \qquad (7.132)$$

In tensor notation from differential geometry:

$$F^a_{\mu\nu} = (d \wedge A^a)_{\mu\nu} + (\omega^a{}_b \wedge A^b)_{\mu\nu}. \qquad (7.133)$$

In the base manifold Eq. (7.133) becomes:

$$F^{\kappa\mu\nu} = \partial^\mu A^{\kappa\nu} - \partial^\nu A^{\kappa\mu} + (\omega^{\kappa\mu}{}_\lambda A^{\lambda\nu} - \omega^{\kappa\nu}{}_\lambda A^{\lambda\mu}) \qquad (7.134)$$

In vector notation Eq. (7.134) splits into two equations, one for the electric field and one for the magnetic field:

$$\boldsymbol{E} = -\boldsymbol{\nabla}\phi - \frac{\partial \boldsymbol{A}}{\partial t} + \phi\boldsymbol{\omega} - \omega\boldsymbol{A} \qquad (7.135)$$

and

$$\boldsymbol{B} = \boldsymbol{\nabla} \times \boldsymbol{A} - \boldsymbol{\omega} \times \boldsymbol{A} \qquad (7.136)$$

For the orbital electric field component of the Coulomb law Eq. (7.135) has the following internal structure:

$$\phi = cA^{00}, \ \boldsymbol{A} = A^{01}\boldsymbol{i} + A^{02}\boldsymbol{j} + A^{03}\boldsymbol{k}, \qquad (7.137)$$

$$\omega = c\omega^{00}{}_0, \ \boldsymbol{\omega} = (\omega^{01}{}_0\boldsymbol{i} + \omega^{02}{}_0\boldsymbol{j} + \omega^{03}{}_0\boldsymbol{k}). \qquad (7.138)$$

This result illustrates that the internal structure of the relation between field and potential is different for each law of electrodynamics in ECE theory. Therefore in a GCUFT such as ECE different types of field and potential exist for each law, and also different types of spin connection.

For the orbital Gauss law of magnetism the internal structure of Eq. (7.136) is:

$$\boldsymbol{A} = A^{01}\boldsymbol{i} + A^{02}\boldsymbol{j} + A^{03}\boldsymbol{k}, \qquad (7.139)$$

$$\boldsymbol{\omega} = -(\omega^{01}{}_0\boldsymbol{i} + \omega^{02}{}_0\boldsymbol{j} + \omega^{03}{}_0\boldsymbol{k}). \qquad (7.140)$$

7.5 Tensor and Vector Laws of Classical Dynamics and Electrodynamics 151

For the Ampère Maxwell law, a spin law, the internal structure of Eqs. (7.135) and (7.136) are again different, and are defined as follows. The structure of Eq. (7.135) is:

$$
\begin{aligned}
\phi &= cA^{00} = cA^{01} = cA^{02} = cA^{03}, \\
A_X &= A^{01} = A^{11} = A^{21} = A^{31}, \\
A_Y &= A^{02} = A^{12} = A^{22} = A^{32}, \\
A_Z &= A^{03} = A^{13} = A^{23} = A^{33}, \\
\omega_X &= \omega^{11}{}_0 = \omega^{11}{}_1 = \omega^{11}{}_2 = \omega^{11}{}_3, \\
\omega_Y &= \omega^{22}{}_0 = \omega^{22}{}_1 = \omega^{22}{}_2 = \omega^{22}{}_3, \\
\omega_Z &= \omega^{33}{}_0 = \omega^{33}{}_1 = \omega^{33}{}_2 = \omega^{33}{}_3, \\
\omega &= c\omega^{10}{}_0 = c\omega^{10}{}_1 = c\omega^{10}{}_2 = c\omega^{10}{}_3 \\
&= c\omega^{20}{}_0 = c\omega^{20}{}_1 = c\omega^{20}{}_2 = c\omega^{20}{}_3 \\
&= c\omega^{30}{}_0 = c\omega^{30}{}_1 = c\omega^{30}{}_2 = c\omega^{30}{}_3
\end{aligned}
\qquad (7.141)
$$

and the structure of Eq. (7.136) is:

$$
\begin{aligned}
B_X &= B^{332} = \frac{\partial A_Z}{\partial Y} - \frac{\partial A_Y}{\partial Z} + \omega_Z A_Y - \omega_Y A_Z, \\
B_Y &= B^{113} = \frac{\partial A_X}{\partial Z} - \frac{\partial A_Z}{\partial X} + \omega_X A_Z - \omega_Z A_X, \\
B_Z &= B^{221} = \frac{\partial A_Y}{\partial X} - \frac{\partial A_X}{\partial Y} + \omega_Y A_X - \omega_X A_Y.
\end{aligned}
\qquad (7.142)
$$

Finally the internal structures are again different for the Faraday law of induction. In arriving at these conclusions the relation between field and potential in the base manifold is:

$$
F^{\kappa\mu\nu} = \partial^\mu A^{\kappa\nu} - \partial^\nu A^{\kappa\mu} + (\omega^{\kappa\mu}{}_\lambda A^{\lambda\nu} - \omega^{\kappa\nu}{}_\lambda A^{\lambda\mu}). \qquad (7.143)
$$

The Hodge dual of this equation is:

$$
\widetilde{F}^{\kappa\mu\nu} = (\partial^\mu A^{\kappa\nu} - \partial^\nu A^{\kappa\mu} + (\omega^{\kappa\mu}{}_\lambda A^{\lambda\nu} - \omega^{\kappa\nu}{}_\lambda A^{\lambda\mu}))_{HD} \qquad (7.144)
$$

and this is needed to give the results for the homogenous laws. An example of taking the Hodge dual is given below:

$$
\begin{aligned}
\widetilde{F}^{001} &= (\partial^0 A^{01} - \partial^1 A^{00} + (\omega^{00}{}_\lambda A^{\lambda 1} - \omega^{01}{}_\lambda A^{\lambda 0}))_{HD} \\
&= \partial^2 A^{03} - \partial^3 A^{02} + (\omega^{02}{}_\lambda A^{\lambda 3} - \omega^{03}{}_\lambda A^{\lambda 2}).
\end{aligned}
\qquad (7.145)
$$

With these rules the overall conclusion is that in a generally covariant unified field theory (GCUFT) such as ECE the four laws of classical electrodynamics can be reduced to the same vector form as the MH laws of un-unified special relativity (nineteenth century), but the four laws are no longer written in a flat, Minkowski spacetime. They are written in a four dimensional space-time with torsion and curvature. This procedure reveals the internal structure of the electric and magnetic fields appearing in each law, for example correctly makes the distinction between a static and radiated electric field, and a static and radiated magnetic field. The relation between field and potential also develops an internal structure which is different for each law, but for each law, the vector relation can be reduced to:

$$\boldsymbol{E} = -\boldsymbol{\nabla}\phi - \frac{\partial \boldsymbol{A}}{\partial t} + \phi\boldsymbol{\omega} - \omega \boldsymbol{A} \qquad (7.146)$$

and

$$\boldsymbol{B} = \boldsymbol{\nabla} \times \boldsymbol{A} - \boldsymbol{\omega} \times \boldsymbol{A}. \qquad (7.147)$$

In a GCUFT, gauge theory is not used because the potential has a physical effect as in the electrotonic state of Faraday. The ECE theory is developed entirely in four dimensions, is entirely self-consistent, and reproduces a range of experimental data [1–12] which the MH theory cannot explain. The ECE theory is also philosophically consistent with the need to apply the philosophy of relativity to the whole of physics. The latter becomes a unified field theory based on geometry. The first attempts by Einstein to develop general relativity were based on Riemann geometry and restricted to the theory of gravitation. In the philosophy of relativity, however, the basic idea that physics is geometry must be used for every equation of physics, and each equation must be part of the same geometrical framework. This is achieved in a GCUFT such as ECE theory by using Cartan's standard differential geometry. This is a self-consistent geometry that recognizes the existence of space-time torsion in the first Cartan structure equation, and space-time curvature in the second. It is also recognized that there is only one Bianchi identity, and that this must always inter-relate torsion and curvature, both are fundamental to the structure of space-time.

7.6 Spin Connection Resonance

One of the most important consequences of general relativity applied to electrodynamics is that the spin connection enters into the relation between the field and potential as described in Section 7.5. The equations of electrodynamics as written in terms of the potential can be reduced to the form of Bernoulli Euler resonance equations. These have been incorporated during the

7.6 Spin Connection Resonance

course of development of ECE theory into the Coulomb law, which is the basic law used in the development of quantum chemistry in for example density functional code. This process has been illustrated [1–12] with the hydrogen and helium atoms. The ECE theory has also been used to design or explain circuits which use spin connection resonance to take power from space-time, notably papers 63 and 94 of the ECE series on www.aias.us. In paper 63, the spin connection was incorporated into the Coulomb law and the resulting equation in the scalar potential shown to have resonance solutions using an Euler transform method. In paper 94 this method was extended and applied systematically to the Bedini motor. The method is most simply illustrated by considering the vector form of the Coulomb law deduced in Section 7.5:

$$\boldsymbol{\nabla} \cdot \boldsymbol{E} = \rho/\epsilon_0 \tag{7.148}$$

and assuming the absence of a vector potential (absence of a magnetic field). The electric field is then described by:

$$\boldsymbol{E} = -(\boldsymbol{\nabla} + \boldsymbol{\omega})\phi \tag{7.149}$$

rather than the standard model's:

$$\boldsymbol{E} = -\boldsymbol{\nabla}\phi. \tag{7.150}$$

Therefore Eq. (7.149) in (7.148) produces the equation

$$\nabla^2 \phi + \boldsymbol{\omega} \cdot \boldsymbol{\nabla}\phi + (\boldsymbol{\nabla} \cdot \boldsymbol{\omega})\phi = -\frac{\rho}{\epsilon_0} \tag{7.151}$$

which is capable of giving resonant solutions as described in paper 63. The equivalent equation in the standard model is the Poisson equation, which is a limit of Eq. (7.151) when the spin connection is zero. The Poisson equation does not give resonant solutions. It is known from the work of Tesla for example that strong resonances in electric power can be obtained with suitable apparatus, and such resonances cannot be explained using the standard model. A plausible explanation of Tesla's well known results is given by the incorporation of the spin connection into classical electrodynamics. Using spherical polar coordinates and restricting consideration to the radial component:

$$\nabla^2 \phi = \frac{\partial^2 \phi}{\partial r^2} + \frac{2}{r}\frac{\partial \phi}{\partial r}, \tag{7.152}$$

$$\boldsymbol{\omega} \cdot \boldsymbol{\nabla}\phi = \omega_r \frac{\partial \phi}{\partial r}, (\boldsymbol{\nabla} \cdot \boldsymbol{\omega})\phi = \frac{\phi}{r^2}\frac{\partial}{\partial r}(r^2 \omega_r), \tag{7.153}$$

so that Eq. (7.151) becomes:

$$\frac{\partial^2 \phi}{\partial r^2} + \left(\frac{2}{r} + \omega\right)\frac{\partial \phi}{\partial r} + \frac{\phi}{r^2}\left(2r\omega_r + r^2\frac{\partial \omega_r}{\partial r}\right) = \frac{-\rho}{\epsilon_0} \qquad (7.154)$$

In paper 63 a spin connection was chosen of the simplest type compatible with its dimensions of inverse meters:

$$\omega_r = -\frac{1}{r} \qquad (7.155)$$

and thus giving the differential equation:

$$\frac{\partial^2 \phi}{\partial r^2} + \frac{1}{r}\frac{\partial \phi}{\partial r} - \frac{1}{r^2}\phi = \frac{-\rho}{\epsilon_0} \qquad (7.156)$$

as a function of r. Eq. (7.156) becomes a resonance equation if the driving term on the right hand side is chosen to be oscillatory, in the simplest instance:

$$\rho = \rho(0)\cos(\kappa_r r). \qquad (7.157)$$

To obtain resonance solutions from Eq. (7.156), an Euler transform [1–12] is needed as follows:

$$\kappa_r r = \exp(i\kappa_r R). \qquad (7.158)$$

This is a standard Euler transform extended to a complex variable. This simple change of variable transforms Eq. (7.156) into:

$$\frac{\partial^2 \phi}{\partial R^2} + \kappa_r^2 \phi = \frac{\rho(0)}{\epsilon_0}\text{Real}(e^{2i\kappa_r R}\cos(e^{i\kappa_r R})) \qquad (7.159)$$

which is an undamped oscillator equation as demonstrated in detail in Eq. (7.63), where the domain of validity of the transformed variable was discussed in detail. It is seen from feedback software to www.aias.us that paper 63 has been studied in great detail by a high quality readership, so we may judge that its impact has been extensive. The concept of spin connection resonance has been extended to gravitational theory and magnetic motors and the theory published in standard model journals [25–27]. In paper 63 the simplest possible form of the spin connection was used, Eq. (7.155) and the resulting Eq. (7.156) was shown to have resonance solutions using a change

7.6 Spin Connection Resonance

of variable. There is therefore resonance in the variable R. In paper 90 of www.aias.us this method was made more general by considering the equation

$$\frac{\partial^2 \phi}{\partial r^2} + \left(\frac{2}{r} + \omega_r\right)\frac{\partial \phi}{\partial r} + \frac{\phi}{r^2}\left(2r\omega_r + r^2 \frac{\partial \omega_r}{\partial r}\right) = \frac{-\rho}{\epsilon_0} \quad (7.160)$$

which is a more general form of Eq. (7.156). When the spin connection is defined as:

$$\omega_r = \omega_0^2 r - 4\beta \log_e r - \frac{4}{r}. \quad (7.161)$$

Eq. (7.160) becomes a simple resonance equation in r itself:

$$\frac{\partial^2 \phi}{\partial r^2} + 2\beta \frac{\partial \phi}{\partial r} + \omega_0^2 \phi = \frac{-\rho}{\epsilon_0}. \quad (7.162)$$

There is freedom of choice of the spin connection. The latter was unknown in electrodynamics prior to ECE theory and must ultimately be determined experimentally. An example of this procedure is given in paper 94, where spin connection resonance (SCR) theory is applied to a patented device. One of the papers published in the standard model literature [26] applies SCR to magnetic motors that are driven by space-time. It is probable that SCR was also discovered and demonstrated by Tesla [28], but empirically before the emergence of relativity theory. SCR has also been applied to gravitation and published in the standard model literature [27]. So a gradual loosening of the ties to the standard model is being observed at present.

In paper 92 of the ECE series (www.aias.us), Eq. (7.160) was further considered and shown to reduce to an Euler Bernoulli resonance equation of the general type:

$$\frac{d^2 x}{dr^2} + 2\beta \frac{dx}{dr} + \kappa_0^2 x = A \cos(\kappa r) \quad (7.163)$$

in which β plays the role of friction coefficient, κ_0 is a Hooke's law wavenumber and in which the right hand side is a cosinal driving term. Eq. (7.160) reduces to Eq. (7.163) when:

$$\omega_r = 2\left(\beta - \frac{1}{r}\right), \kappa_0^2 = \frac{4}{r}\left(\beta - \frac{1}{r}\right) + \frac{\partial \omega_r}{\partial r} \quad (7.164)$$

Therefore the condition udner which the spin connection gives the simple resonance Eq. (7.163) is defined by:

$$\omega_r = \kappa_0^2 - 4\beta \log_e r - \frac{4}{r}. \tag{7.165}$$

Reduction to the standard model Coulomb law occurs when:

$$\beta = \frac{1}{r} \tag{7.166}$$

when

$$\omega_r = 0, \kappa_0^2 = 0. \tag{7.167}$$

In general there is no reason to assume that condition (7.166) always holds. The reason why the standard model Coulomb law is so accurate in the laboratory is that it is tested off resonance. In this off resonant limit the ECE theory has been shown [1–12] to give the Standard Coulomb law as required by a vast amount of accumulated data of two centuries since Coulomb first inferred the law. In general, ECE theory has been shown to reduce to all the known laws of physics, and in addition ECE gives new information. This is a classic hallmark of a new advance in physics. It is probable that Tesla inferred methods of tuning the Coulomb law (and other laws) to spin connection resonance. Many other reports of such surges in power have been made, and it is now known and accepted by the international community of scientists that they come from general relativity applied to classical electrodynamics.

Paper 94 of the ECE series is a pioneering paper in which the theory of SCR is applied to a patented device in order to explain in detail how the patented device takes energy from space-time. No violation of the laws of conservation of energy and momentum occurs in ECE theory or in SCR theory.

7.7 Effects of Gravitation on Optics and Spectroscopy

In the standard model of electrodynamics the electromagnetic sector is described by the nineteenth century Maxwell Heaviside (MH) field theory, which in gauge theory is U(1) invariant and Lorentz covariant in a Minkowski space-time. As such MH theory cannot describe the effect of gravitation on optics and spectroscopy because gravitation requires a non-Minkowski space-time. In ECE theory on the other hand all sectors are generally covariant, and during the course of development of ECE theory several effects of gravitation on optics and spectroscopy have been inferred, notably the effect of gravitation on the Sagnac effect, RFR and on the polarization of

7.7 Effects of Gravitation on Optics and Spectroscopy

light grazing a white dwarf. An explanation for the well known Faraday disk generator has also been given in terms of spinning space-time, an explanation which illustrates the fact that the torsion of space-time produces effects not present in the standard model. Gravitation is the curvature of space-time and in ECE theory the interaction of torsion and curvature is determined by Cartan geometry.

The Faraday disk generator has been explained in ECE theory from the basic assumption that the electromagnetic field is the Cartan torsion within a factor:

$$\boldsymbol{F}_{\text{mech}} = A^{(0)} T_{\text{mech}} \qquad (7.168)$$

where $cA^{(0)}$ is the primordial voltage. The factor $A^{(0)}$ is considered to originate in the magnet of the Faraday disk generator. The Faraday disk generator consists essentially of a spinning disk placed on a magnet, without the magnet no induction is observed, i.e. no p.d.f. is generated between the center and rim of the disk without a magnet being present. The original experiment by Faraday on 26th Dec. 1831 consisted of spinning a disk on top of a static magnet, but an e.m.f. is also observed if both the disk and the magnet are spun about their common vertical axis. There continues to be no explanation for the Faraday disk generator in the standard model, because in the latter there is no connection between the electromagnetic field and mechanical spin, angular momentum and torsion, while ECE makes this connection in Eq. (7.168). The standard model law of induction of Faraday is:

$$\nabla \times \boldsymbol{E} + \frac{\partial \boldsymbol{B}}{\partial t} = \boldsymbol{0} \qquad (7.169)$$

and spinning the magnetic field about its own axis does not produce a non-zero curl of the electric field as required. Clearly, a static magnetic field will not cause induction from Eq. (7.169). So this is a weak point of the standard model, in which induction is caused in the classical textbook description by moving a bar magnet inside a coil, causing a current to appear. In ECE it has been shown [1–12] that the explanation of the Faraday disk generator is simply:

$$\boldsymbol{F} = \boldsymbol{F}_{e/m} + \boldsymbol{F}_{\text{mech}} \qquad (7.170)$$

which in vector notation (section 7.5) produces the law of induction:

$$\nabla \times \boldsymbol{E}_{\text{mech}} + \frac{\partial \boldsymbol{B}_{\text{mech}}}{\partial t} = \boldsymbol{0}. \qquad (7.171)$$

Spinning the disk has the following effect in ECE theory.

In the complex circular basis [1–12] the magnetic flux density in ECE theory is defined by:

$$\boldsymbol{B}^{(1)*} = \boldsymbol{\nabla} \times \boldsymbol{A}^{(1)*} - i\frac{\kappa}{A^{(0)}}\boldsymbol{A}^{(2)} \times \boldsymbol{A}^{(3)} \tag{7.172}$$

$$\boldsymbol{B}^{(2)*} = \boldsymbol{\nabla} \times \boldsymbol{A}^{(2)*} - i\frac{\kappa}{A^{(0)}}\boldsymbol{A}^{(3)} \times \boldsymbol{A}^{(1)} \tag{7.173}$$

$$\boldsymbol{B}^{(3)*} = \boldsymbol{\nabla} \times \boldsymbol{A}^{(3)*} - i\frac{\kappa}{A^{(0)}}\boldsymbol{A}^{(1)} \times \boldsymbol{A}^{(2)} \tag{7.174}$$

where

$$\kappa = \frac{\Omega}{c} \tag{7.175}$$

is a wave-number and Ω is an angular frequency in radians per second. When the disk is stationary the ECE vector potential is [1–12] proportional by fundamental hypothesis to the tetrad:

$$\boldsymbol{A}^{(1)} = A^{(0)}\boldsymbol{q}^{(1)} \tag{7.176}$$

$$\boldsymbol{A}^{(2)} = A^{(0)}\boldsymbol{q}^{(2)} \tag{7.177}$$

$$\boldsymbol{A}^{(3)} = A^{(0)}\boldsymbol{q}^{(3)}. \tag{7.178}$$

In the complex circular basis the tetrads are:

$$\boldsymbol{q}^{(1)} = \frac{1}{\sqrt{2}}(\boldsymbol{i} - i\boldsymbol{j}), \tag{7.179}$$

$$\boldsymbol{q}^{(2)} = \frac{1}{\sqrt{2}}(\boldsymbol{i} + i\boldsymbol{j}), \tag{7.180}$$

$$\boldsymbol{q}^{(3)} = \boldsymbol{k}, \tag{7.181}$$

and have O(3) symmetry as follows:

$$\boldsymbol{q}^{(1)} \times \boldsymbol{q}^{(2)} = i\boldsymbol{q}^{(3)*}, \tag{7.182}$$

$$\boldsymbol{q}^{(2)} \times \boldsymbol{q}^{(1)} = i\boldsymbol{q}^{(1)*}, \tag{7.183}$$

$$\boldsymbol{q}^{(3)} \times \boldsymbol{q}^{(1)} = i\boldsymbol{q}^{(2)*}. \tag{7.184}$$

In the absence of rotation about Z:

$$\boldsymbol{\nabla} \times \boldsymbol{A}^{(1)} = \boldsymbol{\nabla} \times \boldsymbol{A}^{(2)} = \boldsymbol{0}, \tag{7.185}$$

7.7 Effects of Gravitation on Optics and Spectroscopy

$$\boldsymbol{A}^{(3)} = A^{(0)}\boldsymbol{k}. \tag{7.186}$$

In the complex circular basis:

$$\nabla \times \boldsymbol{E}^{(1)} + \partial \boldsymbol{B}^{(1)}/\partial t = \boldsymbol{0}, \tag{7.187}$$

$$\nabla \times \boldsymbol{E}^{(2)} + \partial \boldsymbol{B}^{(2)}/\partial t = \boldsymbol{0}, \tag{7.188}$$

$$\nabla \times \boldsymbol{E}^{(3)} + \partial \boldsymbol{B}^{(3)}/\partial t = \boldsymbol{0}. \tag{7.189}$$

Therefore from Eqs. (7.176) to (7.189) the only field present is:

$$\begin{aligned}\boldsymbol{B}^{(3)*} = \boldsymbol{B}^{(3)} &= -iB^{(0)}\boldsymbol{q}^{(1)} \times \boldsymbol{q}^{(2)} \\ &= B_z^{(3)}\boldsymbol{k} = B_z\boldsymbol{k}\end{aligned} \tag{7.190}$$

which is the static magnetic field of the bar magnet.

Now mechanically rotate the disk at an angular frequency Ω to produce:

$$\boldsymbol{A}^{(1)} = \frac{A^{(0)}}{\sqrt{2}}(\boldsymbol{i} - i\boldsymbol{j})\exp(i\Omega t), \tag{7.191}$$

$$\boldsymbol{A}^{(2)} = \frac{A^{(0)}}{\sqrt{2}}(\boldsymbol{i} + i\boldsymbol{j})\exp(-i\Omega t). \tag{7.192}$$

From Eqs. (7.176) to (7.192) electric and magnetic fields are induced in a direction transverse to Z, i.e. in the XY plane of the spinning disk as observed experimentally. However, the Z axis magnetic flux density is unchanged by physical rotation about the same Z axis. This is again as observed experimentally. The (2) component of the transverse electric field spins around the rim of the disk and is defined from Eq. (7.151) as:

$$\boldsymbol{E}^{(2)} = \boldsymbol{E}^{(1)*} = -\left(\frac{\partial}{\partial t} + i\Omega\right)\boldsymbol{A}^{(2)}. \tag{7.193}$$

It can be seen from section 7.5 that $i\Omega$ is a type of spin connection. The latter is caused by mechanical spin, which in ECE is a spinning of space-time itself. The real and physical part of the induced $E^{(1)}$ is:

$$\text{Real}(\boldsymbol{E}^{(1)}) = \frac{2}{\sqrt{2}}A^{(0)}\Omega(\boldsymbol{i}\sin\Omega t - \boldsymbol{j}\cos\Omega t) \tag{7.194}$$

and is proportional to the product of $A^{(0)}$ and Ω, again as observed experimentally. An electromotive force is set up between the center of the disk and the rotating rim, as first observed experimentally by Faraday. This e. m. f. is measured experimentally with a voltmeter at rest with respect to the rotating disk.

The homogeneous law (7.120) of ECE theory is generally covariant [1–12] by construction, so retains its form in any frame of reference. ECE therefore produces a simple and complete description of the Faraday disk generator in terms of the spinning of space-time, and concomitant spin connection. The latter is therefore demonstrated in classical electrodynamics by the generator. All known experimental features are explained straightforwardly by ECE theory, but cannot be explained by MH theory, in which the spin connection is missing because Minkowski space-time has no connection by construction - it is a "flat" space-time. It is relatively easy to think of electrodynamics as spinning space-time if we think of gravitation as curving space-time. This analysis also gives confidence in the arguments of Section 7.6, where power is obtained from space-time with spin connection resonance.

The same ECE concept just used to explain the Faraday disk generator has been used [1–12] to give a simple explanation of the Sagnac effect (ring laser gyro). Again, the standard model has no satisfactory explanation for the Sagnac effect [1–12]. Consider the rotation of a beam of light of any polarization around a circle of area πr^2 in the XY plane at an angular frequency ω_1. The rotation is a rotation of space-time itself in ECE theory, described by the rotating tetrad:

$$q^{(1)} = \frac{1}{\sqrt{2}}(\boldsymbol{i} - i\boldsymbol{j})e^{i\omega_1 t}. \qquad (7.195)$$

This is rotation around the static platform of the Sagnac interferometer. The fundamental ECE assumption means that this rotation produces the electromagnetic vector potential:

$$\boldsymbol{A}_L^{(1)} = A^{(0)}\boldsymbol{q}^{(1)} \qquad (7.196)$$

for left rotation and:

$$\boldsymbol{A}_R^{(1)} = \frac{A^{(0)}}{\sqrt{2}}(\boldsymbol{i} + i\boldsymbol{j})e^{i\omega_1 t} \qquad (7.197)$$

for right rotation. When the platform is at rest a beam going around leftwise takes the same time to reach its starting point as a beam going around right-wise. The time delay is zero:

$$\Delta t = 2\pi \left(\frac{1}{\omega_1} - \frac{1}{\omega_1}\right) = 0. \qquad (7.198)$$

7.7 Effects of Gravitation on Optics and Spectroscopy

Eqs. (7.196) and (7.197) do not exist in special relativity because in the MH theory electromagnetism is a nineteenth century entity superimposed on a space-time that is flat and static and never rotates.

Now consider the left - wise rotating beam (7.196) and spin the platform mechanically in the same left-wise direction at an angular frequency Ω. The result is an increase in the angular frequency of the rotating tetrad as follows:

$$\omega_1 \to \omega_1 + \Omega. \tag{7.199}$$

Similarly consider the left wise rotating beam (7.196) and spin the platform right-wise. The result is a decrease in the angular frequency of the rotating tetrad:

$$\omega_1 \to \omega_1 - \Omega. \tag{7.200}$$

The time delay between a beam circling left-wise with the platform, and one circling left-wise against the platform is therefore:

$$\Delta t = 2\pi \left(\frac{1}{\omega_1 - \Omega} - \frac{1}{\omega_1 + \Omega} \right) \tag{7.201}$$

which is the Sagnac effect. The angular frequency ω_1 can be calculated from the experimental result [1–12]:

$$\Delta t = \frac{4\Omega}{c^2} Ar = \frac{4\pi\Omega}{\omega_1^2 - \Omega^2} \tag{7.202}$$

If

$$\Omega \ll \omega_1 \tag{7.203}$$

it is found that

$$\omega_1 = \frac{c}{r} = c\kappa \tag{7.204}$$

Q.E.D. Therefore the Sagnac effect is another result of a spin connection, which in this case can be thought of as the potential (7.196) itself.

Similarly, phase effects such as the Tomita Chao effect were also described straightforwardly with the same basic concept during the development of ECE theory.

In order to describe the effects of gravitation on optics and spectroscopy a dielectric version of the ECE theory was developed and implemented to find that the polarization of light is changed by light grazing a very massive object such as a white dwarf, and the dielectric theory was also used to demonstrate the effect of gravitation on the Sagnac effect [1–12]. The standard model is not capable of such descriptions without the use of adjustable parameters in such transient twentieth century artifacts as superstring theory, now being essentially discarded as being untestable experimentally. ECE is far simpler and is also capable of describing data such as the Faraday disk generator and the Sagnac effect straightforwardly. During the course of its development the ECE theory has also been applied to the interaction of three fields [27] and the effect of gravitation on the inverse Faraday effect and its resonance counterpart, known as radiatively induced fermion resonance (RFR).

The interaction of fields in ECE theory is controlled by Cartan geometry, in the particular case of the interaction of gravitation and electromagnetism, there is a very small homogeneous charge current density in the Gauss law and in the Faraday law of induction. For all practical purpose in the laboratory this is not observable. However, it has been shown in ECE theory to result in changes of polarization and other optical properties of light grazing a white dwarf, which is an object many times heavier than the sun. Such changes of polarization are not described by the Einstein Hilbert equation.

7.8 Radiative Corrections in ECE Theory

During the course of development of ECE theory the anomalous g factor of the electron and Lamb shifts in hydrogen and helium have been explained satisfactorily in a far simpler manner than the standard model and using the causal and objective principles of Einsteinian relativity. The usual approach to the radiative corrections in quantum electrodynamics (QED) has been criticized [1–12], especially its claim to accuracy. The QED method of the standard model relies on assumptions that are not present in Einsteinian relativity, and also on adjustable parameters. The Feynman method consists of assuming the existence of virtual particles and on a perturbation method of quantum mechanics which sums thousands of terms of increasing complexity. There is no proof that this sum converges. It is also claimed in standard model QED that the accuracy of the fine structure constant is reproduced theoretically to high precision. However the fine structure constant in S.I. units is:

$$\alpha = \frac{e^2}{4\pi\epsilon\hbar c} \qquad (7.205)$$

and its accuracy is limited by the experimental accuracy of the Planck constant. There is no way that a theory can produce a higher accuracy than experiment, and the theoretical value of the g factor of the electron is based on the value of the fine structure constant. Thus g cannot be known with greater accuracy than that of the fine structure constant. These surprising

7.8 Radiative Corrections in ECE Theory

inconsistencies in the standard model data were discussed in detail [1–12] and a brief summary is given here.

The fundamental constants of physics are agreed upon by treaty and are given on sites such as that of the National Institute for Standards and Technology (www.nist.gov). This site gives:

$$g(\text{exptl.}) = 2.0023193043718 \pm 0.0000000000075 \tag{7.206}$$

$$\hbar(\text{exptl}) = (6.6260693 \pm 0.0000011) \times 10^{-34} Js \tag{7.207}$$

$$e(\text{exptl.}) = (1.60217653 \pm 0.00000014) \times 10^{-19} C \tag{7.208}$$

$$c(\text{exact}) = 2.99792458 \times 10^8 ms^{-1} \tag{7.209}$$

$$\epsilon_0(\text{exact}) = 8.854187817 \times 10^{-12} J^{-1} C^2 m^{-1} \tag{7.210}$$

$$\mu_0(\text{exact}) = 4\pi \times 10^{-7} Js^2 C^{-2} m^{-1} \tag{7.211}$$

with relative standard uncertainties. With a sufficiently precise value of:

$$\pi = 3.141592653590 \tag{7.212}$$

gives, from these data:

$$\alpha = 0.007297(34) \tag{7.213}$$

where the result has been rounded off to the relative standard uncertainty of the Planck constant $\hbar$. This is an experimentally determined uncertainty. The theoretical value of g from ECE theory was found by using Eq. (7.213) in

$$g = 2\left(1 + \frac{\alpha}{4\pi}\right)^2 \tag{7.214}$$

and gives:

$$g(\text{ECE}) = 2.002323(49). \tag{7.215}$$

The experimental value of g is known to a much higher precision than the experimental value of $\hbar$, and is:

$$g(\text{exptl.}) = 2.0023193043718 \pm 0.0000000000075. \tag{7.216}$$

It is seen that:

$$g(\text{ECE}) - g(\text{exptl.}) = 0.000004 \tag{7.217}$$

which is about the same order of magnitude as the experimental uncertainty of h. Therefore it was shown that ECE theory gives g as precisely as the experimental uncertainty in h will allow. The standard model literature was found to be severely self-inconsistent. For example a much used text by Atkins [29] gives h as:

$$h\,(\text{Atkins}) = 6.62818 \times 10^{-34} Js \qquad (7.218)$$

without uncertainty estimates. This is different in the fourth decimal place from the NIST value given above, a discrepancy of four orders of magnitude. Despite this, Atkins gives:

$$\alpha(\text{Atkins}) = 0.00729351 \qquad (7.219)$$

which claims to be different from Eq. (7.213) only in the sixth decimal place. Atkins gives the g factor of the electron as:

$$g(\text{Atkins}) = 2.002319314 \qquad (7.220)$$

which is different from the NIST value in the eighth decimal place, while it is claimed at NIST that $g(\exp)$ from Eq. (7.216) is accurate to the twelfth decimal place. So there is another discrepancy of four orders of magnitude. Ryder on the other hand [18] gives:

$$g(\text{Ryder}) = 2.0023193048 \qquad (7.221)$$

which is different from the NIST value in the tenth decimal place, a discrepancy of two orders of magnitude. One could try to explain these discrepancies by increasing accuracy of experimental method over the years, but there is no way in which QED can reproduce g to the tenth decimal place as claimed by Ryder. This is easily seen from the fact that g is calculated theoretically in QED from the fine structure constant, whose accuracy is limited by h as we have argued. There is also no way in which QED can be a fundamental theory as is often claimed in the standard model literature. This is again easily seen from the fact that QED has several assumptions extraneous to the theory of relativity [1–12]. Examples are virtual particles, acausality (the electron can do what it likes, g backwards in time and so on), dimensional regularization, re-normalization and the hugely elaborate perturbation method known as the Feynman calculus. It is not known whether the series expansion used in the Feynman calculus converges. Its thousands of terms are just worked out by computer in the hope that it converges. In summary:

$$g(\text{Schwinger}) = 2 + \alpha/\pi = 2.002322(8) \qquad (7.222)$$

7.8 Radiative Corrections in ECE Theory

$$g(\text{ECE}) = 2 + \alpha/\pi + \frac{\alpha^2}{8\pi^2} = 2.002323(49) \qquad (7.223)$$

$$g(\text{exptl.}) = 2.0023193043718 \pm 0.0000000000075 \qquad (7.224)$$

$$g(\text{Atkins}) = 2.002319314 \pm (?) \qquad (7.225)$$

$$g(\text{Ryder}) = 2.0023193048 \pm (?) \qquad (7.226)$$

and there is little doubt that other textbooks and sources give further different values of g to add to the confusion in the standard model literature. So where does this analysis leave the claims of QED? The Wolfram site claims that QED gives g using the series

$$g = 2\left(1 + \frac{\alpha}{2\pi} - 0.328\left(\frac{\alpha}{2\pi}\right)^2 + 1.181\left(\frac{\alpha}{\pi}\right)^3 - 1.510\left(\frac{\alpha}{\pi}\right)^4 + \ldots + 4.393 \times 10^{-12}\right) \qquad (7.227)$$

which is derived from thousands of Feynman diagrams (sic). However, the numbers in Eq. (7.227) all come from the various assumptions of QED, none of which are present in Einsteinian relativity. The latter is causal and objective by construction. An even worse internal inconsistency emerges within the NIST site itself, because the fine structure constant is claimed to be:

$$\alpha(\text{NIST}) = (7.297352560 \pm 0.000000024) \times 10^{-3} \qquad (7.228)$$

both experimentally and theoretically. This cannot be true because Eq. (7.228) is different in the eighth decimal place from Eq. (7.213), which is calculated with NIST's OWN data, Eqs. (7.206) to (7.211). So the NIST site is internally inconsistent to several orders of magnitude, because it is at the same time claimed that Eq. (7.228) is accurate to the tenth decimal place. From Eq. (7.207) however it is seen that h at NIST is accurate only to the sixth decimal place, which limits the accuracy of α to this, i.e. four orders of magnitude less precise than claimed.

The theoretical claim for the fine structure constant at NIST comes from QED, which his described as a theory in which an electron emits a virtual photon, which in turn emits virtual electron positron pairs. The virtual positron is attracted and the virtual electron is repelled from the real electron. This process results in a screened charge, a mathematical concept with a limiting value defined as the limit of zero momentum transfer or infinite distance. At high energies the fine structure constant drops to 1/128, and so is not a constant at all. It cannot therefore be claimed to be precise to the relative standard uncertainty of Eq. (7.228), taken directly from the NIST website itself. There is therefore no direct way of proving experimentally the existence of virtual electron positron pairs, or of virtual photons. The experimental claim for the fine structure constant at NIST comes from the quantum Hall effect combined with a calculable cross capacitor to measure standard resistance. The von Klitzing constant:

$$R_\kappa = \frac{\hbar}{e^2} = \frac{\mu_0 c}{2} \text{(sic)} \tag{7.229}$$

is used in this experimental determination. However, this method is again limited by the experimental accuracy of h. The accuracy of e is only ten times better than h from NIST's own data, and R_κ cannot be more accurate than h. If α were really as accurate as claimed in Eq. (7.228), both h and e would have to be this accurate experimentally, and this is obviously not true.

In view of these severe inconsistencies in the standard model and in view of the many ad hoc and indeed unprovable assumptions of QED, it is considered that the so called "precision tests" of QED are of no utility and no meaning. These include the g factor of the electron, the Lamb shift, the Casimir effect, positronium, and so forth.

The ECE theory of these radiative corrections therefore set out to reproduce what is really known experimentally in the simplest way. These methods are of course those of William of Ockham and Francis Bacon. In the non-relativistic quantum approximation to ECE theory the Schrödinger equation was modified as follows [1–12]:

$$-\frac{\hbar^2}{2m} \nabla^2 \left(\frac{\alpha}{2\pi} + \frac{\alpha^2}{16\pi^2} \right) \psi = \frac{e^2}{4\pi\epsilon_0} \left(\frac{1}{r} - \frac{1}{r + r(\text{vac})} \right) \psi \tag{7.230}$$

in which the effect of the vacuum potential is considered to be a shift in the electron to proton distance for each orbital of an atom or molecule, in the simplest case atomic hydrogen (H). Computer algebra was used to show that:

$$\frac{r(\text{vac})(2s)}{r + (r + r(\text{vac}))} - \frac{r(\text{vac})(2\rho_z, \cos\theta = 1)}{r(r + r(\text{vac}))} = \frac{1}{4\pi} \frac{\hbar}{mc} \frac{1}{r^2} \tag{7.231}$$

so that the simple ECE method of Eq. (7.230) gives the correct qualitative result observed first by Lamb in atomic H. This is known as the Lamb shift. Computer algebra was used to show that the ECE Lamb shift is:

$$\Delta E = \left(\frac{1}{16\pi^{3/2}} \frac{\alpha}{a} \frac{\hbar}{mc} \right) \frac{1}{r} = 0.0353 \ cm^{-1} \tag{7.232}$$

in the approximation in which the angular dependence if the Lamb shift is not considered.

7.8 Radiative Corrections in ECE Theory

The potential energy of the unperturbed H atom in wave-numbers is:

$$V_0 = -\frac{\alpha}{r} \tag{7.233}$$

and the vacuum perturbs this as follows:

$$V = -\frac{\alpha}{r + r(\text{vac})}. \tag{7.234}$$

So the change in potential energy due to the vacuum (i.e. the radiative correction) is positive valued as follows:

$$\Delta V = |V - V_0| = \alpha \left(\frac{1}{r} - \frac{1}{r + r(\text{vac})} \right). \tag{7.235}$$

This equation was obtained by assuming that the Schrödinger equation of H in the presence of the radiative correction due to the vacuum is, to first order in α:

$$-\frac{\hbar^2}{2m} \left(1 + \frac{\alpha}{2\pi} \right) \nabla^2 \psi - \frac{e^2}{4\pi\epsilon_0 r} \psi = E\psi \tag{7.236}$$

and that this is equivalent to:

$$-\frac{\hbar^2}{2m} \nabla^2 \psi - \frac{e^2}{4\pi\epsilon_0 (r + r(\text{vac}))} \psi = E\psi. \tag{7.237}$$

It was assumed that $r(\text{vac})$ is small enough to justify using the analytically known unperturbed wave-functions of H (ψ_0) to a good approximation. So:

$$\psi \sim \psi_0 \tag{7.238}$$

and:

$$\nabla^2 \psi_0 = -\frac{4\pi mc}{\hbar} \left(\frac{1}{r} - \frac{1}{r + r(\text{vac})} \right) \psi_0. \tag{7.239}$$

Using computer algebra this approximation gives [1–12]:

$$\frac{1}{r + r_{2p}(\text{vac})} - \frac{1}{r + r_{2s}(\text{vac})} = \frac{1}{2\pi^{3/2}} \frac{\hbar}{mc} \frac{1}{r^2}. \tag{7.240}$$

The change in potential energy due to the radiative correction of the vacuum is thus:

$$\Delta V = \frac{\alpha}{2\pi^{3/2}} \frac{\hbar}{mc} \frac{1}{r^2} \qquad (7.241)$$

and the change in total energy is:

$$\Delta E = \frac{r}{2n^2 a} \Delta V = \left(\frac{1}{16\pi^{3/2}} \frac{\alpha}{a} \frac{\hbar}{mc} \right) \frac{1}{r} = 0.0353 \; cm^{-1} \qquad (7.242)$$

which is the Lamb shift of atomic H. Here:

$$r = 1.69 \times 10^{-7} m \qquad (7.243)$$

From Eq. (240):

$$\frac{r_{2s}(\text{vac}) - r_{2p}(\text{vac})}{(r + r_{2p}(\text{vac}))(r + r_{2s}(\text{vac}))} = \frac{1}{2\pi^{3/2}} \frac{\hbar}{mc} \frac{1}{r^2}. \qquad (7.244)$$

Eq. (238) implies:

$$r \gg r_{2s}(\text{vac}) \sim r_{2p}(\text{vac}) \qquad (7.245)$$

so in this approximation Eq. (7.244) becomes:

$$r_{2s}(\text{vac}) - r_{2p}(\text{vac}) = \frac{1}{2\pi^{3/2}} \frac{\hbar}{mc} \qquad (7.246)$$

i.e.

$$r_{2s}(\text{vac}) - r_{2p}(\text{vac}) = \frac{1}{4\pi^{5/2}} \frac{\hbar}{mc} \qquad (7.247)$$

where the standard Compton wavelength is:

$$\frac{h}{mc} = 2.426 \times 10^{-12} m. \qquad (7.248)$$

Thus we arrive at:

$$r_{2s}(\text{vac}) - r_{2p}(\text{vac}) = 3.48 \times 10^{-13} m. \qquad (7.249)$$

This is a plausible result because the classical electron radius is:

$$r(\text{classical}) = \frac{1}{4\pi\epsilon_0} \frac{e^2}{mc^2} = 2.818 \times 10^{-15} m \qquad (7.250)$$

and the Bohr radius is:

$$a = 5.292 \times 10^{-11} m. \qquad (7.251)$$

So the radiative correction perturbs the electron orbitals by about ten times the classical radius of the electron and by orders less than the Bohr radius. The ECE theory also shows why the Lamb shift is constant as observed because for a given orientation:

$$\cos\theta = 1 \qquad (7.252)$$

the shift is determined completely by $1/r$ within a constant of proportionality defined by:

$$\zeta = \frac{1}{32\pi^{3/2}} \frac{\alpha}{a} \frac{\hbar}{mc}. \qquad (7.253)$$

The angular dependence of the Lamb shift in H was also considered [1–12] and the method extended to the helium atom. Finally, consideration was given to how radiative corrections may be amplified by spin connection resonance.

Therefore in summary, the accuracy of the fine structure constant is determined experimentally by that of the Planck constant h. The LEAST accurately known constant determines the accuracy of the fine structure constant, as should be well known. There is no way that any theory can determine the fine structure constant more accurately than it is known experimentally. Therefore ECE theory sets out to use the experimental accuracy in α. The latter is determined by the accuracy in h as argued. This was done as simply as possible in accordance with Ockham's Razor. QED on the other hand is hugely elaborate, and its claims to be an accurate fundamental theory are unjustifiable. There can be no experimental justification for the existence of virtual particle pairs because of the gross internal inconsistencies in data reviewed in this section. Additionally, there are several ad hoc assumptions in the theory of QED itself.

7.9 Summary of Advances Made by ECE Theory, and Criticisms of the Standard Model

In this section a summary is given of the main advances of ECE theory over the past five years since inception in Spring 2003, and also a summary of implied criticisms of the current model of physics known as the standard model.

The major advantage of ECE theory is that it relies on the original principles of the theory and philosophy of relativity, without any extraneous input. This approach adheres therefore to the Ockham Razor of philosophy, the simpler the better. It also adheres to the principles of Francis Bacon, that every theory is tested experimentally, and not against another theory.

1. The inverse Faraday effect. This is described by the spinning of space-time and the B(3) field (see www.aias.us Omnia Opera) from first principles. In the standard model the effect cannot be described self consistently and cannot be described without an ad hoc conjugate product of non-linear optics. The latter is introduced phenomenologically in the standard model of non-linear optics, a theory of special relativity. In ECE theory the B(3) spin field indicates that optics and spectroscopy are parts of a generally covariant unified field theory (GCUFT).

2. The Aharonov Bohm effects. These are described self consistently in ECE through the spin connection using the principles of general relativity. As shown in this review paper, the standard model description of the Aharonov Bohm (AB) effects is at best controversial and at worst erroneous. A satisfactory description of the AB effects in ECE leads to a new understanding of quantum entanglement and one photon interferometry.

3. The polarization change in light deflected by gravitation. This is not described in the Einstein Hilbert (EH) equation of the standard model because it is a purely kinematic equation relying on the gravitational attraction between a photon and a mass M, for example the solar mass. In ECE all the optical effects of gravitation are developed self consistently from the Bianchi identity of Cartan geometry.

4. The Faraday disk generator. This is described in ECE through the Cartan torsion of space-time introduced by mechanical spin, this concept is missing entirely from the standard model, which still cannot describe the 1831 Faraday disk generator.

5. The Sagnac effect and ring laser gyro. These are described again by the Cartan torsion of space-time introduced by spinning the platform of the Sagnac interferometer. The Sagnac effect is very difficult to understand using Maxwell Heaviside theory, but is easily described in ECE theory. The latter offers a far simpler description than other available attempts at explaining the Sagnac effect of 1913.

6. The velocity curve of a spiral galaxy. This is described straightforwardly and simply in ECE theory by introducing again the concept of constant space-time torsion. The spiral galaxies main features cannot be

7.9 Summary of Advances Made by ECE Theory, and Criticisms of the Standard Model 171

described at all in the standard model. This is because the latter relies on an ad hoc "dark matter" that originates in the EH equation. The latter is self inconsistent as argued in this review paper.

7. The topological phases such as the Berry phase. These are derived in ECE from first principles, and are rigorously inter-related. In the standard model their description is incomplete, and in the case of the electromagnetic phase, erroneous.

8. The electromagnetic phase. This is described self consistently in terms of the B(3) spin field of ECE theory using general relativity. In the standard model the phase is incompletely determined mathematically, and violates parity in simple effects such as reflection.

9. Snell's law, reflection, refraction, diffraction, interferometry and related optical effects. These can be described correctly only in a GCUFT such as ECE. In the standard model the theory of reflection for example, does not fit with parity inversion symmetry due to the neglect of the B(3) spin field.

10. Improvements to the Heisenberg Uncertainty Principle. Various experiments have shown that the principle is incorrect by orders of magnitude, in ECE theory it is developed with causal and objective general relativity and the concept of quantum of action density.

11. The unification of wave mechanics and general relativity. This has been achieved straightforwardly in ECE theory through the use of Cartan geometry. In the standard model it is still not possible to make this basic unification. The Dirac, Proca and other wave equations are limits of the ECE wave equation, which is derived easily from the tetrad postulate of Cartan. So ECE allows the description of the effect of gravitation on such equations, and on such phenomena as the Sagnac effect. This is again not possible in the standard model.

12. Description of particle interaction. This description is achieved with simultaneous ECE equations without assuming the existence either of virtual particles or of the Higgs mechanism. The Higgs boson still has not been verified experimentally, and its energy is not defined theoretically.

13. The photon mass. The Proca equation is derived easily from Cartan geometry using the simple hypothesis that the potential is proportional to the Cartan tetrad. In the standard model the Proca equation is directly incompatible with gauge invariance, a fundamental self-inconsistency of the standard model, one of many self - inconsistencies.

14. Replacement of the gauge principle. The gauge principle is not tenable in a GCUFT such as ECE because the potential in ECE is physically meaningful as in Faraday's original electrotonic state. Abandonment of the gauge principle allows a return to the earlier concepts of relativity without introducing an ad hoc and abstract internal space as in Yang Mills theory. In ECE theory the tetrad postulate is invariant under the general coordinate transform, and this is the principle that governs the potential field in ECE.

15. Description of the electro-weak field without the Higgs mechanism. This becomes possible in a relatively straightforward manner by using simultaneous ECE equations. The Higgs mechanism is ad hoc, and to date unproven experimentally, indeed it is unprovable because an energy cannot be assigned to the Higgs boson. The Higgs boson, having no well defined energy, cannot be proven experimentally by particle collision methods, however powerful the accelerator. No sign of a Higgs boson was found at LEP, and to date no sign at the CERN heavy hadron collider.

16. Description of neutrino oscillations. This is a relatively simple exercise in ECE theory but in the standard model neutrino oscillations remained deeply controversial for years because of adherence to the assumption that the neutrino had no mass. In ECE all particles have mass - a fundamental requirement of relativity.

17. The generally covariant description of the laws of classical electrodynamics. These laws become laws of general relativity and a unified field theory, they are no longer laws of a Minkowski space-time as in the standard model. The concept of spin connection and spin connection resonance make important advances and potentially open up new sources of energy.

18. Derivation of the quark model from general relativity. This has been achieved in ECE theory by using an SU(n) representation space in the wave and field equations. In the standard model the quark theory is one of special relativity. QCD relies on ad hoc concepts such as renormalization, which as argued in section 7.8, are not internally consistent with data. The situation in QCD is worse than that in QED.

19. Derivation of the quantum theory of electrodynamics. This is achieved using the wave equation and the ECE hypothesis, resulting in a generally covariant version of the Proca equation with non-zero photon mass. In so doing a minimum particle volume is always present, so there are no point particles and no need for re-normalization. Feynman's QED is abandoned as described in Section 7.8.

7.9 Summary of Advances Made by ECE Theory, and Criticisms of the Standard Model

20. The origin of particle spin. This is traced to geometry and particle spins of all kinds are successfully incorporated into general relativity. This is not possible with the EH equation, which has been shown to be fundamentally flawed.

21. Development of cosmology. The major advantage of considering the Cartan torsion becomes abundantly clear in cosmology, in particular the explanation of the spiral galaxies. Cosmology based on the EH equation has been shown to be meaningless in several different ways.

22. No Singularities. This is a flawed concept introduced by incorrect solutions of the EH equation. The latter is itself inconsistent with the Bianchi identity. In ECE theory the concept of Big Bang is replaced with the steady state universe with local oscillations. Similarly there are no black holes and no dark matter. Applications of experimentally untestable string theory to these concepts multiplies the heavily criticized obscurantism of modern physics.

23. Explanation of the red shift. This is a simple optical effect in ECE theory, there can be different red shifts in equidistant objects. ECE also offers a new explanation of the background radiation if indeed it is not an artifact of the Earth's atmosphere as some scholars now think.

24. Spin connection resonance. This concept is made possible in ECE and has been offered as an explanation of Tesla's well known giant resonances and similar reports of over a century of work. The latter cannot be explained in the standard model yet is potentially a source of major new energy.

25. Spinning Space-time. This is a key new concept of electrodynamics, akin to curving space-time in gravitation. ECE has made the major discovery that the two concepts are linked ineluctably in relativity, and this has led to the abandonment of the EH equation. A suggested replacement of the equation has been made in recent work.

26. Counter gravitation. It has been shown that this is feasible only by using resonance methods based again on the spin connection and the interaction of gravitation and electromagnetism. It needs a GCUFT such as ECE to begin to describe this interaction of the fundamental fields of force.

27. Gravitational Dynamics. These are developed in ECE in the same way as electrodynamics. For example it is relatively easy to show that there is a gravitational equivalent of the Faraday law of induction, as indeed observed recently. A new approach to the derivation of the acceleration due to gravity has also been made possible, an approach based on the rigorous Bianchi identity given by Cartan.

28. Quantum Entanglement. These well known quantum effects can be understood using the spin connection of ECE in a similar way to the AB effects. Similarly the argument can be extended to such phenomena as one photon Young interferometry. In the standard model they are very difficult to understand because of the use of a Minkowski space-time with no connection. In the standard model these are mysterious effects with many offered explanations, none convincing.

29. Superconductivity and related fields. The equations governing the behavior of classes of materials are all derived in ECE from geometry, so there is an overall self-consistency which is often missing in the standard model. For example plasma, semiconductors, superconductors, and so forth.

30. Quantum Field Theory. This is developed in ECE entirely without he use of string theory or super-symmetry. String theory in particular has been heavily criticized because it cannot be tested experimentally and makes no new predictions at all. Such matters as photon mass theory, canonical quantization, and creation annihilation operator theory are all improved by ECE theory.

31. Radiative Corrections. These are understood in a far simpler way in ECE theory as discussed in Section 7.8. The claims of QED theory have been shown to be false by several orders of magnitude, and the complacency of the standard physics community heavily criticized thereby.

32. Fermion Resonance. New methods of detecting and developing fermion resonance have been developed and it is shown that such resonance can be induced without the use of magnets. This method is known as radiatively induced fermion resonance (RFR). It has been clearly understood to be due to the B(3) field.

33. Ubiquitous B(3) Field. It has been shown that the B(3) field is the one responsible for the general relativistic description of the electromagnetic phase, so it occurs throughout optics and spectroscopy, in everyday phenomena such as reflection.

34. Fundamental Advances in Geometry. In the course of developing ECE theory it has been shown that there is only one Bianchi identity, not two unrelated identities used in the standard model. It has also been shown rigorously in many ways that the Bianchi identity has a Hodge dual. These properties lead to field equations with duality symmetry. Such a symmetry is not present in the standard model.

35. Self Consistency of Cartan's geometry. This has been tested in many ways, and it has been shown that the tetrad postulate is rigorously self consistent and fundamental to physics. Numerous tests of self consistency have been made.

36. Development of Gravitational Relativity. It has been shown that the correct description of gravitation requires the Bianchi identity of Cartan, which links torsion to curvature. The Bianchi identity used by Einstein has been shown to be incomplete, and using computer algebra, it has been shown that the EH equation is inconsistent with the use of a Christoffel connection and symmetric metric. It has also been shown that claimed solutions of the EH equation are often incorrect mathematically. Finally it has been shown that the Ricci flat space-time is incompatible with the Einsteinian equivalence principle. Therefore the standard model literature has to be read with considerable caution. Many claims of the standard model have not stood up to scrutiny, whereas ECE has developed strongly.

Acknowledgments

The British Government is thanked for the award of a Civil List Pension, and the staff of AIAS for five years of voluntary work on ECE theory and websites. The staff of the Telesio-Galilei Association are thanked for appointments and logistical support.

A

Appendix 1: Homogeneous Maxwell Heaviside Equations

In the first of several technical appendices it is shown how to translate the homogeneous Maxwell Heaviside (MH) from tensor to vector notation, giving details that are rarely found in textbooks. In tensor notation the equation is:

$$\partial_\mu \widetilde{F}^{\mu\nu} = 0 \tag{A.1}$$

and involves the Hodge dual of the 4 x 4 field tensor, defined as follows:

$$\widetilde{F}_{\mu\nu} = \frac{1}{2} \epsilon_{\mu\nu\rho\sigma} F^{\rho\sigma}. \tag{A.2}$$

Indices are raised using the Minkowski metric:

$$\widetilde{F}^{\mu\nu} = g^{\mu\kappa} g^{\nu\rho} \widetilde{F}_{\kappa\rho} \tag{A.3}$$

where:

$$g_{\mu\nu} = g^{\mu\nu} = \begin{bmatrix} 1 & 0 & 0 & 0 \\ 0 & -1 & 0 & 0 \\ 0 & 0 & -1 & 0 \\ 0 & 0 & 0 & -1 \end{bmatrix}. \tag{A.4}$$

Therefore the Hodge dual is:

$$\widetilde{F}^{\mu\nu} = \begin{bmatrix} 0 & cB^1 & cB^2 & cB^3 \\ -cB^1 & 0 & -E^3 & E^2 \\ -cB^2 & E^3 & 0 & -E^1 \\ -cB^3 & -E^2 & E^1 & 0 \end{bmatrix} \tag{A.5}$$

For example:

$$\tilde{F}_{01} = \frac{1}{2}(\epsilon_{0123}F^{23} + \epsilon_{0132}F^{32}) = F^{23} \tag{A.6}$$

and

$$\tilde{F}^{01} = g^{00}g^{11}\tilde{F}_{01} = -\tilde{F}_{10}. \tag{A.7}$$

The homogeneous laws of classical electrodynamics are the Gauss law and Faraday law of induction. They are obtained as follows by choice of indices. The Gauss law is obtained by choosing:

$$\nu = 0 \tag{A.8}$$

and so

$$\partial_1 \tilde{F}^{10} + \partial_2 \tilde{F}^{20} + \partial_3 \tilde{F}^{30} = 0. \tag{A.9}$$

In vector notation this is

$$\nabla \cdot \boldsymbol{B} = 0. \tag{A.10}$$

The Faraday law of induction is obtained by choosing:

$$\nu = 1, 2, 3 \tag{A.11}$$

and is three component equations:

$$\partial_0 \tilde{F}^{01} + \partial_2 \tilde{F}^{21} + \partial_3 \tilde{F}^{31} = 0 \tag{A.12}$$

$$\partial_0 \tilde{F}^{02} + \partial_1 \tilde{F}^{12} + \partial_3 \tilde{F}^{32} = 0 \tag{A.13}$$

$$\partial_0 \tilde{F}^{03} + \partial_1 \tilde{F}^{13} + \partial_2 \tilde{F}^{23} = 0. \tag{A.14}$$

These can be condensed into one vector equation, which is

$$\nabla \times \boldsymbol{E} + \frac{\partial \boldsymbol{B}}{\partial t} = \boldsymbol{0}. \tag{A.15}$$

The differential form, tensor and vector notations are summarized as follows:

$$d \wedge F = 0 \to \partial_\mu \widetilde{F}^{\mu\nu} = 0 \to \nabla \cdot B = 0 \quad \text{(A.16)}$$

$$\nabla \times E + \frac{\partial B}{\partial t} = 0$$

The homogeneous laws of classical electrodynamics are most elegantly represented by the differential form notation, but most usefully represented by the vector notation.

B

Appendix 2: The Inhomogeneous Equations

The inhomogeneous laws are the Coulomb law and the Ampère Maxwell law. In tensor notation they are condensed into one equation:

$$\partial_\mu F^{\mu\nu} = \frac{1}{\epsilon_0} J^\nu \tag{B.1}$$

where the charge current density is:

$$J^\nu = \left(\rho, \frac{\boldsymbol{J}}{c}\right) \tag{B.2}$$

and where the partial derivative is:

$$\partial_\mu = \left(\frac{1}{c}\frac{\partial}{\partial t}, \frac{\partial}{\partial X}, \frac{\partial}{\partial Y}, \frac{\partial}{\partial Z}\right) \tag{B.3}$$

The field tensor is:

$$F^{\mu\nu} = \begin{bmatrix} 0 & -E^1 & -E^2 & -E^3 \\ E^1 & 0 & -cB^3 & cB^2 \\ E^2 & cB^3 & 0 & -cB^1 \\ E^3 & -cB^2 & cB^1 & 0 \end{bmatrix} = \begin{bmatrix} 0 & F^{01} & F^{02} & F^{03} \\ F^{10} & 0 & F^{12} & F^{13} \\ F^{20} & F^{21} & 0 & F^{23} \\ F^{30} & F^{31} & F^{32} & 0 \end{bmatrix} \tag{B.4}$$

and in S.I. units:

$$\epsilon_0 \mu_0 = \frac{1}{c^2}. \tag{B.5}$$

In this notation:

$$\left.\begin{array}{l}E_X = E^1 = F^{10}, \\ E_Y = E^2 = F^{20}, \\ E_Z = E^3 = F^{30},\end{array}\right\} \quad (B.6)$$

and so on. The Coulomb law is obtained from choosing:

$$\nu = 0 \quad (B.7)$$

so that:

$$\partial_1 F^{10} + \partial_2 F^{20} + \partial_3 F^{30} = \frac{1}{\epsilon_0} J^0. \quad (B.8)$$

In vector component notation this is:

$$\frac{\partial E_X}{\partial X} + \frac{\partial E_Y}{\partial Y} + \frac{\partial E_Z}{\partial Z} = \frac{1}{\epsilon_0} \rho \quad (B.9)$$

which in vector notation is:

$$\boldsymbol{\nabla} \cdot \boldsymbol{E} = \frac{\rho}{\epsilon_0}. \quad (B.10)$$

The Ampère Maxwell law is obtained from choosing

$$\nu = 1, 2, 3 \quad (B.11)$$

which gives three equations:

$$\partial_0 F^{01} + \partial_2 F^{21} + \partial_3 F^{31} = \frac{1}{\epsilon_0} J^1 \quad (B.12)$$

$$\partial_0 F^{02} + \partial_1 F^{12} + \partial_3 F^{32} = \frac{1}{\epsilon_0} J^2 \quad (B.13)$$

$$\partial_0 F^{03} + \partial_1 F^{13} + \partial_2 F^{23} = \frac{1}{\epsilon_0} J^3. \quad (B.14)$$

In vector component notation these are:

$$-\frac{1}{c} \frac{\partial E_X}{\partial t} + c \left(\frac{\partial B_Z}{\partial Y} - \frac{\partial B_Y}{\partial Z} \right) = \frac{1}{\epsilon_0} J_X \quad (B.15)$$

$$-\frac{1}{c}\frac{\partial E_Y}{\partial t} + c\left(\frac{\partial B_X}{\partial Z} - \frac{\partial B_Z}{\partial X}\right) = \frac{1}{\epsilon_0}J_Y \qquad (B.16)$$

$$-\frac{1}{c}\frac{\partial E_Z}{\partial t} + c\left(\frac{\partial B_Y}{\partial X} - \frac{\partial B_X}{\partial Y}\right) = \frac{1}{\epsilon_0}J_Z. \qquad (B.17)$$

The definition of the vector curl is

$$\boldsymbol{\nabla} \times \boldsymbol{B} = \begin{vmatrix} \boldsymbol{i} & \boldsymbol{j} & \boldsymbol{k} \\ \partial/\partial Z & \partial/\partial Y & \partial/\partial Z \\ B_X & B_Y & B_Z \end{vmatrix} \qquad (B.18)$$

$$= \left(\frac{\partial B_Z}{\partial Y} - \frac{\partial B_Y}{\partial Z}\right)\boldsymbol{i} - \left(\frac{\partial B_Z}{\partial X} - \frac{\partial B_X}{\partial Z}\right)\boldsymbol{j} + \left(\frac{\partial B_Y}{\partial X} - \frac{\partial B_X}{\partial Y}\right)\boldsymbol{k},$$

so it is seen that the three equations (B.15) to (B.17) can be condensed into one vector equation:

$$\boldsymbol{\nabla} \times \boldsymbol{B} - \frac{1}{c^2}\frac{\partial \boldsymbol{E}}{\partial t} = \mu_0 \boldsymbol{J} \qquad (B.19)$$

which is the Ampère Maxwell Law. The differential form, tensor and vector formulations of the inhomogeneous laws of standard model classical electrodynamics are summarized as follows:

$$d \wedge \widetilde{F} = \frac{J}{\epsilon_0} \rightarrow \partial_\mu F^{\mu\nu} = \frac{J^\nu}{\epsilon_0} \rightarrow \boldsymbol{\nabla} \cdot \boldsymbol{E} = \frac{\rho}{\epsilon_0}, \qquad (B.20)$$

$$\boldsymbol{\nabla} \times \boldsymbol{B} - \frac{1}{c^2}\frac{\partial \boldsymbol{E}}{\partial t} = \mu_0 \boldsymbol{J}.$$

C

Appendix 3: Some Examples of Hodge Duals in Minkowski Space-Time

In Minkowski space-time the Hodge dual of a rank two anti-symmetric tensor (two-form) in four dimensions is defined by:

$$\widetilde{F}_{\mu\nu} = \frac{1}{2} \epsilon_{\mu\nu\rho\sigma} F^{\rho\sigma}. \tag{C.1}$$

For example, the B(3) field is defined by:

$$F^{\mu\nu} = \begin{bmatrix} 0 & 0 & 0 & 0 \\ 0 & 0 & -cB^{(3)} & 0 \\ 0 & cB^{(3)} & 0 & 0 \\ 0 & 0 & 0 & 0 \end{bmatrix} \tag{C.2}$$

so its Hodge dual is:

$$\widetilde{F}^{\mu\nu} = \begin{bmatrix} 0 & 0 & 0 & cB^{(3)} \\ 0 & 0 & 0 & 0 \\ 0 & 0 & 0 & 0 \\ -cB^{(3)} & 0 & 0 & 0 \end{bmatrix}. \tag{C.3}$$

It can be seen that the Hodge dual of the B(3) field does not imply the existence of an E(3) field, it is a re-arrangement of matrix elements. There appears to be no experimental evidence for the existence of a radiated E(3) field. In other words there is no electric equivalent of the inverse Faraday effect, and there is no electric equivalent of the Faraday effect.

The radiated B(3) field is generated by the spin connection, the static magnetic field of the standard model is defined without the spin connection as follows:

$$\boldsymbol{B} = \boldsymbol{\nabla} \times \boldsymbol{A}. \tag{C.4}$$

Appendix 3: Some Examples of Hodge Duals in Minkowski Space-Time

In tensor form the static magnetic field is:

$$F^{\mu\nu} = \begin{bmatrix} 0 & 0 & 0 & 0 \\ 0 & 0 & -cB_Z & cB_Y \\ 0 & cB_Z & 0 & -cB_X \\ 0 & -cB_Y & cB_X & 0 \end{bmatrix} \quad (C.5)$$

whose Hodge dual is:

$$\tilde{F}^{\mu\nu} = \begin{bmatrix} 0 & cB_X & cB_Y & cB_Z \\ -cB_X & 0 & 0 & 0 \\ -cB_Y & 0 & 0 & 0 \\ -cB_Z & 0 & 0 & 0 \end{bmatrix}. \quad (C.6)$$

Again, the Hodge dual does not generate an electric field. In ECE theory the magnetic field in vector notation always includes the spin connection vector as follows:

$$\boldsymbol{B} = \nabla \times \boldsymbol{A} - \boldsymbol{\omega} \times \boldsymbol{A} \quad (C.7)$$

and this is true for all types of magnetic field.

D

Appendix 4: Standard Tensorial Formulation of the Homogeneous Maxwell Heaviside Field Equations

The standard tensorial formulation developed in this appendix is:

$$\partial_\mu \widetilde{F}^{\mu\nu} = \partial^\mu \widetilde{F}_{\mu\nu} = 0 \qquad (D.1)$$

and is needed as a baseline for the development of ECE theory. The field tensor is defined as:

$$F^{\mu\nu} = \begin{bmatrix} 0 & cB^1 & cB^2 & cB^3 \\ -cB^1 & 0 & -E^3 & E^2 \\ -cB^2 & E^3 & 0 & -E^1 \\ -cB^3 & -E^2 & E^1 & 0 \end{bmatrix}. \qquad (D.2)$$

where, in standard covariant - contravariant notation and in S.I. units:

$$\partial_\mu = \left(\frac{1}{c}\frac{\partial}{\partial t}, \frac{\partial}{\partial X}, \frac{\partial}{\partial Y}, \frac{\partial}{\partial Z}\right), \qquad (D.3)$$

$$\partial^\mu = \left(\frac{1}{c}\frac{\partial}{\partial t}, -\frac{\partial}{\partial X}, -\frac{\partial}{\partial Y}, -\frac{\partial}{\partial Z}\right), \qquad (D.4)$$

$$x^\mu = (ct, X, Y, Z), \qquad (D.5)$$

$$x_\mu = (ct, -X, -Y, -Z). \qquad (D.6)$$

The metric and inverse metric tensors in Minkowski space-time are equal, and are given by:

$$g_{\mu\nu} = g^{\mu\nu} = \begin{bmatrix} 1 & 0 & 0 & 0 \\ 0 & -1 & 0 & 0 \\ 0 & 0 & -1 & 0 \\ 0 & 0 & 0 & -1 \end{bmatrix}. \qquad (D.7)$$

Appendix 4: Standard Tensorial Formulation of the Homogeneous Maxwell

Indices are raised and lowered with the metric, for example:

$$\widetilde{F}_{\mu\nu} = g_{\mu\rho}\, g_{\nu\sigma}\, \widetilde{F}^{\rho\sigma} \tag{D.8}$$

where

$$g_{00} = 1,\, g_{11} = g_{22} = g_{33} = -1 \tag{D.9}$$

and so on. Therefore:

$$\widetilde{F}_{01} = g_{00}\, g_{11} F^{01} = -\widetilde{F}^{01},\ \widetilde{F}_{02} = -\widetilde{F}^{02},\, \widetilde{F}_{03} = -\widetilde{F}^{03} \tag{D.10}$$

and so on. Therefore:

$$\widetilde{F}^{\mu\nu} = \begin{bmatrix} 0 & cB_X & cB_Y & cB_Z \\ -cB_x & 0 & -E_Z & E_Y \\ -cB_Y & E_Z & 0 & -E_X \\ -cB_Z & -E_Y & E_X & 0 \end{bmatrix},\ \widetilde{F}_{\mu\nu} = \begin{bmatrix} 0 & -cB_X & -cB_Y & -cB_Z \\ cB_X & 0 & -E_Z & E_Y \\ cB_Y & E_Z & 0 & -E_X \\ cB_Z & -E_Y & E_X & 0 \end{bmatrix}. \tag{D.11}$$

If the field tensor is defined with raised indices then the Gauss law is given by:

$$\partial_1 \widetilde{F}^{10} + \partial_2 \widetilde{F}^{20} + \partial_3 \widetilde{F}^{30} = 0 \tag{D.12}$$

i.e.:

$$-\nabla \cdot \boldsymbol{B} = 0 \tag{D.13}$$

and the Faraday law of induction is given by

$$\partial_0 \widetilde{F}^{01} + \partial_2 \widetilde{F}^{21} + \partial_3 \widetilde{F}^{31} = 0 \tag{D.14}$$

$$\partial_0 \widetilde{F}^{02} + \partial_1 \widetilde{F}^{12} + \partial_3 \widetilde{F}^{32} = 0 \tag{D.15}$$

$$\partial_0 \widetilde{F}^{03} + \partial_1 \widetilde{F}^{13} + \partial_2 \widetilde{F}^{23} = 0 \tag{D.16}$$

i.e.

$$\nabla \times \boldsymbol{E} + \frac{\partial \boldsymbol{B}}{\partial t} = \boldsymbol{0}. \tag{D.17}$$

In almost all textbooks the Gauss law is written as:

$$\nabla \cdot \boldsymbol{B} = 0, \tag{D.18}$$

but the above is the rigorously correct result.

Similarly if the field tensor is written with lowered indices, : i.e.:

$$\partial^\mu \widetilde{F}_{\mu\nu} = 0 \tag{D.19}$$

the rigorously correct result is:

$$-\nabla \cdot \mathbf{B} = 0 \tag{D.20}$$

$$-\left(\nabla \times \mathbf{E} + \frac{\partial \mathbf{B}}{\partial t}\right) = 0$$

The minus signs are always omitted in textbook material.

If the field tensor is defined with indices raised:

$$\partial_\mu F^{\mu\nu} = \frac{J^\nu}{\epsilon_0} \tag{D.21}$$

where:

$$F^{\mu\nu} = \frac{1}{2} \epsilon^{\mu\nu\rho\sigma} \widetilde{F}_{\rho\sigma}. \tag{D.22}$$

The totally anti-symmetric unit tensor in four-dimensions has elements:

$$\begin{aligned}
\epsilon^{0123} &= -\epsilon^{1230} = \epsilon^{2301} = -\epsilon^{3012} = 1 \\
\epsilon^{1023} &= -\epsilon^{2130} = \epsilon^{3201} = -\epsilon^{0312} = -1 \\
\epsilon^{1032} &= -\epsilon^{2103} = \epsilon^{3210} = -\epsilon^{0321} = 1 \\
\epsilon^{1302} &= -\epsilon^{2013} = \epsilon^{3120} = -\epsilon^{0231} = -1
\end{aligned} \tag{D.23}$$

So for example:

$$\begin{aligned}
F^{01} &= \frac{1}{2}\left(\epsilon^{0123}\widetilde{F}_{23} + \epsilon^{0132}\widetilde{F}_{32}\right) = \widetilde{F}_{23} = -E_X \\
F^{02} &= \frac{1}{2}\left(\epsilon^{0231}\widetilde{F}_{31} + \epsilon^{0213}\widetilde{F}_{13}\right) = \widetilde{F}_{31} = -E_Y \\
F^{03} &= \frac{1}{2}\left(\epsilon^{0312}\widetilde{F}_{12} + \epsilon^{0321}\widetilde{F}_{21}\right) = \widetilde{F}_{12} = -E_Z \\
F^{23} &= \frac{1}{2}\left(\epsilon^{2301}\widetilde{F}_{01} + \epsilon^{2310}\widetilde{F}_{10}\right) = \widetilde{F}_{01} = -cB_X \\
F^{13} &= \frac{1}{2}\left(\epsilon^{1302}\widetilde{F}_{02} + \epsilon^{1320}\widetilde{F}_{20}\right) = -\widetilde{F}_{02} = cB_Y \\
F^{12} &= \frac{1}{2}\left(\epsilon^{1230}\widetilde{F}_{30} + \epsilon^{1203}\widetilde{F}_{03}\right) = \widetilde{F}_{03} = -cB_Z
\end{aligned} \tag{D.24}$$

Appendix 4: Standard Tensorial Formulation of the Homogeneous Maxwell

Therefore:

$$F^{\mu\nu} = \begin{bmatrix} 0 & -E_X & E_Y & -E_Z \\ E_X & 0 & -cB_Z & cB_Y \\ E_Y & cB_Z & 0 & -cB_X \\ E_Z & -cB_Y & cB_X & 0 \end{bmatrix} = \begin{bmatrix} 0 & -E^1 & -E^2 & -E^3 \\ E^1 & 0 & -cB^3 & cB^2 \\ E^2 & cB^3 & 0 & -cB^1 \\ E^3 & -cB^2 & cB^1 & 0 \end{bmatrix}. \tag{D.25}$$

The charge current density is:

$$J^\nu = \left(\rho, \frac{J}{c}\right). \tag{D.26}$$

The Coulomb law is:

$$\partial_1 F^{10} + \partial_2 F^{20} + \partial_3 F^{30} = \frac{1}{\epsilon_0} J^0 = \frac{\rho}{\epsilon_0} \tag{D.27}$$

which in vector notation is:

$$\boldsymbol{\nabla} \cdot \boldsymbol{E} = \frac{\rho}{\epsilon_0}. \tag{D.28}$$

The Ampère Maxwell law is:

$$\partial_0 F^{01} + \partial_2 F^{21} + \partial_3 F^{31} = J^1/\epsilon_0 \tag{D.29}$$

$$\partial_0 F^{02} + \partial_1 F^{12} + \partial_3 F^{32} = J^2/\epsilon_0 \tag{D.30}$$

$$\partial_0 F^{03} + \partial_1 F^{13} + \partial_2 F^{23} = J^3/\epsilon_0 \tag{D.31}$$

i.e.:

$$-\frac{1}{c}\frac{\partial E_X}{\partial t} + c\left(\frac{\partial B_Z}{\partial Y} - \frac{\partial B_Y}{\partial Z}\right) = \frac{1}{\epsilon_0} J_X \tag{D.32}$$

$$-\frac{1}{c}\frac{\partial E_Y}{\partial t} + c\left(\frac{\partial B_X}{\partial Z} - \frac{\partial B_Z}{\partial X}\right) = \frac{1}{\epsilon_0} J_Y \tag{D.33}$$

$$-\frac{1}{c}\frac{\partial E_Z}{\partial t} + c\left(\frac{\partial B_Y}{\partial X} - \frac{\partial B_X}{\partial Y}\right) = \frac{1}{\epsilon_0} J_Z \tag{D.34}$$

which is:

$$\boldsymbol{\nabla} \times \boldsymbol{B} - \frac{1}{c^2}\frac{\partial \boldsymbol{E}}{\partial t} = \mu_0 \, \boldsymbol{J}. \tag{D.35}$$

Therefore the standard adopted is:

$$\partial_\mu F^{\mu\nu} = \frac{1}{\epsilon_0} J^\nu \to \nabla \cdot \mathbf{E} = \rho/\epsilon_0 \qquad (D.36)$$

$$\nabla \times \mathbf{B} - \frac{1}{c^2} \frac{\partial \mathbf{E}}{\partial t} = \mu_0 \, \mathbf{J}.$$

To be precisely correct therefore, the tensorial formulation of the four laws of electrodynamics is:

$$\partial_\mu F^{\mu\nu} = \frac{1}{\epsilon_0} J^\nu \qquad (D.37)$$

$$-\partial^\mu \tilde{F}_{\mu\nu} = 0 \qquad (D.38)$$

where:

$$F^{\mu\nu} = \begin{bmatrix} 0 & -E^1 & -E^2 & -E^3 \\ E^1 & 0 & -cB^3 & cB^2 \\ E^2 & cB^3 & 0 & -cB^1 \\ E^3 & -cB^2 & cB^1 & 0 \end{bmatrix} \qquad (D.39)$$

and

$$\tilde{F}^{\mu\nu} = \begin{bmatrix} 0 & -cB^1 & -cB^2 & -cB^3 \\ cB^1 & 0 & -E^3 & E^2 \\ cB^2 & E^3 & 0 & -E^1 \\ -cB^3 & -E^2 & E^1 & 0 \end{bmatrix}. \qquad (D.40)$$

In free space:

$$\partial_\mu F^{\mu\nu} = 0, \qquad (D.41)$$

$$-\partial^\mu \tilde{F}_{\mu\nu} = 0. \qquad (D.42)$$

The free space equations are duality invariant under:

$$F^{\mu\nu} \leftrightarrow \tilde{F}_{\mu\nu} \qquad (D.43)$$

i.e.:

$$E_X \leftrightarrow cB_X, E_Y \leftrightarrow cB_Y, E_Z \leftrightarrow cB_Z. \qquad (D.44)$$

Appendix 4: Standard Tensorial Formulation of the Homogeneous Maxwell   189

The Hodge dual transform is:

$$F^{\mu\nu} = \frac{1}{2}\epsilon^{\mu\nu\rho\sigma}\widetilde{F}_{\rho\sigma} \qquad (D.45)$$

and can be summarized as:

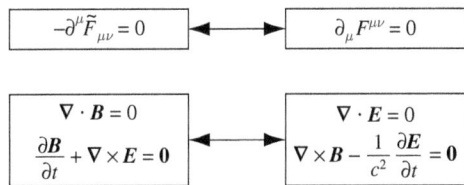

$$\boxed{-\partial^{\mu}\widetilde{F}_{\mu\nu} = 0} \quad \longleftrightarrow \quad \boxed{\partial_{\mu}F^{\mu\nu} = 0}$$

$$\boxed{\begin{array}{c}\nabla \cdot \mathbf{B} = 0 \\ \dfrac{\partial \mathbf{B}}{\partial t} + \nabla \times \mathbf{E} = \mathbf{0}\end{array}} \quad \longleftrightarrow \quad \boxed{\begin{array}{c}\nabla \cdot \mathbf{E} = 0 \\ \nabla \times \mathbf{B} - \dfrac{1}{c^2}\dfrac{\partial \mathbf{E}}{\partial t} = \mathbf{0}\end{array}}$$

The presence of matter and charge-current density breaks the duality symmetry, or duality invariance.

E

Appendix 5: Illustrating the Meaning of the Connection with Rotation in a Plane

Consider the clockwise rotation in a plane of a vector V^1 to V^2 as in Fig. (7.E1). This rotation is carried out by moving the vector and keeping the frame of reference fixed. This process is equivalent to keeping the vector fixed and rotating the frame of reference anti-clockwise through an equal angle θ. In Cartesian coordinates:

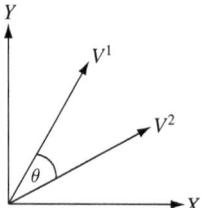

Fig. E.1. Rotation of a Vector in a Plane.

and

$$\boldsymbol{V}^1 = V_X^1 \boldsymbol{i} + V_Y^1 \boldsymbol{j} \tag{E.1}$$

$$\boldsymbol{V}^2 = V_X^2 \boldsymbol{i} + V_Y^2 \boldsymbol{j} \tag{E.2}$$

where:

$$|\boldsymbol{V}^1| = |\boldsymbol{V}^2|, \tag{E.3}$$

$$|\boldsymbol{V}^1| = (V_X^{1\ 2} + V_Y^{1\ 2})^{\frac{1}{2}}, \tag{E.4}$$

$$|\boldsymbol{V}^2| = (V_X^{2\ 2} + V_Y^{2\ 2})^{\frac{1}{2}}. \tag{E.5}$$

Appendix 5: Illustrating the Meaning of the Connection with Rotation in a Plane

This is a rotation in which the frame is fixed, i.e. the Cartesian unit vectors i and j do not change. The rotation could equally well be represented by:

$$\boldsymbol{V}^1 = V_X \boldsymbol{i}_1 + V_Y \boldsymbol{j}_1, \tag{E.6}$$

$$\boldsymbol{V}^2 = V_X \boldsymbol{i}_2 + V_Y \boldsymbol{j}_2, \tag{E.7}$$

and in this case the vector is fixed and the frame rotated anti-clockwise. We now have:

$$|\boldsymbol{V}^1| = |\boldsymbol{V}^2| = (V_X^2 + V_Y^2)^{\frac{1}{2}} \tag{E.8}$$

because:

$$\left.\begin{array}{c} \boldsymbol{i}_1 \cdot \boldsymbol{i}_1 = \boldsymbol{i}_2 \cdot \boldsymbol{i}_2 = 1 \\ \boldsymbol{j}_1 \cdot \boldsymbol{j}_1 = \boldsymbol{j}_2 \cdot \boldsymbol{j}_2 = 1 \end{array}\right\}. \tag{E.9}$$

The invariance under rotation of the complete vector field is true in both cases:

$$\begin{array}{l} \text{a)} \quad V^{12} = V_X^{12} + V_Y^{12} = V_X^{22} + V_Y^{22} = V^{22}, \\ \text{b)} \quad V^{12} = V_X^2 + V_Y^2 = V^{22}. \end{array} \tag{E.10}$$

The rotation can also be represented by:

$$\begin{bmatrix} V_X^1 \\ V_Y^1 \\ V_Z^1 \end{bmatrix} = \begin{bmatrix} \cos\theta & \sin\theta & 0 \\ -\sin\theta & \cos\theta & 0 \\ 0 & 0 & 1 \end{bmatrix} \begin{bmatrix} V_X^2 \\ V_Y^2 \\ V_Z^2 \end{bmatrix} \tag{E.11}$$

i.e.:

$$V_X^1 = V_X^2 \cos\theta + V_Y^2 \sin\theta \tag{E.12}$$

$$V_Y^1 = -V_X^2 \sin\theta + V_Y^2 \cos\theta \tag{E.13}$$

$$V_Z^1 = V_Z^2. \tag{E.14}$$

These equations are usually interpreted as the vector rotated clockwise with fixed frame. However they are also true for a fixed vector and frame rotated anti-clockwise. So this is an example of the frame itself moving. Therefore a connection can be defined because the connection determines how the frame itself moves. The general rule for covariant derivative is:

$$D_\nu V^\mu = \partial_\nu V^\mu + \Gamma^\mu{}_{\lambda\nu} V^\lambda. \tag{E.15}$$

This equation means that D_ν acting on V^μ is the four derivative ∂_ν plus the term $\Gamma^\mu{}_{\lambda\nu} V^\lambda$. The three index symbol is referred to as "the connection", and describes the movement of the frame itself. The latter produces, for a given ν:

$$U^\mu = \Gamma^\mu{}_\lambda V^\lambda. \tag{E.16}$$

It is seen that Eq. (E.11) is an example of Eq. (E.16) in three dimensions, X, Y, and Z. So for a rotation of the frame anti-clockwise in three dimensions about the Z axis the matrix is the rotation matrix:

$$\Gamma^\mu{}_\lambda = \begin{bmatrix} \cos\theta & \sin\theta & 0 \\ -\sin\theta & \cos\theta & 0 \\ 0 & 0 & 1 \end{bmatrix}. \tag{E.17}$$

Thus:

$$\left.\begin{array}{l} \Gamma^1{}_1 = \cos\theta, \Gamma^1{}_2 = \sin\theta, \Gamma^1{}_3 = 0, \\ \Gamma^2{}_1 = -\sin\theta, \Gamma^2{}_2 = \cos\theta, \Gamma^2{}_3 = 0, \\ \Gamma^3{}_1 = 0, \Gamma^3{}_2 = 0, \Gamma^3{}_3 = 1 \end{array}\right\}. \tag{E.18}$$

for each ν. Summation over repeated indices is used in Eq. (E.16) so:

$$\left.\begin{array}{l} U^1 = \Gamma^1{}_1 V^1 + \Gamma^1{}_2 V^2 + \Gamma^1{}_3 V^3, \\ U^2 = \Gamma^2{}_1 V^1 + \Gamma^2{}_2 V^2 + \Gamma^2{}_3 V^3, \\ U^3 = \Gamma^3{}_1 V^1 + \Gamma^3{}_2 V^2 + \Gamma^3{}_3 V^3, \end{array}\right\} \tag{E.19}$$

for each ν. These equations (E.19) are the same as Eqs. (E.12) to (E.14).
The covariant derivative of Eq. (E.15) in this case is therefore:

$$D_\nu V^\mu = (\partial + \Gamma^\mu{}_\lambda)_\nu V^\lambda. \tag{E.20}$$

For example:

$$\begin{aligned} D_\nu V^1 &= (\partial + \Gamma^1{}_1)_\nu V^1 + \Gamma^1{}_{2\nu} V^2 \\ D_\nu V^1 &= (\partial + \cos\theta)_\nu V^1 + (\sin\theta)_\nu V^2 \\ D_\nu V^1 &= \partial_\nu V^1 + (\cos\theta)_\nu V^1 + (\sin\theta)_\nu V^2 \end{aligned} \tag{E.21}$$

Appendix 5: Illustrating the Meaning of the Connection with Rotation in a Plane

Thus:

$$\Gamma^1{}_{1\nu} = (\cos\theta)_\nu, \Gamma^1{}_{2\nu} = (\sin\theta)_\nu. \tag{E.22}$$

These connections must have the units of inverse meters and must operate in the same way as the four derivative ∂_ν. So it is reasonable to assume:

$$\Gamma^1_{1\nu} = \frac{1}{2}\cos\theta\,\partial_\nu,\ \Gamma^1_{2\nu} = \frac{1}{2}\sin\theta\,\partial_\nu \tag{E.23}$$

and

$$D_\nu V^1 = \frac{1}{2}((1+\cos\theta)\partial_\nu V^1 + \sin\theta\partial_\nu V^2) \tag{E.24}$$

If there is no frame rotation:

$$\theta = 0 \tag{E.25}$$

and

$$D_\nu V^1 = \partial_\nu V^1. \tag{E.26}$$

This method regards the connection as an operator. It is well known that the set is a basis set in Riemann geometry. Others possibilities consistent with the correct dimensions of the connection are

$$(\cos\theta)_\nu = \frac{\cos\theta}{r}, (\sin\theta)_\nu = \frac{\sin\theta}{r}. \tag{E.27}$$

References

[1] M. W. Evans, "Generally Covariant Unified Field Theory: the Geometrization of Physics" (Abramis Academic, 2005 to present), vols. 1 to 4, vol. 5 in prep. (Papers 71 to 93 on www.aias.us).

[2] L. Felker, "The Evans Equations of Unified Field Theory" (Abramis Academic, 2007).

[3] K. Pendergast, "Crystal Spheres" (preprint on www.aias.us, Abramis to be published).

[4] Omnia Opera Section of www.aias.us.

[5] H. Eckardt, L. Felker, D. Indranu and S. Crothers, articles on www.aias.us.

[6] M. W. Evans, (ed.), "Modern Non-linear Optics", a special topical issue of I. Prigogine and S. A. Rice, "Advances in Chemical Physics" (Wiley Interscience, New York, 2001, second edition), vols. 119(1) to 119(3).

[7] M. W. Evans and S. Kielich (eds.), ibid., first edition (Wiley Interscience, New York, 1992, 1993, 1997), vols. 85(1) to 85(3).

[8] M. W. Evans and L. D. Crowell, "Classical and Quantum Electrodynamics and the B(3) Field" (World Scientific, 2001).

[9] M. W. Evans and J.-P. Vigier, "The Enigmatic Photon" (Kluwer, 1994 to 2002), in five volumes.

[10] M. W. Evans and A. A. Hasanein, "The Photomagneton in Quantum Field Theory" (World Scientific, 1994).

[11] M. W. Evans, "The Photon's Magnetic Field, Optical NMR" (World Scientific, 1992).

[12] M. W. Evans, Physica B, **182**, 227 and 237 (1992).

[13] S. P. Carroll, "Spacetime and Geometry, an Introduction to General Relativity" (Addison Wesley, New York, 2004), chapter three.

[14] M. W. Evans, Omnia Opera Section of www.aias.us (1992 to present).

[15] J. B. Marion and S. T. Thornton, "Classical Dynamics of Particles and Systems" (Harcourt Brace College Publishers, 1988, third edition).

[16] See for example a review by Zawodny in ref. (7).

[17] The first papers on the B(3) field are ref. (12), available in the Omnia Opera section of www.aias.us.

[18] L. H. Ryder, "Quantum Field Theory" (Cambridge University Press, 1996, 2nd ed.).

[19] S. Crothers, papers and references therein on the www.aias.us site and papers 93 of the ECE series on www.aias.us.
[20] P. W. Atkins, "Molecular Quantum Mechanics" (Oxford University Press, 1983, 2nd edition and subsequent editions).
[21] J. R. Croca, "Towards a Non-linear Quantum Physics" (World Scientific, 2001).
[22] This claim is made in ref. (18) and corrected in ref. (1).
[23] E. G. Milewski (Chief Editor), "The Vector Analysis Problem Solver" (Research and Education Association, New York, 1987, revised printing).
[24] J. D. Jackson, "Classical Electrodynamics" (Wiley, New York, 1999, third edition).
[25] M. W. Evans, Acta Phys. Polonica, **38**, 2211 (2007).
[26] M. W. Evans and H. Eckardt, Physica B, **400**, 175 (2007).
[27] M. W. Evans, Physica B, **403**, 517 (2008).
[28] Michael Krause (Director), "All About Tesla" (a film available on DVD that premiered in 2007).
[29] P. W. Atkins, frontispiece of ref. (20).

8

The Incompatibility of the Christoffel Connection with the Bianchi Identity

by

Myron W. Evans,
Alpha Institute for Advanced Study, Civil List Scientist.
(emyrone@aol.com and www.aias.us)

Abstract

It is shown that the use of the Christoffel connection in general relativity is inconsistent with the Bianchi identity of differential geometry. This finding means that the Einstein Hilbert (EH) theory is not a valid theory of physics because it is based on the Ricci cyclic equation in which the Cartan torsion is missing. When the Cartan torsion is correctly re-instated in the Bianchi identity, it is found that the Christoffel connection is not a valid solution in the theory of gravitation. More generally, Einstein Cartan Evans (ECE) theory is needed for a self consistent development of the gravitational field unified with other fundamental fields such as electromagnetism.

Keywords: Incompatibility of the Christoffel connection and Bianchi identity; Einstein Hilbert field theory, Einstein Cartan Evans (ECE) field theory.

8.1 Introduction

Recently a relatively straightforward and generally covariant unified field theory has been suggested [1–10] based on the well known Cartan geometry [11]. This is known as Einstein Cartan Evans (ECE) field theory because it is based directly on the well known correspondence between Einstein and Cartan in the early part of the twentieth century. It is internationally accepted [12] that ECE is a valid and well tested field theory which has essentially supplanted the 1915 Einstein Hilbert (EH) field theory. In Section 8.2 one of the basic weaknesses of EH theory is proven using the well known [11] Bianchi identity of differential geometry. This was developed by Cartan and is a cyclic sum of definitions of the Riemann tensor for any connection. The Riemann and torsion tensors are in turn derived from the commutator of covariant derivatives acting on the general four vector [1–11]. Both tensors are well defined for any connection irrespective of metric compatibility. One tensor cannot exist without the other. The basic weakness of the EH theory is the complete neglect of Cartan torsion. The so-called first Bianchi identity used in EH theory is in fact the Ricci cyclic equation, in which torsion does not appear at all. The correct Bianchi identity of differential geometry must include a non-zero torsion. In Section 8.2 this is developed in tensor notation and it is shown that the Christoffel connection cannot be a valid solution of the Bianchi identity. There is only one Bianchi identity [1–11], the so called first and second Bianchi identities of EH theory are approximations that neglect the Cartan torsion. In Section three a method is illustrated in which the interaction of two fields is considered in ECE theory, and an equation developed in which this basic weakness of EH theory is approximately circumvented.

8.2 Development of the Bianchi Identity and its Incompatibility with the Christoffel Connection

The Bianchi identity of Cartan's standard differential geometry [11] is expressed most succinctly and in indexless notation [1–10] as:

$$D \wedge T := R \wedge q. \tag{8.1}$$

It is seen that it is the balance of torsion (T) and curvature (R). Here $D \wedge$ is a well defined covariant exterior derivative, T is Cartan's torsion form, R is Cartan's curvature form, and q is Cartan's tetrad. The condensed notation (8.1) reveals the basic structure of the identity the most clearly. The symbol := denotes that it is an exact identity. One side is identically equal to the other side. Restoring the indices of Cartan's tangent space-time [1–11], Eq. (8.1) becomes:

$$d \wedge T^a + \omega^a{}_b \wedge T^b := R^a{}_b \wedge q^b \tag{8.2}$$

where the spin connection $\omega^a{}_b$ has been written out in full. Translating to tensor notation Eq. (8.2) becomes [1–11]:

$$\begin{aligned}\partial_\mu T^a_{\nu\rho} + \partial_\rho T^a_{\mu\nu} + \partial_\nu T^a_{\rho\mu} + \omega^a{}_{\mu b} T^b_{\nu\rho} + \omega^a{}_{\rho b} T^b_{\mu\nu} + \omega^a{}_{\nu b} T^b_{\rho\mu} \\ := R^a{}_{\mu\nu\rho} + R^a{}_{\rho\mu\nu} + R^a{}_{\nu\rho\mu}\end{aligned} \tag{8.3}$$

8.2 Development of the Bianchi Identity and its Incompatibility

using the well known rules [1–11] for the wedge product. Now define the Hodge dual tensors [1–11]:

$$\widetilde{R}^a{}_\rho{}^{\alpha\beta} = \frac{1}{2}\|g\|^{1/2}\epsilon^{\alpha\beta\mu\nu}R^a{}_{\rho\mu\nu} \tag{8.4}$$

and

$$\widetilde{T}^{a\alpha\beta} = \frac{1}{2}\|g\|^{1/2}\epsilon^{\alpha\beta\mu\nu}T^a_{\mu\nu} \tag{8.5}$$

in four dimensions. The rule for Hodge duals [1–11] means that the square root of the determinant of the metric $g_{\mu\nu}$ must premultiply the 4-D Levi-Civita symbol of the Hodge dual in a base manifold with curvature and torsion. If metric compatibility is assumed then:

$$D_\rho g_{\mu\nu} = D_\rho g^{\mu\nu} = 0 \tag{8.6}$$

and the cyclic sums in Eq. (8.3) translate into [1–11]:

$$D_\mu \widetilde{T}^{a\mu\nu} := \widetilde{R}^a{}_\mu{}^{\mu\nu} \tag{8.7}$$

This is the tensorial version of the condensed identity (8.1). In the tensorial version appear the Hodge duals of the Cartan torsion tensor and the curvature tensor $\widetilde{R}^a{}_\mu{}^{\mu\nu}$. A particular solution of Eq. (8.7) in the base manifold is [1–11]

$$D_\mu \widetilde{T}^{\kappa\mu\nu} := \widetilde{R}^\kappa{}_\mu{}^{\mu\nu}. \tag{8.8}$$

Eq. (8.1) may also be written [1–11] as:

$$\begin{aligned}R^\lambda{}_{\rho\mu\nu} + R^\lambda{}_{\mu\nu\rho} + R^\lambda{}_{\nu\rho\mu} &:= \partial_\mu\Gamma^\lambda_{\nu\rho} - \partial_\nu\Gamma^\lambda_{\mu\rho} + \Gamma^\lambda_{\mu\sigma}\Gamma^\sigma_{\nu\rho} - \Gamma^\lambda_{\nu\sigma}\Gamma^\sigma_{\mu\rho} \\ &+ \partial_\nu\Gamma^\lambda_{\rho\mu} - \partial_\rho\Gamma^\lambda_{\nu\mu} + \Gamma^\lambda_{\nu\sigma}\Gamma^\sigma_{\rho\mu} - \Gamma^\lambda_{\rho\sigma}\Gamma^\sigma_{\nu\mu} \\ &+ \partial_\rho\Gamma^\lambda_{\mu\nu} - \partial_\mu\Gamma^\lambda_{\rho\nu} + \Gamma^\lambda_{\rho\sigma}\Gamma^\sigma_{\mu\nu} - \Gamma^\lambda_{\mu\sigma}\Gamma^\sigma_{\rho\nu}\end{aligned} \tag{8.9}$$

i.e. as a cyclic sum of definitions of the Riemann tensor:

$$\begin{aligned}R^\lambda{}_{\rho\mu\nu} &:= \partial_\mu\Gamma^\lambda_{\nu\rho} - \partial_\nu\Gamma^\lambda_{\mu\rho} + \Gamma^\lambda_{\mu\sigma}\Gamma^\sigma_{\nu\rho} - \Gamma^\lambda_{\nu\sigma}\Gamma^\sigma_{\mu\rho}, \\ R^\lambda{}_{\mu\nu\rho} &:= \partial_\nu\Gamma^\lambda_{\rho\mu} - \partial_\rho\Gamma^\lambda_{\nu\mu} + \Gamma^\lambda_{\nu\sigma}\Gamma^\sigma_{\rho\mu} - \Gamma^\lambda_{\rho\sigma}\Gamma^\sigma_{\nu\mu}, \\ R^\lambda{}_{\nu\rho\mu} &:= \partial_\rho\Gamma^\lambda_{\mu\nu} - \partial_\mu\Gamma^\lambda_{\rho\nu} + \Gamma^\lambda_{\rho\sigma}\Gamma^\sigma_{\mu\nu} - \Gamma^\lambda_{\mu\sigma}\Gamma^\sigma_{\rho\nu}\end{aligned} \tag{8.10}$$

for any connection. The Riemann and torsion tensors are furthermore related by the basic equation [1–11]:

$$[D_\mu, D_\nu]V^\rho = R^\rho{}_{\sigma\mu\nu}V^\sigma - T^\lambda_{\mu\nu}D_\lambda V^\rho. \tag{8.11}$$

One tensor cannot exist without the other and Eq. (8.11) shows that they are always defined by:

$$R^\lambda_{\rho\mu\nu} := \partial_\mu \Gamma^\lambda_{\nu\rho} - \partial_\nu \Gamma^\lambda_{\mu\rho} + \Gamma^\lambda_{\mu\sigma}\Gamma^\sigma_{\nu\rho} - \Gamma^\lambda_{\nu\sigma}\Gamma^\sigma_{\mu\rho}, \tag{8.12}$$

and

$$T^\lambda_{\mu\nu} := \Gamma^\lambda_{\mu\nu} - \Gamma^\lambda_{\nu\mu}. \tag{8.13}$$

In general:

$$R^\lambda_{\rho\mu\nu} + R^\lambda_{\mu\nu\rho} + R^\lambda_{\nu\rho\mu} \neq 0. \tag{8.14}$$

Eq. (8.9) shows that Eq (8.1) is indeed an exact identity. One side is a restatement of the other side by definition. Eq. (8.11) shows that both tensors must be anti-symmetric in their last two indices:

$$R^\lambda_{\rho\mu\nu} = -R^\lambda_{\rho\nu\mu}, \tag{8.15}$$
$$T^\lambda_{\mu\nu} = -T^\lambda_{\nu\mu}, \tag{8.16}$$

and so have well defined Hodge duals in four dimensions [1–11]:

$$\widetilde{R}^\lambda_{\rho\mu\nu} = \frac{1}{2}\|g\|^{1/2}\epsilon_{\mu\nu\alpha\beta}R^\lambda{}_\rho{}^{\alpha\beta}, \tag{8.17}$$

$$\widetilde{T}^\lambda_{\mu\nu} = \frac{1}{2}\|g\|^{1/2}\epsilon_{\mu\nu\alpha\beta}T^{\lambda\alpha\beta}. \tag{8.18}$$

The Hodge dual of the torsion tensor is the same as the Hodge dual of the difference of anti-symmetric connections, so:

$$\widetilde{T}^\lambda_{\mu\nu} = (\Gamma^\lambda_{\mu\nu} - \Gamma^\lambda_{\nu\mu})_{HD}. \tag{8.19}$$

There also exists a Hodge dual of the commutator of covariant derivatives, because this is also an anti-symmetric operator:

$$[D_\mu, D_\nu]_{HD} = \frac{1}{2}\|g\|^{1/2}\epsilon_{\mu\nu\alpha\beta}[D^\alpha, D^\beta]. \tag{8.20}$$

8.2 Development of the Bianchi Identity and its Incompatibility

So we obtain the Hodge dual of Eq. (8.11):

$$[D_\mu, D_\nu]_{HD} V^\rho = \tilde{R}^\rho{}_{\sigma\mu\nu} V^\sigma - \tilde{T}^\lambda_{\mu\nu} D_\lambda V^\rho \qquad (8.21)$$

where the Hodge dual curvature and torsion tensors are defined by the Hodge dual commutator acting on the arbitrary four vector V^ρ.

In general the Hodge dual curvature and torsion are defined by the connection appearing in:

$$[D_\mu, D_\nu]_{HD} V^\rho := (D_\mu(D_\nu V^\rho) - D_\nu(D_\mu V^\rho))_{HD}. \qquad (8.22)$$

Using Eq. (8.21) in Eq. (8.9) it follows that:

$$\tilde{R}^\lambda_{\rho\mu\nu} := (\partial_\mu \Gamma^\lambda_{\nu\rho} - \partial_\nu \Gamma^\lambda_{\mu\rho} + \Gamma^\lambda_{\mu\sigma}\Gamma^\sigma_{\nu\rho} - \Gamma^\lambda_{\nu\sigma}\Gamma^\sigma_{\mu\rho})_{HD}, \qquad (8.23a)$$

$$\tilde{T}^\lambda_{\mu\nu} := (\Gamma^\lambda_{\mu\nu} - \Gamma^\lambda_{\nu\mu})_{HD}, \qquad (8.23b)$$

and it also follows that:

$$D \wedge \tilde{T} := \tilde{R} \wedge q. \qquad (8.24)$$

This equation in tensor notation is:

$$\partial_\mu \tilde{T}^a_{\nu\rho} + \partial_\rho \tilde{T}^a_{\mu\nu} + \partial_\nu \tilde{T}^a_{\rho\mu} + \omega^a{}_{\mu b}\tilde{T}^b_{\nu\rho} + \omega^a{}_{\rho b}\tilde{T}^b_{\mu\nu} + \omega^a{}_{\nu b}\tilde{T}^b_{\rho\mu}$$
$$:= \tilde{R}^a{}_{\mu\nu\rho} + \tilde{R}^a{}_{\rho\mu\nu} + \tilde{R}^a{}_{\nu\rho\mu} \qquad (8.25)$$

and this sum is equivalent to:

$$D_\mu T^{a\mu\nu} := R^a{}_\mu{}^{\mu\nu}. \qquad (8.26)$$

A particular solution of Eq. (8.26) is:

$$D_\mu T^{\kappa\mu\nu} := R^\kappa{}_\mu{}^{\mu\nu}. \qquad (8.27)$$

This equation is the most convenient form of the Bianchi identity and shows that the covariant derivative of the torsion tensor on the left hand side is identically equal to the Ricci type tensor appearing on the right hand side of Eq. (8.26). It has been shown by computer algebra for many different metrics that the tensor $R^\kappa{}_\mu{}^{\mu\nu}$ is in general non-zero for a Christoffel connection:

$$\Gamma^\lambda_{\mu\nu} = \Gamma^\lambda_{\nu\mu}. \qquad (8.28)$$

However, for the connection (8.28) the torsion must be zero:

$$T^\lambda_{\mu\nu} = \Gamma^\lambda_{\mu\nu} - \Gamma^\lambda_{\nu\mu} = 0. \tag{8.29}$$

Therefore the Bianchi identity and Christoffel connection are incompatible in general. The Christoffel connection can be a solution of the Bianchi identity if and only if the metric defines a Ricci flat space-time where all elements of the Ricci tensor vanish [1–12]. However, for such a space-time the EH field equation means that the $T_{\mu\nu}$ tensor also vanishes. This is the canonical energy-momentum density tensor as is well known. Crothers has shown [13] that the Ricci flat space-time is incompatible with the Einstein equivalence principle and in such a space-time there can be no fields and no mass/energy density. Such a space-time has no physical meaning therefore. Several other conceptual errors in the EH theory have been pointed out by Crothers [13, 14].

There is no way in which the EH theory can escape its inherent weaknesses. Furthermore it is almost entirely based on the use of the Christoffel symbol, which as we have shown, violates the fundamental geometry (8.1). It is concluded that all cosmologies based on EH theory are physically meaningless, notably Big Bang, black hole theory and dark matter. None can exist in nature. The Christoffel symbol can be used if and only if

$$R^\kappa{}_\mu{}^{\mu\nu} = 0 \tag{8.30}$$

when Eq. (8.1) reduces to:

$$R \wedge q = 0 \tag{8.31}$$

This in tensor notation is:

$$R^\lambda{}_{\rho\mu\nu} + R^\lambda{}_{\nu\rho\mu} + R^\lambda{}_{\mu\nu\rho} = 0 \tag{8.32}$$

and is incorrectly referred to as the "first Bianchi identity". In fact it was first given by Ricci and Levi-Civita and later developed by Cartan into Eq. (8.1). The so called "second Bianchi identity" is in form notation:

$$D \wedge R = 0 \tag{8.33}$$

but it has been shown during the course of the development of ECE theory [1–11] that the "second Bianchi identity" is a restatement of Eq. (8.1) as follows:

$$D \wedge (D \wedge T) := D \wedge (R \wedge q). \tag{8.34}$$

So there is only one Bianchi identity, Eq. (8.1).

The incompatibility of the Christoffel connection with Eq. (8.1) was discovered in paper 93 of the ECE series (www.aias.us) when it was found that the tensor $R^\kappa{}_\mu{}^{\mu\nu}$ is non-zero in general for a Christoffel connection, and zero only for physically meaningless Ricci flat space-times. This property of the Christofel connection is incompatible with the fact that for such a connection, the torsion tensor must be zero. The Christoffel connection cannot be a solution of Eq. (8.27) in general. To calculate $R^\kappa{}_\mu{}^{\mu\nu}$ required computer algebra, because the calculation of $R^\kappa{}_\mu{}^{\mu\nu}$ by hand is exceedingly intricate. So this fatal flaw in EH theory has remained hidden for more than ninety years. The EH equation itself is valid if and only if it does not use a Christoffel connection and does not use a Ricci flat space-time. However, ECE shows that the EH field equation is only one part of the whole.

8.3 Interaction of Gravitation and Electromagnetism

It has been shown in Section 8.2 that the interpretation of the well known EH theory with a Christoffel connection is basically incorrect geometrically. This may perhaps be surprising to the uninitiated, but it has been known for some years to scholars like Crothers [13–14] for example that EH is deeply flawed in several ways. The dogma of EH is unfortunately adhered to in the cosmological literature, and this has become harmful to science. In this section the basic equation of paper 93 of www.aias.us is derived by considering Eq. (8.27) for the interaction of two fields. It is found that in this case the calculation of $R^\kappa{}_\mu{}^{\mu\nu}$ with the Christoffel symbol can be approximately valid, but if and only if one field interacts with another in the context of ECE theory.

The ECE theory asserts that the electromagnetic field tensor is the Cartan torsion within a proportionality:

$$F^a_{\mu\nu} = A^{(0)} T^a_{\mu\nu} \tag{8.35}$$

so there exists an electromagnetic torsion, denoted T (e/m), as well as a gravitational torsion denoted T (grav). Similarly there exists an electromagnetic curvature R (e / m) as well as a gravitational curvature R (grav). The connection in ECE theory is not the symmetric Christoffel connection. The electromagnetic field in ECE theory is for all practical purposes defined by the balance condition:

$$R \wedge q = \omega \wedge T \tag{8.36}$$

which is based on experimental data, notably that the homogenous field equation must be:

$$d \wedge F^a = 0 \tag{8.37}$$

for all practical purposes in the laboratory. The condition (8.36) then means that there is no interaction between gravitation and electromagnetism in this well defined sense. From Eq. (8.36) it follows that:

$$\widetilde{R} \wedge q = \omega \wedge \widetilde{T} \tag{8.38}$$

so that there is no contribution from electromagnetism to the charge current density of the inhomogeneous field equation [1–10] of ECE. Under these circumstances the basic equation of paper 93 follows:

$$\partial_\mu F^{\kappa\mu\nu}(e/m) \doteq A^{(0)} R^\kappa{}_\mu{}^{\mu\nu}(\text{grav}) \tag{8.39}$$

where the electromagnetic torsion on the left hand side is balanced by the gravitational curvature on the right hand side. Under these circumstances the Christoffel connection can be used to compute $R^\kappa{}_\mu{}^{\mu\nu}$, and this automatically implies that there is no gravitational torsion present.

Acknowledgments

The British Government is thanked for the award of a Civil List Pension and the Telesio Galieli Association of generous funding of this work. Colleagues worldwide are thanked for many interesting discussions.

References

[1] M. W. Evans, "Generally Covariant Unified Field Theory" (Abramis Academic, Suffolk, 2005–2007), volumes one to four, volume five in prep. (www.aias.us, papers 71–93).

[2] L. Felker, "The Evans Equations of Unified Field Theory" (Abramis Academic, Suffolk, 2007).

[3] Papers and articles on www.aias.us and www.atomicprecision.com.

[4] K. Pendergast, "Mercury as Crystal Spheres" (preprint on www.aias.us and Abramis in prep).

[5] M. W. Evans et al., Omnia Opera section of www.aias.us, 1992 to present.

[6] M. W. Evans and L. B. Crowell, "Classical and Quantum Electrodynamics and the B(3) Field" (World Scientific, 2001).

[7] M. W. Evans (ed.), "Modern Non-Linear Optics", a special topical issue in three parts of I. Prigogine and S. A. Rices (series eds.), Advances in Chemical Physics", (Wiley Interscience, New York, 2001, second edition), vols 119(1) to 119(3); ibid., M. W. Evans and S. Kielich (eds.), (Wiley Interscience, New York, 1992, 1993 and 1997, first edition), vols. 85(1) to 85(3).

[8] M. W. Evans and J-P. Vigier, "The Enigmatic Photon" (Kluwer, Dordrecht, 1994 to 2002), hardback and softback), vols. 1 to 5.

[9] M. W. Evans and A. A. Hasanein, "The Photomagneton in Quantum Field Theory" (World Scientific, 1994).

[10] M. W. Evans, Physica B, **182**, 227, 237 (1992).

[11] S. P. Carroll, "Space-time and Geometry: an Introduction to General Relativity" (Addison Wesley, New York, 2004 and online 1997 notes).

[12] Intense interest in www.aias.us worldwide recorded daily for 3.5 years using feedback software.

[13] S. J. Crothers in paper 93 of www.aias.us.

[14] Critical articles by S. J. Crothers on the www.aias.us website.

9

The Fundamental Origin of the Bianchi Identity of Cartan Geometry and ECE Theory

by

Myron W. Evans,
Alpha Institute for Advanced Study, Civil List Scientist.
(emyrone@aol.com and www.aias.us)

Abstract

The fundamental origin of the Bianchi identity of Cartan geometry is shown to be the commutator of covariant derivatives acting on a four vector in a space-time with curvature and torsion. The Hodge dual of this operation results in a Hodge dual Bianchi identity. In tensorial form the Bianchi identity and its Hodge dual are the field equations of electrodynamics in Einstein Cartan Evans (ECE) theory. These field equations reduce to the same vector formulation as the well known Maxwell Heaviside field equations, but are generally covariant and unified with other fundamental force fields.

Keywords: Cartan geometry, Bianchi identity, Einstein Cartan Evans (ECE) field theory, Hodge duality, field equations.

9.1 Introduction

It is well known [1] that Cartan's geometry is a self consistent geometry that generalizes Riemann geometry in an elegant manner based on the Cartan structure equations and Bianchi identity. It has been shown recently [2–11] that there is only one Bianchi identity, which must always relate the curvature form to the torsion form. To neglect the latter is arbitrary and inadmissible, yet this is what happens in the standard model theory of general relativity. This flaw has persisted for ninety years, so the only valid theory of general relativity is the well known Einstein Cartan Evans (ECE) theory introduced

208   9 The Fundamental Origin of the Bianchi Identity

in 2003 on the basis of Cartan geometry with torsion and curvature. Any self consistent geometry may be used in general relativity, whose fundamental assertion is that natural philosophy is geometry. In ECE theory the philosophy of general relativity is adhered to in the manner of Einstein and Hilbert, but the torsion is re-instated following Cartan. The internal consistency of Cartan geometry depends on the use both of curvature and torsion, and also on the tetrad postulate [1–11]. The latter is the requirement that the complete vector field be invariant under the general coordinate transformation. There is no situation in natural philosophy where this is not true. There may be abstract and abstruse geometries in which the tetrad postulate is not true, but they are not Cartan's geometry as taught for eighty years [1] since first proposed by Cartan circa 1922.

In Section 9.2 the Bianchi identity of Cartan is proven from the commutator of covariant derivatives operating on a four vector in a space-time with curvature and torsion. It is well known [1] that the commutator operating in this manner produces the fundamental definitions of the Riemann and torsion tensors in terms of connections. These definitions are fundamental and are true irrespective of the postulate of metric compatibility [1]. The definitions are therefore true for any type of line element, metric and connection. The Riemann geometry used by Einstein and Hilbert [1] (EH theory) is based on the Christoffel connection, which is symmetric in its lower two indices. This assumption is the basis of EH theory and is based on the arbitrary neglect of the torsion tensor. The Christoffel connection is related to a symmetric metric in EH theory using the postulate of metric compatibility [1]. All line elements and solutions of the EH equation use these arbitrary assumptions. In 1915, when the EH theory was proposed, the existence of the Cartan torsion of 1922 was obviously unknown. In Section 9.2 it is proven that the Bianchi identity of Cartan is a re-expression of the commutator of covariant derivatives acting on the four vector. The Bianchi identity of Cartan is a cyclic sum of fundamental definitions of the curvature tensor. It is therefore an exact identity which is true, however, if and only if the torsion tensor is defined from the same commutator of covariant derivatives acting on the same four vector.

In Section 9.3 the Hodge dual of the Bianchi identity is proven in the same way, by using the Hodge dual of the commutator of covariant derivatives acting on the same four vector. This operation produces the Hodge dual of the Riemann or curvature tensor and the Hodge dual of the torsion tensor. A well defined Hodge dual of the Bianchi identity follows from this proof.

In Section 9.4 it is proven that the use of the Christoffel symbol or connection is incompatible fundamentally with the Bianchi identity of Cartan. This is clear just by considering the torsion tensor, which is the difference of connections as follows:

$$T^{\kappa}_{\mu\nu} = \Gamma^{\kappa}_{\mu\nu} - \Gamma^{\kappa}_{\nu\mu}. \tag{9.1}$$

For the Christoffel connection:

$$\Gamma^\kappa_{\mu\nu} = \Gamma^\kappa_{\nu\mu} \qquad (9.2)$$

and the torsion tensor vanishes. This is incompatible with the fundamental operation of a commutator of covariant derivatives acting on a four vector. In ECE theory the torsion is correctly re-instated, and the general connection used instead of the Christoffel connection. The ECE theory is internally consistent whereas the EH theory is not.

9.2 Proof of the Bianchi Identity

In previous work [1–11] it has been proven that the Bianchi identity of Cartan is the cyclic sum of definitions of the curvature or Riemann tensor. This result is true if and only if the torsion tensor is defined as in Eq. (9.1). In this section the Bianchi identity is proven from the well known equation [1–11]:

$$[D_\mu, D_\nu] V^\rho = R^\rho{}_{\sigma\mu\nu} V^\sigma - T^\lambda_{\mu\nu} D_\lambda V^\rho \qquad (9.3)$$

in which the commutator of covariant derivatives acts on a four vector. In Eq. (9.3) no assumption is made concerning metric compatibility or symmetry of the metric or connection, so Eq. (9.3) is a fundamental and general result. In Eq. (9.3) the commutator is an operator defined by:

$$[D_\mu, D_\nu] = D_\mu D_\nu - D_\nu D_\mu \qquad (9.4)$$

where the covariant derivative is defined by:

$$D_\mu V^\nu = \partial_\mu V^\nu + \Gamma^\nu_{\mu\lambda} V^\lambda. \qquad (9.5)$$

Here V^ν is any four-vector in a space-time with both torsion and curvature. In Eq. (9.3) the curvature tensor is:

$$R^\lambda{}_{\rho\mu\nu} := \partial_\mu \Gamma^\lambda_{\nu\rho} - \partial_\nu \Gamma^\lambda_{\mu\rho} + \Gamma^\lambda_{\mu\sigma} \Gamma^\sigma_{\nu\rho} - \Gamma^\lambda_{\nu\sigma} \Gamma^\sigma_{\mu\rho} \qquad (9.6)$$

and the torsion tensor is:

$$T^\lambda_{\mu\nu} := \Gamma^\lambda_{\mu\nu} - \Gamma^\lambda_{\nu\mu}. \qquad (9.7)$$

There is no a priori reason for assuming that the connection must be a Christoffel connection as defined in Eq. (9.2). In almost all standard model

210   9 The Fundamental Origin of the Bianchi Identity

general relativity the torsion is arbitrarily neglected by using the Christoffel connection and the symmetric metric:

$$g_{\mu\nu} = g_{\nu\mu}. \tag{9.8}$$

To recast Eq. (9.3) in the form of Cartan's Bianchi identity these assumptions are not used. The only additional equation needed is the tetrad postulate [1–11]:

$$D_\mu q^a_\nu = 0. \tag{9.9}$$

where q^a_ν is the well known Cartan tetrad. The latter is a vector valued one-form of the standard differential geometry introduced by Cartan. The latter's Bianchi identity is:

$$d \wedge T^a + \omega^a{}_b \wedge T^b := R^a{}_b \wedge q^b \tag{9.10}$$

where T^a is the torsion form, a vector valued two-form [1–11], $R^a{}_b$ is the curvature form, a tensor valued two-form, and $\omega^a{}_b$ is the spin connection. The tetrad postulate relates the spin connection and gamma connection as follows [1–11]:

$$\partial_\mu q^a_\sigma + \omega^a_{\mu b} q^b_\sigma = \Gamma^\lambda_{\mu\sigma} q^a_\lambda. \tag{9.11}$$

It is proven as follows that Eqs. (9.3) and (9.10) are the same equation, provided that the tetrad postulate (9.11) is used. First translate Eq. (9.10) into tensor notation using the rules for the wedge product $\wedge$ of differential geometry [1]. This translation from form to tensor notation produces the result:

$$\partial_\mu T^a_{\nu\rho} + \omega^a_{\mu b} T^b_{\nu\rho} + \partial_\rho T^a_{\mu\nu} + \omega^a_{\rho b} T^b_{\mu\nu} + \partial_\nu T^a_{\rho\mu} + \omega^a_{\nu b} T^b_{\rho\mu}$$
$$:= (R^\lambda_{\mu\nu\rho} + R^\lambda_{\rho\mu\nu} + R^\lambda_{\nu\rho\mu}) q^a_\lambda \tag{9.12}$$

where:

$$T^a_{\nu\rho} = (\Gamma^\lambda_{\nu\rho} - \Gamma^\lambda_{\rho\nu}) q^a_\lambda \quad \text{etc.}, \tag{9.13}$$

$$T^b_{\nu\rho} = (\Gamma^\lambda_{\nu\rho} - \Gamma^\lambda_{\rho\nu}) q^b_\lambda \quad \text{etc.} \tag{9.14}$$

Using the Leibnitz Theorem:

$$\partial_\mu T^a_{\nu\rho} = (\partial_\mu \Gamma^\lambda_{\nu\rho} - \partial_\mu \Gamma^\lambda_{\rho\nu}) q^a_\lambda$$
$$+ (\Gamma^\lambda_{\nu\rho} - \Gamma^\lambda_{\rho\nu}) \partial_\mu q^a_\lambda \quad \text{etc.}, \tag{9.15}$$

so Eq. (9.12) becomes:

$$(\partial_\mu \Gamma^\lambda_{\nu\rho} - \partial_\mu \Gamma^\lambda_{\rho\nu})q^a_\lambda + (\Gamma^\lambda_{\nu\rho} - \Gamma^\lambda_{\rho\nu})(\partial_\mu q^a_\lambda + \omega^a_{\mu b} q^b_\lambda) + \ldots$$
$$:= (R^\lambda_{\mu\nu\rho} + R^\lambda_{\rho\mu\nu} + R^\lambda_{\nu\rho\mu})q^a_\lambda. \qquad (9.16)$$

Now re-label dummy (i.e. repeated) indices in the second term on the left hand side:

$$\lambda \to \sigma \qquad (9.17)$$

to obtain:

$$(\partial_\mu \Gamma^\lambda_{\nu\rho} - \partial_\mu \Gamma^\lambda_{\rho\nu})q^a_\lambda + (\Gamma^\sigma_{\nu\rho} - \Gamma^\sigma_{\rho\nu})(\partial_\mu q^a_\sigma + \omega^a_{\mu b} q^b_\sigma) + \ldots$$
$$:= (R^\lambda_{\mu\nu\rho} + R^\lambda_{\rho\mu\nu} + R^\lambda_{\nu\rho\mu})q^a_\lambda \qquad (9.18)$$

Use the tetrad postulate (9.11) to obtain the cyclic sum:

$$\partial_\mu \Gamma^\lambda_{\nu\rho} - \partial_\mu \Gamma^\lambda_{\rho\nu} + \Gamma^\lambda_{\mu\sigma}(\Gamma^\sigma_{\nu\rho} - \Gamma^\sigma_{\rho\nu}) + \partial_\rho \Gamma^\lambda_{\mu\nu} - \partial_\rho \Gamma^\lambda_{\nu\mu}$$
$$+ \Gamma^\lambda_{\rho\sigma}(\Gamma^\sigma_{\mu\nu} - \Gamma^\sigma_{\nu\mu}) + \partial_\nu \Gamma^\lambda_{\rho\mu} - \partial_\nu \Gamma^\lambda_{\mu\rho} + \Gamma^\lambda_{\nu\sigma}(\Gamma^\sigma_{\rho\mu} - \Gamma^\sigma_{\mu\rho}) \qquad (9.19)$$
$$:= R^\lambda_{\mu\nu\rho} + R^\lambda_{\rho\mu\nu} + R^\lambda_{\nu\rho\mu}.$$

Re-arrange this cyclic sum as follows:

$$R^\lambda_{\rho\mu\nu} + R^\lambda_{\mu\nu\rho} + R^\lambda_{\nu\rho\mu}$$
$$:= \partial_\mu \Gamma^\lambda_{\nu\rho} - \partial_\nu \Gamma^\lambda_{\mu\rho} + \Gamma^\lambda_{\mu\sigma}\Gamma^\sigma_{\nu\rho} - \Gamma^\lambda_{\nu\sigma}\Gamma^\sigma_{\mu\rho}$$
$$+ \partial_\nu \Gamma^\lambda_{\rho\mu} - \partial_\rho \Gamma^\lambda_{\nu\mu} + \Gamma^\lambda_{\nu\sigma}\Gamma^\sigma_{\rho\mu} - \Gamma^\lambda_{\rho\sigma}\Gamma^\sigma_{\nu\mu} \qquad (9.20)$$
$$+ \partial_\rho \Gamma^\lambda_{\mu\nu} - \partial_\mu \Gamma^\lambda_{\rho\nu} + \Gamma^\lambda_{\rho\sigma}\Gamma^\sigma_{\mu\nu} - \Gamma^\lambda_{\mu\sigma}\Gamma^\sigma_{\rho\nu}.$$

It is seen that this is a cyclic sum of three definitions of the curvature tensor:

$$R^\lambda_{\rho\mu\nu} := \partial_\rho \Gamma^\lambda_{\mu\nu} - \partial_\mu \Gamma^\lambda_{\rho\nu} + \Gamma^\lambda_{\rho\sigma}\Gamma^\sigma_{\mu\nu} - \Gamma^\lambda_{\mu\sigma}\Gamma^\sigma_{\rho\nu}, \qquad (9.21)$$

$$R^\lambda_{\mu\nu\rho} := \partial_\mu \Gamma^\lambda_{\nu\rho} - \partial_\nu \Gamma^\lambda_{\mu\rho} + \Gamma^\lambda_{\mu\sigma}\Gamma^\sigma_{\nu\rho} - \Gamma^\lambda_{\nu\sigma}\Gamma^\sigma_{\mu\rho}, \qquad (9.22)$$

$$R^\lambda_{\nu\rho\mu} := \partial_\nu \Gamma^\lambda_{\rho\mu} - \partial_\rho \Gamma^\lambda_{\nu\mu} + \Gamma^\lambda_{\nu\sigma}\Gamma^\sigma_{\rho\mu} - \Gamma^\lambda_{\rho\sigma}\Gamma^\sigma_{\nu\mu}. \qquad (9.23)$$

These definitions come from Eq. (9.3) as does the definition of the torsion tensor needed to obtain the result (9.20), Q.E.D.

It has been proven that Cartan's Bianchi identity (9.10) is the same as Eq. (9.3) given the tetrad postulate (9.11). The identity is exact, because its left hand side is identically the same as its right hand side. One takes three definitions (9.21)–(9.23) and adds them. In general:

$$R^\lambda_{\rho\mu\nu} + R^\lambda_{\mu\nu\rho} + R^\lambda_{\nu\rho\mu} \neq 0 \tag{9.24}$$

because the Bianchi identity is:

$$D_\mu T^a_{\nu\rho} + D_\rho T^a_{\mu\nu} + D_\nu T^a_{\rho\mu} := R^a_{\mu\nu\rho} + R^a_{\rho\mu\nu} + R^a_{\nu\rho\mu} \tag{9.25}$$

and there is no reason for assuming that the torsion is zero, assuming that the connection is symmetric or assuming that the metric is symmetric, not for assuming metric compatibility.

What is usually done in EH theory is to make all these assumptions and to describe the resulting geometry as Riemann geometry. This is arbitrary and unjustifiable. The resulting physics of general relativity is all based on these arbitrary assumptions. The correct Bianchi identity is Eq. (9.25), which can be re-written as:

$$D_\mu \tilde{T}^{a\,\mu\nu} := \tilde{R}^a_{\ \mu}{}^{\mu\nu} \tag{9.26}$$

where the tilde denotes Hodge duality. In deriving Eq. (9.26) from Eq. (9.25) the following Hodge duals are used [1–11]:

$$\tilde{T}^{a\,\alpha\beta} = \frac{1}{2}\|g\|^{\frac{1}{2}}\epsilon^{\alpha\beta\mu\nu} T^a_{\mu\nu}, \tag{9.27}$$

$$\tilde{R}^a_{\ b}{}^{\alpha\beta} = \frac{1}{2}\|g\|^{\frac{1}{2}}\epsilon^{\alpha\beta\mu\nu} R^a_{\ b\mu\nu}. \tag{9.28}$$

Here $\|g\|^{\frac{1}{2}}$ is the square root of the positive value of the metric determinant, and $\epsilon^{\alpha\beta\mu\nu}$ is the four dimensional Levi-Civita symbol of Minkowski spacetime. Since two-forms are anti-symmetric by definition [1–11]:

$$T^a_{\mu\nu} = -T^a_{\nu\mu}, \tag{9.29}$$

$$R^a_{\ b\mu\nu} = -R^a_{\ b\nu\mu}, \tag{9.30}$$

their Hodge duals are also anti-symmetric, and are also two-forms. A particular solution of Eq. (9.26) is the base manifold equation:

$$D_\mu \tilde{T}^{\kappa\mu\nu} := \tilde{R}^\kappa_{\ \mu}{}^{\mu\nu} \tag{9.31}$$

which is the basis [2–11] of the homogeneous ECE field equation. Note carefully that Eq. (9.26) is less general than Eq. (9.25) because in deriving Eq. (9.26) metric compatibility is used as follows:

$$D_\mu \left(\|g\|^{\frac{1}{2}} \right) = 0. \tag{9.32}$$

However there is no situation in natural philosophy in which metric compatibility is not true, because the metric is defined by:

$$g_{\mu\nu} = q_\mu^a q_\nu^b \eta_{ab}. \tag{9.33}$$

Here η_{ab} is the Minkowski metric [1–11]. The tetrad postulate (9.11) then implies metric compatibility, which is the equation [1–11]:

$$D_\rho g_{\mu\nu} = 0. \tag{9.34}$$

As we have argued, the tetrad postulate is the fundamental requirement that the complete vector field be invariant under general coordinate transformation, and this is always true in natural philosophy.

9.3 Hodge Dual of the Bianchi Identity

This identity of geometry is the basis for the inhomogeneous field equation of ECE theory [2–11]. It is proven by taking the Hodge dual term by term of Eq. (9.3) to give:

$$[D_\mu, D_\nu]_{HD} V^\rho = \tilde{R}^\rho{}_{\sigma\mu\nu} V^\sigma - \tilde{T}^\lambda{}_{\mu\nu} D_\lambda V^\rho \tag{9.35}$$

where the subscript HD denotes Hodge dual. The Hodge duals of the curvature and torsion tensors are evidently:

$$\tilde{R}^\lambda{}_{\rho\mu\nu} := \left(\partial_\mu \Gamma^\lambda_{\nu\rho} - \partial_\nu \Gamma^\lambda_{\mu\rho} + \Gamma^\lambda_{\mu\sigma}\Gamma^\sigma_{\nu\rho} - \Gamma^\lambda_{\nu\sigma}\Gamma^\sigma_{\mu\rho} \right)_{HD}, \tag{9.36}$$

$$\tilde{T}^\lambda{}_{\mu\nu} := \left(\Gamma^\lambda_{\mu\nu} - \Gamma^\lambda_{\nu\mu} \right)_{HD}. \tag{9.37}$$

Therefore, following the methods of Section 9.2, Eq. (9.35) is a re-statement of:

$$d \wedge \tilde{T}^a + \omega^a{}_b \wedge \tilde{T}^b := \tilde{R}^a{}_b \wedge q^b. \tag{9.38}$$

The same connections occur in Eqs. (9.10) and (9.37) and Eqs. (9.3) and (9.35) can be inter-converted by the Hodge dual transformation. Therefore they are duality invariant. The tensor formulation of Eq. (9.37) is:

$$D_\mu \tilde{T}^a{}_{\nu\rho} + D_\rho \tilde{T}^a{}_{\mu\nu} + D_\nu \tilde{T}^a{}_{\rho\mu} := \tilde{R}^a{}_{\mu\nu\rho} + \tilde{R}^a{}_{\rho\mu\nu} + \tilde{R}^a{}_{\nu\rho\mu} \qquad (9.39)$$

which is the same as:

$$D_\mu T^{a\mu\nu} := R^a{}_\mu{}^{\mu\nu} \qquad (9.40)$$

a special case of which is:

$$D_\mu T^{\kappa\mu\nu} = R^\kappa{}_\mu{}^{\mu\nu} \qquad (9.41)$$

This equation is the basis of the inhomogeneous field equation of ECE theory [1–11].

In deriving the Hodge dual (9.35) the $\|g\|^{\frac{1}{2}}$ factor cancels out because it is the same on both sides. Therefore the Hodge duality can be carried out with the $\epsilon^{\mu\nu\rho\sigma}$ tensor of Minkowski space-time. This is the totally anti-symmetric unit tensor in four dimensions. It is important to note that the same connections occur in the Bianchi identity and also in its Hodge dual (9.37), In concise, index-less notation they can be written as:

$$D \wedge T := R \wedge q \qquad (9.42)$$

and

$$D \wedge \tilde{T} := \tilde{R} \wedge q \qquad (9.43)$$

and so are clearly interchangeable under the Hodge dual transforms:

$$T \to \tilde{T}; \quad R \to \tilde{R}. \qquad (9.44)$$

This is what is meant by duality invariance. The latter forms the basis for topics in physics such as Montonen-Olive duality, topological magnetic monopoles and so forth [12]. The usual Maxwell Heaviside (MH) field equations are not duality invariant because there is no magnetic monopole. The MH equations in form notation are:

$$d \wedge F = 0, \qquad (9.45)$$

$$d \wedge \tilde{F} = \tilde{j}/\epsilon_0, \qquad (9.46)$$

and it is seen that under the Hodge dual transformation of F to $\widetilde{F}$ they are not duality invariant. In contrast it is shown in the next section that the ECE field equations are duality invariant and have a fundamental symmetry that is missing in the MH theory of special relativity. This symmetry is possible only in a generally covariant unified field theory.

9.4 Incompatibility of the Christoffel Connection

The tensorial formulations of the Bianchi identity and its Hodge dual are duality invariant equations which in the base manifold are as follows:

$$D_\mu \widetilde{T}^{\kappa\mu\nu} := \widetilde{R}^\kappa{}_\mu{}^{\mu\nu}, \tag{9.47}$$

$$D_\mu T^{\kappa\mu\nu} := \widetilde{R}^\kappa{}_\mu{}^{\mu\nu}. \tag{9.48}$$

Indices can be raised and lowered on the torsion and curvature tensors in these expressions by use of the metric. For example:

$$T^\kappa_{\mu\nu} = g_{\mu\rho} g_{\nu\sigma} T^{\kappa\rho\sigma}. \tag{9.49}$$

By use of computer algebra [2-11] the tensor $R^\kappa{}_\mu{}^{\mu\nu}$ has been evaluated for various metrics and line elements based on the Christoffel connection (see for example paper 93 of the www.aias.us series). It was found by Maxima that the tensor is non-zero in general for a Christoffel connection. It vanishes only when the line element is constructed from a Ricci flat space-time. Crothers has argued recently on www.aias.us that the use of a Ricci flat space-time is incompatible with the equivalence principle of Einstein. From the point of view of ECE theory such a space-time implies that there is no electromagnetic field because the canonical energy momentum density vanishes. The canonical angular energy momentum density is a rank three tensor as is well known [12] and also vanishes in a Ricci flat space-time. In ECE the electromagnetic field is directly proportional to this rank three tensor density, and this is justified experimentally through the well known fact that the electromagnetic field has angular momentum as observed in the Beth effect. So a Ricci flat space-time is one in which there are no fields and no energy momentum density.

For all line elements that consider a finite energy momentum density the Christoffel connection is always used in Einstein Hilbert field theory. In this case the tensor $R^\kappa{}_\mu{}^{\mu\nu}$ is non-zero, but for a Christoffel connection the tensor $T^{\kappa\mu\nu}$ is zero. The Christoffel connection is therefore incompatible with the Bianchi identity and its Hodge dual. This is intuitively clear from the fact that one cannot neglect torsion, as argued in sections 9.2 and 9.3. This is an irretrievable flaw in EH theory and progress must be made by solving the Cartan equations without discarding torsion. This is precisely what ECE theory sets out to do and so the well accepted ECE theory is the only self-consistent theory of general relativity and of the generally covariant unified field.

Acknowledgments

The British Government is thanked for a Civil List Pension and the Telesio Galilei Association for generous funding of this work. The staff of AIAS and Telesio Galilei are thanked for many interesting discussions.

References

[1] S. P. Carroll, "Space-time and Geometry : an Introduction to General Relativity" (Addison Wesley, New York, 2004 and online notes), notably chapter three.

[2] M. W. Evans, "Generally Covariant Unified Field Theory" (Abramis Academic, Suffolk, 2005 to present), vols. 1–4, volume five in prep. (Papers 71 to 93 on www.aias.us).

[3] L. Felker, "The Evans Equations of Unified Field Theory" (Abramis Academic, Suffolk, 2007).

[4] K. Pendergast, "Crystal Spheres" (Abramis Academic, Suffolk in prep., preprint on www.aias.us), an account in historical context of ECE theory and the British Civil List Scientists from Newton to present.

[5] M. W. Evans et al., Omnia Opera Section of www.aias.us, precursor theories of ECE, which was initiated in Spring 2003 (1992 to 2003).

[6] M. W. Evans (ed.), "Modern Non-Linear Optics", a special topical issue in three parts of I. Prigogine and S. A. Rice (series editors), "Advances in Chemical Physics" (Wiley, New York, 2001, second edition), vols 119(1) to 119(3); ibid., M. W. Evans and S. Kielich (eds.), (Wiley, New York, 1992, 1993 and 1997, first edition), vols. 85(1) to 85(3).

[7] M. W. Evans and L. B. Crowell, "Classical and Quantum Electrodynamics and the B(3) Field" (World Scientific, 2001).

[8] M. W. Evans and J.-P. Vigier, "The Enigmatic Photon" (Kluwer, 1994 to 2002, hardback and softback), in five volumes.

[9] M. W. Evans and A. A. Hasanein, "The Photomagneton in Quantum Field Theory" (World Scientific, 1994).

[10] M. W. Evans, "The Photon's Magnetic Field" (World Scientific, 1992).

[11] M. W. Evans, Physica B, 182, 227 and 237 (1992), the first B(3) papers.

[12] L. H. Ryder, "Quantum Field Theory" (Cambridge University Press, 1996, second edition).

Development of the Einstein Hilbert Field Equation into the Einstein Cartan Evans (ECE) Field Equation

by

Myron W. Evans,
Alpha Institute for Advanced Study, Civil List Scientist.
(emyrone@aol.com and www.aias.us)

Abstract

Recently in this series of papers it has been shown that the Einstein Hilbert (EH) field equation is self-inconsistent because the use of the Christoffel connection is inconsistent with the one true Bianchi identity of Cartan geometry. The EH field equation is extended in this paper to include the effect of the Cartan torsion through the angular energy - momentum density tensor. The canonical energy - momentum density tensor of the EH equation is recognized as having an internal structure based on geometry. Therefore the ECE field equation is based on geometry rather than on the lagrangian method leading to the conventional Noether Theorem used in the derivation by Einstein of the EH equation.

Keywords: Einstein Cartan Evans (ECE) field equation, Einstein Hilbert (EH) field equation, Cartan torsion, angular energy momentum density tensor, energy momentum density tensor.

10.1 Introduction

Recently in this series of papers on the Einstein Cartan Evans (ECE) field theory [1–10] it has been shown that there is only one true Bianchi identity of Cartan geometry, and that the use of the Christoffel symbol in general relativity is inconsistent with this one true identity because of the neglect of the Cartan torsion in conventional general relativity [11, 12]. There are many criticisms available [12] of conventional general relativity but this newly discovered [1–10] incompatibility is an irretrievable flaw in the Einstein Hilbert (EH) field equation itself. The ECE theory retains the basic idea of general relativity, that physics or natural philosophy is geometry, but bases the subject of general relativity on the standard and well known Cartan geometry [11] of 1922. The latter is well taught and well known to be rigorously self-consistent [11]. However, this self consistency relies on the recognition of the torsion form of Cartan, as well as the curvature form of Cartan. The one true Bianchi identity is an equation which is a rigorously correct identity, one side of the equation being identically the same as the other [1–11]. In tensor notation the one true Bianchi identity is the cyclic sum of definitions of the Riemann tensor. The Riemann and torsion tensors are defined by the commutator of covariant derivatives acting on the four vector [1–11]. The one true Bianchi identity is a re-expression of this fundamental and well known [1–11] operation. The latter generates the Riemann (or curvature) and torsion tensors simultaneously. One is ineluctably linked to the other, and there is no a priori reason to assert that one or the other must vanish. This is a self-consistent conclusion of Riemann geometry, and Cartan geometry is equivalent to Riemann geometry through the well known and accepted tetrad postulate [1–11]. The latter is the fundamental requirement that the complete vector field be invariant under the general coordinate transformation. This is always true in natural philosophy.

The EH field equation was inferred independently by Einstein and Hilbert in 1915 and was based on the development by Einstein of general relativity from about 1907 onwards using a form of Riemann geometry in which torsion is implicitly zero. The existence of the torsion tensor was unknown in this era, and the Riemann geometry used by Einstein neglects torsion. Einstein inferred the EH equation by making the "second Bianchi identity" torsionless and thus self inconsistent Riemann geometry proportional to the Noether Theorem applied to energy momentum density only, neglecting spin. The covariant derivative of this linear momentum density tensor is zero, but no angular motion is considered. The proportionality constant is k, the Einstein constant. The torsion form was inferred by Cartan in 1922 and it can be translated into the torsion tensor [1–11] using the tetrad postulate. The latter is essential for the compatibility of Cartan and general Riemann geometry. When the torsion tensor is zero the connection must be the Christoffel connection, which is symmetric in its lower two indices. More generally the connection does not have this property, and in general the metric tensor is not symmetric. In conventional general relativity an equation is derived defining the Christoffel connection in terms of the symmetric metric by using the equation of metric compatibility [1–11]. However, the curvature tensor and torsion tensor are found by the operation of the commutator of covariant derivatives on the four vector, as argued already, and this operation does not assume any

10.2 Self-Inconsistency of Standard Model Riemann Geometry

particular symmetry of the connection or metric. Neither does it assume metric compatibility [1–11]. So conventional or standard model general relativity is based on entirely arbitrary assumptions concerning the symmetry of the metric and connection, and also assumes metric compatibility. Recently it has been found that this inherent arbitrariness leads to a fundamental internal inconsistency in standard model general relativity - the Bianchi identity is fundamentally inconsistent with the Christoffel connection as conventionally defined via the symmetric metric.

In Section 10.2 the geometrical reason for this inconsistency is given. In Section 10.3 a new field equation is inferred geometrically, and this is referred to as the ECE field equation. The latter uses the angular energy momentum density tensor as well as the energy momentum tensor, and shows that the two forms of energy momentum are always inter-convertible. Total energy momentum is conserved only for the whole universe, or for a well defined experimental situation. The curvature tensor can represent both angular or linear motion, and the torsion tensor can represent both angular and linear motion. They are two tensors which are bound together ineluctably by the commutator of covariant derivatives acting on the four vector in a spacetime with curvature and torsion. The history of general relativity is such that the torsion tensor was not known in the era leading to the EH equation of 1915. When the torsion form and tensor were inferred by Cartan in 1922 the EH equation remained as it was in 1915. So all claims to the existence of Big Bang, black holes and dark matter are based on an internally incorrect geometry and are therefore unscientific. There are many other criticisms [12] of conventional general relativity which must be addressed if progress is to be made. These criticisms have been made by many leading scientists [12] and ECE is a suggestion for progress.

10.2 Self-Inconsistency of Standard Model Riemann Geometry

The self-inconsistency is most clearly shown by considering the Hodge dual [1–11] of the one true Bianchi identity. Such a consideration leads to a simple tensor equation

$$D_\mu T^{\kappa\mu\nu} = R^\kappa{}_\mu{}^{\mu\nu} \tag{10.1}$$

which links the covariant derivative of the torsion tensor $T^{\kappa\mu\nu}$ to the Ricci type tensor $R^\kappa{}_\mu{}^{\mu\nu}$ on the right hand side. Summation over repeated internal indices means that this is a two index tensor. Summation over repeated indices occurs also on the left hand side of Eq. (10.1) so indices are balanced as required in tensor algebra. The Ricci-type tensor on the right hand side of

Eq. (10.1) was evaluated by computer algebra [1–11] using the Christoffel connection as defined in the standard model of general relativity:

$$\Gamma^\rho{}_{\mu\alpha} = \frac{1}{2}g^{\rho\lambda}(\partial_\mu g_{\lambda\alpha} + \partial_\alpha g_{\mu\lambda} - \partial_\lambda g_{\alpha\mu}) \tag{10.2}$$

and was found to be non-zero in general (papers 93 onwards of www.aias.us). It vanishes only when the conventional Ricci tensor is assumed to be zero by construction - the so called vacuum or Ricci flat solutions of EH. The Christoffel connection (10.2), as conventionally defined, is symmetric in its lower two indices, because the metric in standard model general relativity is assumed to be symmetric:

$$g_{\mu\nu} = g_{\nu\mu}. \tag{10.3}$$

So the torsion tensor in standard model general relativity is zero:

$$T^\kappa{}_{\mu\nu} = \Gamma^\kappa{}_{\mu\nu} - \Gamma^\kappa{}_{\nu\mu} = 0. \tag{10.4}$$

Therefore this leads to a fundamental internal self-inconsistency in standard model general relativity: the left hand side of Eq. (10.1) is always zero by construction (use of Eq. (10.2)) but the right hand side is not zero in general for the SAME Eq. (10.2).

The origin of this self-inconsistency is the arbitrary neglect of torsion. The one true Bianchi identity can be written most clearly in an indexless notation [1–10]:

$$D \wedge T := R \wedge q \tag{10.5}$$

where $D\wedge$ represents a type of exterior derivative $d\wedge$ supplemented by the spin connection, i.e. a type of covariant exterior derivative [11]. The wedge product is $\wedge$ and T, R and q represent respectively the torsion, curvature and tetrad forms of Cartan geometry [11]. If T is taken arbitrarily to be zero, Eq. (10.5) becomes:

$$R \wedge q = 0 \tag{10.6}$$

which is referred to in the literature as "the first Bianchi identity". In tensor notation Eq. (10.6) is:

$$R_{\kappa\mu\nu\rho} + R_{\kappa\rho\mu\nu} + R_{\kappa\nu\rho\mu} = 0 \tag{10.7}$$

and is true if and only if the metric is symmetric [1–11]. It is not an identity therefore, and was inferred not by Bianchi but by Ricci and Levi-Civita.

For these reasons it is referred to in ECE theory as the Ricci cyclic equation. It is clearly inconsistent with the true Bianchi identity (10.5) due to arbitrary neglect of torsion and this self inconsistency carries through to Eq. (10.1). Similarly the "second Bianchi identity" of standard model general relativity is:

$$D \wedge R = 0 \qquad (10.8)$$

and in tensor notation this becomes:

$$D_\rho R^\kappa{}_{\sigma\mu\nu} + D_\mu R^\kappa{}_{\sigma\nu\rho} + D_\nu R^\kappa{}_{\sigma\rho\mu} = 0. \qquad (10.9)$$

Again this neglects torsion arbitrarily, and is a special case of [1–10]:

$$D \wedge (D \wedge T) := D \wedge (R \wedge q). \qquad (10.10)$$

There is only one true Bianchi identity therefore, Eq. (10.5). The torsionless "second Bianchi identity" Eq. (10.9) can be re-expressed [1–11] as:

$$D^\mu \left(R_{\mu\nu} - \frac{1}{2} R g_{\mu\nu} \right) = 0 \qquad (10.11)$$

where $R_{\mu\nu}$ is the torsionless Ricci tensor, R is the torsionless scalar curvature and $g_{\mu\nu}$ is the torsionless metric. The torsionless Einstein tensor, finally, is defined as:

$$G_{\mu\nu} := R_{\mu\nu} - \frac{1}{2} R g_{\mu\nu}. \qquad (10.12)$$

The torsionless Noether Theorem is [11] is:

$$D^\mu T_{\mu\nu} = 0 \qquad (10.13)$$

and the torsionless EH equation is obtained from:

$$D^\mu G_{\mu\nu} = k D^\mu T_{\mu\nu} = 0 \qquad (10.14)$$

as the special case:

$$G_{\mu\nu} = k T_{\mu\nu}. \qquad (10.15)$$

This is the well known EH equation which is used as the basis for unscientific assertions such as the existence of black holes (also severely criticized on www.aias.us by Crothers), the existence of an unscientific Big Bang, and the existence of unscientific dark matter. Not only is the basic geometry of EH flawed, but also its methods of solution. Crothers has also criticized the concept of Ricci flat space-times, and has shown that the so called Schwarzschild metric is not due to Schwarzschild. The latter's original solution of 1916 does not contain the mass M, but a parameter alpha unrelated in general to M. Santilli [12] has summarized numerous criticisms of EH theory, criticisms made by well known scientists since inception of EH theory in 1915. Contemporary standard model proponents appear to be unaware of this scholarship, or if they are aware of it, neglect it in the same arbitrary manner in which they neglect Cartan's torsion. So the subject is unscientific and this must be recognized for progress. The following section is a suggestion for progress.

10.3 ECE Field Equation

The ECE equation is based on the proportionality of the Cartan torsion and the three index angular energy momentum density tensor [1–10]:

$$T^{\kappa\mu\nu} = kJ^{\kappa\mu\nu} \tag{10.16}$$

as inferred in previous work. The structure of Eq. (10.1) then suggests that a similar proportionality exists between the Ricci type tensor $R^{\kappa}{}_{\mu}{}^{\mu\nu}$ and an energy momentum density tensor:

$$R^{\kappa}{}_{\mu}{}^{\mu\nu} = kT^{\kappa}{}_{\mu}{}^{\mu\nu}. \tag{10.17}$$

Therefore we arrive at the field equation:

$$D_{\mu}J^{\kappa\mu\nu} = T^{\kappa}{}_{\mu}{}^{\mu\nu} \tag{10.18}$$

on this geometrical basis. This field equation suggests in turn that there exists a conservation of energy equation:

$$D_{\kappa}(D_{\mu}J^{\kappa\mu\nu}) = D_{\kappa}T^{\kappa}{}_{\mu}{}^{\mu\nu} \tag{10.19}$$

which is a balance between linear and angular energy momentum. In general both sides of Eq. (10.19) are non-zero. Therefore Eq. (10.19) represents a balance of linear and angular energy - momentum densities, both being always non-zero in general. There is no a priori reason for assuming that one or the other should be zero. As a matter of semantics it is possible to assert that "pure translation" is represented by:

$$J^{\kappa\mu\nu} = 0 \tag{10.20}$$

and

$$D_\kappa T^\kappa{}_\mu{}^{\mu\nu} = 0. \tag{10.21}$$

This is the Noether Theorem used in the EH equation if the following identification is made:

$$\begin{aligned} T^{\kappa\nu} &:= T^\kappa{}_\mu{}^{\mu\nu} \\ &= T^\nu{}_\mu{}^{\mu\kappa} \end{aligned} \tag{10.22}$$

This symmetry follows from Eq. (10.17) if the Christoffel connection is used to calculate the Riemann tensor leading to the Ricci type tensor $R^\kappa{}_\mu{}^{\mu\nu}$. Similarly, it is possible to assert that "pure rotation" is given by:

$$T^\kappa{}_\mu{}^{\mu\nu} = 0 \tag{10.23}$$

and

$$D_\mu J^{\kappa\mu\nu} = 0. \tag{10.24}$$

This equation is conservation of angular energy momentum density using the covariant derivative. It is the covariant version of an equation given in chapter three of Ryder [13] but it does not appear in Einstein's general relativity [13] because the latter did not consider torsion and did not consider angular energy momentum density. Eq. (10.18) asserts that energy momentum can be interconverted between linear and angular forms, but it is arbitrary to assert a state of pure translation or pure rotation. Similarly, it is arbitrary to assert that torsion and curvature can exist irrespective of each other in geometry. Total energy momentum is conserved as in Eq. (10.19), but this total energy momentum must be carefully defined for each situation under consideration.

Einstein derived the Poisson equation leading to the Newton inverse square law by consideration of a weak field limit of the EH equation. In this derivation the Christoffel symbol is used in the limit of slowly moving fields. However the underlying geometrical entity is the curvature tensor. In the weak field limit the space-time is approximated by a Minkowski space-time in which the curvature becomes infinitesimally small, so the distance between two point masses becomes not curve but a straight line. Using the same ideas in Eq. (10.18), the covariant derivative is replaced in the weak field limit by the ordinary derivative:

$$D_\mu \to \partial_\mu \tag{10.25}$$

so we obtain:

$$\partial_\mu T^{\kappa\mu\nu} \doteq R^\kappa{}_\mu{}^{\mu\nu}. \tag{10.26}$$

Using $\nu = 0$, Eq. (10.26) reduces to:

$$\nabla \cdot \boldsymbol{g} = c^2 R \tag{10.27}$$

where R has the units of inverse meters squared, i.e. of scalar curvature. Eq. (10.17) gives:

$$R = k\rho_m \tag{10.28}$$

where ρ_m is mass density in kilograms per meters cubed. So:

$$\nabla \cdot \boldsymbol{g} = kc^2 \rho_m \tag{10.29}$$

which can be compared with the Coulomb law of ECE theory [1–11]:

$$\nabla \cdot \boldsymbol{E} = \rho_e/\epsilon_0 \tag{10.30}$$

where ρ_e is charge density in coulombs per meters cubed. It is well known that Eq. (10.30) is a form of the Coulomb inverse square law, so Eq. (10.29) is the Newton inverse square law, derived straightforwardly in the weak field limit Eq. (10.25) from the tensor equation (10.1) that originates in the Hodge dual of the Bianchi identity.

Acknowledgments

The British Government is thanked for a Civil List Pension and the Telesio-Galilei Association for funding this work. The staff of AIAS and others are thanked for many interesting discussions.

References

[1] M. W. Evans, "Generally Covariant Unified Field Theory" (Abramis Academic, Suffolk, 2005 onwards), volumes 1–4, ibid., volume five in prep. (www.aias.us, papers 71–93).

[2] L. Felker, "The Evans Equations of Unified Field Theory" (Abramis, Suffolk, 2007).

[3] K. Pendergast, "Crystal Spheres" preprint on www.aias.us, (Abramis, in prep.).

[4] M. W. Evans, et al, over one hundred papers, 2003 to present on www.aias.us, and educational articles on ECE theory.

[5] M. W. Evans (ed.), "Modern Nonlinear Optics", a special topical issue in three parts of I. Prigogine and S. A. Rice (series eds.,), "Modern Nonlinear Optics" (Wiley, New York, 2001, second edition), vols. 119(1) to 119(3); M. W. Evans and S. Kielich (eds.), ibid., first edition (Wiley, New York, 1992, 1993, 1997), vols. 85(1) to 85(3).

[6] M. W. Evans and L. B. Crowell, "Classical and Quantum Electrodynamics and the B(3) Field" (World Scientific, 2001).

[7] M. W. Evans and J.-P. Vigier, "The Enigmatic Photon" (Kluwer, 1994 to 2002 in hardback and softback), in five volumes.

[8] M. W. Evans et al., Omnia Opera section of www.aias.us (1992 to 2003), precursor theories of ECE theory in circa twenty five journals.

[9] M. W. Evans and A. A. Hasanein, "The Photomagneton in Quantum Field Theory" (World Scientific, 1994).

[10] S. P. Carroll, "Space-time and Geometry: an Introduction to General Relativity" (Addison Wesley, New York, 2004).

[11] R. Santilli et al., on www.telesio-galilei.com.

[12] L. H. Ryder, "Quantum Field Theory" (Cambridge Univ. Press, 1996).

[13] A. Einstein, "The Meaning of Relativity" (Princeton Univ. Press, 1921–1954).

11

A Rigorous Proof of the Hodge Dual of the Bianchi Identity of Cartan

by

Myron W. Evans,
Alpha Institute for Advanced Study, Civil List Scientist.
(emyrone@aol.com and www.aias.us)

Abstract

Using the action of the commutator of covariant derivatives on the four vector in a four dimensional space-time, it is proven that there exists a well defined Hodge dual of the Bianchi identity of Cartan for any metric and any connection.

Keywords: Bianchi identity of Cartan, Hodge dual, ECE theory, four-dimensional space-time.

11.1 Introduction

It is well known that the Bianchi identity as developed by Cartan [1] relates the torsion form and curvature form. It has been shown during the development [2–10] of ECE theory that this Bianchi identity is a rigorous identity that states that the cyclically symmetric sum of three curvature tensors is identically equal to the same cyclically symmetric sum of fundamental definitions of the same curvature tensors. The definitions of the curvature and torsion tensors arise [1] from the action of the commutator of covariant derivatives on the four vector. The curvature and torsion tensors are defined by the same commutator, and therefore the one tensor cannot exist without the other tensor. The Bianchi identity is derivable from the commutator given the tetrad postulate [2–10], and the Bianchi identity must always link the torsion to the curvature. In Section 2 a well defined Hodge dual of the same

Bianchi identity is derived rigorously by considering the Hodge dual of the commutator of covariant derivatives acting on the four vector. The resulting Hodge dual identity, when developed in tensor notation in the base manifold, shows that there is a fundamental self inconsistency in the Einstein Hilbert (EH) theory of general relativity and in the Einstein Hilbert field equation. The Christoffel connection of the EH theory is fundamentally incompatible with the Bianchi identity as developed by Cartan. In previous work [2–10] a development of the EH equation has been suggested, a development based on the Bianchi identity. Furthermore, the Einstein Cartan Evans (ECE) field equations of dynamics and electrodynamics are based on the Bianchi identity of Cartan and its Hodge dual, proven rigorously in Section 11.2.

11.2 Proof of the Hodge Dual Bianchi Identity

Consider the commutator of covariant derivatives acting on the four vector V^ρ in a four-dimensional space-time:

$$[D_\mu, D_\nu]V^\rho = R^\rho{}_{\sigma\mu\nu}V^\sigma - T^\lambda_{\mu\nu}D_\lambda V^\rho \qquad (11.1)$$

Here $R^\rho{}_{\sigma\mu\nu}$ is the curvature tensor [1], $T^\lambda_{\mu\nu}$ is the torsion tensor, and D_λ is the covariant derivative. It has been shown [2–10] that Eq. (11.1) is the basis for the Bianchi identity of Cartan [1–10]:

$$D \wedge T := R \wedge q \qquad (11.2)$$

where a shorthand notation has been adopted suppressing indices for structural clarity. In this notation T is the torsion form of Cartan's differential geometry, R is the Riemann form, q is the tetrad form, $\wedge$ is the wedge product and $D\wedge$ is the covariant exterior derivative of Cartan's differential geometry. In order for Eq. (11.2) to be true, the basic definitions generated by Eq. (11.1) must be used in Eq. (11.2), and the tetrad postulate [1–10] must also be used. The Bianchi identity (11.2) is the basic structure used for the homogeneous field equation of ECE theory, both in dynamics and electrodynamics.

It is proven that there exists the following identity:

$$D \wedge \widetilde{T} := \widetilde{R} \wedge q \qquad (11.3)$$

which is the basis for the inhomogeneous field equation of ECE theory, both in dynamics and electrodynamics. The proof is as follows.

Consider the Hodge dual transformations [1–10]:

$$[D^\mu, D^\nu]_{HD} = \frac{1}{2} \parallel g \parallel^{\frac{1}{2}} \epsilon^{\mu\nu\alpha\beta}[D_\alpha, D_\beta], \qquad (11.4)$$

11.2 Proof of the Hodge Dual Bianchi Identity

$$\tilde{R}^\rho{}_\sigma{}^{\mu\nu} = \frac{1}{2} \parallel g \parallel^{\frac{1}{2}} \epsilon^{\mu\nu\alpha\beta} R^\rho{}_{\sigma\alpha\beta}, \tag{11.5}$$

$$\tilde{T}^{\lambda\mu\nu} = \frac{1}{2} \parallel g \parallel^{\frac{1}{2}} \epsilon^{\mu\nu\alpha\beta} T^\lambda{}_{\alpha\beta}. \tag{11.6}$$

Here $\epsilon^{\mu\nu\alpha\beta}$ is the four dimensional totally anti-symmetric unit tensor of Minkowski space-time [1], and the Hodge dual transformations are weighted by definition [1] by the square root of the modulus of the determinant of the metric, denoted by $\|g\|^{1/2}$. The Hodge dual is denoted by ν. In four dimensions, the Hodge dual of an anti-symmetric tensor (i.e. differential two-form [1–10]) is another anti-symmetric tensor. Using Eqs. (11.4) to (11.6) in Eq. (11.1):

$$[D^\mu, D^\nu]_{HD} V^\rho = \tilde{R}^\rho{}_\sigma{}^{\mu\nu} V^\sigma - \tilde{T}^{\lambda\mu\nu} D_\lambda V^\rho \tag{11.7}$$

where the subscript HD on the left hand side denotes Hodge dual of the commutator, which is anti-symmetric in μ and ν. Eq. (11.7) proves Eq. (11.3) with raised indices μ and ν. In order to obtain the final form of the ECE field equations it is necessary to prove Eq. (11.7) with lowered indices μ and ν. To lower indices requires the use of the metric by definition [1–10]. Therefore the three anti-symmetric tensors in Eq. (11.7) are expressed in terms of their equivalents with lowered indices as follows:

$$[D^\mu, D^\nu]_{HD} = g^{\mu\alpha} g^{\nu\beta} [D_\alpha, D_\beta]_{HD}, \tag{11.8}$$

$$\tilde{R}^\rho{}_\sigma{}^{\mu\nu} = g^{\mu\alpha} g^{\nu\beta} \tilde{R}^\rho{}_{\sigma\alpha\beta}, \tag{11.9}$$

$$\tilde{T}^{\lambda\mu\nu} = g^{\mu\alpha} g^{\nu\beta} \tilde{T}^\lambda{}_{\alpha\beta}. \tag{11.10}$$

It follows by use of Eqs. (11.8) to (11.10) in Eq. (11.7) that:

$$g^{\mu\alpha} g^{\nu\beta} [D_\alpha, D_\beta]_{HD} V^\rho = g^{\mu\alpha} g^{\nu\beta} (\tilde{R}^\rho{}_{\sigma\alpha\beta} V^\sigma - \tilde{T}^\lambda{}_{\alpha\beta} D_\lambda V^\rho) \tag{11.11}$$

a particular solution of which is:

$$[D_\alpha, D_\beta]_{HD} V^\rho := \tilde{R}^\rho{}_{\sigma\alpha\beta} V^\sigma - \tilde{T}^\lambda{}_{\alpha\beta} D_\lambda V^\rho. \tag{11.12}$$

This is the required identity with lowered indices. Eq. (11.12) means that:

$$(D \wedge \tilde{T}^a)_{\mu\nu} := (\tilde{R}^a{}_b \wedge q^b)_{\mu\nu} \tag{11.13}$$

with lowered μ and ν indices, and it is seen that the metric has been eliminated from consideration provided that the particular solution (11.12) is used. In tensor notation Eq. (11.13) is:

$$D_\mu \widetilde{T}^a_{\nu\rho} + D_\rho \widetilde{T}^a_{\mu\nu} + D_\nu \widetilde{T}^a_{\rho\mu} := \widetilde{R}^a_{\mu\nu\rho} + \widetilde{R}^a_{\rho\mu\nu} + \widetilde{R}^a_{\nu\rho\mu} \qquad (11.14)$$

where the rules for wedge product and covariant exterior derivative [1–10] have been used. Eq. (11.14) is the same as:

$$D_\mu T^{a\mu\nu} := R^a{}_\mu{}^{\mu\nu} \qquad (11.15)$$

because the four dimensional totally anti-symmetric unit tensor is the same by definition [1–10] as that used in Minkowski or flat space-time. The factor $\|g\|^{1/2}$ has been cancelled out on both sides of Eq. (11.15). A particular solution of Eq. (11.15) is the base manifold equation:

$$D_\mu T^{\kappa\mu\nu} = R^\kappa{}_\mu{}^{\mu\nu} \qquad (11.16)$$

which is the inhomogeneous field equation of ECE theory, Q.E.D.

By compute algebra [2–10] it has been shown that for the Christoffel connection:

$$\Gamma^\lambda_{\mu\nu} = \Gamma^\lambda_{\nu\mu} \qquad (11.17)$$

the right hand side of Eq. (11.16) is not zero in general. The computer algebra shows that it is zero only in a Ricci flat space-time. For the same Christoffel connection the left hand side of Eq. (11.16) is always zero because:

$$T^\lambda_{\mu\nu} = \Gamma^\lambda_{\mu\nu} - \Gamma^\lambda_{\nu\mu}. \qquad (11.18)$$

So the use of the Christoffel connection is incompatible with fundamental geometry.

This conclusion signals the collapse of the EH theory of general relativity and conclusions based thereon, because all so called "exact solutions" of EH are based on the Christoffel symbol and line elements deduced therefrom. Other severe limitations and internal inconsistencies of EH are given by Crothers on www.aias.us and Santilli on www.telesio-galilei.com.

Similarly, the Bianchi identity (11.2) translates into:

$$D_\mu T^a_{\nu\rho} + D_\rho T^a_{\mu\nu} + D_\nu T^a_{\rho\mu} := R^a_{\mu\nu\rho} + R^a_{\rho\mu\nu} + R^a_{\nu\rho\mu} \qquad (11.19)$$

which in form notation is:

$$(D \wedge T^a)_{\mu\nu} := (R^a{}_b \wedge q^b)_{\mu\nu}. \qquad (11.20)$$

11.2 Proof of the Hodge Dual Bianchi Identity

This equation is the same as:

$$D_\mu \widetilde{T}^{a\mu\nu} := \widetilde{R}^a{}_\mu{}^{\mu\nu}. \qquad (11.21)$$

A particular solution of Eq. (11.21) is:

$$D_\mu \widetilde{T}^{\kappa\mu\nu} = \widetilde{R}^\kappa{}_\mu{}^{\mu\nu} \qquad (11.22)$$

which is the homogeneous ECE field equation's structure. Eqs. (11.16) and (11.22) are the ECE field equations for dynamics, and using the fundamental hypothesis:

$$F^a_{\mu\nu} = A^{(0)} T^a_{\mu\nu} \qquad (11.23)$$

become the ECE field equations for electrodynamics. The homogeneous electro-dynamical equation is:

$$D_\mu \widetilde{F}^{\kappa\mu\nu} = A^{(0)} \widetilde{R}^\kappa{}_\mu{}^{\mu\nu} \qquad (11.24)$$

and the inhomogeneous electro-dynamical equation is:

$$D_\mu F^{\kappa\mu\nu} = A^{(0)} R^\kappa{}_\mu{}^{\mu\nu}. \qquad (11.25)$$

The covariant derivative may be expanded in terms of the spin connection to give:

$$\partial_\mu \widetilde{F}^{\kappa\mu\nu} = A^{(0)} (\widetilde{R}^\kappa{}_\mu{}^{\mu\nu} - \omega^\kappa_{\mu\lambda} \widetilde{T}^{\lambda\mu\nu}), \qquad (11.26)$$

and

$$\partial_\mu F^{\kappa\mu\nu} = A^{(0)} (R^\kappa{}_\mu{}^{\mu\nu} - \omega^\kappa_{\mu\lambda} T^{\lambda\mu\nu}), \qquad (11.27)$$

Experimental data show [2–10] that for all practical purposes (F. A. P. P.) in the laboratory:

$$\widetilde{R}^\kappa{}_\mu{}^{\mu\nu} = \omega^\kappa_{\mu\lambda} \widetilde{T}^{\lambda\mu\nu} \qquad (11.28)$$

but:

$$R^\kappa{}_\mu{}^{\mu\nu} \neq \omega^\kappa_{\mu\lambda} T^{\lambda\mu\nu}. \qquad (11.29)$$

Using these data, Eqs. (11.26) and (11.27) reduce to:

$$\partial_\mu \widetilde{F}^{\kappa\mu\nu} = 0, \tag{11.30}$$

$$\partial_\mu F^{\kappa\mu\nu} = J^{\kappa\nu}/\epsilon_0. \tag{11.31}$$

It has been shown [2–10] that in vector form these become the same as the Maxwell Heaviside field equations as follows:

$$\boldsymbol{\nabla} \cdot \boldsymbol{B} = 0, \tag{11.32}$$

$$\boldsymbol{\nabla} \times \boldsymbol{E} + \frac{\partial \boldsymbol{B}}{\partial t} = \boldsymbol{0}, \tag{11.33}$$

$$\boldsymbol{\nabla} \cdot \boldsymbol{E} = \rho/\epsilon_0, \tag{11.34}$$

$$\boldsymbol{\nabla} \times \boldsymbol{B} - \frac{1}{c^2}\frac{\partial \boldsymbol{E}}{\partial t} = \mu_0 \boldsymbol{J}. \tag{11.35}$$

but written in a more general space-time with curvature and torsion. The relation between the electric and magnetic fields and the potentials are developed as follows in terms of the spin connection vector $\boldsymbol{\omega}$, and scalar:

$$\boldsymbol{E} = -\boldsymbol{\nabla}\phi - \frac{\partial \boldsymbol{A}}{\partial t} + \phi\boldsymbol{\omega} - \omega\boldsymbol{A}, \tag{11.36}$$

$$\boldsymbol{B} = \boldsymbol{\nabla} \times \boldsymbol{A} - \boldsymbol{\omega} \times \boldsymbol{A}. \tag{11.37}$$

In standard S.I. units **B** is the magnetic flux density, **E** is the electric field strength, ϵ_0 is the vacuum permittivity, ρ is the electric charge density, μ_0 is the vacuum permeability, and **J** is the electric current density. In Eqs. (11.36) and (11.37) is the scalar potential, **A** is the vector potential, $\boldsymbol{\omega}$ is the spin connection vector and ω is the spin connection scalar. It is seen that all the equations are derived from the Bianchi identity.

Acknowledgments

The British Government is thanked for the high honour of a Civil List Pension and the staff of AIAS and Telesio-Galilei for many interesting discussions.

References

[1] S. P. Carroll, "Space-time and Geometry: an Introduction to General Relativity" (Addison Wesley, New York, 2004, and lecture notes of 1997 downloadable from the web).

[2] M. W. Evans, "Generally Covariant Unified Field Theory : the Geometrization of Physics" (Abramis, 2005 onwards), in four volumes to date, volume five in prep. (Papers 71 to 93 of www.aias.us).

[3] L. Felker, "The Evans Equations of Unified Field Theory" (Abramis, 2007).

[4] K. Pendergast, "Crystal Spheres" (Abramis in preps. Preprint on www.aias.us).

[5] M. W. Evans and H. Eckardt, Physica B, **400**, 175 (2007).

[6] M. W. Evans, Acta Phys. Polonica B, **38**, 2211 (2007).

[7] M. W. Evans, Physica B, **403**, 517 (2008).

[8] M. W. Evans et al., Omnia Opera section of www.aias.us, 1992 to present.

[9] M. W. Evans and J.-P. Vigier, "The Enigmatic Photon" (Kluwer Dordrecht, 1994 to 2002, hardback and softback), in five volumes.

[10] M. W. Evans, "Modern Non-Linear Optics", in "Advances in Chemical Physics" (Wiley Interscience, 1992 to 2001, first and second editions, hardback and softback), vols. 85 and 119; M. W. Evans and L. B. Crowell, "Classical and Quantum Electrodynamics and the B(3) Field" (World scientific, 2001).

A New Theory of Light Deflection Due to Gravitation

by

Myron W. Evans,
Alpha Institute for Advanced Study, Civil List Scientist.
(emyrone@aol.com and www.aias.us)

Abstract

A new theory of light deflection due to gravitation is proposed based on Einstein Cartan Evans (ECE) unified field theory. It is shown that the Schwarzschild parameter of 1916 can be expressed as:

$$\alpha = -T/R$$

where T/R is the ratio of torsion to curvature, c is the vacuum speed of light and G is Newton's constant. In the Einstein Hilbert theory alpha is identified with the mass M of the object that deflects light. However this identification of alpha with M was not made by Schwarzschild and conflicts with the concept of Ricci flat vacuum used in the EH equation. In the new theory the deflection of light is governed by the ratio R/T and the theory suggests a simple explanation for the Pioneer anomaly, in which a small and anomalous gravitational attraction towards the sun has been found within experimental error.

Keywords: Light deflection due to gravitation, ECE theory, Pioneer anomaly.

12.1 Introduction

Recently a generally covariant unified field theory (GCUFT) has been developed [1–10] and accepted as providing one means of unifying physics with geometry, the aim of relativity theory. During the course of development of the theory it was found that the geometry of the Einstein Hilbert field equation is incompatible with the Hodge dual of the Bianchi identity. This incompatibility was discovered through the use of computer algebra, which showed that the Ricci type tensor $R^\kappa{}_\mu{}^{\mu\nu}$ is non-zero for all space-times that are not Ricci flat space-times. This result is incompatible with the fact that the inhomogeneous field equation of ECE theory is

$$D_\mu T^{\kappa\mu\nu} = R^\kappa{}_\mu{}^{\mu\nu} \tag{12.1}$$

On the left hand side occurs the covariant derivative of the torsion tensor $T^{\kappa\mu\nu}$, and for all Christoffel symbols this must be zero [11]. On the right hand side occurs the tensor $R^\kappa{}_\mu{}^{\mu\nu}$ which is non-zero in general as argued.

Therefore the Einstein Hilbert equation cannot be used to correctly describe the phenomenon of light deflection by an object of mass M. In order to achieve this aim this paper sets out to develop an expression for the parameter α used by Schwarzschild in 1916 [1–10] in two exact solutions of the EH field equation. It is known from this work by Schwarzschild that a line element of the type:

$$ds^2 = -(1-\alpha)c^2 dt^2 + (1-\alpha)^{-1} dr^2 + r^2 d\Omega^2 \tag{12.2}$$

with

$$\alpha = \frac{2GM}{c^2 r} \tag{12.3}$$

describes the light deflection due to gravitation to an accuracy of 1: 100,000 [1–10]. Here G is the Newton constant, M is the mass of the gravitating object, c is the speed of light and r is the distance between the center of mass of the gravitating object and a point mass m called the particle. In the case of light being deflected by M the particle m is the photon. In the EH theory however, the Schwarzschild solution (12.2) is obtained with a Christoffel connection and by using a Ricci flat assumption. The Christoffel connection is incompatible with the Hodge dual of the Bianchi identity 1-10 and Crothers [1–10] has shown that the Ricci flat assumption is incompatible with the equivalence principle. Also, the Ricci flat assumption means a vanishing Ricci tensor, so that there is no energy momentum density present, no source present, and therefore no M. This is why the Schwarzschild solution is known as a vacuum solution. Computer algebra shows that the tensor $R^\kappa{}_\mu{}^{\mu\nu}$ is zero only for a vacuum solution, otherwise it is non-zero, but the torsion tensor is always zero. This is self-inconsistent and is the result of assuming a Christoffel connection. In the presence of torsion, the latter cannot be used to produce line elements, and conversely, a line element such as (12.2) cannot be used to compute $R^\kappa{}_{\mu}{}^{\mu\nu}$ through the use of a Christoffel symbol and a symmetric metric. These are all procedures of the EH equation, which arbitrarily and incorrectly neglects torsion. In the correct procedure adopted in this paper, the tensor and the torsion tensor are always non-zero, there is always a source for gravitation, and both the metric and connection are asymmetric. In this paper no assumption is made concerning their symmetry.

There are several reasons therefore for discarding the EH explanation of light bending due to gravitation, primarily because it neglects torsion, and also neglects the fact that a finite source M requires a finite energy momentum density. In general, the Christoffel connection must be asymmetric for finite torsion, and both the tensors $T^{\kappa\mu\nu}$ and $R^\kappa{}_{\mu}{}^{\mu\nu}$ must be non-zero. These fundamental geometrical requirements of Cartan geometry are reviewed in Section 12.2, leading to a definition of **g** in terms of elements of the torsion tensor. In Section 12.3 a new explanation of the experimental result for light bending is given, and it is shown that the Schwarzschild parameter is:

$$\alpha = -\frac{T}{R} \qquad (12.4)$$

where T/R is a ratio of a well defined scalar torsion to a well defined scalar curvature. The experimental result of NASA Cassini is therefore produced when:

$$\alpha = -\frac{T}{R} = \frac{2GM}{c^2} \cdot \frac{1}{r} \qquad (12.5)$$

but the small Pioneer anomaly [11, 12] requires alpha to be different from this. This is straightforwardly explained in ECE theory when the ratio T/R is slightly different from that needed to give M, the mass of the object that deflects light. There is no explanation for the Pioneer anomaly in the EH theory, as is well known. The calculation leading to (12.4) does not make any assumptions concerning the metric or connection, and self consistently accounts for the existence of torsion as well as curvature in the general four dimensional space-time. The result (12.4) is also compatible with both the Bianchi identity and the Hodge dual of the Bianchi identity.

12.2 The Fundamental Geometry

It has been shown in previous work that the Bianchi identity as developed by Cartan is, in an index-less notation [1–10]:

$$D \wedge T := R \wedge q \qquad (12.6)$$

where $D\wedge$ denotes the covariant derivative, T denotes the torsion form, R denotes the curvature form, $\wedge$ denotes the wedge product and q denotes the tetrad form. In the base manifold, Eq. (12.6) is equivalent to the tensorial expression:

$$D_\mu \tilde{T}^{\kappa\mu\nu} = \tilde{R}^\kappa{}_\mu{}^{\mu\nu} \tag{12.7}$$

where D_μ is the covariant derivative, $\tilde{T}^{\kappa\mu\nu}$ is the Hodge dual of the torsion tensor, and where $\tilde{R}^\kappa{}_\mu{}^{\mu\nu}$ is the Hodge dual of the curvature tensor $R^\kappa{}_{\mu\alpha\beta}$. It has been shown [1–10] that there also exists the Hodge dual of Eq. (12.6), denoted:

$$D \wedge \tilde{T} := \tilde{R} \wedge q \tag{12.8}$$

and that the tensorial expression of Eq. (12.8) is:

$$D_\mu T^{\kappa\mu\nu} = R^\kappa{}_\mu{}^{\mu\nu}. \tag{12.9}$$

Eqs. (12.6) to (12.9) are the equations of gravitation in the presence of torsion. In general, neither the metric nor the connection in these equations can be symmetric, because a symmetric metric produces:

$$T^{\kappa\mu\nu} = 0, \quad R^\kappa{}_\mu{}^{\mu\nu} \neq 0 \tag{12.10}$$

which is incompatible with Eq. (12.10). The geometry of the Einstein Hilbert (EH) equation produces Eq. (12.10), and so a new approach to the subject of gravitation is needed, one that is fully compatible with the Bianchi identity (12.6) and with its Hodge dual (12.8).

In the EH theory:

$$\tilde{R}^\kappa{}_\mu{}^{\mu\nu} = 0 \tag{12.11}$$

because:

$$R^\kappa{}_{\mu\nu\rho} + R^\kappa{}_{\rho\mu\nu} + R^\kappa{}_{\nu\rho\mu} = 0. \tag{12.12}$$

This is the same as the so-called [11] "first Bianchi identity":

$$R \wedge q = q \wedge R = 0. \tag{12.13}$$

Eqs. (12.11) and (12.12) are true if and only if:

$$g_{\mu\nu} = g_{\nu\mu} \tag{12.14}$$

and

$$\Gamma^\kappa_{\mu\nu} = \Gamma^\kappa_{\nu\mu} \tag{12.15}$$

which is the Christoffel connection. Under the geometrically arbitrary assumptions (12.14) and (12.15) the torsion tensor vanishes:

$$T^\kappa_{\mu\nu} = \Gamma^\kappa_{\mu\nu} - \Gamma^\kappa_{\nu\mu} = 0. \tag{12.16}$$

It has also been shown that there is only one true Bianchi identity (12.6) in Cartan geometry. The so called "second Bianchi identity" of EH theory is a special case of Eq. (12.6) when there is no torsion. This was shown [1–10] by taking the covariant exterior derivative on both sides of Eq. (12.6) to give:

$$D \wedge (D \wedge T) := D \wedge (R \wedge q). \tag{12.17}$$

This reduces to

$$D \wedge R = 0 \tag{12.18}$$

when the torsion is arbitrarily neglected. Reinstating the indices in Eq. (12.18) gives:

$$D \wedge R^\kappa{}_{\mu\nu\rho} = 0 \tag{12.19}$$

i.e:

$$D_\sigma R^\kappa{}_{\mu\nu\rho} + D_\rho R^\kappa{}_{\mu\sigma\nu} + D_\nu R^\kappa{}_{\mu\rho\sigma} = 0 \tag{12.20}$$

which is almost always referred to in the EH literature as "the second Bianchi identity", whereas the true identity is Eq. (12.6) and its Hodge dual, Eq. (12.8).

The usual EH approach to light deflection in the solar system is therefore geometrically incorrect. It also suffers from the assumption that the Ricci tensor is zero by construction [1–10], and from the assumption that the Schwarzschild parameter α of 1916 is assumed arbitrarily and incorrectly [1–10] to be determined by the mass M of the attracting object (the solar mass for example). So the EH method is self-inconsistent fundamentally. Furthermore, Crothers [1–10] has shown that the method of solution used in EH field theory is also fundamentally flawed, notably the proper radius and radius of curvature of a line element is confused [1–10]. This confusion means that an initial singularity cannot be assumed in the EH theory, and cannot be deduced geometrically. Therefore theories that depend on an initial singularity such as Big Bang, cannot be correct geometrically. They have apparently developed uncritically for ninety years. There are several other well known criticisms and limitations of the EH theory on a website such as www.telesio-galilei.com.

The geometrically correct equations of Cartan geometry are (12.6) and (12.8). In principle these must be solved without any a priori assumption about the symmetry of the metric and connection. Therefore this solution in general must be a numerical one, using a supercomputer. In an analytical approximation however the weak field limit may be considered. The latter can be defined as the limiting approach to Minkowski or flat space-time. In this limit it is assumed that the spin connection goes to zero, so Eq. (12.9) is approximated by:

$$\partial_\mu T^{\kappa\mu\nu} \doteq R^\kappa{}_\mu{}^{\mu\nu}. \tag{12.21}$$

In vector notation [1–10], Eq. (12.21) is:

$$\boldsymbol{\nabla} \cdot \boldsymbol{T}_1 = R_1, \tag{12.22}$$

$$\boldsymbol{\nabla} \times \boldsymbol{T}_2 - \frac{1}{c}\frac{\partial \boldsymbol{T}_3}{\partial t} = \boldsymbol{R}_2, \tag{12.23}$$

were the subscripts denote particular types of torsion and curvature defined by:

$$\boldsymbol{T}_1 = T^{010}\boldsymbol{i} + T^{020}\boldsymbol{j} + T^{030}\boldsymbol{k}, \tag{12.24}$$

$$R_1 = R^0{}_1{}^{10} + R^0{}_2{}^{20} + R^0{}_3{}^{30}, \tag{12.25}$$

$$\boldsymbol{T}_2 = T^{332}\boldsymbol{i} + T^{113}\boldsymbol{j} + T^{221}\boldsymbol{k}, \tag{12.26}$$

$$\boldsymbol{T}_3 = T^{110}\boldsymbol{i} + T^{220}\boldsymbol{j} + T^{330}\boldsymbol{k}, \tag{12.27}$$

and where $\boldsymbol{R}_2$ is defined by Eqs. (12.26) and (12.27). The Newton inverse square law is obtained in this weak field limit as:

$$\boldsymbol{\nabla} \cdot \boldsymbol{g} = c^2 R. \tag{12.28}$$

The structures of Eqs. (12.22) and (12.28) are the same so:

$$\boldsymbol{g} = c^2 \boldsymbol{T}_1, R = R_1. \tag{12.29}$$

The acceleration due to gravity in the weak field limit is therefore:

$$\boldsymbol{g} = c^2(T^{010}\boldsymbol{i} + T^{020}\boldsymbol{j} + T^{030}\boldsymbol{k}) \tag{12.30}$$

and the mass density is defined by [1–11]:

$$\rho_m = \frac{1}{k}(R^0{}_1{}^{10} + R^0{}_2{}^{20} + R^0{}_3{}^{30}) \qquad (12.31)$$

where k is Einstein's constant.

In a Ricci flat space-time, and using a symmetric metric and connection, it is found from Eq. (12.31) using computer algebra [1–10] that:

$$\rho_m(\text{Ricci flat}) = 0. \qquad (12.32)$$

In such a space-time there is therefore no acceleration due to gravity because:

$$\mathbf{g} = \mathbf{0}. \qquad (12.33)$$

The elements T^{010}, T^{020}, and T^{030} are proportional to elements of the angular energy momentum density tensor. In other words if there is no source, i.e. no angular energy momentum density, there is no mass density ($\rho_m = 0$) and no field ($\mathbf{g} = \mathbf{0}$). Therefore the Ricci flat approach to solutions of the EH equation is self-inconsistent. Light bending due to gravitation is governed by Eq. (12.28), and by six components $T^{010}, T^{020}, T^{030}, R^0{}_1{}^{10}, R^0{}_2{}^{20}$ and $R^0{}_3{}^{30}$ in general. If $\mathbf{g}$ is restricted to one axis, e.g. $\mathbf{k}$, then the problem reduces to two components, T^{030} and $R^0{}_3{}^{30}$, i.e.:

$$\mathbf{g} = c^2 T^{030} \mathbf{k}, \quad \rho_m = \frac{1}{k} R^0{}_3{}^{30}. \qquad (12.34)$$

12.3 The General Geodesic Method

The basic method used in EH theory to calculate the light deflection due to gravitation is the null geodesic method based on the following condition on the general line element. Cartesian coordinates are used in the following:

$$ds^2 = -g_{00}c^2 dt^2 + g_{11}dX^2 + g_{22}dY^2 + g_{33}dZ^2$$
$$= 0. \qquad (12.35)$$

This method and the geodesic equation remain valid in ECE theory provided no assumptions are made concerning the symmetry of the metric and connection. The general geodesic equation is [11]:

$$\epsilon = -g_{\mu\nu}\frac{dx^\mu}{d\lambda}\frac{dx^\nu}{d\lambda} = \text{constant} \qquad (12.36)$$

where $g_{\mu\nu}$ is the general asymmetric metric. In spherical polar coordinates Eq. (12.36) is

$$-g_{00}c^2\left(\frac{dt}{d\lambda}\right)^2 + g_{11}\left(\frac{dr}{d\lambda}\right)^2 + g_{22}r^2\left(\frac{d\phi}{d\lambda}\right)^2 = -\epsilon \tag{12.37}$$

if the metric matrix is diagonal. The conserved quantities may be defined in general [11] by:

$$E = g_{00}c\frac{dt}{d\lambda}, \quad L = g_{22}r^2\frac{d\phi}{d\lambda} \tag{12.38}$$

so that Eq. (12.37) becomes:

$$-\frac{E^2}{2} + \frac{1}{2}g_{00}g_{11}\left(\frac{dr}{d\lambda}\right)^2 + \frac{g_{00}}{2}\left(g_{22}r^2\left(\frac{d\phi}{d\lambda}\right)^2 + \epsilon\right) = 0. \tag{12.39}$$

Define the potential in reduced units by:

$$V := \frac{1}{2}g_{00}\left(g_{22}r^2\left(\frac{d\phi}{d\lambda}\right)^2 + \epsilon\right). \tag{12.40}$$

where

$$\lambda = c\tau \tag{12.41}$$

and where τ is the proper time. Define the potential energy in units of joules by:

$$V := \frac{1}{2}mc^2 g_{00}\left(g_{22}r^2\left(\frac{d\phi}{d\lambda}\right)^2 + \epsilon\right). \tag{12.42}$$

To consider light deflected by mass the null geodesic is used, so:

$$\epsilon = 0 \tag{12.43}$$

and in this case:

$$V = \frac{1}{2}mg_{00}g_{22}r^2\left(\frac{d\phi}{d\tau}\right)^2. \tag{12.44}$$

12.3 The General Geodesic Method

Without loss of generality it may be assumed that:

$$g_{22} = 1 \qquad (12.45)$$

so that the potential energy in joules is:

$$V = \frac{1}{2} m g_{00} r^2 \left(\frac{d\phi}{d\tau}\right)^2. \qquad (12.46)$$

This expression must now be related to the potential Φ of the ECE theory, and to the curvature, spin connection and torsion. From this, an explanation of may g_{00} be found from first principles. The ECE theory gives in general:

$$\boldsymbol{\nabla} \cdot \boldsymbol{g} = c^2 (R - \omega T) \qquad (12.47)$$

where R, omega and T are defined by:

$$R := R^0{}_1{}^{10} + R^0{}_2{}^{20} + R^0{}_3{}^{30}, \qquad (12.48)$$

$$\omega T := (\omega^0{}_{1\lambda} T^{\lambda 10} + \omega^0{}_{2\lambda} T^{\lambda 20} + \omega^0{}_{3\lambda} T^{\lambda 30}). \qquad (12.49)$$

Here the Einstein and Newton constants in S. I. units are:

$$k = 1.86595 \times 10^{-26} N s^2 kg^{-2}, \qquad (12.50)$$

$$G = 6.6726 \times 10^{-11} N m^2 kg^{-2}. \qquad (12.51)$$

Firstly the Newtonian limit may be defined by:

$$R \gg \omega T \qquad (12.52)$$

i.e.

$$\omega \to 0. \qquad (12.53)$$

In this limit:

$$\boldsymbol{g} = -\boldsymbol{\nabla}\Phi = \frac{Gm_2}{r^2} \boldsymbol{r} \qquad (12.54)$$

and the force is:

$$\boldsymbol{F} = m_1 \boldsymbol{g} = -\left(\frac{Gm_1 m_2}{r^2}\right) \boldsymbol{r}. \qquad (12.55)$$

Therefore:

$$\nabla \cdot g = 2G \frac{m_2}{r^3} \tag{12.56}$$

and

$$\rho_m = \frac{m_2}{4\pi r^3} := \frac{m_2}{3V_0}. \tag{12.57}$$

Therefore the mass m_2 is defined by a curvature:

$$R = \frac{km_2}{3V_0} = R^0{}_1{}^{10} + R^0{}_2{}^{20} + R^0{}_3{}^{30}. \tag{12.58}$$

If the spin connection is fully considered, the acceleration due to gravity is defined in a manner analogous to the electric field strength in the Coulomb law of ECE theory [1–10]

$$g = -(\nabla + \omega)\Phi. \tag{12.59}$$

Here

$$V = m_2 \Phi. \tag{12.60}$$

The basic equations of the system are therefore:

$$V = m_2 g_{00} r^2 (d\phi/d\tau)^2, \tag{12.61}$$

$$g = -(\nabla + \omega)\Phi, \tag{12.62}$$

$$\nabla \cdot g = c^2 (R - \omega T), \tag{12.63}$$

$$V = m_2 \Phi, \tag{12.64}$$

and the mathematical problem is to solve this system of equations to give an expression for g_{00} in terms of R, ω, and T. The expression for g_{00} may then be used with the structure of Schwarzschild's solution to define his α parameter in terms of R and T.

From Eq. (12.63):

$$g = c^2 \int (R - \omega T) \, dr. \tag{12.65}$$

12.3 The General Geodesic Method

From Eq. (12.62):

$$g = -\frac{\partial \Phi}{\partial r} - \omega \Phi \quad (12.66)$$

where

$$\Phi = g_{00} r^2 \left(\frac{d\phi}{d\tau}\right)^2. \quad (12.67)$$

From Eqs. (12.66) and (12.67):

$$g = -g_{00} r \left(\frac{d\phi}{d\tau}\right)^2 (2 + \omega r). \quad (12.68)$$

If it is assumed as in paper 63 of www.aias.us that the radial component of the spin connection vector is:

$$\omega = -\frac{1}{r} \quad (12.69)$$

then:

$$g = -g_{00} r \left(\frac{d\phi}{d\tau}\right)^2 = c^2 \int \left(R + \frac{T}{r}\right) dr. \quad (12.70)$$

Differentiating both sides of this equation gives:

$$-g_{00} \left(\frac{d\phi}{d\tau}\right)^2 = c^2 \left(R + \frac{T}{r}\right) \quad (12.71)$$

and

$$g_{00} = c^2 \left(R + \frac{T}{r}\right) / \left(\frac{d\phi}{d\tau}\right)^2. \quad (12.72)$$

This is the required expression for g_{00} in terms of R and T.

The experimental result of NASA Cassini is that the light deflection by the sun is twice the Newtonian value within 1 : 100,000. Such a result is given by [1–11]:

$$g_{00} = -\left(1 - \frac{2GM}{c^2 r}\right) \quad (12.73)$$

where M is the mass of the sun. It is seen that the structures of Eqs. (12.72) and (12.73) are the same. Comparing the equations:

$$R = \frac{1}{c^2}\left(\frac{d\phi}{d\tau}\right)^2 \tag{12.74}$$

and

$$T = -\frac{2GM}{c^4}\left(\frac{d\phi}{d\tau}\right)^2 \tag{12.75}$$

i.e.

$$T = -\frac{2GM}{c^2}R. \tag{12.76}$$

In the original solution of EH for a vacuum, given by Schwarzschild in 1916, a parameter α was used, defined by:

$$\alpha = \frac{2GM}{rc^2}\text{(Cassini result)} \tag{12.77}$$

so

$$\alpha = -\frac{T}{R}. \tag{12.78}$$

Therefore the structure of α has been deduced from the true Bianchi identity without making any assumptions concerning either the metric or the spin connection and also without assuming a Ricci flat condition. The EH result coincides with:

$$-\frac{T}{R} = \frac{2GM}{rc^2} \tag{12.79}$$

but as the Pioneer anomaly [12] shows, this may not be the case in general, there is a small anomalous acceleration of $(8.74 \pm 1.33) \times 10^{-10}\text{ms}^{-2}$ towards the sun which cannot be explained in EH because M cannot be adjusted. However it is explained in ECE as a small deviation from the result (12.79), i.e. a small increment in T/R.

Acknowledgments

The British Government is thanked for a Civil List Pension for pre-eminent service to Britain in science, and the staff of AIAS and the Telesio-Galilei Association for many interesting discussions.

References

[1] M. W. Evans, "Generally Covariant Unified Field Theory" (Abramis Academic, Suffolk, 2005 onwards), volumes one to four, vol. 5 in prep (papers 71 to 93 on www.aias.us).

[2] L. Felker, "The Evans Equations of Unified Field Theory" (Abramis Academic, Suffolk, 2007).

[3] K. Pendergast, "Crystal Spheres" (Abramis Academic, to be published, preprint on www.aias.us, this book contains a unique description of the history of the British Civil List and Order of Merit scientists).

[4] M. W. Evans, Omnia Opera section of www.aias.us from 1992 to present, hyperlinked papers on the ECE theory and precursor gauge and B(3) theories.

[5] M. W. Evans and H. Eckardt, Physica B, **400**, 175 (2007).

[6] M. W. Evans, Acta Physica Polonica, B, **38**, 2211 (2007).

[7] M. W. Evans, Physica B, **403**, 517 (2008

[8] M. W. Evans (ed.), second edition of "Modern Non-linear Optics", in Advances in Chemical Physics, vol. 119 (2001); M. W. Evans and S. Kielich (eds.), first edition, vol. 85 (1992, 1993, 1997).

[9] M. W. Evans and L. B. Crowell, "Classical and Quantum Electrodynamics and the B(3) Field" (World Scientific, 2001).

[10] M. W. Evans and J.-P. Vigier, "The Enigmatic Photon" (Kluwer, 1994 to 2002, hardback and softback), in five volumes.

[11] S. P. Carroll, "Spacetime and Geometry: an Introduction to General Relativity" (Addison-Wesley, New York, 2004, and freely available notes of 1997 on the internet).

[12] Descriptions of the well known Pioneer anomaly on the internet.

13

ECE Theory of the Orbit of Binary Pulsars

by

Myron W. Evans,
Alpha Institute for Advanced Study, Civil List Scientist.
(emyrone@aol.com and www.aias.us)

Abstract

Recently in this series of papers on Einstein Cartan Evans (ECE) unified field theory it has been shown that the Einstein Hilbert (EH) field equation is incompatible with the fundamental Bianchi identity of Cartan geometry. This means that new explanations must be sought for the experimental tests of gravitational relativity theory now available. One of these is the 3 millimeter decrease per revolution in the orbit of the Hulse Taylor binary pulsar. It is no longer valid to attempt to explain this with EH theory and a straightforward qualitative explanation is given in terms of ECE theory without using the flawed EH postulate of gravitational radiation. The explanation is based on the experimentally observed r dependent ratio of torsion (T) to curvature (R). If this ratio is not precisely constant, but increases with the radial co-ordinate r, the orbit decays as observed.

Keywords: ECE theory of the Hulse Taylor binary pulsar, r dependent ratio of torsion to curvature.

13.1 Introduction

Recently it has been demonstrated in this series of papers [1–10] that the Einstein Hilbert (EH) field equation is incompatible with the fundamental Bianchi identity of Cartan geometry, the incompatibility was first observed by computer algebra in paper 93 of the ECE series on www.aias.us. and shows up in the Hodge dual of the Bianchi identity. The latter has been proven in several ways during the course of development of the ECE series of papers, of which this is paper 106. This means that the EH field equation must be rejected, because it is based on a torsion-less geometry. Cartan showed in 1922 that there are two structure equations that determine the geometry of a four-dimensional space-time such as that used in general relativity. The first defines torsion in terms of the spin connection and tetrad, and the second defines curvature in terms of the spin connection. The Bianchi identity due to Cartan ineluctably relates torsion to curvature. The Hodge dual of this identity was proven in this series of papers, and leads to a tensor equation in which the covariant derivative of the three index torsion tensor $T^{\kappa\mu\nu}$ is equated to a curvature tensor $R^{\kappa}{}_{\mu}{}^{\mu\nu}$

$$D_\mu T^{\kappa\mu\nu} = R^{\kappa}{}_{\mu}{}^{\mu\nu}. \tag{13.1}$$

The curvature tensor was evaluated by computer algebra in paper 93 for several well used line elements, all based on the Christoffel connection. It was found by computer algebra that in general the tensor $R^{\kappa}{}_{\mu}{}^{\mu\nu}$ is non zero. However, for the same Christoffel connection the torsion tensor is always zero by definition so the covariant derivative of the torsion is also zero. Therefore the use of a Christoffel connection is incompatible with Cartan geometry and there is no way out of this "ECE Paradox" for the EH equation. The latter is always based on a Christoffel connection.

Therefore all inferences based on the EH equation must be re-evaluated and made consistent with the Bianchi identity and its newly inferred Hodge dual. In paper 103 a new field equation was suggested-one that includes torsion and curvature self consistently, in paper 104 the Hodge dual of the Bianchi identity was proven for any metric and spin connection, and in paper 105 a new explanation of the precision tests of general relativity was given in terms of the ratio of a well defined scalar torsion T to a well defined scalar curvature R. In this paper the theory of paper 105 is applied to the Hulse Taylor binary pulsar [11] in which a decrease in orbit of about 3 millimeters per revolution is observed. In the standard model the now obsolete Einstein Hilbert (EH) field equation is used in an attempt to explain this decrease in terms of quadrupole gravitational radiation [12] in the weak field limit. However, no direct observation of gravitational radiation has been made. In Section 13.2 a much simpler explanation of this decrease in orbit is given. When the ratio of T/R is not exactly constant, but increases with the radial coordinate r, the orbit spirals inwards as observed. Therefore the data of the Hulse Taylor binary pulsar and another binary pulsar recently discovered by Joddrell Bank [13] must be used to determine the precise dependence of T/R upon r. No postulate of gravitational radiation is needed. The correct method of developing gravitational radiation is based on the well known ECE wave equation:

$$(\Box + kT)q^a_\mu = 0 \tag{13.2}$$

where the gravitational wave is the tetrad q_μ^a. The eigenvalues of this gravitational wave equation are kT, where k is Einstein's constant and T in this context is a well defined scalar canonical energy momentum density, not to be confused with the symbol T for scalar torsion. The eigenvalues are related to a well defined scalar curvature by a generalization of the Einstein postulate to all fields [1–10]:

$$R = -kT. \tag{13.3}$$

Therefore if gravitational radiation is ever detected, it is due to Eq. (13.2), and not due to the EH field equation. The latter is geometrically incorrect, so no physical inference can be based upon it. There are no black holes, no Big Bang, and no dark matter. These are all based on the incorrect EH field equation. Additionally, the so called Schwarzschild metric was not inferred [14] by Schwarzschild in 1916. The so called Schwarzschild radius is not due to Schwarzschild. The latter published two papers in 1916, one of which gave a vacuum solution of the EH equation in terms of a parameter α. The mass M of the gravitating source was not used by Schwarzschild. In paper 105 we redefined the so called Schwarszchild radius r_s as:

$$r_s = -\frac{T}{R} = \frac{2mG}{c^2} \tag{13.4}$$

where G is Newton's constant and c the vacuum speed of light. Therefore the standard model approach to gravitational relativity must be finally discarded. It is already highly controversial and effectively obsolete, and is being replaced by the internationally accepted ECE theory. In these papers we are applying ECE to the experimental data directly, without being influenced by the EH equation or by flawed dogma, however oft repeated. A theory is accepted or rejected upon logic and the Ockham/Bacon scientific method, not upon the transient subjective opinion of any era.

13.2 Orbit of the Hulse Taylor Binary Pulsar

The first binary pulsar was found in 1974 by Hulse and Taylor [11]. It consists of a pulsar (a neutron star) with a pulsation period of 59 milliseconds, and a second neutron star in an elongated orbit of period 7.75 hours. Each neutron star is 1.4 million times the mass of the sun in the solar system. The orbit is observed to be gradually decreasing by 3.1 mm per orbit. The orbital precession is 4.2° a year. In ECE theory (paper 105) a constant orbital precession is given by a constant:

$$r_s = -\frac{T}{R}. \tag{13.5}$$

However, the orbit of the binary pulsar is a spiral inwards, so experimentally the ratio of T to R in Eq. (13.5) is not constant. Universal gravitation means a constant G for a given constant M, but more generally, G depends on T/R when the gravitational fields are enormous, such as in a binary pulsar. The well known Pioneer anomalies [15] are due in ECE theory to the same ratio of T/R. In other words gravitation is not universal in general.

It is well known that the orbit in standard model general relativity is calculated from the line element and the latter in turn is taken to be the so called Schwarzschild metric. In paper 105 an entirely new method of calculating the orbit was proposed. The result is that the line element is expressed as:

$$c^2 d\tau^2 = \left(1 + \frac{T}{Rr}\right) c^2 dt^2 - \left(1 + \frac{T}{Rr}\right)^{-1} dr^2 - r^2 d\phi^2 \qquad (13.6)$$

where:

$$\frac{dr}{d\phi} = \frac{dr}{d\tau}\frac{d\tau}{d\phi}. \qquad (13.7)$$

Here τ is the proper time, r is the radial coordinate and ϕ an angular coordinate of the spherical polar system. For a constant perihelion advance per orbit, the experimental data are reproduced by a constant as in Eq. (13.4). In the weak field limit this condition is the universal gravitation of Newton, in which G is a universal constant. In all field theories of relativity, including ECE theory, c is a universal constant. The constant c is taken to be exact in standard laboratories, but G is among the least precisely known of the constants of physics. Therefore Eq. (13.4) means that G is determined by T/R for a given M and given c. In the weak field limit, T/R is constant, and this is the limit of ECE that defines the universal gravitation of Newton. The Hulse Taylor binary pulsar shows that more generally, gravitation is not universal, in the huge gravitational fields of a binary pulsar G is not the same as in the solar system, where the sun's gravitational field is a millionth of that of a neutron star. In the solar system itself, the Pioneer anomaly suggests that there are also small departures from universal gravitation of the order of one part in 10^{-10}. In ECE these are small departures from the constancy of T/R.

From the line element (13.6) two constants of motion are defined [12]:

$$E = mc^2 \left(1 + \frac{T}{Rr}\right) \frac{dt}{d\tau} \qquad (13.8)$$

and

$$L = mr^2 \frac{d\phi}{d\tau} \qquad (13.9)$$

in S.I. units. Therefore the orbital equation is:

$$\left(\frac{dr}{d\tau}\right)^2 = \frac{E}{m^2c^2} - \left(1 + \frac{T}{Rr}\right)\left(c^2 + \frac{L^2}{m^2r^2}\right). \tag{13.10}$$

For a planet such as Mercury in the solar system the advance of the perihelion is only a few arc-seconds per century and is essentially constant. This is due to the relatively weak gravitational forces in the solar system in comparison with the Hulse Taylor binary pulsar, where the advance in the perihelion is 4.2° per year. Eq. (13.10) is for a mass m orbiting a mass M and may be written as a function of T/R as follows:

$$\frac{1}{2}m\left(\frac{dr}{d\tau}\right)^2 = \left(\frac{E^2}{2mc^2} - \frac{1}{2}mc^2\right) - \frac{c^2}{2}\frac{T}{rR}m \\ - \frac{L^2}{2mr^2} - \frac{T}{2R}\frac{L^2}{mr^3}. \tag{13.11}$$

The potential energy of the orbital equation is:

$$V(r) = \frac{c^2}{2}\frac{T}{rR}m + \frac{L^2}{2mr^2} + \frac{T}{2R}\frac{L^2}{mr^3}, \tag{13.12}$$

and consists of the $1/r$ dependent gravitational attraction, the $1/r^2$ dependent centripetal repulsion, and the $1/r^3$ dependent relativistic attraction. It is seen that the negative attraction terms depend on T/R, but the centripetal repulsion does not. The perihelion advance is given [1–10] by:

$$\delta\phi = -\left(\frac{3\pi r}{A(1-e^2)}\right)\frac{T}{R} \tag{13.13}$$

where A is the semi-major axis and where e is the orbital eccentricity. When T/R is constant the perihelion advance is constant, as appears to be the case for Mercury within experimental precision at present.

In the Hulse Taylor binary pulsar the perihelion advance is not constant, because the orbit decreases by 3.1 mm per revolution. Therefore r in Eq. (13.12) decreases by 3.1 mm each revolution. In consequence the attractive $1/r^3$ term begins to dominate over the other two terms as

$$r \to 0. \tag{13.14}$$

Therefore the attraction between the two objects of mass M and m will increase per revolution of orbit as observed. Using the experimental data the ratio T/R can be found for the Hulse Taylor binary pulsar, or any other binary pulsar such as the one discovered at Joddrell Bank in 2003/2004 [13].

In the solar system the data from the Pioneer anomaly can be used to find the appropriate T/R for the solar system, and in general for every system that is observed to have anomalous behavior. More generally T/R may depend on the three coordinates of the spherical polar system (r, θ, ϕ). Therefore a systematic astronomical survey may be carried out to describe each system in terms of its characteristic T/R parameter. This is precisely constant only if the perihelion advance is constant and the orbit is stable.

Similarly, light deflection due to gravity is precisely twice the Newtonian value only if the ratio T/R is precisely constant. Light deflection is calculated by eliminating the proper time as follows:

$$\left(\frac{dr}{d\phi}\right)^2 = \left(\frac{dr}{d\tau}\right)^2 \left(\frac{d\tau}{d\phi}\right)^2 = \left(\frac{dr}{d\tau}\right)^2 \left(\frac{mr^2}{L}\right)^2 \tag{13.15}$$

which implies that:

$$\left(\frac{dr}{d\phi}\right)^2 = \frac{r^4}{b^2} - \left(1 + \frac{T}{rR}\right)\left(\frac{r^4}{a^2} + r^2\right) \tag{13.16}$$

where a and b are constants [12]:

$$a := \frac{L}{mc}, \quad b := \frac{cL}{E}. \tag{13.17}$$

This is the orbital equation in a form where proper time has been eliminated. It is seen that the orbital equation also depends on the ratio T/R. All orbits are defined by this ratio. From Eq. (13.16):

$$\phi = \int \left(r^2 \left(\frac{1}{b^2} - \left(1 + \frac{T}{rR}\right)\right)\left(\frac{1}{a^2} + \frac{1}{r^2}\right)\right)^{-1} dr. \tag{13.18}$$

The formula for light deflection due to gravitation is obtained as:

$$m \to 0 \tag{13.19}$$

and so:

$$\phi \to \int \left(r^2 \frac{1}{b^2} - \left(1 + \frac{T}{rR}\right)\frac{1}{r^2}\right)^{-1} dr. \tag{13.20}$$

Expanding in power of $\frac{T}{rR}$ gives:

$$\delta\phi \sim -\frac{2T}{Rb}. \tag{13.21}$$

Here b is identified with the distance of closest approach. Any deviation from twice the Newtonian value means that the ratio T/R is not precisely constant, as in the problem of the perihelion advance. In addition to light deflection, ECE theory predicts that gravitation will change all the electro-dynamical properties of light, notably polarization. This phenomenon could be looked for in the newly discovered Joddrell Bank binary pulsar or other objects with mass much greater than the sun of the solar system. The greater the mass the greater the ratio T/R from the formula:

$$-\frac{T}{R} = \frac{2mG}{c^2} \qquad (13.22)$$

and the greater the angle of deflection from the formula (13.21).

In future work a search will be made for a theoretical method that may give T/R from first principles, and also modeling of T/R with various kinds of r dependence may be made with the use of computer graphics and algebra. This is a much simpler explanation than gravitational radiation, which cannot exist according to the EH equation because the latter is fundamentally flawed, i.e. is not compatible with the Bianchi identity as given by Cartan. Therfore gravitational radiation from the EH equation cannot be a precise test of gravitational relativity. Unless torsion is correctly incorporated as in ECE theory, light deflection and perihelion advance cannot be regarded as precise tests of the EH equation. It appears that the claims of the standard model in this respect are due to modeling with many parameters.

Acknowledgments

The British Government is thanked for a Civil List Pension for pre-eminent contributions in science, and the staff of AIAS and the Telesio-Galilei Association for many interesting discussions.

References

[1] M. W. Evans, "Generally Covariant Unified Field Theory" (Abramis, Suffolk, 2005 onwards), volumes one to four, volume five in prep. (Papers 71 to 93 on www.aias.us).

[2] L. Felker, "The Evans Equations of Unified Field Theory" (Abramis, 2007).

[3] K. Pendergast, "Crystal Spheres" (preprint on www.aias.us, Abramis in prep.).

[4] M. W. Evans, Omnia Opera section of www.aias.us from 1992 to present, hyperlinked papers on ECE theory and precursor gauge theories.

[5] M. W. Evans and H. Eckardt, Physica B, **400**, 175 (2007).

[6] M. W. Evans, Physica B, **403**, 517 (2008).

[7] M. W. Evans, Acta Physica Polonica, B, **38**, 2211 (2007).

[8] M. W. Evans (ed.), second edition of "Modern Non-Linear Optics", in Advances in Chemical Physics, vol. 119 (2001); M. W. Evans and S. Kielich (eds.), first editions, vol. 85 (1992, 1993, 1997).

[9] M. W. Evans and L. B. Crowell, "Classical and Quantum Electrodynamics and the B(3) Field" (World Scientific 2001).

[10] M. W. Evans and J.-P. Vigier, "The Enigmatic Photon" (Kluwer, 1994 to 2002, hardback and softback), in five volumes.

[11] The first binary pulsar to be discovered.

[12] S. P. Carroll, "Space-time and Geometry: an Introduction to General Relativity" (Addison-Wesley, New York, 2004, and downloadable noted of 1997).

[13] Discovered circa 2003/2004 and consisting of two pulsars whose orbit decreases by about 7 millimeters a revolution, see ref. (3) for more details.

[14] See section by S. Crothers in paper 93 of ECE theory on www.aias.us.

[15] These show a small but systematic non-Newtonian attraction towards the sun, and new anomalies are being discovered, suggesting a dependence of $-T/R$ on all three spherical polar coordinates in general.

Spin Connection Resonance in the Faraday Disk Generator

by

Myron W. Evans,
Alpha Institute for Advanced Study, Civil List Scientist.
(emyrone@aol.com and www.aias.us)

and

F. Amador and Horst Eckardt,
Alpha Institute for Advanced Studies (AIAS).

Abstract

Using Einstein Cartan Evans (ECE) unified field theory the conditions are deduced under which a Faraday disk generator may be used to demonstrate a resonant peak of power due to the spin connection used in ECE theory (spin connection resonance or SCR). The analytical analysis is supported by a Faraday disk design with variable spin speeds which has recently demonstrated the existence of SCR experimentally. Three principal types of resonances have been identified.

Keywords: ECE theory, spin connection resonance, Faraday disk generator.

14.1 Introduction

During the course of development [1–10] of ECE theory the phenomenon of spin connection resonance (SCR) has been developed (for example papers 63 and 94 of the ECE series on www.aias.us) and shown to be important in the acquisition of electric power from space-time through the Cartan torsion. This source of electric power is well known experimentally and was demonstrated for example by Tesla [11] in several devices. Other groups have observed such effects [1–10] for over a hundred years, but as in the case of the Faraday disk generator [12] the standard Maxwell Heaviside (MH) theory does not have an explanation for them. Therefore there has been a tendency to under-implement these potentially important devices despite their obvious importance for the generation of electric power. In papers 43 and 44 of the ECE series a straightforward explanation for the Faraday disk generator was given in terms of the spinning of space-time and in Section 14.2 this explanation is adapted to demonstrate analytically the possibility in the Faraday disk generator of spin connection resonance induced by varying the speed of the spinning disk. Further details and numerical evaluations are given in Section 14.3. In Sections 14.4 and 14.5 such a device is described experimentally and suggestion is made for the improved control and engineering of devices that take electric power from space-time using the Faraday disk design.

14.2 Analytical Theory

The theory of the Faraday disk was first developed with ECE theory in papers 43 and 44. They were based on the fundamental idea of ECE theory:

$$F = A^{(0)} T \tag{14.1}$$

in short-hand index-less notation [1–10]. Here F denotes the electromagnetic field form and T the Cartan torsion form [13, 14]. The quantity $cA^{(0)}$ is a fundamental voltage [1–10]. In the Faraday disk the torsion T is set up by mechanical rotation. So the basic equations of the generator are:

$$F = A^{(0)} T(\text{mechanical}) = d \wedge A + \omega \wedge A \tag{14.2}$$

where $\wedge$ denotes wedge product and $d\wedge$ denotes exterior derivative. Here A is the potential form of ECE theory [1–10] and ω is its spin connection form [1–10, 13, 14]. The field equations of the system are based on the Bianchi identity as developed by Cartan and are:

$$d \wedge F + \omega \wedge F = R \wedge A, \tag{14.3}$$

$$d \wedge \tilde{F} + \omega \wedge \tilde{F} = \tilde{R} \wedge A. \tag{14.4}$$

The second equation is the Hodge dual of the Bianchi identity and was developed during the course of development of ECE theory. The field equations can be reduced [1–10] to vector notation as used in standard electrical engineering. They then become the following set of six equations for all practical purposes in the laboratory.

$$\nabla \cdot \mathbf{B} = 0 \tag{14.5}$$

$$\nabla \times \mathbf{E} + \frac{\partial \mathbf{B}}{\partial t} = 0 \tag{14.6}$$

$$\nabla \cdot \mathbf{E} = \frac{\rho}{\epsilon_0} \tag{14.7}$$

$$\nabla \times \mathbf{B} - \frac{1}{c^2}\frac{\partial \mathbf{E}}{\partial t} = \mu_0 \mathbf{J} \tag{14.8}$$

$$\mathbf{B} = \nabla \times \mathbf{A} - \boldsymbol{\omega} \times \mathbf{A} \tag{14.9}$$

$$\mathbf{E} = -\nabla \phi - \frac{\partial \mathbf{A}}{\partial t} + \phi \boldsymbol{\omega} - \omega \mathbf{A}. \tag{14.10}$$

The first four are the Gauss law, the Faraday law of induction, the Coulomb law and the Ampère Maxwell law, the fifth and sixth are the equations expressing fields in terms of the potentials and spin connection scalar and vector. In these vector equations, expressed in S.I. units, $\mathbf{B}$ is the magnetic flux density, $\mathbf{E}$ is the electric field strength, ρ is the charge density, ϵ_0 is the vacuum permittivity, $\mathbf{J}$ is the current density, $\mathbf{A}$ is the vector potential, ω is the scalar connection, ϕ is the scalar potential and $\boldsymbol{\omega}$ is the vector connection. Details of this derivation are given in the ECE series of papers and books (www.aias.us), notably in review paper 100.

The Bianchi identity (14.3) gives the homogeneous field equation in tensor notation, and the Hodge dual identity (14.4) gives the inhomogeneous field equation in tensor notation. The tensor equations are then written in the base manifold, which is a four dimensional space-time with torsion and curvature. The latter is expressed in the original field equations through the curvature form R in index-less notation, the link between geometry and the electromagnetic field being expressed by the basic relation (14.2). The classical field equations of electrodynamics therefore become field equations of general relativity, not field equations of special relativity, in which both torsion and curvature are absent, and in which the space-time is a Minkowski space-time. The MH field theory, in which the electromagnetic field is a nineteenth century concept defined on a Minkowski frame of reference, is one of special relativity. In ECE theory the electromagnetic field is the space-time geometry itself within a factor $A^{(0)}$, where $cA^{(0)}$ is a primordial voltage. Finally the two tensor equations in the base manifold are developed as four vector equations, and the tensor relation between field and potential developed into two further vector equations.

In paper 44 a complex circular basis [1–10] was used to define a rotating potential set up by mechanically rotating the Faraday disk at an angular frequency Ω in radians per second:

$$\mathbf{A}^{(1)} = \frac{A^{(0)}}{\sqrt{2}}(\mathbf{i} - i\mathbf{j})e^{i\Omega t}. \tag{14.11}$$

Its complex conjugate is denoted:

$$\mathbf{A}^{(2)} = \frac{A^{(0)}}{\sqrt{2}} \left(\mathbf{i} + i\mathbf{j}\right) e^{-i\Omega t}. \tag{14.12}$$

This is the key concept of the ECE explanation of the Faraday disk generator. The real parts of $\mathbf{A}^{(1)}$ and $\mathbf{A}^{(2)}$ are the same and can be worked out with de Moivre's Theorem:

$$e^{i\Omega t} = \cos \Omega t + i \sin \Omega t. \tag{14.13}$$

The ECE concept is based on:

$$A = A^{(0)} q \tag{14.14}$$

where q is the Cartan tetrad [1–10]. The tetrad relevant to the Faraday disk is:

$$\mathbf{q}^{(1)} = \mathbf{q}^{(2)*} = \frac{1}{\sqrt{2}} \left(\mathbf{i} - i\mathbf{j}\right) e^{i\Omega t}. \tag{14.15}$$

This concept is one of rotational general relativity, whereas the Maxwell Heaviside (MH) theory is one of special relativity in a flat or Minkowski space-time. It is well known that MH is unable to explain the Faraday disk generator, whereas ECE explains it straightforwardly. It is clear therefore that electrodynamics is part of ECE theory, a generally covariant unified field theory (www.aias.us). Classical and quantum electrodynamics have been extensively developed within ECE theory, and unified with other fundamental fields, notably gravitation.

In the Faraday disk the mechanical spin sets up a rotational tetrad, which is a rotation of space-time ITSELF. In paper 44 a special case of Eq. (14.10) was used:

$$\mathbf{E} = -\frac{\partial \mathbf{A}}{\partial t} - \omega \mathbf{A} \tag{14.16}$$

in which ϕ was assumed to be zero and where $\mathbf{A}$ is generated by the magnet of the Faraday disk generator, essentially as used by Faraday and reported in his diary on Dec 26th 1831. The scalar spin connection in paper 44 was assumed to be proportional to Ω, so the electric field strength is:

$$\mathbf{E}^{(2)} = \mathbf{E}^{(1)*} = -\left(\frac{\partial}{\partial t} + i\Omega\right) \mathbf{A}^{(2)}. \tag{14.17}$$

The real part of this expression is worked out with Eq. (14.13) and is:

$$\mathbf{E} = \frac{2}{\sqrt{2}} A^{(0)} \Omega \left(\mathbf{i} \sin \Omega t - \mathbf{j} \cos \Omega t \right). \qquad (14.18)$$

This electric field strength (in volts per meter) spins around the rim of the rotating disk. As observed experimentally it is proportional to the product of $A^{(0)}$ and Ω. An electromotive force is set up between the center of the disk and its rim, as first observed by Faraday, and this emf is measured by a voltmeter at rest with respect to the spinning disk.

Recently [15] there have been reports of a Faraday disk generator exhibiting a powerful resonance effect hitherto unknown. The onset of this surge of electric power occurs when the angular frequency of the spinning disk is time dependent. At a sharply defined Ω the apparatus was observed to disintegrate (explode). There is no explanation for this in standard electrical engineering, which is based on the MH theory. In ECE theory it can be explained by spin connection resonance provided that the rate of spin of the disk is time dependent, i.e. its RPM increases so that:

$$\frac{\partial \Omega}{\partial t} \neq 0. \qquad (14.19)$$

Use Eq. (14.8) with

$$\nabla \times \mathbf{B} = \mathbf{0} \qquad (14.20)$$

because in the Faraday disk generator:

$$\mathbf{B} = B^{(0)} \mathbf{k}. \qquad (14.21)$$

Therefore

$$\frac{\partial \mathbf{E}}{\partial t} = -\epsilon_0 \mathbf{J} \qquad (14.22)$$

where:

$$\mathbf{E} = \frac{2}{\sqrt{2}} A^{(0)} \Omega \left(\mathbf{i} \sin \Omega t - \mathbf{j} \cos \Omega t \right). \qquad (14.23)$$

and in complex circular notation the electric field strength is:

$$\mathbf{E}^{(2)} = -\frac{\partial \mathbf{A}^{(2)}}{\partial t} - i\Omega \mathbf{A}^{(2)}. \qquad (14.24)$$

Differentiating this equation with respect to time:

$$\frac{\partial \mathbf{E}^{(2)}}{\partial t} = -\frac{\partial^2 \mathbf{A}^{(2)}}{\partial t^2} - i\Omega \frac{\partial \mathbf{A}^{(2)}}{\partial t} - i\frac{\partial \Omega}{\partial t}\mathbf{A}^{(2)} = -\frac{\mathbf{J}}{\epsilon_0}, \quad (14.25)$$

so using Eq. (14.24) the equation for the potential is:

$$\frac{\partial^2 \mathbf{A}^{(2)}}{\partial t^2} + i\Omega \frac{\partial \mathbf{A}^{(2)}}{\partial t} + i\frac{\partial \Omega}{\partial t}\mathbf{A}^{(2)} = \frac{\mathbf{J}}{\epsilon_0}. \quad (14.26)$$

This is an Euler Bernoulli resonance equation [1–10, 16] under the condition:

$$\frac{\partial \Omega}{\partial t} \neq 0 \quad (14.27)$$

i.e. that the RPM of the spinning disk increases. Thus:

$$\frac{\partial^2 \mathbf{A}^{(2)}}{\partial t^2} + i\frac{\partial \left(\Omega \mathbf{A}^{(2)}\right)}{\partial t} = \frac{\mathbf{J}}{\epsilon_0}. \quad (14.28)$$

This is an undamped resonator equation if J is designed experimentally to be periodic, for example:

$$\mathbf{J}^{(2)} = J^{(0)} \cos(\Omega_0 t)\, \mathbf{e}^{(2)}. \quad (14.29)$$

Finally if the engineering design is such that:

$$\frac{\partial \Omega}{\partial t} \gg \Omega \quad (14.30)$$

we obtain the equation:

$$\frac{\partial^2 \mathbf{A}^{(2)}}{\partial t^2} + i\frac{\partial \Omega}{\partial t}\mathbf{A}^{(2)} = J^{(0)} \cos(\Omega_0 t)\, \mathbf{e}^{(2)}. \quad (14.31)$$

At resonance [1–10, 16]:

$$\mathbf{A}^{(2)} \to \infty. \quad (14.32)$$

The observed explosion of the Faraday disk generator [15] may be explained in this way, i.e. the design must be a rapidly varying Ω and a periodic current density coming from the emf set up between the center and rim of the rotating disk.

14.3 Numerical Results

The resonance equation (14.26) has to be solved nuerically. First we rewrite the equation to two equations for the real and imaginary part of $\mathbf{A}^{(2)}$, denoted by A_r and A_i (considering only the time dependence):

$$\frac{d^2 A_r}{dt^2} - A_i \left(\frac{d\Omega}{dt}\right) - \left(\frac{dA_i}{dt}\right)\Omega = \frac{J}{\epsilon_0}, \qquad (14.33)$$

$$\frac{d^2 A_i}{dt^2} + A_r \left(\frac{d\Omega}{dt}\right) + \left(\frac{dA_r}{dt}\right)\Omega = 0. \qquad (14.34)$$

In case of vanishing Ω these equations pass into

$$\frac{d^2 A_r}{dt^2} = \frac{J}{\epsilon_0}, \qquad (14.35)$$

$$\frac{d^2 A_i}{dt^2} = 0. \qquad (14.36)$$

These simple equations then have the general solutions

$$A_r = \frac{J}{2\epsilon_0} t^2 + k_1 t + k_2, \qquad (14.37)$$

$$A_i = k_3 t + k_4. \qquad (14.38)$$

These are growing solutions in t, even if the constants k_1 and k_2 are chosen as zero. Therefore the real part of the electric Field (see Eq. (14.24))

$$E_r = \mathrm{Re}\left(-\frac{\partial A}{\partial t} - i\Omega A\right) = A_i \Omega - \frac{dA_r}{dt}. \qquad (14.39)$$

will grow linearly by the mechanical rotation. The effect of Ω will be to overlay an oscillatory structure to the simple solutions (14.37), (14.38).

For the numerical solution of Eqs. (14.33, 14.34) we define

$$\Omega = \alpha_0 \cos \omega_0 t, \qquad (14.40)$$

$$J = J_0 \cos \omega_J t. \qquad (14.41)$$

To make the simulations more realistic we have added a conductivity term according to Ohm's law:

$$\mathbf{J}_{cond} = \sigma \, \mathbf{E} \qquad (14.42)$$

266   14 Spin Connection Resonance in the Faraday Disk Generator

with a suitable conductivity value σ. Replacing $\mathbf{J}$ by $\mathbf{J} + \mathbf{J}_{cond}$ in Eq. (14.26) and selecting the real and imaginariy part again leads to

$$\frac{d^2 A_r}{dt^2} - A_i \left(\frac{d\Omega}{dt}\right) - \left(\frac{dA_i}{dt}\right)\Omega = \frac{J}{\epsilon_0} + \frac{\sigma}{\epsilon_0}\left(A_i \Omega - \frac{dA_r}{dt}\right), \quad (14.43)$$

$$\frac{d^2 A_i}{dt^2} + A_r \left(\frac{d\Omega}{dt}\right) + \left(\frac{dA_r}{dt}\right)\Omega = -\frac{\sigma}{\epsilon_0}\left(A_r \Omega + \frac{dA_i}{dt}\right). \quad (14.44)$$

Equations (14.33,14.34) have been solved numerically. The results for a moderate amplitude of mechanical rotation are shown in Figs. (14.1)–(14.3). In Fig. (14.1) the real and imaginary part of the vector potential are graphed. The real part grows quadratically as predicted by Eq. (14.37), with a superimposed oscillatory structure by the mechanical rotation Ω. The imaginary part of the potential is purely oscillatory as are both time derivatives (Fig. (14.2)). The most relevant quantity is the electric field, Eq. (14.39), which describes the induction effect due to rotation. As can be seen from Fig. (14.3), The real part of E grows linearly, despite of the oscillating potentials, only in right-most part there is a very low undulation.

For a modified parameter set we chose similar values of mechanical and current frequency, leading to a heterodyne effect in the time behaviour of the potential (Figs. (14.4), (14.5)). The electrical field (Fig. (14.6)) is oscillatory now. The imaginary part is practically zero because the model does not contain any damping or energy dissipation terms.

There should be a resonance behaviour according to section 14.1. In Fig. (14.7) the maximum electric field value within $60s$ simulated time has been

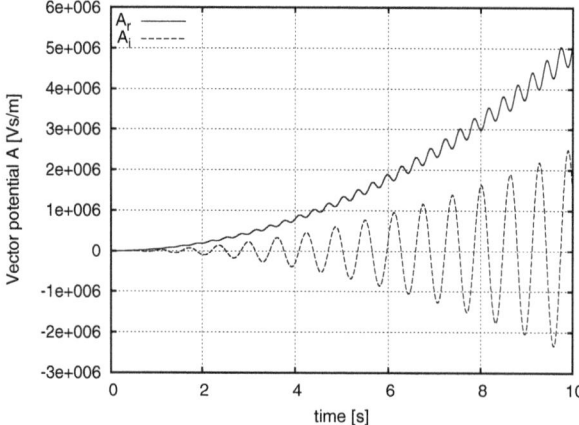

Fig. 14.1. Vector potental for $\alpha_0 = 5$, $\omega_0 = 10$, $\omega_J = 0$.

14.3 Numerical Results   267

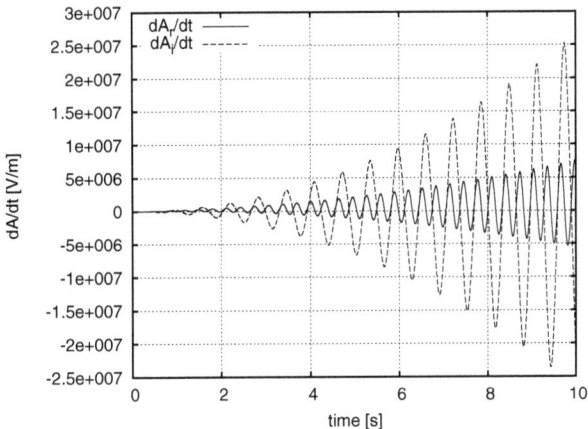

Fig. 14.2. Derivative of Vector potental for $\alpha_0 = 5$, $\omega_0 = 10$, $\omega_J = 0$.

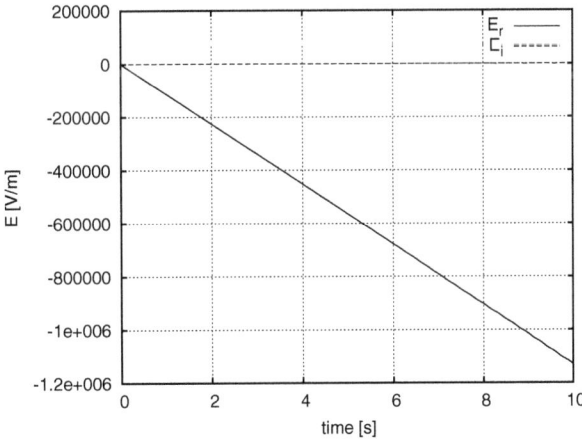

Fig. 14.3. Electric field for $\alpha_0 = 5$, $\omega_0 = 10$, $\omega_J = 0$.

268   14 Spin Connection Resonance in the Faraday Disk Generator

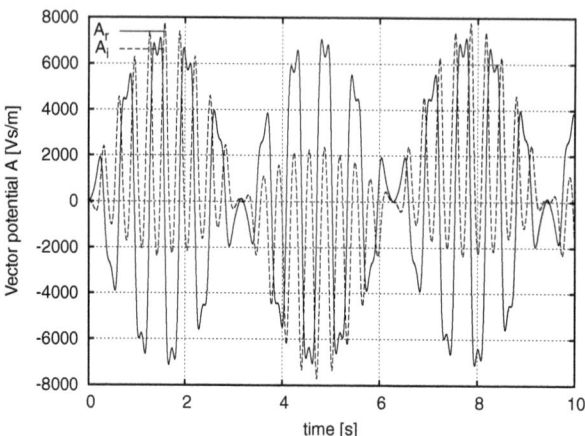

Fig. 14.4. Vector potental for $\alpha_0 = 20$, $\omega_0 = 10$, $\omega_J = 9$.

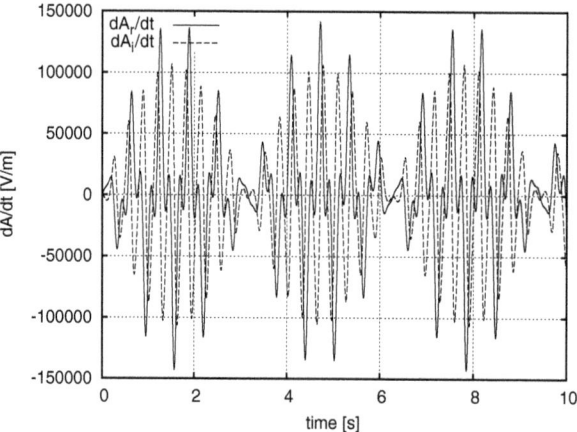

Fig. 14.5. Derivative of Vector potental for $\alpha_0 = 20$, $\omega_0 = 10$, $\omega_J = 9$.

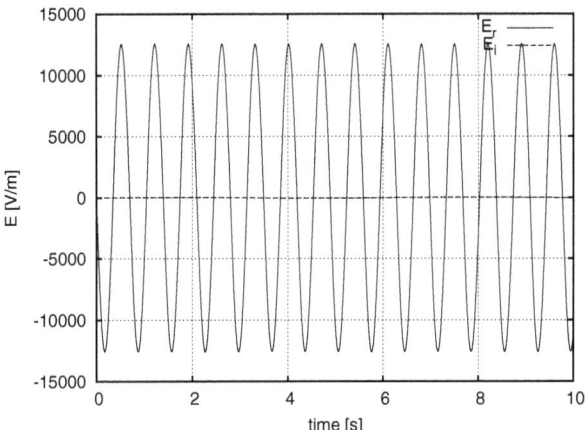

Fig. 14.6. Electric field for $\alpha_0 = 20$, $\omega_0 = 10$, $\omega_J = 9$.

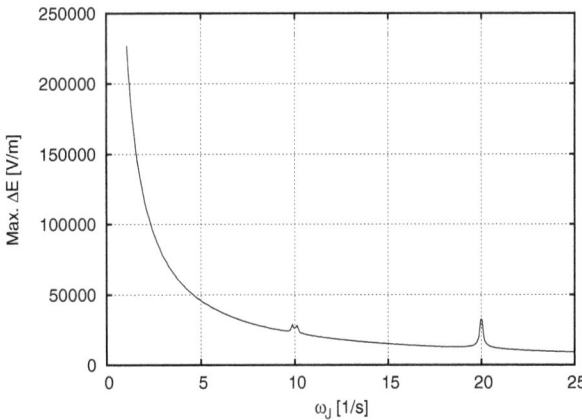

Fig. 14.7. Resonance curve of E field for variable current frequency ω_J.

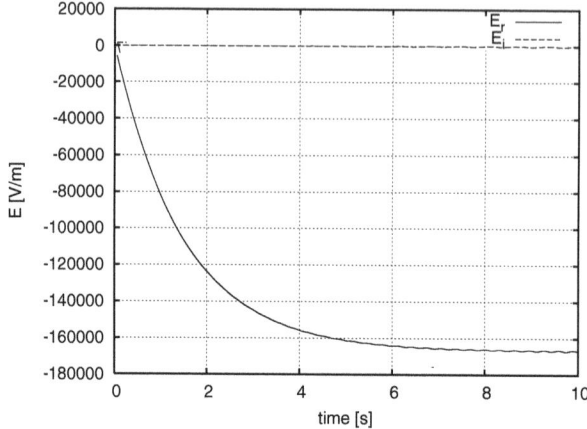

Fig. 14.8. Electric field for $\alpha_0 = 5$, $\omega_0 = 10$, $\omega_J = 0$ with conductivity term.

plotted in dependence of the current frequency ω_J. There are minor maxima for ω_J values of 10/s and 20/s which correspond to the given rotational frequency of $\omega_0 = 10/s$. Most significant is the increase of the maximum amplitude for $\omega_J \to 0$. This means that a direct current gives the highest E field value.

From electrical engineering it is known that the feedback of the electric field to a current can be modeled by Ohm's law, see Eq. (14.42). The resulting Eqs. (14.43, 14.44) therefore lead to a reduction in current. We have repeated the calculation for Fig. (14.3) with a conductivity term. The result (Fig. (14.8)) shows that there is a limit of the E field strength, making the result more realistic.

14.4 Dynamics of the Homopolar Generator

As described in ECE [1–10], torsion in space-time is electromagnetism. It has further been proposed that Spin Connection Resonance (SCR) is also a critical part of ECE. The connecting evidence between both concepts is present when a varying $d\mathbf{\Omega}/dt$ is applied to a Faraday disk (Homopolar Generator). Moreover, it will be shown that the varying spin connection (identified with $d\mathbf{\Omega}/dt \neq 0$) produces additional accelerations to electrons hereto assumed to be ignored under random collisions and generalized to drift velocity.

The accelerative state of a rotating reference frame is analyzed to the application of moving electrons. As shown in a standard mathematical dictionary of equations [17], a body in acceleration with a rotating reference frame experiences the rotational operator

14.4 Dynamics of the Homopolar Generator

$$\mathbf{R} \equiv \left(\frac{d}{dt}\right)_{body} + \mathbf{\Omega} \times . \tag{14.45}$$

The velocity in the rotating frame of space is

$$\mathbf{v}_{space} = \mathbf{R}\mathbf{r} = \frac{d\mathbf{r}}{dt} + \mathbf{\Omega} \times \mathbf{r} \tag{14.46}$$

and the equation expanded to acceleration in space is

$$\mathbf{a}_{space} = \mathbf{R}^2\mathbf{r} = \left(\frac{d}{dt} + \mathbf{\Omega}\times\right)^2 \mathbf{r} \tag{14.47}$$

where $\mathbf{r}$ is the coordinate vector in the rotating system. Eq. (14.47) gives a simplified result of

$$\begin{aligned}\mathbf{a}_{space} &= \left(\frac{d}{dt} + \mathbf{\Omega}\times\right)\left(\frac{d\mathbf{r}}{dt} + \mathbf{\Omega} \times \mathbf{r}\right) \\ &= \frac{d^2\mathbf{r}}{dt^2} + \frac{d}{dt}(\mathbf{\Omega} \times \mathbf{r}) + \mathbf{\Omega} \times \frac{d\mathbf{r}}{dt} + \mathbf{\Omega} \times (\mathbf{\Omega} \times \mathbf{r}) \\ &= \frac{d^2\mathbf{r}}{dt^2} + \mathbf{\Omega} \times \frac{d\mathbf{r}}{dt} + \frac{d\mathbf{\Omega}}{dt} \times \mathbf{r} + \mathbf{\Omega} \times \frac{d\mathbf{r}}{dt} + \mathbf{\Omega} \times (\mathbf{\Omega} \times \mathbf{r}) .\end{aligned} \tag{14.48}$$

When grouping for Velocity and Angular Velocity,

$$\mathbf{v} \equiv \frac{d\mathbf{r}}{dt}, \tag{14.49}$$

$$\boldsymbol{\alpha} \equiv \frac{d\mathbf{\Omega}}{dt}, \tag{14.50}$$

we get

$$\mathbf{a}_{space} = \frac{d^2\mathbf{r}}{dt^2} + 2\mathbf{\Omega} \times \mathbf{v} + \mathbf{\Omega} \times (\mathbf{\Omega} \times \mathbf{r}) + \boldsymbol{\alpha} \times \mathbf{r}. \tag{14.51}$$

As a result, we get four terms

$$\mathbf{a}_{body} = \frac{d^2\mathbf{r}}{dt^2}, \tag{14.52}$$

$$\mathbf{a}_{coriolis} = 2\mathbf{\Omega} \times \mathbf{v}, \tag{14.53}$$

$$\mathbf{a}_{centrifugal} = \mathbf{\Omega} \times (\mathbf{\Omega} \times \mathbf{r}), \qquad (14.54)$$

$$\mathbf{a}_{angular} = \mathbf{\alpha} \times \mathbf{r} \qquad (14.55)$$

for all accelerative states for the electron particle. The final acceleration of a body in space is summarized as follows:

$$\mathbf{a}_{space} = \mathbf{a}_{body} + \mathbf{a}_{coriolis} + \mathbf{a}_{centrifugal} + \mathbf{a}_{angular} \qquad (14.56)$$

where

1. conventional analysis normally ignores the $d\mathbf{\Omega}/dt$ when the assumption of an uniformly rotating frame of reference is given,
2. Coriolis acceleration is assumed to vanish in the background through random collisions within either gas, liquids, or solids, and
3. does not link any of these ignored variables to resonance.

Furthermore, the Lozenz Force Law equation in the coordinates $\mathbf{r}'$ of the rest system is

$$\mathbf{F_L}' = q(\mathbf{E}' + \mathbf{v}' \times \mathbf{B}'). \qquad (14.57)$$

We have to transform this equation to the rotating system according to Eq. (14.46):

$$\mathbf{F_L} = q\left(\mathbf{E} + \mathbf{v} \times \mathbf{B} + (\mathbf{\Omega} \times \mathbf{r}) \times \mathbf{B}\right). \qquad (14.58)$$

To re-study the Homopolar Generator in the context of ECE, we use the Newtonian limit and obtain the equation of motion

$$m\mathbf{a}_{space} = \mathbf{F_L} \qquad (14.59)$$

which can be rewritten with aid of Eq. (14.56) to

$$m\mathbf{a}_{body} = \mathbf{F_L} - m\left(\mathbf{a}_{coriolis} + \mathbf{a}_{centrifugal} + \mathbf{a}_{angular}\right). \qquad (14.60)$$

With substitution of Eqs. (14.52–14.55) and (14.58) follows

$$\frac{d^2\mathbf{r}}{dt^2} = \frac{q}{m}\left(\mathbf{E} + \mathbf{v} \times \mathbf{B} + (\mathbf{\Omega} \times \mathbf{r}) \times \mathbf{B}\right) - 2\mathbf{\Omega} \times \mathbf{v} - \mathbf{\Omega} \times (\mathbf{\Omega} \times \mathbf{r}) - \mathbf{\alpha} \times \mathbf{r}. \qquad (14.61)$$

14.4 Dynamics of the Homopolar Generator

We assume that $\mathbf{\Omega}$ and $\mathbf{B}$ point into the direction of the z coordinate,

$$\mathbf{\Omega} = \begin{pmatrix} 0 \\ 0 \\ \Omega \end{pmatrix}, \tag{14.62}$$

$$\mathbf{B} = \begin{pmatrix} 0 \\ 0 \\ B \end{pmatrix}, \tag{14.63}$$

and E is oscillating in radial direction,

$$\mathbf{E} = E_0 \cos(\omega_J t)\,\widehat{\mathbf{r}} = E_0 \cos(\omega_J t)\,\frac{\mathbf{r}}{r}. \tag{14.64}$$

Then we obtain for the single terms in Eq. (14.61)

$$\mathbf{v} \times \mathbf{B} = \begin{pmatrix} v_2 B \\ -v_1 B \\ 0 \end{pmatrix}, \quad \mathbf{\Omega} \times \mathbf{v} = \begin{pmatrix} -v_2 \Omega \\ v_1 \Omega \\ 0 \end{pmatrix},$$

$$\mathbf{\Omega} \times \mathbf{r} = \begin{pmatrix} -r_2 \Omega \\ r_1 \Omega \\ 0 \end{pmatrix}, \quad \mathbf{\Omega} \times (\mathbf{\Omega} \times \mathbf{r}) = \begin{pmatrix} -r_1 \Omega^2 \\ -r_2 \Omega^2 \\ 0 \end{pmatrix}, \tag{14.65}$$

$$(\mathbf{\Omega} \times \mathbf{r}) \times \mathbf{B} = \begin{pmatrix} r_1 \Omega B \\ r_2 \Omega B \\ 0 \end{pmatrix}, \quad \boldsymbol{\alpha} \times \mathbf{r} = \begin{pmatrix} -r_2 \frac{d\Omega}{dt} \\ r_1 \frac{d\Omega}{dt} \\ 0 \end{pmatrix}.$$

Inserting these terms into Eq. (14.61) shows that the particle moves in the x-y plane exclusively. The two coupled equations of motion in the rotating frame are

$$\frac{d^2 r_1}{dt^2} = \frac{q}{m}\left(E_0 \cos(\omega_J t)\,\frac{r_1}{\sqrt{r_1^2 + r_2^2}} + v_2 B + r_1 \Omega B\right) + 2v_2 \Omega + r_1 \Omega^2 + r_2 \frac{d\Omega}{dt}, \tag{14.66}$$

$$\frac{d^2 r_2}{dt^2} = \frac{q}{m}\left(E_0 \cos(\omega_J t)\,\frac{r_2}{\sqrt{r_1^2 + r_2^2}} - v_1 B + r_2 \Omega B\right) - 2v_1 \Omega + r_2 \Omega^2 - r_1 \frac{d\Omega}{dt}. \tag{14.67}$$

This is a model for the dynamics within the Homopolar Generator. As already stated, we cannot expect that electrons move in a solid in such a way, but it may be a hint for certain effects incurred by mechanical rotation. The Eqs. (14.66,14.67) have been solved numerically similar to the previously discussed results. In order to make the single force contributions visible we have solved the equations with omitting particular terms at the right-hand side. First we studied the Coriolis force being present as the only force. The result is shown in Fig. (14.9) for a constant rotation speed Ω. The orbit of a "free" electron is a spiral as is known from classical mechanics. Fig. (14.10) shows the same for an oscillating Ω. In the rotating coordinate system the particle is pushed back and forth as an effect of the variation of Ω. This is superimposed to the spiralling behaviour. The Lorentz force term in action is graphed in Fig. (14.11), for better graphical representation with a factor of $q/m = -1$. It consists of a force in radial direction (from E field) and a circular orbit (from $v \times B$ term). The result is an open rosette orbit. Near to the center there is a sharp edge where direction is changed abruptly. A detailed analysis showed that the velocity in this point is zero, it is like a classical turning point. The electric and magnetic term alone (not shown) give a linear oscillating orbit and a circular orbit.

All these effect computed together lead to a quite chaotic behaviour (Fig. (14.12)). The radius of the orbit is bound due to the Lorentz force term. The maximum radius taken over a certain time interval can be considered as an indicator for a certain resonance behaviour. In our interpretation this means that the charged particle crosses a certain range faster, leading to a higher current. In Fig. (14.13) this criterion is shown in dependence of a variable "driving force" term ω_J. There are indeed certain resonances which depend in a complicated way on the term $q/m \cdot B$.

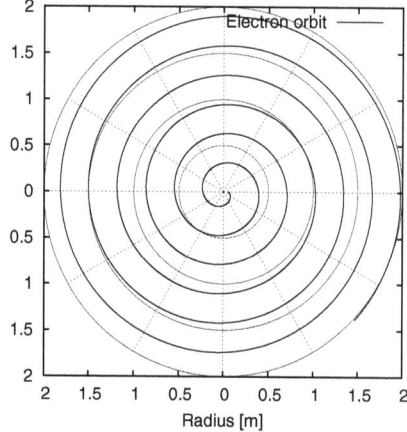

Fig. 14.9. Electron orbit for Coriolis force.

14.4 Dynamics of the Homopolar Generator   275

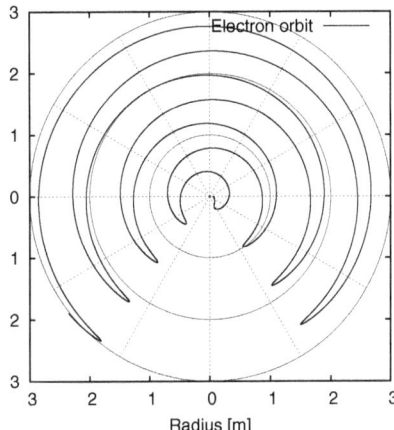

Fig. 14.10. Electron orbit for variable rotation frequency Ω.

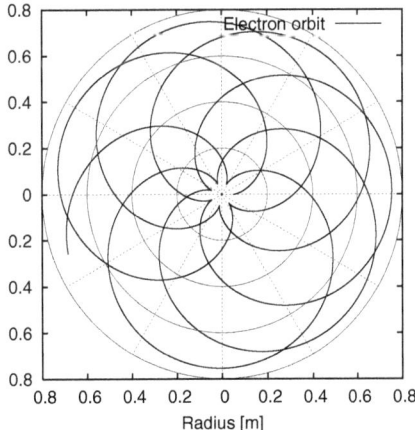

Fig. 14.11. Electron orbit for Lorentz force (electric and magnetic).

276   14 Spin Connection Resonance in the Faraday Disk Generator

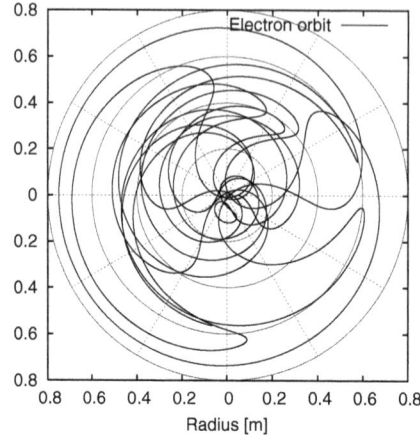

Fig. 14.12. Electron orbit for all force components.

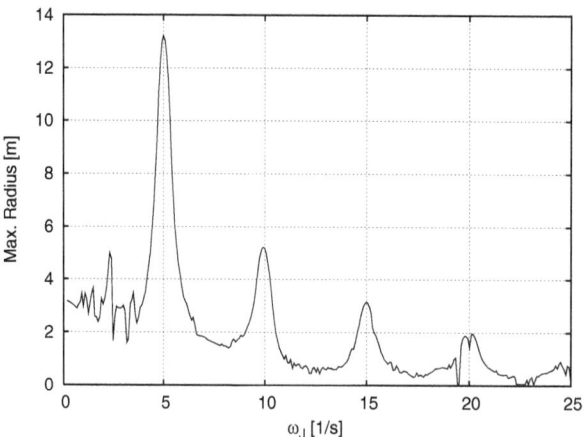

Fig. 14.13. Resonance curve of max. radius for variable current frequency ω_J.

14.5 Discussion of Design

As we have shown in the last section, a varying rotation speed adds sophisticated trajectories to the electron particles within the Homopolar Generator. It is proposed that the relationships (harmonics/standing waves) between Coriolis, centrifugal, and angular accelerations are of key importance; in which, either the geometrical energy being carried by the electrons or the exposed spaces around nucleuses creates additional unrecognized forces.

A detailed prototype of this nature is needed for further study. It's important to notice that additional experiments are needed besides SCR in copper plates; for instance, using aluminum, mercury, carbon nanotubes, superconducting wires, doped semiconducting materials, Stainless Steel, and additional alloys are also good candidates. Moreover, a close study of Weber's Force [18] application to the study of the Homopolar Generator is also needed. A great piece of work concerning classical forces and torque was already done by Guala-Valverde and coworkers (see [25] and references therein).

14.5.1 Magnetic Flux Path Configuration

It is important to mention that a similar path of the accelerated electron just mentioned can be taken by magnetic flux lines by configuring a transformer core away from the usual one pass architecture to many flux path architecture (Fig. (14.14)). A transformer with a coil core design will give flux paths a 1:n relationship amplifying the flux lines through simple geometry. Technically the new architecture would make this the first B(3) Transformer in existence. Engineers using magnets and combination with transformer cores would find this design useful.

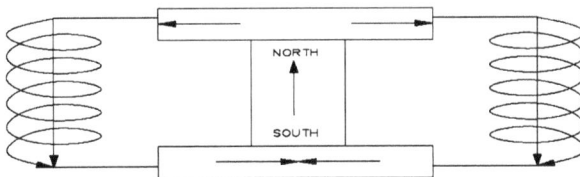

Fig. 14.14. Magnetic flux line amplification by multiple paths.

14.5.2 Proposed Prototype

14.5.2.1 Dynamo Electric Machine

As a result, it is proposed in this document that when SCR & ECE is established, the imposed drift velocities no longer generalize a particles' trajectory and the proposed harmonic/standing waves thru SCR create powerful resonances causing new anomalous results.

The proposed AIAS Homopolar Generator will take on a similar design to Nikola Tesla's work [20], see Fig. (14.15). In his design he proposed a means to extract the energy produced from the spinning disks from the least point on interruptions, the central axis, to both disks. This made his design simple and effective to keep his contacts from experiencing too much wear. Moreover, his design included two disks where a common spin direction produced current from the center-to-periphery of one disk and from the periphery-to-center of the other disk. His arrangements of the magnets made this simple design effective as he tied both disks' periphery with a copper belt band.

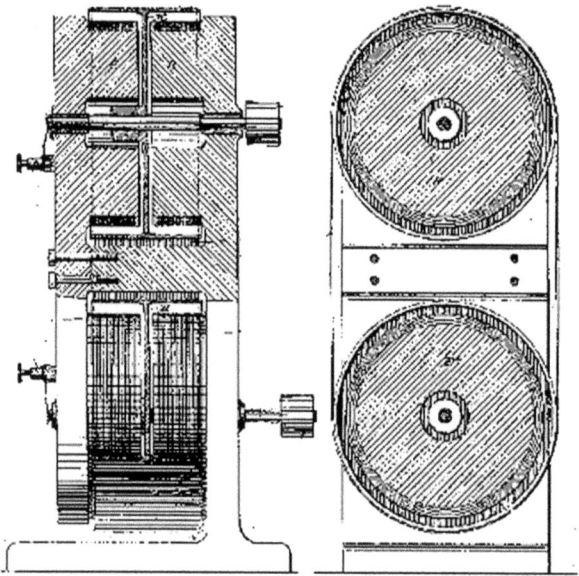

Fig. 14.15. Homopolar generator design of Nikola Tesla.

14.5.2.2 P&ID Detail Control

In our design (Fig. (14.16)) we decouple the outside belts for independent spin control for each Homopolar generator; however, they will have a common coupling point for the amps to flow through between the central axes of both disks. The power will be measured and extracted at the periphery of both disk with contacts going to terminal blocks (V1 and V2) for load connection. The independent spin control of both disks will superimpose two output waves that would give further study to ECE and SCR connection.

The basic AIAS Homopolar Process and Instrumentation Diagram (Fig. (14.16)) shows the general arrangement with Programmable Logic Controllers (PLC) with a Supervisory Control and Data Acquisition (SCADA) package. Standard network protocols (Ethernet and Devicenet) are used for control of

14.5 Discussion of Design   279

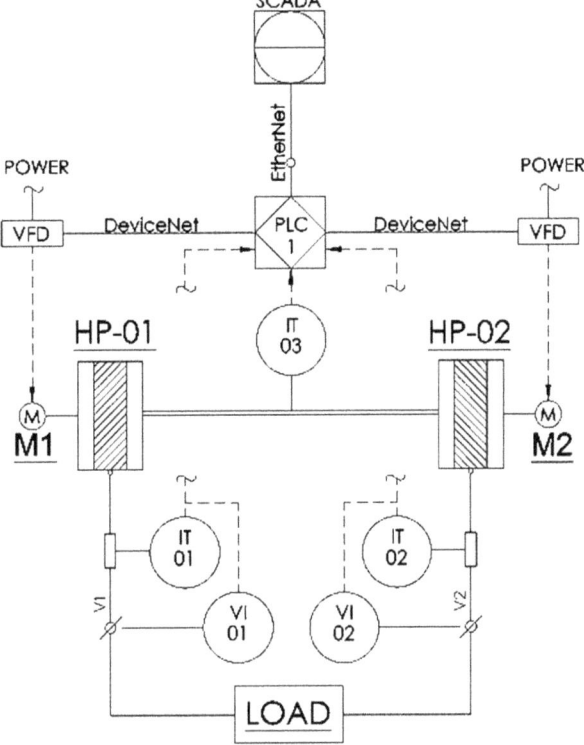

Fig. 14.16. AIAS Homopolar Process & Instrumentation Diagram.

Homopolar generators through variable frequency drives (VFDs) and logging data from instruments and trending of experimental data (SQL Server) are stored in a centralized server. Each instrument shown is for proper data analysis and control of said experiment with Proportional, Integral and Derivative (PID) software.

The final controls design will give both the engineers and scientist the flexibility to either expand or refine the necessary $d\Omega/dt$ control for proper ECE and SCR connections.

14.5.2.3 Circuit Analysis of Current

A circuit analysis of the Homopolar Generator reveals a current source-current output network architecture.

On expansion of this layout it is possible to introduced both positive and negative feedback through an Inductor by adding or subtracting magnetic fields to the static field of the magnets. The inductor would take the arbitrary current produced by the varying current. Since the inductor will be coiled around the homopolar generator, the end result will be additional input of energy into the system.

As mentioned before, an aligned winding of the inductor to the magnetic fields of the magnet would constitute a positive feedback arrangement. Likewise, if the aligned winding of the inductor is in the opposite direction to the magnetic fields of the magnet then it would constitute a negative feedback. As a result,

$$V(t) = \frac{1}{2}\Omega(t) r^2 B_t \tag{14.68}$$

where

$$B_t = B_{\text{magnet}} + B_{\text{inductor}} \tag{14.69}$$

for positive feedback. Moreover, equation (14.68) becomes

$$V(t) = \frac{1}{2}\Omega(t) r^2 (B_{\text{magnet}} + B_{\text{inductor}}), \tag{14.70}$$

where

$$B_{\text{inductor}} = \mu N \frac{I(t)}{l} \tag{14.71}$$

and

$$V(t) = I(t)\, R. \tag{14.72}$$

Hence, substituting (14.70) and (14.71) into (14.72),

$$I(t)\, R = \frac{1}{2}\left(\left(\Omega(t)\, r^2 \left(B_{\text{magnet}} + \mu N \frac{I(t)}{l}\right)\right)\right) \tag{14.73}$$

or

$$I(t)\left(R - \Omega(t)\, r^2\, \mu N \frac{I(t)}{2l}\right) = \frac{1}{2}\Omega(t)\, r^2\, B_{\text{magnet}}, \tag{14.74}$$

we finally get

$$I(t) = \frac{\Omega(t)\, r^2\, B_{\text{magnet}}}{2\left(R - \Omega(t)\, r^2 \frac{\mu N}{2l}\right)} \tag{14.75}$$

where the known voltages of the system are:

$$V_{\text{homopolar}}(t) = I(t)\,R, \tag{14.76}$$

$$V_{\text{inductor}}(t) = L\frac{dI(t)}{dt} \tag{14.77}$$

so that the total voltage is

$$V_{\text{total}}(t) = I(t)R + L\frac{dI(t)}{dt} \tag{14.78}$$

and total power created by the system is

$$\text{Power}(t) = V_{\text{total}}(t)\,I(t). \tag{14.79}$$

With substitution of (14.78) into (14.79) we get

$$\text{Power}(t) = \left(I(t)R + L\frac{dI(t)}{dt}\right)I(t) \tag{14.80}$$

or

$$\text{Power}(t) = I^2(t)R + L\frac{dI(t)}{dt}I(t). \tag{14.81}$$

For negative feedback, Eq. (14.70) would have a minus before the magnetic field on the inductor indicating that it has been wound in the opposite direction of the magnet's field. Again, this feedback introduces additional standard electrical results which would make the system react much different from the standard Homopolar Generators. We also point out that Tesla did use such inductors on his design to add additional energy into the system.

We conclude with the statement that Eq. (14.75) represent a further type of resonance. The current goes to inifity if the design is chosen in a way that the denominator of this equation tends to zero:

$$R - \Omega(t)\,r^2\frac{\mu N}{2l} = 0. \tag{14.82}$$

Assuming the harmonic time behaviour of Eq. (14.40) this gives a resonant current as shown in Fig. (14.17) (in arbitrary units). A time dependent Ω is even not required in this case. Variation of Ω being constant in time leads to the pole-like resonance graphed in Fig. (14.18).

In total we have shown in this paper that in the homopolar generator there are three types of resonances possible: A resonance of potential, a resonance

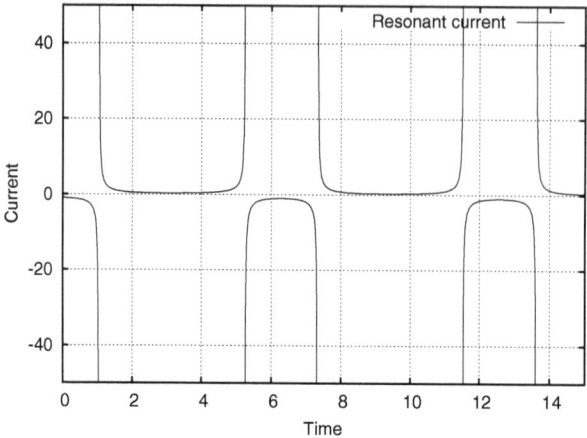

Fig. 14.17. Current resonance according to positive feedback design, periodic $\Omega(t)$.

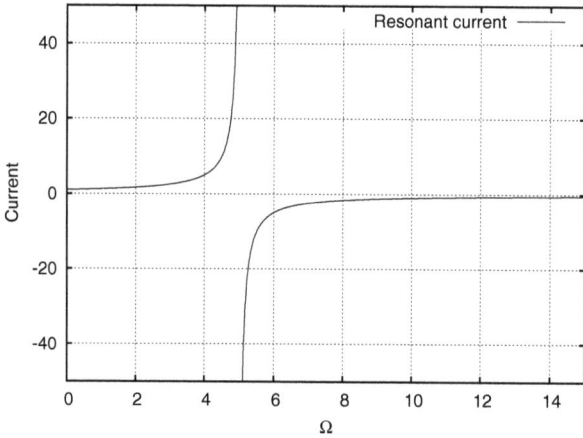

Fig. 14.18. Resonance curve for current resonance, no time-dependent Ω.

due to movement of charge carriers, and a resonance of the current by positive feedback.

Acknowledgments

The British Government is thanked for a Civil List Pension and the staff of AIAS and the Telesio Galilei Association for many interesting discussions.

References

[1] M. W. Evans, "Generally Covariant Unified Field Theory" (Abramis, 2005 onwards), in five volumes to date (see www.aias.us).

[2] L. Felker, "The Evans Equations of Unified Field Theory" (Abramis 2007).

[3] K. Pendergast, "Crystal Spheres" (preprint on www.aias.us, Abramis in prep.).

[4] M. W. Evans and H. Eckardt, Physica B, 400, 175 (2007).

[5] M. W. Evans, Physica B, 403, 517 (2008).

[6] M. W. Evans, Acta Phys. Polonica B, 38, 2211 (2007).

[7] M. W. Evans (ed.), Adv. Chem. Phys., vol. 119 (2001); ibid., M. W. Evans and S. Kielich, vol. 85 (1992, 1993, 1997).

[8] M. W. Evans, Omnia Opera section of www.aias.us, 1992 to present.

[9] M. W. Evans and L. B. Crowell, "Classical and Quantum Electrodynamics and the B(3) Field" (World Scientific, 2001).

[10] M. W. Evans and J.-P. Vigier, "The Enigmatic Photon" (Kluwer, 1994 to 2002), in five volumes.

[11] M. Krause, Director, "All about Tesla", on DVD, Berlin 2007.

[12] T. Valone, "The Homopolar Handbook, A Definite Guide to Faraday Disk and n Machine Technologies"(Integrity Research Institute, 2001, 3rd. Edition).

[13] S. P. Carroll, "Space-time and Geometry, an Introduction to General Relativity" (Addison Wesley, New York, 2004, and 1997 downloadable notes).

[14] R. M. Wald, "General Relativity" (Univ. of Chicago Press, 1984)

[15] Private communication of Walter Thurner to H. Eckardt.

[16] J. B. Marion and S. T. Thornton, "Classical Dynamics" (HBC, New York, 1988, 3rd. Ed.).

[17] E. W. Weisstein, "CRC Concise Encyclopedia of Mathematics" (Chapman & Hall/CRC, 1999)

[18] A. K. T. Assis, "Weber's Electrodynamics" (Kluwer Academic Publishers, 1994)

[19] J. A. Edminister, "Theory and Problems of Electromagnetics" (McGraw Hill Schaum's Outlines Series, 1995, 2nd Edition)

[20] N. Tesla, "My Inventions : The Autobiography of Nikola Tesla", (Barnes & Noble, 1995, reprint)

[21] Programmable Logic Controllers (www.ab.com)

[22] FactoryHMI and FactorySQL SCADA package (www.inductiveautomation.com)
[23] ODVA Devinet protocol (www.odva.org)
[24] PowerFlex Variable Frequency Drives (www.ab.com)
[25] J. Guala-Valverde and R. Achilles, "Ampère: The Avis Phoenix of Electrodynamics", Apeiron, Vol. 15, No. 3, July 2008

Orbital Einstein Cartan Evans (ECE) Theory and Non Einstein Hilbert (EH) Orbits in Astronomy and Cosmology

by

Myron W. Evans,
Alpha Institute for Advanced Study, Civil List Scientist.
(emyrone@aol.com and www.aias.us)

and

Horst Eckardt and K. Pendergast,
Alpha Institute for Advanced Studies (AIAS).
www.aias.us and www.atomicprecision.com

Abstract

It is shown that orbits of all kinds are determined in ECE theory by a well defined ratio of scalar torsion (T) to scalar curvature (R). Non-relativistic and relativistic circular orbits are considered as examples. The orbit of a binary pulsar is reproduced by using the complete theory with T/R as a parameter. It is shown that the precessing ellipse with decreasing orbital radius can be explained in ECE theory without assuming gravitational radiation. The latter is a flawed concept based on the geometrically incorrect Einstein Hilbert (EH) equation. In general orbits cannot be described by the EH theory, but must be described by the geometrically self-consistent ECE theory. Several examples of non-EH orbits are now known, such as binary pulsars and the various Pioneer/Cassini anomalies. These are explained straightforwardly with ECE theory, which also gives a simple explanation for the equivalence principle.

Keywords: Non- EH orbits, ECE theory of orbits, equivalence principle.

15.1 Introduction

It has been shown recently in this series of papers on ECE theory [1–10] that the Einstein Hilbert (EH) field equation is geometrically self-inconsistent because of its neglect of the Cartan torsion. Therefore conclusions based on the EH theory must be discarded and the theory of orbits developed with the geometrically self-consistent ECE theory, a generally covariant unified field theory rigorously based on Cartan geometry [11]. In order to demonstrate the ability of ECE theory to explain orbits straightforwardly, we consider in this paper the various examples of orbits now known not to be describable by the EH theory. Many criticisms of the latter have been made down the years [12] and it is well known to be a deeply flawed theory. It was finally discarded when it was shown using computer algebra [1–10] that the Christoffel connection is incompatible with the fundamental Bianchi identity as given by Cartan [1–11].

In Section 15.2 the general ECE orbital theory is developed in terms of a well defined ratio [1–10] of scalar torsion (denoted T) to scalar curvature (denoted R). This ratio is used as a parameter with which to describe non - EH orbits such as those of binary pulsars, or those observed in the Pioneer/Cassini anomalies. The limit of non-relativistic and relativistic circular orbits is considered as an example. In Section 15.3 the equivalence principle is explained straightforwardly with ECE orbital theory, and in Section 15.4 a perturbatin model of the binary pulsar orbits is developed. Graphical results and discussion are given in section 15.5 for non - EH orbits in binary pulsars and in the solar system, where non-EH orbits are now known from Pioneer and Cassini data. In section 15.6 the Hulse-Taylor binary pulsar is discussed.

15.2 General ECE Orbital Theory

In S.I. units and for a planar orbit, ECE orbital theory is based on the line element:

$$-c^2 d\tau^2 = -\left(1 - \frac{r_S}{r}\right) c^2 dt^2 + \left(1 - \frac{r_S}{r}\right)^{-1} dr^2 + r^2 d\phi^2 \qquad (15.1)$$

in spherical polar coordinates. Here [1–10]:

$$r_S = -\frac{T}{R}, \quad |r_S| = \frac{T}{R}, \qquad (15.2)$$

where T/R is a parameter determined by the data. An excellent description of the great majority of orbits is found in the limit:

$$r_S \to \frac{2MG}{c^2} \qquad (15.3)$$

15.2 General ECE Orbital Theory

where G is Newton's constant, and where a mass m is attracted to a gravitating mass M. In the solar system M is the mass of the Sun. In general however Eq. (15.3) does not apply. Examples of observed orbits in which Eq. (15.3) does not apply are those of binary pulsars and the Pioneer/Cassini anomalies. The standard model attempts to explain these orbits must be discarded because they are based on the geometrically incorrect Einstein Hilbert (EH) theory [1–12]. From Eq. (15.1):

$$
-\epsilon = -c^2 \left(\frac{d\tau}{d\lambda}\right)^2 = -\left(1 - \frac{r_S}{r}\right) c^2 \left(\frac{dt}{d\lambda}\right)^2 \\
+ \left(1 - \frac{r_S}{r}\right)^{-1} \left(\frac{dr}{d\lambda}\right)^2 + r^2 \left(\frac{d\phi}{d\lambda}\right)^2
\quad (15.4)
$$

where ϵ is a constant of motion [13, 14]. From Killing vector analysis [11, 13, 14] the following are also constants of motion for all gravitational fields:

$$
E_r = \left(1 - \frac{r_S}{r}\right) \left(\frac{dt}{d\lambda}\right), \quad L_r = r^2 \left(\frac{d\phi}{d\lambda}\right)
\quad (15.5)
$$

These constants of motion remain so under all conditions, including the conditions of a binary pulsar. Therefore the first type of ECE orbital equation is:

$$
\frac{1}{2}\left(\frac{dr}{d\lambda}\right)^2 + V_r(r) = \frac{1}{2} E_r^2.
\quad (15.6)
$$

This equation has the units of energy if:

$$
\lambda = \tau
\quad (15.7)
$$

where τ is the proper time [11, 13, 14] and if both sides of Eq. (15.6) are in S.I. units of joules. This is achieved by multiplying both sides by m to obtain:

$$
\frac{1}{2} m \left(\frac{dr}{d\tau}\right)^2 + V = E = \frac{1}{2} mc^2 E_r^2.
\quad (15.8)
$$

The potential energy of the system in joules is:

$$
V = mV_r = \frac{1}{2} m \left(\epsilon - \epsilon \frac{r_S}{r} + \frac{L^2}{r^2} - \frac{r_S L^2}{r^3}\right)
\quad (15.9)
$$

and the ECE orbital equation of type one, in S.I. units of joules is:

$$\frac{1}{2} m \left(\frac{dr}{d\tau}\right)^2 = E - V \qquad (15.10)$$

with:

$$\epsilon = c^2. \qquad (15.11)$$

Therefore the potential energy in joules is:

$$V = \frac{1}{2} m \left(1 - \frac{r_S}{r}\right) \left(c^2 + \frac{L}{r^2}\right) \qquad (15.12)$$

with the constant of motion:

$$L = r^2 \frac{d\phi}{d\tau}. \qquad (15.13)$$

Various types of orbits can be found by integrating Eq. (15.10) numerically. In the standard model the radius r_S is always assumed to be:

$$r_S = \frac{2MG}{c^2} \qquad (15.14)$$

and is incorrectly attributed [1–10] to Schwarzschild. This attribution is due to poor scholarship, Schwarzchild in 1916 produced a parameter α which was not identified with Eq. (15.14). The identification of α with the so called Schwarzschild radius is in fact arbitrary, as shown by Crothers [1–10]. The identification follows the data, and does not predict the data as claimed in the standard model. The observational fact is that there are orbits which do not obey Eq. (15.14). This is now understood [1–10] to be due to the fact that the EH equation is geometrically self-inconsistent for reasons carefully developed recently [1–10] in a series of proofs and arguments, and using computer algebra.

In a binary pulsar [11] for example the mean distance between the two component stars is decreasing per revolution. This decrease cannot be described by Eq.(15.14). In ECE theory the observed decrease per revolution is explained by using the fact that in general:

$$|r_S| = \frac{T}{R}(r, \theta, \phi). \qquad (15.15)$$

15.2 General ECE Orbital Theory

A simple and well known [11] example of orbital theory is the limit of circular orbits:

$$\frac{\partial V}{\partial r} = 0. \tag{15.16}$$

From Eqs. (15.12) and (15.16):

$$r = \frac{1}{c^2 r_S}\left(L^2 \pm L\left(L^2 - 12c^2 r_S^2\right)^{\frac{1}{2}}\right). \tag{15.17}$$

In the Newtonian limit:

$$r_S \to 0 \tag{15.18}$$

and:

$$r \to \frac{L^2}{c^2 r_S}. \tag{15.19}$$

Therefore:

$$r \to \frac{L^2}{c^2}\frac{R}{T}. \tag{15.20}$$

In the Newtonian limit:

$$L \to rv \tag{15.21}$$

where v is the orbital velocity, so:

$$\frac{R}{T} \to \left(\frac{c}{v}\right)^2 \frac{1}{r}. \tag{15.22}$$

In the Newtonian limit it is known from observation that:

$$\frac{R}{T} = \frac{c^2}{2GM} \tag{15.23}$$

to an excellent approximation, so:

$$rv^2 = GM. \tag{15.24}$$

Therefore in the Earth to Sun system for example these Newtonian limits apply. Some numerical results and discussion on the self-consistency of this analysis are given in Section 15.6. It is shown there that for a large enough T/R the orbital radius will decrease to zero, without assuming gravitational radiation. In a binary pulsar the orbit is elliptical, the masses of the two stars are about equal, and the ecliptic precesses a great deal every revolution. On top of this the distance between the two stars decreases. All these features are described by Eq. (15.10) and other types of ECE orbital equation, given the fact that T/R is in general a function of the spherical polar coordinates, and not a constant. This conclusion is also shown by the Pioneer/Cassini anomalies.

The binary pulsar is a two particle problem in dynamics, governed by:

$$\mathbf{r} = \mathbf{r}_1 - \mathbf{r}_2 \tag{15.25}$$

and

$$\mathbf{r}_1 = \frac{m_2}{m_1 + m_2}\mathbf{r}, \quad \mathbf{r}_2 = -\frac{m_1}{m_1 + m_2}\mathbf{r}, \tag{15.26}$$

where m_1 and m_2 are the masses of the two stars and where $\mathbf{r}_1$ and $\mathbf{r}_2$ are the distances between each star and the center of mass of the two star system. If the reduced mass is defined by:

$$\mu = \frac{m_1 m_2}{m_1 + m_2} \tag{15.27}$$

the lagrangian [13] is:

$$L = \frac{1}{2}\mu\left(\dot{r}^2 + r^2\dot{\phi}^2\right) - U(r) \tag{15.28}$$

in spherical polar coordinates. The following is a constant of motion:

$$l = \mu r^2 \dot{\phi} \tag{15.29}$$

and this is Kepler's second law as is well known. The total energy of the system is constant:

$$E = T + U \tag{15.30}$$

and so the orbital law may be described by:

$$\phi(r) = \int \frac{l}{r^2}\left(2\mu\left(E - U - \frac{l^2}{2\mu r^2}\right)\right)^{-1/2} dr. \tag{15.31}$$

15.2 General ECE Orbital Theory

This law may be rewritten [13] in terms of the force:

$$F(r) = -\frac{\partial U}{\partial r} \tag{15.32}$$

as

$$\frac{d^2 u}{dr^2} + u = -\frac{\mu}{l^2}\frac{1}{u^2} F(u) \tag{15.33}$$

where

$$u = \frac{1}{r}. \tag{15.34}$$

The centrifugal force is identified as [13]:

$$F_c = -\frac{\partial U_c}{\partial r} = \frac{l^2}{\mu r^3} = \mu r \dot{\phi}^2 \tag{15.35}$$

and the effective potential is:

$$V(r) = U(r) + \frac{l^2}{2\mu r^2}. \tag{15.36}$$

In Newtonian dynamics:

$$F(r) = -\frac{mMG}{r^2} \tag{15.37}$$

so:

$$U(r) = -\int F(r) dr = -\frac{mMG}{r}. \tag{15.38}$$

In Newtonian dynamics therefore:

$$V(r) = -\frac{mMG}{r} + \frac{l^2}{2\mu r^2}. \tag{15.39}$$

Eq. (15.33) may be integrated to give:

$$\frac{\alpha}{r} = 1 + \epsilon \cos(\theta) \tag{15.40}$$

where:

$$\alpha = \frac{l^2}{\mu k}, \quad \epsilon = \left(1 + \frac{2El^2}{\mu k^2}\right)^{\frac{1}{2}}, \quad (15.41)$$

and

$$k = mMG. \quad (15.42)$$

The quantity ϵ is known as the eccentricity of the orbit.

From these equation it is inferred that a Newtonian orbit is an exactly closed ellipse whose perihelion does not advance, i.e. in a Newtonian orbit:

$$\frac{d^2u}{d\phi^2} + u = \frac{Gm^2M}{l^2} \quad (15.43)$$

which is the orbit of a mass m attracted by a mass M through Newton's inverse square law. In the standard model the relativistic orbital equation becomes [13]:

$$\frac{d^2u}{d\phi^2} + u = \frac{Gm^2M}{l^2} + 3\frac{GM}{c^2}u^2 \quad (15.44)$$

i.e.

$$\frac{d^2u}{d\phi^2} + u = \frac{1}{\alpha} + \delta u^2 \quad (15.45)$$

using the notation:

$$\frac{1}{\alpha} = \frac{Gm^2M}{l^2}, \quad \delta = 3\frac{GM}{c^2}. \quad (15.46)$$

In ECE theory the orbital equation is Eq. (15.45) with:

$$GM = \frac{c^2}{2}\left|\frac{T}{R}\right| = \frac{c^2}{2}r_S. \quad (15.47)$$

Therefore orbits may also be described by numerically integrating the type two ECE orbital equation (15.45). The orbit of a binary pulsar is therefore described by finding u from Eq. (15.45) and using Eqs. (15.25) and (15.26) to find the individual orbits of the stars of the binary pulsar as a function of T/R. The latter is adjusted to fit the observed perihelion advance of $4°$ per revolution in the Hulse Taylor binary pulsar for example, and the observed decrease of 3.1 mm a revolution in the inter-star separation $r_1 + r_2$.

A third type of ECE orbital equation may be found by using:

$$\frac{dr}{d\phi} = \frac{dr}{d\tau}\frac{d\tau}{d\phi} = \frac{r^2}{L}\frac{dr}{d\tau} \tag{15.48}$$

so:

$$\frac{dr}{d\phi} = \frac{r^2}{L}\left(\frac{2}{m}(E-V)\right)^{\frac{1}{2}} \tag{15.49}$$

and:

$$\phi = \int \frac{L}{r^2}\left(\frac{2}{m}(E-V)\right)^{-\frac{1}{2}} dr. \tag{15.50}$$

This equation may be integrated numerically as a function of T/R to give such effects as light bending by gravitation as a function of T/R. So in general the light bending is not precisely twice the Newtonian value - this is an ECE prediction that can be looked for experimentally in for example the Joddrell Bank binary pulsar in which the two stars are each pulsars.

15.3 The Equivalence Principle

The inertial mass is that mass that determines the acceleration of a particle under the action of a given force [13]. The gravitational mass is that mass that determines the gravitational force between two particles. The two types of mass are the same within one part in 10^{12}. This is known as the equivalence principle. Thus:

$$F = mg = -\frac{mMG}{r^2} \tag{15.51}$$

and there exists an acceleration due to gravity that is independent of m. Thus two objects of different m fall to the surface of the Earth of mass M at the same time for a given r. This fact is usually attributed to Galileo but was known to the ancients, for example John Philoponus in the sixth century. The gravitational potential defines g as follows:

$$\mathbf{g} = -\boldsymbol{\nabla}\Phi \tag{15.52}$$

where:

$$\Phi = -\frac{GM}{r} \tag{15.53}$$

and the potential energy in joules is:

$$U = m\Phi. \tag{15.54}$$

In ECE the equivalence principle is a consequence of Cartan geometry, as might be expected. In ECE theory [1–10]:

$$g = Tc^2 \tag{15.55}$$

and:

$$\left|\frac{T}{R}\right| \to \frac{2MG}{c^2}. \tag{15.56}$$

Therefore Eq. (15.51) follows from Eqs. (15.55) and (15.56) if:

$$Tc^2 = \frac{c^2}{2}\left|\frac{T}{R}\right|\frac{1}{r^2} \tag{15.57}$$

i.e.:

$$R = \frac{1}{2r^2}. \tag{15.58}$$

Now use:

$$\left|\frac{T}{R}\right| = r_S = \frac{2MG}{c^2} \tag{15.59}$$

so from Eqs. (15.58) and (15.59):

$$|T| = \frac{r_S}{2r^2}. \tag{15.60}$$

Therefore the fundamental geometrical reason why m should be the same on both sides of Eq. (15.51) is that the curvature and torsion are given by Eqs. (15.58) and (15.60). The Pioneer/Cassini anomalies and the orbits of binary pulsars are due to the fact that r_S deviates from $2MG/c^2$, but the equivalence principle in ECE is always true, as for any theory of relativity.

15.4 Perturbation Theory of the Orbit of a Binary Pulsar System

In the simplest instance it is shown in this section that the main features of the orbit of a binary pulsar system can be explained with:

$$r_S = -\frac{T}{R} = \frac{2MG}{c^2} + \frac{a}{r} \tag{15.61}$$

where a is a parameter of the system. Therefore the potential energy of the binary pulsar is given by:

$$V = \frac{1}{2}m\left(1 - \frac{r_S}{r}\right)\left(c^2 + \frac{L^2}{r^2}\right) \tag{15.62}$$

with r_S defined as in Eq. (15.61). The latter produces an additional force of attraction:

$$\Delta F = -\frac{\partial \Delta V}{\partial r} = -\frac{2am}{r^3}\left(c^2 + \frac{2L^2}{r^2}\right). \tag{15.63}$$

It may be shown straightforwardly using Eq. (15.33) that this additional force of attraction results in a logarithmic spiral orbit of the type:

$$r = k^{\frac{1}{3}} \exp\left(\frac{\alpha}{3}\phi\right) \tag{15.64}$$

so that the complete orbit of the binary pulsar is a precessing ellipse superimposed on a logarithmic spiral. This result has been confirmed using computer simulation and is illustrated in Fig. (15.12). It can be seen that the observed decrease in orbital radius of 3.1 mm a revolution in a binary pulsar such as the Hulse Taylor binary pulsar can be reproduced by the simple assumption of Eq. (15.61). The ratio of torsion to curvature is assumed therefore to be of the form (15.61), in which a can be considered to be a small perturbation. In the solar system it is responsible for the Pioneer/Cassini anomalies, which is a small but finite extra gravitational attraction not given by the Einstein Hilbert theory. Therefore Eq. (15.61) gives a simple and consistent explanation of non EH orbits of all kinds without the assumption of gravitational radiation and by self consistently incorporating the Cartan torsion missing in the EH equation. The total potential of the binary pulsar system may be expressed as:

$$V = V_0 + \Delta V \tag{15.65}$$

where ΔV is an extra attractive potential due to Eq. (15.61). In an initially circular orbit:

$$\frac{\partial V_0}{\partial r} = 0 \qquad (15.66)$$

and it is clear that the extra force of attraction due to ΔV will cause the orbit to spiral inwards on a logarithmic spiral trajectory of type (15.64). In the earth to sun system in the solar system the extra ΔV is very small, as can be seen from the Pioneer/Cassini anomaly, and the earth's orbit is essentially circular and non-relativistic. The advance of the perihelion for earth is very small per orbital revolution of one year. However in a binary pulsar the extra force of attraction causes the orbit to spiral inwards on a logarithmic spiral trajectory by an average of 3.1 mm a revolution. The orbit of the Hulse Taylor binary pulsar for example is very elliptical, and its perihelion advances by 4° per revolution. The advance of the perihelion can be described by V_0 alone, but the decrease of the orbit needs for its explanation the additional ΔV. All binary pulsar systems currently catalogued can be described in this way, without the need to assume gravitational radiation.

One of the major discoveries of ECE theory [1–10] is that the EH equation is self inconsistent because of its neglect of torsion. Therefore all the physical predictions of the EH equation must be re-explained anew using ECE theory, in which the torsion is correctly incorporated. These include light deflection due to gravitation, perihelion advance, the orbits of binary pulsars, the Pioneer/Cassini anomalies, frame dragging, the Shapiro delay, and in general all the precision tests of relativity. Concepts such as Big Bang, dark matter theory, black hole theory, and so on must be rejected because they are based on a self-inconsistent geometry, Riemann geometry without torsion. Careful scholarship [1–10] has revealed that the Schwarzschild vacuum solutions of 1916 do not contain the parameter $2MG/c^2$, but a parameter α. Therefore α in Eq. (15.61) is modeled to give the experimentally observed orbits. In ECE theory the α parameter is recognized as the ratio of $-T/R$. The orbit of a binary pulsar has been described without assuming gravitational radiation. The latter is again a false concept based on the flawed EH equation, and gravitational radiation has never been directly observed. It is merely assumed to exist because EH theory cannot describe the decrease in the orbital radius of a binary pulsar. In ECE theory [1–10], gravitational radiation exists in principle from the ECE wave equation based on the tetrad postulate of Cartan geometry. However gravitational radiation is exceedingly difficult to observe, and has not yet been observed in a quarter century of effort. Therefore the exact solutions of the EH equation are obsolete because they are solutions of an inconsistent geometry. It has been shown in paper 93 onwards of ECE theory [1–10] that the Christoffel or symmetric connection is inconsistent with the Bianchi identiy as developed by Cartan. All exact solutions of the EH equation are based on the Christoffel symbol and so must be rejected. During

15.4 Perturbation Theory of the Orbit of a Binary Pulsar System

the course of development of ECE theory, new and self consistent methods of relativity theory have been devised.

In order to calculate non Einstein Hilbert orbits self-consistently the dependence of r_S on r and on the constants of motion is needed. Consider the potential energy of the relativistic Kepler problem using the notation of section 15.1:

$$V(r) = \frac{1}{2} m \left(1 - \frac{r_S}{r}\right) \left(c^2 + \frac{L^2}{r^2}\right) \tag{15.67}$$

It is found that:

$$\frac{\partial V}{\partial r_S} = -\frac{m \left(\frac{L^2}{r^2} + c^2\right)}{2r} \tag{15.68}$$

The rate of change of V with r is:

$$\frac{\partial V}{\partial r} = \frac{m r_S}{2 r^2} \left(\frac{L^2}{r^2} + c^2\right) - \frac{mL^2}{r^3} \left(1 - \frac{r_S}{r}\right) \tag{15.69}$$

Now use:

$$\frac{\partial V}{\partial r_S} = \frac{\partial V}{\partial r} \frac{\partial r}{\partial r_S} \tag{15.70}$$

to find:

$$\frac{\partial r_S}{\partial r} = -\frac{3 r_S L^2 - 2 r L^2 + c^2 r^2 r_S}{r (L^2 + c^2 r^2)} \tag{15.71}$$

and

$$\frac{\partial r}{\partial r_S} = -\frac{r (L^2 + c^2 r^2)}{3 r_S L^2 - 2 r L^2 + c^2 r^2 r_S} \tag{15.72}$$

Therefore by integration, the required dependence of r_S on r and L is found:

$$r_S = \int -\frac{3 r_S L^2 - 2 r L^2 + c^2 r^2 r_S}{r (L^2 + c^2 r^2)} dr \tag{15.73}$$

Conversely, the dependence of r on r_S is found by the following integration:

$$r = \int -\frac{r (L^2 + c^2 r^2)}{3 r_S L^2 - 2 r L^2 + c^2 r^2 r_S} dr_S. \tag{15.74}$$

The integral of Eq. (15.73) can be worked out analytically:

$$r_S(r) = \frac{(L^2 + c^2 r^2)}{r^3} \left(2L^2 \left(\frac{\log(L^2 + c^2 r^2)}{2c^4} + \frac{L^2}{2c^4 L^2 + 2c^6 r^2} \right) + b \right) \quad (15.75)$$

while the integral of Eq. (15.74) is not solvable analytically. Solution (15.75) contains the integration constant b which leads to a threefold classification of the results:

$$r_S \to \begin{cases} +\infty \\ -\infty \\ 0 \end{cases} \text{ for } r \to 0.$$

Examples of the first two classes are graphed in Section 5.2. Self-consistency of the perturbation model is shown there by deriving an expression for the perturbation parameter a which is compatible with the r dependence of r_S.

In Section 15.6 a review of known binary pulsar systems is given from a recent observational survey. These orbits are all likely to be non EH orbits and in future work they will be catalogued using Eq. (15.61), so that each can be assigned an a parameter. Furthermore, orbits of all kinds can be described by Eq. (15.61), for example the equation shows that the earth's orbit will very slowly spiral into the sun due to the non EH attractive force discovered by both the Pioneer and Cassini spacecraft as they escape the solar system. This is equivalent to a g of order 10^{-12}ms^{-2}. So Eq. (15.61) is a suggestion for a self consistent cosmology, one that explains all orbits without gravitational radiation, and one which correctly takes account of the Cartan torsion in relativity theory.

15.5 Graphical Results and Discussion

15.5.1 Schwarzschild Radius and Potential

The results of the analytical and numerical calcualtions are presented in this section. First the effect of a variable Schwarzschild radius r_S was studied. A parametric form

$$r_S = \gamma \, r_{S0} \quad (15.76)$$

was used where r_{S0} is the value from standard theory:

$$r_{S0} = \frac{2GM}{c^2} \quad (15.77)$$

Table 15.1 Simulation parameters (in SI units).

	Earth/Sun	Hulse-Taylor
m_1	$5.9742 \cdot 10^{24}$ kg	$2.86629 \cdot 10^{30}$ kg
m_2	$1.9891 \cdot 10^{30}$ kg	$2.75888 \cdot 10^{30}$ kg
r_1	$1.4960 \cdot 10^{11}$ m	$3.15357 \cdot 10^{9}$ m
L	$4.4580 \cdot 10^{15}$ m^2/s	$3.4 \cdot 10^{14}$ m^2/s

and γ is an adjustable parameter. In the Newtonian limit the orbital radius r depends inversely linear on r_S, see Eq. (15.19). This is graphed in Fig. (15.1) for the earth radius in the solar system. For $\gamma = 1$ it takes the standard value of about $1.5 \cdot 10^{11}$ m but falls hyperbolically to zero for a largely increased r_S.

In Fig. (15.2) the Newtonian limit of Eq. (15.19) is compared with the relativistic case described by Eq. (15.17). With the parameters of the solar system (Table (15.1)) both approaches lead to indistinguishable values with exception of a very large r_S where the relativistic curve drops to a vertical tangent. Beyond this range, the square root expression of Eq. (15.17) is imaginary. It can be seen that both equations are self consistent because they result in the same curve over a wide range of r_S.

Next the relativistic potential of Eq. (15.12) is described and graphed by some examples. In the Newtonian case the term proportional to $1/r^3$ is absent, i.e. r_S is effectively set to zero for this part. From Fig. (15.3) the principle difference between the Newtonian and relativistic case can be seen (the term $1/2\, mc^2$ has been omitted). In the Newtonian case the potential goes to plus infinity for small radii, while it falls to minus infinity in the relativistic case. This does not change significantly if parameters are altered. For example in Fig. (15.4) we have increased the mass of the earth artificially to that of the sun. Theoretically the relativistic curve could have a local maximum, but this does not show up in the parameter range considered here. The behaviour of the Newtonian potential changes if the angular momentum of the earth would be considerably lower as indicated in Fig. (15.5). However, all this would happen inside the radius of the sun and therefore is beyond reality.

The change in the potential for a varying r_S can be studied from Fig. (15.6). For increasing r_S, there is no visible change in the Newtonian potential,

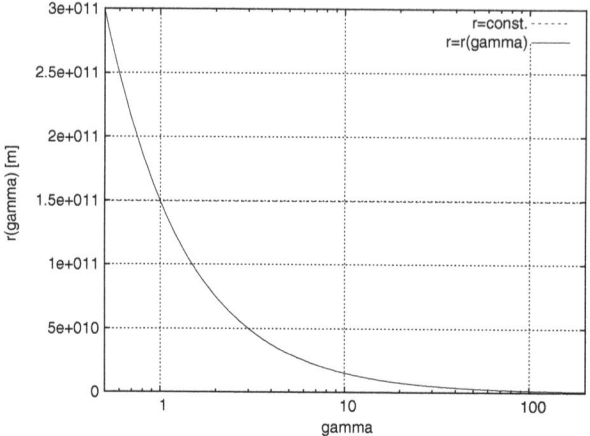

Fig. 15.1. Orbital radius r of Earth in dependence of variable Schwarzschild radius $r_S = \gamma \cdot r_{S0}$.

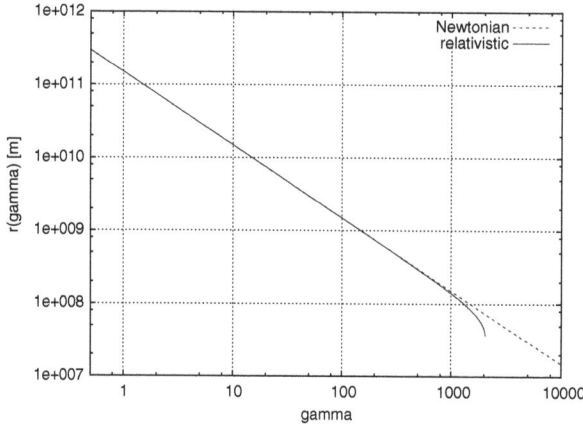

Fig. 15.2. Orbital radius of Earth in dependence of variable Schwarzschild in Newtonian and relativistic limit.

but a tendency to an earlier dropping in the relativistic case. The behaviour of the Newtonian potential is different for a reduced angular momentum (Fig. (15.7)). Increasing the Schwarzschild radius leads to a significant lowering of the potential near to the center.

15.5.2 Self-Consistency of Schwarzschild Radius Calculations

According to the results of section 15.4, r_S can be obtained as a function of the radius by integration (Eqs. 15.73, 15.75). The solution (15.75) contains the

15.5 Graphical Results and Discussion

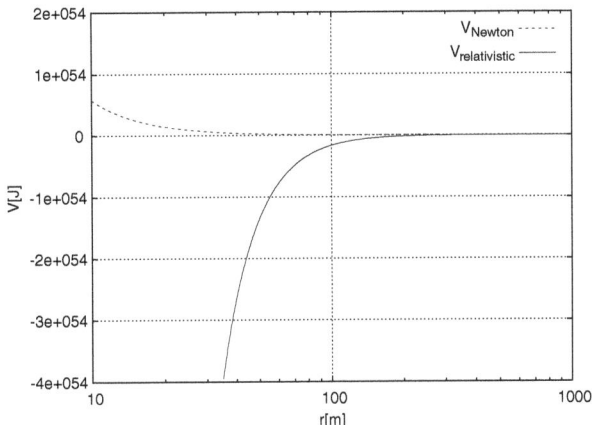

Fig. 15.3. Potential of Sun for Earth, $r_S = 2954$ km, m = m(Earth), L = L(Earth).

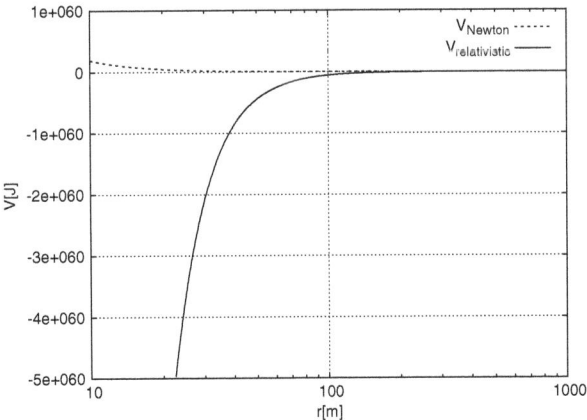

Fig. 15.4. Potential of Sun for enhanced earth mass, $r_S = 2954$ km, m = m(Sun), L = L(Earth).

302   15 Orbital Einstein Cartan Evans (ECE) Theory and Non Einstein Hilbert

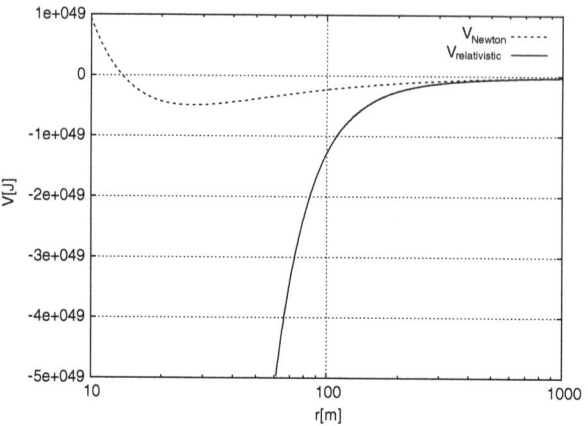

Fig. 15.5. Potential of Sun for enhanced earth mass, $r_S = 2954$ km, m=m(Sun), lowered angular momentum: $L = 6 \cdot 10^{10} \mathrm{m}^2/\mathrm{s}$ (reduced).

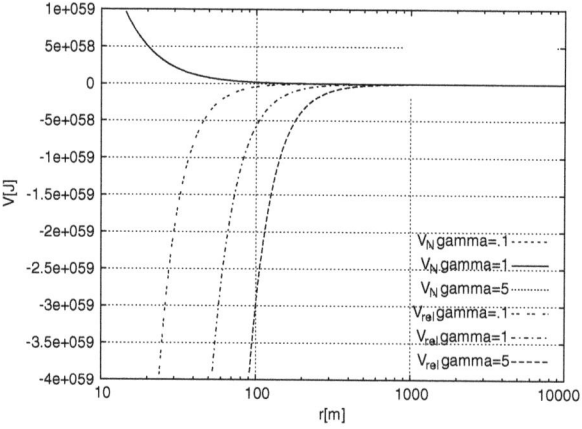

Fig. 15.6. Dependence of Potential of Sun from $|T/R|$, for $|T/R| = \gamma \cdot r_S$ with $r_S = 2954$ km, m=m(Sun), L = L(Earth).

15.5 Graphical Results and Discussion

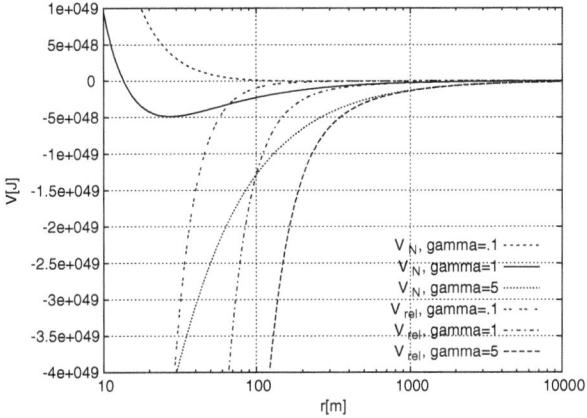

Fig. 15.7. Dependence of Potential of Sun from $|T/R|$, for $|T/R| = \gamma \cdot r_S$ with $r_S = 2954$ km, m = m(Sun), $L = 6 \cdot 10^{10} \text{m}^2/\text{s}$ (reduced).

integration constant b which leads to a threefold classification of the results: $r_S \to +\infty, -\infty, 0$. Examples of the first two classes are graphed in Fig. (15.8). The third class of solutions is defined by $r_S(r)$ hitting the coordinate origin:

$$r_S(0) = 0.$$

This is the type of solution which is physical because it has the right asymptotic behaviour

$$r_S \to 0 \quad \text{for} \quad r \to 0.$$

This is a bound solution while the other two classes are unbound solutions. Setting

$$r_S(r) = 0 \tag{15.78}$$

defines the constant b for which this condition is met, in dependence of r:

$$b = -\frac{\left(L^4 + c^2 r^2 L^2\right) \log\left(L^2 + c^2 r^2\right) + L^4}{c^4 L^2 + c^6 r^2}. \tag{15.79}$$

We obtain the desired value of the constant by setting $r = 0$:

$$b = -\frac{L^2 \left(2 \log(L) + 1\right)}{c^4}. \tag{15.80}$$

This expression depends sensitively on the constant of motion L. By construction, r_S should have the known value of 2954 m for the earth orbit. For $L(Earth)$ we obtain the upper curve in Fig. (15.9) which does not meet this condition properly. Therefore we have adopted the value of L to $L = 1.42 \cdot 10^{15} \mathrm{m}^2/\mathrm{s}$ so that the curve goes through the point

$$r_S(1.49 \cdot 10^{11} \mathrm{m}) \approx r_{S0} = 2954 \, \mathrm{m}.$$

This leads to the lower curve in Fig. (15.9). The numerical analysis shows that the formula for r_S depends sensitively on the value of b. Since we are operating with very large numbers here (up to 10^{60}) the numerical stability is not very high even in evaluating analytical formulae. This point has to be further investigated in subsequent work, perhaps by switching to astronomical units. Here we restrict to showing up the basic properties of a variable $r_S(r)$.

To circumvent this problem we try another approach for model 3. The integration constant b is defined now by the condition

$$r_S(r_{Earth}) = r_{S0}.$$

This gives the result graphed in Fig. (15.10). r_S takes a maximum at the earth radius. This approach is not sufficient since r_S takes negative values for small r.

As a last point in this subsection we inspect the behaviour of the perturbation model Eq. (15.61) in section 15.4. For reasons of self-consistency, the parameter a must be a function of r so that both models can coincide. Therefore we have, with Eq. (15.75):

$$r_{s0} + \frac{a}{r} = \frac{(L^2 + c^2 r^2)}{r^3} \left(2 L^2 \left(\frac{\log(L^2 + c^2 r^2)}{2 c^4} + \frac{L^2}{2 c^4 L^2 + 2 c^6 r^2} \right) + b \right). \tag{15.81}$$

and after resolving for a:

$$a = \frac{(L^4 + c^2 r^2 L^2) \log(L^2 + c^2 r^2) + L^4 + b c^4 L^2 - c^4 r^3 r_{s0} + b c^6 r^2}{c^4 r^2}. \tag{15.82}$$

Using the three models of Figs. (15.9) and (15.10) we obtain three distinct values for b and therefore three different solutions for Eq. (15.81), graphed in Fig. (15.11). Models 1 and 3 have a maximum at the earth orbit. Models 2 and 3 lead to $a = 0$ at the earth orbit by construction. It can be seen from the graphs that this condition is fulfilled. Model 3 gives negative values for a and therefore seems to be the worst one. Model 2 is the only model with

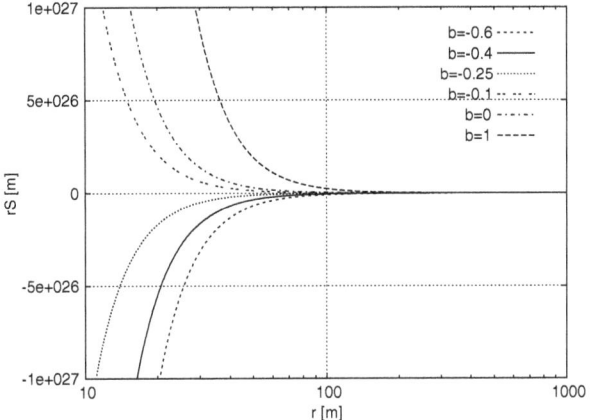

Fig. 15.8. Dependence of Potential of Sun from $|T/R|$, for $|T/R| = \gamma \cdot r_S$ with $r_S = 2954$ km, m = m(Sun), L = $6 \cdot 10^{10}$ m^2/s (reduced).

increasing, non-negative a for $r < r(Earth)$ and seems to be most compatible with the perturbation model.

We can conclude that the self-consistent calculation of r_S is in principle compatible with the perturbation model, but it is quite difficult to find the "physical" integration constant because the results depend sensitively on it.

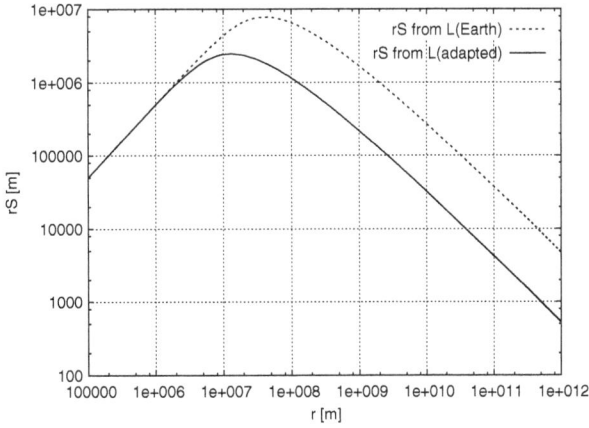

Fig. 15.9. Solutions of third type for $r_S(r)$ going through the coordinate origin, for original L(Earth) and adapted L.

306   15 Orbital Einstein Cartan Evans (ECE) Theory and Non Einstein Hilbert

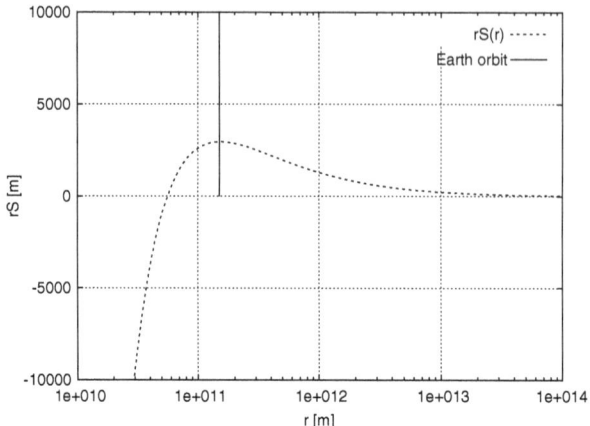

Fig. 15.10. $r_S(r)$ for integration constant b defined from condition $r_S(r_{Earth}) = r_{S0}$.

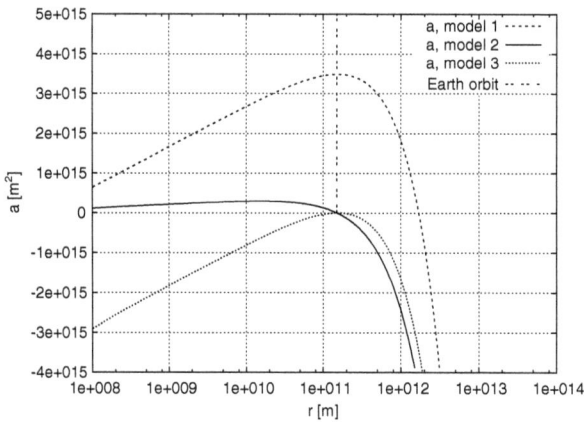

Fig. 15.11. Self-consistent parameter a for perturbation term of r_S, for three models.

15.5.3 Simulations of Orbits

The computation of orbits is based on Eqs. (15.44–15.46). Since these are not soluble analytically, a simulation program was written on base of the Runge-Kutta algorithm. With this method, coupled ordinary differential equations of first order can be solved in 4th order precision. The set of equations is

$$\frac{du}{d\phi} = w, \qquad (15.83)$$

$$\frac{dw}{d\phi} = \frac{1}{\alpha} + \delta u^2 - u \qquad (15.84)$$

with

$$\frac{1}{\alpha} = \frac{G\mu^2 m_2}{l^2} = \frac{\mu^2 c^2}{l^2}\frac{r_S}{2}, \qquad (15.85)$$

$$\delta = \frac{3Gm_2}{c^2} = \frac{3}{2}r_S^2, \qquad (15.86)$$

μ being the reduced mass, $l = L\mu^2$ the angular momentum and w an auxiliary variable (inverse radial velocity). Without the δu^2 term, the orbits $r(\phi) = 1/u(\phi)$ have the form of ellipses. The effect of the relativistic δu^2 is to prevent the elliptic orbits from being closed curves, there is an advance of the perihelion per revolution. With a constant r_S the minimum and maximum radius would not change, but they do if r_S depends on the radius. Both effects (relativistic term and a variable r_S) are inlcuded in the results shown in Fig. (15.12) where they can be observed nicely. The shrinking of radius is made more visisble in Fig. (15.13). One sees that the outer radius shrinks faster than the inner radius, thus stabilizing the orbit for a long time. Correspondingly, the inverse orbital velocities, described by the function $w(\phi)$, shrink in the same way (Fig. (15.14)) so that the total energy remains constant.

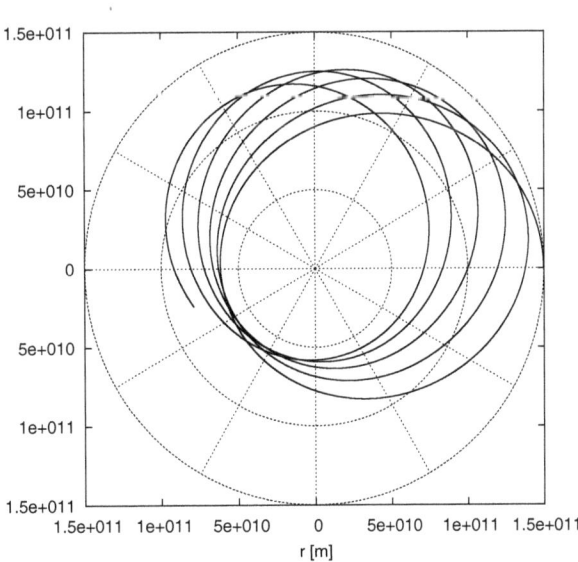

Fig. 15.12. Orbit $r(\phi)$ for a relativistic potential with Schwarzschild radius perturbation a/r.

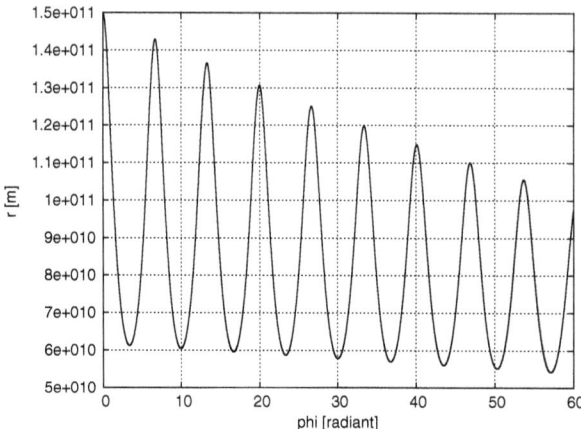

Fig. 15.13. Angular dependence of orbital radius r of Fig. 15.12.

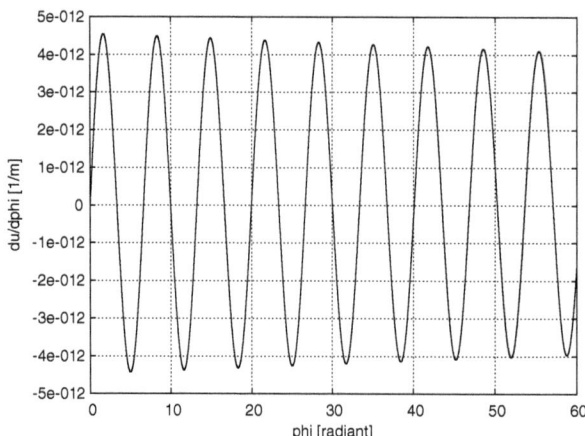

Fig. 15.14. Angular dependence of inverse velocity w of Fig. 15.12.

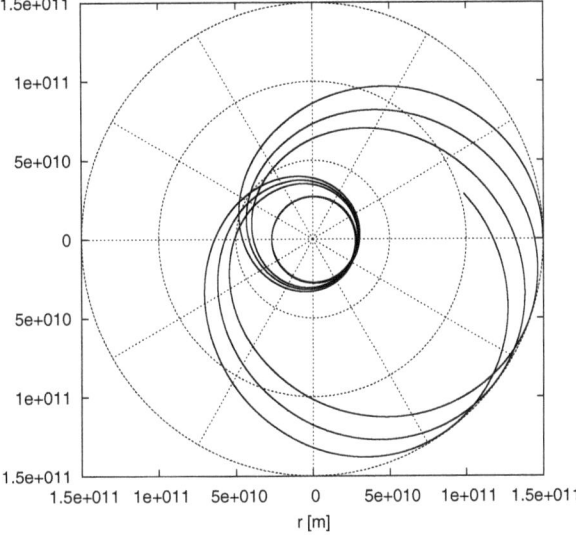

Fig. 15.15. Orbit $r(\phi)$ for a relativistic potential with Schwarzschild radius dependence according to model 2.

For the more complicated models of $r_S(r)$ other types of orbits are to be expected. Applying the model of Fig. (15.9), there is a maximum of $r_S(r)$ which leads to two stable orbital radii and the mass body oscillates between both (Fig. (15.15)). This is in accordance with orbits of periodic central fields as described in [13].

15.6 Pulsars and Double Star Systems

15.6.1 History of Discovery

Sir Arthur Eddington famously led an expedition to observe the total eclipse of the Sun in 1919 to see if light grazing the Sun was deflected by gravity according to the Einstein-Hilbert equation of general relativity. The conclusion that the Einstein-Hilbert equation described the bending of light better that Newtons work, led to Einstein becoming the world's first global superstar scientist and introduced the public to Einstein's concept of gravity as the curvature of space by massive objects. Here, mass tells space how to bend and space tells masses how to move!

Looking further into space than the Sun astronomers have sought to locate objects in the space with much greater gravitational fields than the Sun in order to improve the accuracy of the Edddington experiment, which was carried out close to the limit of experimental accuracy. The objects which are frequently studied for this purpose are white dwarfs, neutron stars and pulsars. The intense beam of light from pulsars make them particularly suited to testing out general relativity.

The first pulsar to be discovered was by the postgraduate student Jocelyn Bell Burnell at Cambridge University in the nineteen sixties. Pulsars are stars that have lived their lives, gone supernova and collaped to produce neutron stars which have most of the mass of the original star, but concentrated into a much smaller volume. The concentration of the mass increases the gravity of the collapsed star in its neighborhood and this leads to a much greater curving of space, as predicted by Einstein's theory of general relativity.

In 1974 the first binary double star to have a pulsar as one of the components was discovered by Hulse and Taylor 21,000 light years away in the constellation of Hercules using the Arecibo radio antenna. The binary pulsar is in orbit with another star, with both components having a mass 1.4 times that of the Sun and the binary nature of the pulsar being given away by the 7.75 hour periodic variations in the arrival times of the radio pulses due to the pulsar approaching and receeding, which corresponds with the time taken for the system to complete one orbit. The Hulse Taylor binary pulsar was widely studied as a test bed for general relativity and gave its discoverers the 1993 Nobel Prize for Physics. About one hundred double pulsars have been found so far, giving researchers a range of orbits to and masses to study. However, the disovery of a binary double star in which both components were pulsars and in which both pulsar beams were reaching Erth would be not just an amazing stoke of luck, but would also be the system of choice to use to study general relativity!

The first and only double pulsar system ever found is called PSR J0737-3039A, B and was discovered by the Jodrell Bank radio-observatory in Manchester in January 2004.

The PSR prefix stands for pulsar and the letters A and B refer to the two component pulsars that are trapped by the immense gravitational field into mutually orbiting one another as a binary double star. Binary double stars are common, but for one of the components of a double star to be a pulsar is rare, because pulsars are comparatively rare and for both components to be a pulsar is a godsend, one of the greatest discoveries in all of science.

15.6.2 Numerical Results for the Hulse Taylor Double Star System

The doubls star system was simulated as described in section 5.3. The orbit of both masses is depicted in Fig. (15.16). Both stars move on ellipses with a common focal point. The relations between the geometric parameters are

$$e = \sqrt{a^2 - b^2}, \tag{15.87}$$

$$\epsilon = \frac{e}{a} = \frac{\sqrt{a^2 - b^2}}{a}, \tag{15.88}$$

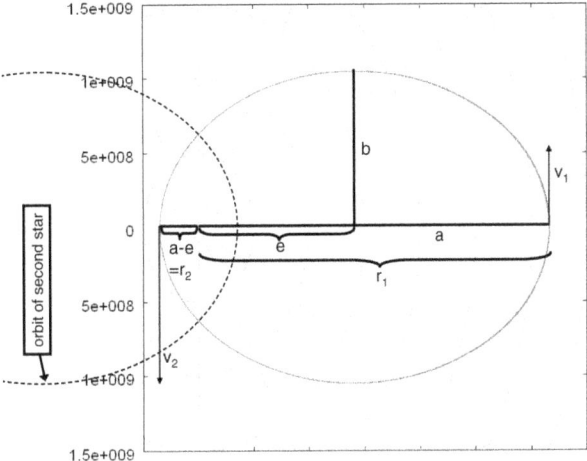

Fig. 15.16. Hulse-Taylor pulsar double star system, orbit and parameters.

$$b = a\sqrt{1 - \epsilon^2} \tag{15.89}$$

with excentricity ϵ. The angular momentum parameter L can be determined in a simple way at the aphelion and perihelion points where the orbital velocity v is perpendicular to the radius, i.e. its radial component vanishes:

$$L = (a + e)v_1 = (a - e)v_2. \tag{15.90}$$

The kinetic, potential and total energy can be derived from the computed variables u and w in the following way. For w we have

$$w = \frac{du}{d\phi} = \frac{du}{dr}\frac{dr}{d\phi} = -\frac{1}{r^2}\frac{dr}{d\phi} \tag{15.91}$$

or

$$\frac{dr}{d\phi} = -wr^2. \tag{15.92}$$

With this equation we have

$$\dot{r} = \frac{dr}{d\phi}\frac{l}{\mu r^2} = -\frac{wl}{\mu}. \tag{15.93}$$

Therefore the kinetic energy becomes

$$T = \frac{1}{2}\left(\mu \dot{r}^2 + \frac{l^2}{\mu r^2}\right) = \frac{l^2}{2\mu}\left(w^2 + u^2\right). \tag{15.94}$$

Note that the centrifugal potential belongs to the kinetic energy. The potential energy is

$$U = \frac{1}{2}\mu\left(c^2 - \frac{r_S}{r}c^2 - r_S\frac{l^2}{\mu^2 r^3}\right) \quad (15.95)$$

and the total energy

$$E_{tot} = T + U. \quad (15.96)$$

These energies have been plotted in Fig. (15.17). It can be seen that the total energy is conserved throughout the orbit. Kinetic and potential energy are highest in the region of maximal closeness of the stars, i.e. in the perihelion.

The relativistic part of the potential is compared to the Newtonian part (see Eq. (15.12)) in Fig. (15.18). The distance of the stars, even being closer than the sun-earth system, is large enough to make the relativistic effects small compared to the Newtonian part of the potential. It can nicely be seen that relativistic effects are highest in the perihelion, while they tend to zero in the aphelion.

Finally the effect of the parameter a of the perturbation model of the Schwarzschild radius is shown (Fig. (15.19)). The decrease in radius per revolution, Δr, is graphed against a. Obviously the dependence is linear over a range of several orders of magnitude. The physical value of Δr is some millimeters per revolution, so a has to be chosen in the range of 10^{-4}m^2. The

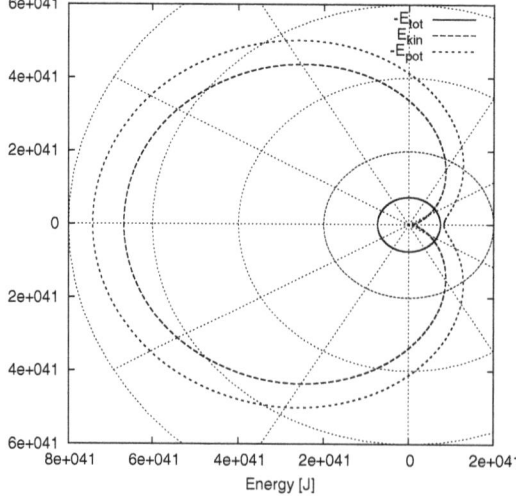

Fig. 15.17. Hulse-Taylor pulsar double star system, energy.

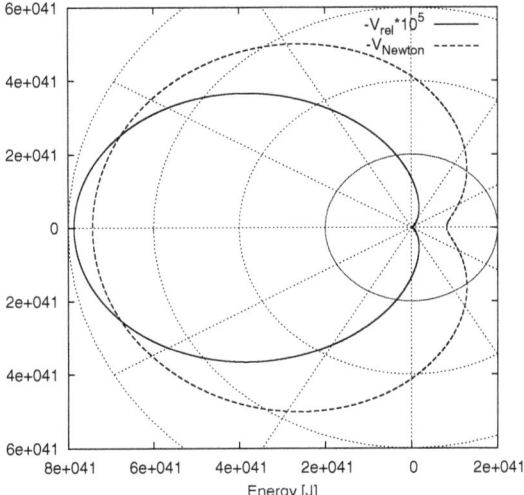

Fig. 15.18. Hulse-Taylor pulsar double star system, relativistic effects (enhanced by 10^5).

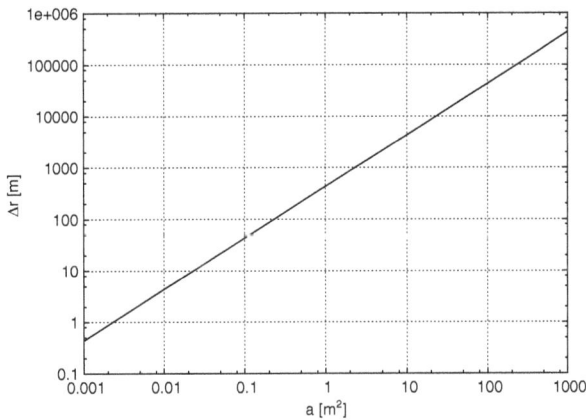

Fig. 15.19. Hulse-Taylor pulsar double star system, radius change per revolution Δr in dependence of perturbation a.

exact value is beyond the precision of the calculation. Although the Runge-Kutta scheme is precise to the order 4, the computed orbit was not stable enough to account for such small differences compared to the ellipse radius. We leave the exact determination of a to later numerical studies.

References

[1] M. W. Evans, "Generally Covariant Unified Field Theory" (Abramis, Suffolk, 2005 onwards), in five volumes to date (see also www.aias.us).

[2] K. Pendergast, "Crystal Spheres", preprint on www.aias.us.

[3] L. Felker, "The Evans Equations of Unified Field Theory" (Abramis, 2007).

[4] M. W. Evans, Omnia Opera Section of www.aias.us (1992 to present).

[5] M. W. Evans, (ed.), Adv. Chem. Phys., vol. 119 (2001), ibid., M. W. Evans and S. Kielich, vol. 85 (1992, 1993, 1997).

[6] M. W. Evans and L. B. Crowell, "Classical and Quantum Electrodynamics and the B(3) Field" (World Scientific, 2001).

[7] M. W. Evans and J.-P. Vigier, "The Enigmatic Photon" (Kluwer, 1994 to 2002), in five volumes.

[8] M. W. Evans and H. Eckardt, Physica B, 400, 175 (2007).

[9] M. W. Evans, Physica B, 403, 517 (2008).

[10] M. W. Evans, Acta Physica Polonica, B, 38, 2211 (2007).

[11] S. P. Carroll, "Spacetime and Geometry: an Introduction to General Relativity" (Addison Wesley, New York, 2004, and online 1997 notes).

[12] These are reviewed on www.telesio-galilei.com.

[13] J. B. Marion and S. T. Thornton, "Classical Dynamics" (HBC, New York, 1988, 3rd. ed.).

[14] R. M. Wald, "General Relativity" (Univ. Chicago Press, 1984).

16

Generalized Cartan Bianchi Identity and a New Theorem of the Cartan Torsion

by

Myron W. Evans,
Alpha Institute for Advanced Study, Civil List Scientist.
(emyrone@aol.com and www.aias.us)

Abstract

It is shown that the commutator of covariant derivatives acting on the four vector produces in general two tensors which are inter-related by a generalized form of the well known Bianchi identity of differential geometry as developed by Cartan. Examples of this theorem are the original Cartan/Bianchi identity itself, its Hodge dual identity, and its derivative identity. By using the well known rule for covariant derivative of a rank three tensor a new cyclic identity of the Cartan torsion is proven. In general these two tensors always exist in four dimensional space-time, and their existence is a direct consequence of the definition of covariant derivative in four dimensions. Therefore any self-consistent theory of relativity must be based on this self-consistent geometry.

Keywords: Generalized Cartan/Bianchi identity, new theorem of the Cartan torsion, theory of relativity.

16.1 Introduction

It is well known that the differential geometry of Cartan [1] is developed in terms of two structure equations and the Bianchi identity. The structure equations define the Cartan torsion in terms of the tetrad, and the Cartan curvature in terms of the spin connection. The Cartan/Bianchi identity relates the torsion and curvature. Recently the Einstein Cartan Evans (ECE) theory has been developed into a self-consistent and generally covariant unified field theory based directly on these structure equations and Bianchi identity [2–10]. The ECE theory produces all the equations of physics from the Cartan geometry, and unifies relativity and wave mechanics using the tetrad postulate. The latter is the fundamental theorem that links Cartan to Riemann geometry. In this paper it is proven that the Cartan/Bianchi identity can be generalized to produce two tensors whose existence depends only on the fundamental definition of the covariant derivative itself in four dimensions. This theorem is proven in Section 16.2. Examples of the theorem are the Cartan/Bianchi identity itself, its Hodge dual identity, and its derivative identity. This self-consistent theorem of geometry is fundamental to any valid theory of relativity in physics, or natural philosophy. Any theory that arbitrarily assumes that one tensor is zero is self-inconsistent geometrically and cannot produce a correct description of physics. An example of a self-inconsistent theory is the Einstein Hilbert field theory of gravitation, which is based on an incorrect geometry, so cannot produce correct physics. In Section 16.3 a new theorem of the Cartan torsion is proven from the general theorem of Section 16.2.

16.2 Generalized Cartan/Bianchi Identity

Define any two anti-symmetric tensors by:

$$[D_\mu, D_\nu]V^\kappa := A^\kappa{}_{\sigma\mu\nu}V^\sigma - B^\lambda_{\mu\nu}D_\lambda V^\kappa \tag{16.1}$$

i.e. by the action of the commutator of covariant derivatives:

$$[D_\mu, D_\nu] = -[D_\nu, D_\mu] \tag{16.2}$$

on the four vector V^κ. The two tensors are always related by the generalized Cartan/Bianchi identity:

$$D \wedge B^a = A^a{}_b \wedge q^b \tag{16.3}$$

where in Eq. (16.3) the standard notation of differential geometry [1–10] has been used.

The proof of this theorem relies only on the fundamental definition of the covariant derivative:

$$D_\mu V^\nu = \partial_\mu V^\nu + \Gamma^\nu_{\mu\lambda}V^\lambda \tag{16.4}$$

where Γ is the connection. Eq. (16.4) implies that the structure of the two tensors must be:

$$A^\kappa{}_{\sigma\mu\nu} := \partial_\mu \Gamma^\kappa_{\nu\sigma} - \partial_\nu \Gamma^\kappa_{\mu\sigma} + \Gamma^\kappa_{\mu\lambda}\Gamma^\lambda_{\nu\sigma} - \Gamma^\kappa_{\nu\lambda}\Gamma^\lambda_{\mu\sigma} \tag{16.5}$$

16.2 Generalized Cartan/Bianchi Identity

and

$$B^\kappa_{\mu\nu} := \Gamma^\kappa_{\mu\nu} - \Gamma^\kappa_{\nu\mu} \tag{16.6}$$

and that the two tensors must be anti-symmetric as follows:

$$A^\kappa{}_{\sigma\mu\nu} = -A^\kappa{}_{\sigma\nu\mu}, \tag{16.7}$$

$$B^\kappa_{\mu\nu} = -B^\kappa_{\nu\mu}. \tag{16.8}$$

Eqs. (16.5) to (16.8) follow directly from Eq. (16.1) without any assumption other than Eq. (16.4). Therefore any two tensors with the structures (16.5) and (16.6) will obey the generalized Cartan/Bianchi identity (16.3). Both tensors are generated directly from the commutator (16.2) and it is incorrect to assert that one tensor must vanish. In general both tensors must be non-zero and both must exist in the same space-time. Given the structure (16.5), the structure (16.6) follows. This result is true for all metrics and all connections, and is true irrespective of any postulate of metric compatibility [1–10]. The result depends only on the anti-symmetry (16.2) of the commutator of covariant derivatives. The latter is basic to field theory as is well known [11, 12].

It is necessary to prove that if the anti-symmetric tensors (16.5) and (16.6) exist from the commutator (16.2), then they always obey Eq. (16.3). In tensor notation, Eq. (16.3) is:

$$D_\mu B^a_{\nu\rho} + D_\rho B^a_{\mu\nu} + D_\nu B^a_{\rho\mu} = A^a_{\mu\nu\rho} + A^a_{\rho\mu\nu} + A^a_{\nu\rho\mu}. \tag{16.9}$$

The covariant derivatives on the left hand side of this equation are expanded [1–10] with the spin connection as follows:

$$\partial_\mu B^a_{\nu\rho} + \omega^a_{\mu b} B^b_{\nu\rho} + \partial_\rho B^a_{\mu\nu} + \omega^a_{\rho b} B^b_{\mu\nu} + \partial_\nu B^a_{\rho\mu} + \omega^a_{\nu b} B^b_{\rho\mu}$$
$$= (A^\lambda_{\mu\nu\rho} + A^\lambda_{\rho\mu\nu} + A^\lambda_{\nu\rho\mu}) q^a_\lambda \tag{16.10}$$

where:

$$B^a_{\nu\rho} = (\Gamma^\lambda_{\nu\rho} - \Gamma^\lambda_{\rho\nu}) q^a_\lambda \quad \text{etc.,} \tag{16.11}$$

$$B^b_{\nu\rho} = (\Gamma^\lambda_{\nu\rho} - \Gamma^\lambda_{\rho\nu}) q^b_\lambda \quad \text{etc.} \tag{16.12}$$

The Cartan tetrad is defined by the tetrad postulate [1–10]:

$$D_\mu q^a_\sigma = \partial_\mu q^a_\sigma + \omega^a_{\mu b} q^b_\sigma - \Gamma^\lambda_{\mu\sigma} q^a_\lambda = 0 \tag{16.13}$$

318   16 Generalized Cartan Bianchi Identity

Therefore the tetrad postulate links the spin connection and gamma connection:

$$\partial_\mu q^a_\sigma + \omega^a_{\mu b} q^b_\sigma = \Gamma^\lambda_{\mu\sigma} q^a_\lambda. \tag{16.14}$$

Using the Leibnitz Theorem:

$$\partial_\mu B^a_{\nu\rho} = (\partial_\mu \Gamma^\lambda_{\nu\rho} - \partial_\mu \Gamma^\lambda_{\rho\nu}) q^a_\lambda + (\Gamma^\lambda_{\nu\rho} - \Gamma^\lambda_{\rho\nu}) \partial_\mu q^a_\lambda \quad \text{etc.,} \tag{16.15}$$

so Eq. (16.10) becomes:

$$(\partial_\mu \Gamma^\lambda_{\nu\rho} - \partial_\mu \Gamma^\lambda_{\rho\nu}) q^a_\lambda + (\Gamma^\lambda_{\nu\rho} - \Gamma^\lambda_{\rho\nu})(\partial_\mu q^a_\lambda + \omega^a_{\mu b} q^b_\lambda) + \ldots$$
$$= (A^\lambda_{\mu\nu\rho} + A^\lambda_{\rho\mu\nu} + A^\lambda_{\nu\rho\mu}) q^a_\lambda. \tag{16.16}$$

Now re-label the summation indices in the second term on the left hand side as follows:

$$\lambda \to \sigma. \tag{16.17}$$

These are the repeated indices or dummy indices of the covariant - contravariant notation, and are summed over by definition. They can therefore take any label and after this re-labeling Eq. (16.16) becomes:

$$(\partial_\mu \Gamma^\lambda_{\nu\rho} - \partial_\mu \Gamma^\lambda_{\rho\nu}) q^a_\lambda + (\Gamma^\sigma_{\nu\rho} - \Gamma^\sigma_{\rho\nu})(\partial_\mu q^a_\sigma + \omega^a_{\mu b} q^b_\sigma) + \ldots$$
$$= (A^\lambda_{\mu\nu\rho} + A^\lambda_{\rho\mu\nu} + A^\lambda_{\nu\rho\mu}) q^a_\lambda. \tag{16.18}$$

Finally use the tetrad postulate (16.14) to obtain:

$$\begin{aligned} A^\lambda_{\mu\nu\rho} + A^\lambda_{\rho\mu\nu} + A^\lambda_{\nu\rho\mu} &= \partial_\mu \Gamma^\lambda_{\nu\rho} - \partial_\mu \Gamma^\lambda_{\rho\nu} + \Gamma^\lambda_{\mu\sigma}(\Gamma^\sigma_{\nu\rho} - \Gamma^\sigma_{\rho\nu}) \\ &\quad + \partial_\rho \Gamma^\lambda_{\mu\nu} - \partial_\rho \Gamma^\lambda_{\nu\mu} + \Gamma^\lambda_{\rho\sigma}(\Gamma^\sigma_{\mu\nu} - \Gamma^\sigma_{\nu\mu}) \\ &\quad + \partial_\mu \Gamma^\lambda_{\rho\mu} - \partial_\nu \Gamma^\lambda_{\mu\rho} + \Gamma^\lambda_{\nu\sigma}(\Gamma^\sigma_{\rho\mu} - \Gamma^\sigma_{\mu\rho}) \end{aligned} \tag{16.19}$$

Re-arrange this cyclic sum as follows:

$$\begin{aligned} A^\lambda_{\rho\mu\nu} &+ A^\lambda_{\mu\nu\rho} + A^\lambda_{\nu\rho\mu} \\ &= \partial_\mu \Gamma^\lambda_{\nu\rho} - \partial_\nu \Gamma^\lambda_{\mu\rho} + \Gamma^\lambda_{\mu\sigma} \Gamma^\sigma_{\nu\rho} - \Gamma^\lambda_{\nu\sigma} \Gamma^\sigma_{\mu\rho} \\ &\quad + \partial_\nu \Gamma^\lambda_{\rho\mu} - \partial_\rho \Gamma^\lambda_{\nu\mu} + \Gamma^\lambda_{\nu\sigma} \Gamma^\sigma_{\rho\mu} - \Gamma^\lambda_{\rho\sigma} \Gamma^\sigma_{\nu\mu} \\ &\quad + \partial_\rho \Gamma^\lambda_{\mu\nu} - \partial_\mu \Gamma^\lambda_{\rho\nu} + \Gamma^\lambda_{\rho\sigma} \Gamma^\sigma_{\mu\nu} - \Gamma^\lambda_{\mu\sigma} \Gamma^\sigma_{\rho\nu}. \end{aligned} \tag{16.20}$$

16.2 Generalized Cartan/Bianchi Identity

This is precisely the cyclic sum of three definitions (16.5). In order to obtain this result the definition (16.6) must be used. So the two tensors $A^\kappa_{\sigma\mu\nu}$ and $B^\kappa_{\mu\nu}$ are related by Eq. (16.3), Q.E.D.

Note that Eq. (16.20) is the cyclic sum of three tensors (16.5) on one side, and the cyclic sum of the definitions of these same tensors on the other side. So Eq. (16.20) and therefore Eq. (16.3) are exact identities which are obeyed by ANY two tensors with the structures (16.5) and (16.6). As seen from Eq. (16.1) these two tensors always exist in four dimensions. Multiply both sides of Eq. (16.9) by the tetrad q^κ_a:

$$(D_\mu B^a_{\nu\rho} + D_\rho B^a_{\mu\nu} + D_\nu B^a_{\rho\mu})q^\kappa_a = (A^a_{\mu\nu\rho} + A^a_{\rho\mu\nu} + A^a_{\nu\rho\mu})q^\kappa_a \quad (16.21)$$

to find that the following is a particular solution of Eq. (16.3):

$$D_\mu B^\kappa_{\nu\rho} + D_\rho B^\kappa_{\mu\nu} + D_\nu B^\kappa_{\rho\mu} = A^\kappa_{\mu\nu\rho} + A^\kappa_{\rho\mu\nu} + A^\kappa_{\nu\rho\mu}. \quad (16.22)$$

Eq. (16.22) is expressed in the base manifold (a four dimensional space-time) and eliminates the tangent index a of Cartan geometry [1–10].

The Cartan/Bianchi identity is recovered if:

$$A^\kappa_{\mu\nu\rho} = R^\kappa_{\mu\nu\rho}, \quad B^\kappa_{\nu\rho} = T^\kappa_{\nu\rho} \quad (16.23)$$

where $R^\kappa_{\mu\nu\rho}$ is the curvature tensor and $T^\kappa_{\nu\rho}$ the torsion tensor. Another important example of Eq. (16.22) is found by using the Hodge dual of the commutator operator as follows:

$$[D^\mu, D^\nu]_{\text{HD}} = \frac{1}{2} \| g \|^{\frac{1}{2}} \epsilon^{\mu\nu\alpha\beta} [D_\alpha, D_\beta] \quad (16.24)$$

Here $\| g \|^{\frac{1}{2}}$ is the square root of the determinant of the metric [1–10], and $\epsilon^{\mu\nu\alpha\beta}$ is the totally anti-symmetric unit tensor in four dimensions. The latter is defined in the same way [1–10] as in Minkowski space-time. From Eqs. (16.1) and (16.24):

$$[D_\mu, D_\nu]_{\text{HD}} V^\kappa = \widetilde{A}^\kappa{}_{\sigma\mu\nu} V^\sigma - \widetilde{B}^\lambda_{\mu\nu} D_\lambda V^\kappa \quad (16.25)$$

where:

$$\widetilde{A}^\kappa{}_{\sigma\mu\nu} := (\partial_\mu \Gamma^\kappa_{\nu\sigma} - \partial_\nu \Gamma^\kappa_{\mu\sigma} + \Gamma^\kappa_{\mu\lambda}\Gamma^\lambda_{\nu\sigma} - \Gamma^\kappa_{\nu\lambda}\Gamma^\lambda_{\mu\sigma})_{\text{HD}}. \quad (16.26)$$

and

$$\widetilde{B}^\kappa_{\mu\nu} := (\Gamma^\kappa_{\mu\nu} - \Gamma^\kappa_{\nu\mu})_{\text{HD}}. \quad (16.27)$$

16 Generalized Cartan Bianchi Identity

The Hodge dual tensors (16.26) and (16.27) are examples of the tensors $A^\kappa{}_{\sigma\mu\nu}$ and $B^\kappa_{\mu\nu}$ and so are related by:

$$D \wedge \widetilde{B}^a = \widetilde{A}^a{}_b \wedge q^b. \tag{16.28}$$

For example:

$$\widetilde{A}^\kappa{}_\sigma{}^{01} = \partial_2 \Gamma^\kappa_{3\sigma} - \partial_3 \Gamma^\kappa_{2\sigma} + \Gamma^\kappa_{2\lambda}\Gamma^\lambda_{3\sigma} - \Gamma^\kappa_{3\lambda}\Gamma^\lambda_{2\sigma} \tag{16.29}$$

and:

$$\widetilde{B}^{\kappa 01} = \Gamma^\kappa_{23} - \Gamma^\kappa_{32} \tag{16.30}$$

and these are related to each other in the same way as Eqs. (16.5) and (16.6), using the same connections.

Therefore we have proven rigorously that if:

$$D \wedge B^a = A^a{}_b \wedge q^b \tag{16.31}$$

then:

$$D \wedge \widetilde{B}^a = \widetilde{A}^a{}_b \wedge q^b. \tag{16.32}$$

Eq. (16.32) is referred to as the Hodge dual of the Cartan/Bianchi identity and was first inferred during the development of ECE theory [2–10]. It is clear that there also exists the derivative identity:

$$D \wedge (D \wedge B^a) := D \wedge (A^a{}_b \wedge q^b) \tag{16.33}$$

which is the correct form of the so called "second Bianchi identity" of the standard model.

In paper 93 of the ECE theory (www.aias.us) it was shown using computer algebra that curvature tensors of the type $R^\kappa{}_\mu{}^{\mu\nu}$ are non-zero in general if calculated from exact solutions of the Einstein Hilbert (EH) field equation of gravitation. All these solutions use the well known Christoffel connection:

$$\Gamma^\kappa_{\mu\nu} = \Gamma^\kappa_{\nu\mu} \tag{16.34}$$

and symmetric metric:

$$g_{\mu\nu} = g_{\nu\mu}. \tag{16.35}$$

16.2 Generalized Cartan/Bianchi Identity

So for example:

$$R^0{}_\mu{}^{\mu 0} \neq 0 \tag{16.36}$$

and summing over repeated μ indices:

$$R^0{}_1{}^{10} + R^0{}_2{}^{20} + R^0{}_3{}^{30} \neq 0. \tag{16.37}$$

From anti-symmetry:

$$R^0{}_1{}^{10} + R^0{}_2{}^{20} + R^0{}_3{}^{30} \neq 0. \tag{16.38}$$

By definition, the Hodge duals in four dimensions [1–10] of these tensor elements are also anti-symmetric in their last two indices and are defined by:

$$\left. \begin{array}{l} \widetilde{R}^0{}_{123} = \| g \|^{\frac{1}{2}} R^0{}_1{}^{01} \\ \widetilde{R}^0{}_{231} = \| g \|^{\frac{1}{2}} R^0{}_2{}^{02} \\ \widetilde{R}^0{}_{312} = \| g \|^{\frac{1}{2}} R^0{}_3{}^{03}. \end{array} \right\} \tag{16.39}$$

Therefore Eq. (16.38) is:

$$\widetilde{R}^0{}_{123} + \widetilde{R}^0{}_{231} + \widetilde{R}^0{}_{312} \neq 0 \tag{16.40}$$

in general for exact solutions of the EH field equation of gravitation. Eq. (16.40) is an example of:

$$\widetilde{R}^\kappa{}_{\mu\nu\rho} + \widetilde{R}^\kappa{}_{\rho\mu\nu} + \widetilde{R}^\kappa{}_{\nu\rho\mu} \neq 0 \tag{16.41}$$

in general. Therefore by Eq. (16.22):

$$D_\mu \widetilde{T}^\kappa{}_{\nu\rho} + D_\rho \widetilde{T}^\kappa{}_{\mu\nu} + D_\mu \widetilde{T}^\kappa{}_{\rho\mu} = \widetilde{R}^\kappa{}_{\mu\nu\rho} + \widetilde{R}^\kappa{}_{\rho\mu\nu} + \widetilde{R}^\kappa{}_{\nu\rho\mu} \neq 0 \tag{16.42}$$

which is the same as the equation:

$$D_\mu T^{\kappa\mu\nu} = R^\kappa{}_\mu{}^{\mu\nu} \neq 0. \tag{16.43}$$

Therefore the right hand side of Eq (16.43) is not zero in general for exact solutions of the EH equation of gravitation, but the left hand side of Eq. (16.43) is always zero for these same exact solutions because they all use the Christoffel symbol (16.34). The torsion tensor $T^{\kappa\mu\nu}$ is always zero for the Christoffel symbol but as we have shown using computer algebra, the curvature tensor is not zero in general for the same Christoffel symbols or connections.

Therefore the EH field equation is self-inconsistent at a fundamental level and must be regarded as obsolete. The reason for the self-inconsistency is that the EH equation was developed in 1915 before the existence of the torsion tensor was realized by Cartan in 1922. The only self-consistent theory of relativity that has been developed and applied to all physics is ECE theory [2–10]. Inferences drawn from the EH theory must be re-evaluated using the correct ECE theory. Example claims of EH theory are Big Bang, the existence of black holes and gravitational radiation, the existence of dark matter, and similar claims that are all based on an incorrect geometry. They cannot therefore be correct physics. Similarly, the so called precision tests of general relativity must be re-evaluated and re-explained with ECE theory, and in this series of papers that re-evaluation has been initiated. Similarly, all known solutions of EH must be tested with the curvature tensor $R^\kappa{}_\mu{}^{\mu\nu}$, and this is also work in progress. Clearly this is a re-evaluation of a large part of modern physics, the so-called "standard model". The latter can never be a correct unified field theory for these and many other well known reasons [2–10]. Prior to 2007 (paper 93), the curvature tensor $R^\kappa{}_\mu{}^{\mu\nu}$ was unknown.

In paper 93 it was checked with the same code that for all exact solutions of EH:

$$R^\kappa{}_{\mu\nu\sigma} + R^\kappa{}_{\sigma\mu\nu} + R^\kappa{}_{\nu\sigma\mu} = 0 \tag{16.44}$$

This result is known in the now obsolete standard model as "the first Bianchi identity". From Eq. (16.31) it is seen that Eq. (16.44) is equivalent to:

$$R^a{}_b \wedge q^b = 0 \tag{16.45}$$

and therefore incorrectly omits the torsion tensor. Therefore it is geometrically incorrect. It was again developed before the existence of torsion was inferred by Cartan in 1922. It is neither an identity nor was it first inferred by Bianchi. It was actually inferred by Ricci and Levi-Civita. Similarly the so-called "second Bianchi identity" of the standard model:

$$D_\mu R^\kappa{}_{\sigma\nu\rho} + D_\rho R^\kappa{}_{\sigma\mu\nu} + D_\nu R^\kappa{}_{\sigma\rho\mu} = 0 \tag{16.46}$$

is incorrect, again because it incorrectly neglects torsion. Eq. (16.46) in form notation is:

$$D \wedge R^a{}_b = 0 \tag{16.47}$$

whereas the correct version of Eq. (16.47) must be the derivative identity (16.33). The latter was again inferred (in paper 88) during the development of ECE theory.

An example of Eq. (16.44) is:

$$R^0{}_{123} + R^0{}_{312} + R^0{}_{231} = 0. \tag{16.48}$$

Using Eq. (16.39), Eq. (16.48) is the same as:

$$\tilde{R}^0{}_1{}^{01} + \tilde{R}^0{}_2{}^{02} + \tilde{R}^0{}_3{}^{03} = 0 \tag{16.49}$$

which is an example of:

$$\tilde{R}^\kappa{}_\mu{}^{\mu\nu} = 0 \tag{16.50}$$

for solutions of the EH field equation of gravitation. Therefore for these solutions:

$$D_\mu \tilde{T}^{\kappa\mu\nu} = \tilde{R}^\kappa{}_\mu{}^{\mu\nu} = 0 \tag{16.51}$$

or in form notation:

$$D \wedge T^a = R^a{}_b \wedge q^b = 0. \tag{16.52}$$

Therefore:

$$\tilde{T}^{\kappa\mu\nu} = 0, \quad \tilde{R}^\kappa{}_\mu{}^{\mu\nu} = 0 \tag{16.53}$$

is fortuitously obeyed by solutions of the EH field equation of gravitation, but the Hodge dual, Eq. (16.43) is NOT obeyed by these same solutions of the EH field equation of gravitation. This result means the logical end of the EH theory and was discovered in 2007. Since then ECE theory has been developed to take account correctly of both the Bianchi identity and its Hodge dual identity. These identities provide the basis of the ECE field equations [2–10].

16.3 A New Cyclic Identity of the Cartan Torsion

This identity is inherent in the Bianchi identity in the form (16.22), but has not been hitherto proven. The proof is as follows.

Use the definition of the covariant derivative of a rank three tensor [1–10] to find that:

$$D_\sigma T^\kappa{}_{\mu\nu} = \partial_\sigma T^\kappa{}_{\mu\nu} + \Gamma^\kappa{}_{\sigma\lambda} T^\lambda{}_{\mu\nu} - \Gamma^\lambda{}_{\sigma\mu} \Gamma^\kappa{}_{\lambda\nu} - \Gamma^\lambda{}_{\sigma\nu} \Gamma^\kappa{}_{\mu\lambda}, \tag{16.54}$$

$$D_\mu T^\kappa{}_{\nu\sigma} = \partial_\mu T^\kappa{}_{\nu\sigma} + \Gamma^\kappa{}_{\mu\lambda} T^\lambda{}_{\nu\sigma} - \Gamma^\lambda{}_{\mu\nu} T^\kappa{}_{\lambda\sigma} - \Gamma^\lambda{}_{\mu\sigma} T^\kappa{}_{\nu\lambda} \tag{16.55}$$

$$D_\nu T^\kappa{}_{\sigma\mu} = \partial_\nu T^\kappa{}_{\sigma\mu} + \Gamma^\kappa{}_{\nu\lambda} T^\lambda{}_{\sigma\mu} - \Gamma^\lambda{}_{\nu\sigma} \Gamma^\kappa{}_{\lambda\mu} - \Gamma^\lambda{}_{\nu\mu} T^\kappa{}_{\sigma\lambda} \tag{16.56}$$

Therefore using Eqs. (16.54) to (16.56) in Eq. (16.22):

$$D_\sigma T^\kappa_{\mu\nu} + D_\mu T^\kappa_{\nu\sigma} + D_\mu T^\kappa_{\sigma\mu}$$
$$= (\partial_\sigma T^\kappa_{\mu\nu} + \partial_\mu T^\kappa_{\nu\sigma} + \partial_\nu T^\kappa_{\sigma\mu} + \Gamma^\kappa_{\sigma\lambda} T^\lambda_{\mu\nu} + \Gamma^\kappa_{\mu\lambda} T^\lambda_{\nu\sigma} + \Gamma^\kappa_{\nu\lambda} T^\lambda_{\sigma\mu})$$
$$- (\Gamma^\lambda_{\sigma\mu} T^\kappa_{\lambda\nu} + \Gamma^\lambda_{\mu\sigma} T^\kappa_{\nu\lambda} + \Gamma^\lambda_{\sigma\nu} T^\kappa_{\mu\lambda} + \Gamma^\lambda_{\nu\sigma} T^\kappa_{\lambda\mu} + \Gamma^\lambda_{\mu\nu} T^\kappa_{\lambda\sigma} + \Gamma^\lambda_{\nu\mu} T^\kappa_{\sigma\lambda})$$
(16.57)

Using the definition of the torsion tensor [1–10]:

$$T^\kappa_{\mu\nu} = \Gamma^\kappa_{\mu\nu} - \Gamma^\kappa_{\nu\mu} \tag{16.58}$$

the first bracket in Eq. (16.57) gives the Bianchi identity (16.3), i.e. gives:

$$R^\kappa{}_{\sigma\mu\nu} + R^\kappa{}_{\mu\nu\sigma} + R^\kappa{}_{\nu\sigma\mu}$$
$$= \partial_\mu \Gamma^\kappa_{\nu\sigma} - \partial_\nu \Gamma^\kappa_{\mu\sigma} + \Gamma^\kappa_{\mu\lambda} \Gamma^\lambda_{\nu\sigma} - \Gamma^\kappa_{\nu\lambda} \Gamma^\lambda_{\mu\sigma}$$
$$+ \partial_\nu \Gamma^\kappa_{\sigma\mu} - \partial_\sigma \Gamma^\kappa_{\nu\mu} + \Gamma^\kappa_{\nu\lambda} \Gamma^\lambda_{\sigma\mu} - \Gamma^\kappa_{\sigma\lambda} \Gamma^\lambda_{\nu\mu}$$
$$+ \partial_\sigma \Gamma^\kappa_{\mu\nu} - \partial_\mu \Gamma^\kappa_{\sigma\nu} + \Gamma^\kappa_{\sigma\lambda} \Gamma^\lambda_{\mu\nu} - \Gamma^\kappa_{\mu\lambda} \Gamma^\lambda_{\sigma\nu}$$
(16.59)

Therefore the second bracket in Eq. (16.57) must be identically zero:

$$\Gamma^\lambda_{\sigma\mu} T^\kappa_{\lambda\nu} + \Gamma^\lambda_{\nu\sigma} T^\kappa_{\lambda\mu} + \Gamma^\lambda_{\mu\nu} T^\kappa_{\lambda\sigma} + \Gamma^\lambda_{\sigma\nu} T^\kappa_{\mu\lambda} + \Gamma^\lambda_{\mu\sigma} T^\kappa_{\nu\lambda} + \Gamma^\lambda_{\nu\mu} T^\kappa_{\sigma\lambda} = 0. \tag{16.60}$$

Now use the anti-symmetry:

$$T^\kappa_{\lambda\mu} = -T^\kappa_{\mu\lambda} \tag{16.61}$$

to prove a new cyclic identity obeyed by the Cartan torsion:

$$T^\kappa_{\lambda\nu} T^\lambda_{\sigma\mu} + T^\kappa_{\lambda\mu} T^\lambda_{\nu\sigma} + T^\kappa_{\lambda\sigma} T^\lambda_{\mu\nu} = 0 \tag{16.62}$$

In form notation Eq. (16.62) is the wedge product:

$$T^\kappa{}_\lambda \wedge T^\lambda = 0 \tag{16.63}$$

which in short-hand notation may be denoted:

$$T \wedge T = 0. \tag{16.64}$$

Here $T^\kappa{}_\lambda$ is defined as the tensor valued one-form [1–10] of index ν:

$$T^\kappa{}_\lambda := T^\kappa{}_{\lambda\nu} \tag{16.65}$$

and T^λ is defined as the vector valued two-form of index $\sigma\mu$:

$$T^\lambda := T^\lambda{}_{\sigma\mu}. \tag{16.66}$$

The short-hand notation (16.64) explains the basic structure of the tensor identity (16.62) as being akin, approximately writing, to the cross product of a vector with itself, or akin to the Poincare Lemma in structure. The new theorem (16.62) is inherent in the Bianchi identity itself and is another way of revealing the rigorous self-consistency of Cartan geometry. If an attempt is made to assert a geometry that is not consistent with the Cartan geometry, the inconsistency will sooner or later reveal itself. In the case of EH theory it finally revealed itself through the curvature tensor $R^\kappa{}_\mu{}^{\mu\nu}$. The latter is essentially impossible to compute by hand, because of the intricate complexity of its internal structure (see paper 93 of www.aias.us). Therefore $R^\kappa{}_\mu{}^{\mu\nu}$ was never computed prior to 2007, and was obviously unknown to Einstein himself. It was also unknown to the instigators of Big Bang theory, notably Hawking and Penrose, and the instigator of black hole theory, Wheeler. Therefore the standard model claims have become elaborate, but are all based on a fundamentally incorrect geometry. The ECE theory is therefore the only self-consistent theory of relativity.

Acknowledgments

The British Government is thanked for a Civil List pension and the Telesio-Galilei Association and AIAS for many interesting discussions.

References

[1] S. P. Carroll, "Space-time and Geometry: an Introduction to General Relativity" (Addison Wesley, New York, 2004, also 1997 lecture notes online).

[2] M. W. Evans, "Generally Covariant Unified Field Theory" (Abramis, Suffolk, 2005 to present), vols. 1–4, vol. 5 in prep. (Papers 71 to 93 on www.aias.us).

[3] L. Felker, "The Evans Equations of Unified Field Theory" (Abramis, 2007).

[4] K. Pendergast, "Crystal Spheres" (Abramis to be published, preprint on www.aias.us).

[5] M. W. Evans, Omnia Opera section of www.aias.us (1992 to present on ECE and precursor theories).

[6] M. W. Evans, (ed.) Adv. Chem. Phys., vol. 119 (2001), ibid., M. W. Evans and S. Kielich (eds.), vol. 85 (1992, 1993, 1997, first edition of vol. 119).

[7] M. W. Evans and L. B. Crowell, "Classical and Quantum Electrodynamics and the B(3) Field" (World Scientific, 2001).

[8] M. W. Evans and J.-P. Vigier, "The Enigmatic Photon" (Kluwer, 1994 to 2002 hardback and softback), in five volumes.

[9] M. W. Evans and A. A. Hasanein, "The Photomagneton in Quantum Field Theory" (World Scientific, 1994).

[10] M. W. Evans, Acta Phys. Polon. B, **38**, 2211 (2007); M. W. Evans and H. Eckardt, Physica B, **400**, 175 (2007); **403**, 517 (2008).

[11] L. H. Ryder, "Quantum Field Theory" (Cambridge Univ. Press, 2nd. Ed., 1996).

[12] R. M. Wald, "General Relativity" (Chicago Univ. Press, 1984).

Derivation of the Thomas Precession in Terms of the Infinitesimal Torsion Generator

by

Myron W. Evans,
Alpha Institute for Advanced Study, Civil List Scientist.
(emyrone@aol.com and www.aias.us)

Abstract

The Thomas precession is shown to be due to the rotation of Minkowski space-time, a rotation which is described by the infinitesimal generator of the Cartan torsion. Rotation of the Minkowski space-time defines the Thomas angular velocity and the Thomas time dilation observed in spin-orbit coupling in atomic and molecular spectroscopy. It is shown that the Thomas angular velocity is generated directly by the commutator of two Lorentz boost generators. The infinitesimal Cartan torsion generator for a rotation about the Z axis is shown to be the time derivative of the infinitesimal rotation generator of the Lorentz or Poincaré groups. Within a factor $\hbar$ this is the angular momentum operator of quantum mechanics. Therefore the Cartan torsion generator is observed throughout quantum physics. Thomas precession is observed in pendulum precession and is due to the infinitesimal generator of Cartan torsion.

Keywords: Thomas precession, infinitesimal generator of Cartan torsion, ECE theory.

17.1 Introduction

In the well known differential geometry of Cartan [1–11] the torsion is derived by the first structure equation in terms of the tetrad and spin connection. The curvature is defined by the second structure equation in terms of the spin connection. Finally the torsion and curvature are linked by the development of the Bianchi identity by Cartan. This is a self-consistent geometry based on the existence of space-time torsion and curvature. It is incorrect to assert arbitrarily that torsion vanishes, so the meaning of Cartan torsion in relativity must be evaluated systematically. This has been done in the Einstein Cartan Evans series of papers on a suggested unified field theory based on the principle of general covariance (see ECE papers on the www.aias.us website). In this paper the well known Thomas precession is shown to be due to the infinitesimal generator of a Cartan torsion caused by the rotation of the Minkowski frame. In Section 17.2 the Thomas angular velocity and proper time are obtained straightforwardly by rotating the Minkowski metric. In a static Minkowski space-time all Christoffel connections are zero by definition, so there is no Christoffel torsion, no Cartan torsion, no Riemann curvature and no Cartan curvature. On rotation of the Minkowki frame however, a Cartan torsion and non-zero connection are generated as shown in Section 17.3. This Cratan torsion due to rotation is the time derivative of the rotation matrix, and so it is possible to define the infinitesimal generator of Cartan torsion. By considering the commutator of two Lorentz boost generators, it is shown that this generator of torsion is produced directly by the commutator, together with the Thomas angular velocity. The Thomas factor is produced by time dilation (again introduced by rotation of Minkowski space-time) and is observed in spin orbit coupling in atomic and molecular spectroscopy. The latter observes the generator of torsion directly as the angular momentum operator of quantum mechanics, which within h is the infinitesimal rotation generator. The infinitesimal generator of Cartan torsion is therefore proportional directly to the angular momentum operator [12] of quantum mechanics.

17.2 Rotation of the Minkowski Space-Time

The static Minkowski line element in cylindrical polar co-ordinates is well known to be:

$$ds^2 = c^2 dt^2 - dr^2 - r^2 d\phi^2 - dZ^2. \tag{17.1}$$

The cylindrical polar and Cartesian coordinates are related by:

$$\left.\begin{array}{r} X = r\cos\phi, \\ Y = r\sin\phi, \\ Z = Z. \end{array}\right\} \tag{17.2}$$

Here c is the speed of light and t the time. Now consider a Minkowski frame denoted by:

$$ds'^2 = c^2 dt^2 - dr^2 - r^2 d\phi'^2 - dZ^2 \tag{17.3}$$

where:

$$\phi' = \phi + \omega t \tag{17.4}$$

represents a rotation of the ϕ coordinate defined by the angular velocity ω. Therefore the infinitesimals are related by:

$$d\phi' = d\phi + \omega dt \tag{17.5}$$

and the squares of the infinitesimals are related by:

$$d\phi'^2 = d\phi^2 + 2\omega d\phi dt + \omega^2 dt^2. \tag{17.6}$$

Therefore the rotating Minkowski line element is:

$$ds'^2 = (c^2 - r^2\omega^2)dt^2 - 2\omega r^2 d\phi dt - dr^2 - r^2 d\phi^2 - dZ^2 \tag{17.7}$$

Now define the linear velocity v by:

$$v = r\omega \tag{17.8}$$

to find that the rotating line element is:

$$ds'^2 = \left(1 - \frac{v^2}{c^2}\right)(c^2 dt^2 - 2r^2 \Omega d\phi dt) - dr^2 - r^2 d\phi^2 - dZ^2 \tag{17.9}$$

where:

$$\Omega = \omega\left(1 - \frac{v^2}{c^2}\right)^{-1} \tag{17.10}$$

is a relativistically corrected angular velocity. It is known as the Thomas angular velocity and is produced directly by the rotation of the Minkowski space-time. The infinitesimal of time is changed by this rotation to:

$$d\tau = \left(1 - \frac{v^2}{c^2}\right)^{1/2} dt \tag{17.11}$$

and this is the time dilation effect responsible for the well known Thomas factor in spin-orbit coupling. The effects of spin orbit coupling are well known in atomic and molecular spectroscopy. Finally the Thomas precession observed in a pendulum for a rotation of 2π radians:

$$\omega dt = 2\pi \tag{17.12}$$

is defined as:

$$\alpha = \Omega d\tau - \omega dt = 2\pi \left(\left(1 - \frac{v^2}{c^2}\right)^{1/2} - 1 \right). \tag{17.13}$$

These are well known effects demonstrated straightforwardly by the simple rotation of a Minkowski space-time as defined in Eq. (17.4). The original demonstrations by Thomas in 1927 and Sommerfeld in 1931 are much more complicated.

17.3 Infinitesimal Generator of the Cartan Torsion

In this section it is shown that the well known phenomena derived in Section 17.2 are due to the Cartan torsion, specifically the infinitesimal torsion generator. Consider the rotation of a three-vector about the Z axis [13, 14]:

$$\begin{bmatrix} V'_X \\ V'_Y \\ V'_Z \end{bmatrix} = \begin{bmatrix} \cos\theta & \sin\theta & 0 \\ -\sin\theta & \cos\theta & 0 \\ 0 & 0 & 1 \end{bmatrix} \begin{bmatrix} V_X \\ V_Y \\ V_Z \end{bmatrix} \tag{17.14}$$

where θ is the angle of rotation. The tetrad is always defined in Cartan geometry as the matrix linking two vectors, so the tetrad for this rotation is defined as:

$$q^a_\mu = \begin{bmatrix} \cos\theta & \sin\theta & 0 \\ -\sin\theta & \cos\theta & 0 \\ 0 & 0 & 1 \end{bmatrix} \tag{17.15}$$

The angular velocity is defined as:

$$\omega = \frac{d\theta}{dt}. \tag{17.16}$$

The spin connection for rotation is the matrix representation of the tetrad [1–11]:

$$\omega^a{}_b = \frac{\omega}{c} \epsilon^a{}_{bc} q^c \tag{17.17}$$

where $\epsilon^a{}_{bc}$ is the totally anti-symmetric unit tensor in three dimensions. So the Cartan torsion for this rotation:

$$T^a_{\mu\nu} = \partial_\mu q^a_\nu - \partial_\nu q^a_\mu + \omega^a{}_{\mu b} q^b_\nu - \omega^a{}_{\nu b} q^b_\mu \tag{17.18}$$

17.3 Infinitesimal Generator of the Cartan Torsion

is fully defined. The elements of the tetrad are:

$$\left.\begin{array}{l} q_1^1 = \cos\theta, q_2^1 = \sin\theta, q_3^1 = 0, \\ q_1^2 = -\sin\theta, q_2^2 = \cos\theta, q_3^2 = 0, \\ q_1^3 = 0, q_2^3 = 0, q_3^3 = 1. \end{array}\right\} \quad (17.19)$$

These are components of tetrad vectors in the Cartesian coordinate system:

$$\left.\begin{array}{l} \boldsymbol{q}^1 = \cos\theta \boldsymbol{i} + \sin\theta \boldsymbol{j}, \\ \boldsymbol{q}^2 = -\sin\theta \boldsymbol{i} + \cos\theta \boldsymbol{j}, \\ \boldsymbol{q}^3 = \boldsymbol{k}. \end{array}\right\} \quad (17.20)$$

The spin connection and tetrad components are related by:

$$\left.\begin{array}{l} \omega^1{}_2 = \dfrac{\omega}{c} q^3, \\ \omega^2{}_3 = \dfrac{\omega}{c} q^1, \\ \omega^3{}_1 = \dfrac{\omega}{c} q^2. \end{array}\right\} \quad (17.21)$$

Therefore the non-zero spin connection components are:

$$\left.\begin{array}{l} \omega^2{}_{13} = \dfrac{\omega}{c} q_1^1, \omega^3{}_{11} = \dfrac{\omega}{c} q_1^2, \\ \omega^2{}_{23} = \dfrac{\omega}{c} q_2^1, \omega^3{}_{21} = \dfrac{\omega}{c} q_2^2, \\ \omega^1{}_{32} = \dfrac{\omega}{c} q_3^3. \end{array}\right\} \quad (17.22)$$

The angle θ depends on time as in Eq. (17.16), which defines the angular velocity. Therefore the non-zero torsion elements are:

$$T^a_{0\mu} = \frac{1}{c}\frac{\partial q^a_\mu}{\partial t} = \frac{\omega}{c}\begin{bmatrix} -\sin\theta & \cos\theta & 0 \\ -\cos\theta & -\sin\theta & 0 \\ 0 & 0 & 0 \end{bmatrix}. \quad (17.23)$$

Define the infinitesimal generator of the Cartan torsion by:

$$T_Z = \frac{1}{i}\frac{\omega}{c}\begin{bmatrix} -\sin\theta & \cos\theta & 0 \\ -\cos\theta & -\sin\theta & 0 \\ 0 & 0 & 0 \end{bmatrix}_{\theta \to 0} \quad (17.24)$$

to find that it is directly proportional to the well known infinitesimal rotation generator:

$$T_Z = \frac{\omega}{c} J_Z \qquad (17.25)$$

where:

$$J_Z = -i \begin{bmatrix} 0 & 1 & 0 \\ -1 & 0 & 0 \\ 0 & 0 & 0 \end{bmatrix} \qquad (17.26)$$

The angular momentum operator of quantum mechanics is the rotation generator within a factor $\hbar$, the reduced Planck constant. Therefore the infinitesimal generator of torsion is directly proportional to the angular momentum operator which is basically important to all quantum mechanics [1–14]. Therefore the Cartan torsion is observed in all quantum mechanics.

It is also well known 1-14 that the Lorentz and Poincaré groups contain commutator equations such as:

$$[\kappa_X, \kappa_Y] = -iJ_Z \qquad (17.27)$$

where κ_X and κ_Y are Lorentz boost generators. Lorentz boosts in different directions cause a rotation, and this is the origin of the Thomas factor and Thomas precession. In Section 2 the Minkowski frame was rotated to illustrate these relativistic phenomena in a simple way. Consider a Lorentz boost in the X axis:

$$X' = \gamma(X + vt), \ Y' = Y, t' = \gamma(t + vX/c^2),$$
$$\gamma = (1 - v^2/c^2)^{-1/2}, \ \beta = v/c, \ X^0 = ct, \ X' = X, \qquad (17.28)$$
$$r = \cosh\phi, \gamma\beta = \sinh\phi, \frac{v}{c} = \tanh\phi.$$

The Lorentz boost in matrix form is therefore:

$$\begin{bmatrix} X^{0'} \\ X^{1'} \\ X^{2'} \\ X^{3'} \end{bmatrix} = \begin{bmatrix} \gamma & \gamma\beta & 0 & 0 \\ \gamma\beta & \gamma & 0 & 0 \\ 0 & 0 & 1 & 0 \\ 0 & 0 & 0 & 1 \end{bmatrix} \begin{bmatrix} X^0 \\ X^1 \\ X^2 \\ X^3 \end{bmatrix} \qquad (17.29)$$

where the matrix in Eq. (17.29) is a Lorentz boost matrix. This is also a type of tetrad in Cartan geometry. Therefore in Cartan geometry the group

17.3 Infinitesimal Generator of the Cartan Torsion

generators of the well known Lorentz and Poincaré groups of special relativity are all tetrads. Similarly the Lorentz boost matrix in Y is [1–14]:

$$B_Y = \begin{bmatrix} \gamma & 0 & \gamma\beta & 0 \\ 0 & \gamma & 0 & 0 \\ \gamma\beta & 0 & 1 & 0 \\ 0 & 0 & 0 & 1 \end{bmatrix} \tag{17.30}$$

The infinitesimal boost generators are:

$$\kappa_X = -i\frac{\partial B_X}{\partial \phi}\bigg|_{\phi \to 0} = -i \begin{bmatrix} 0 & 1 & 0 & 0 \\ 1 & 0 & 0 & 0 \\ 0 & 0 & 0 & 0 \\ 0 & 0 & 0 & 0 \end{bmatrix},$$

$$\kappa_Y = -i\frac{\partial B_Y}{\partial \phi}\bigg|_{\phi \to 0} = -i \begin{bmatrix} 0 & 0 & 1 & 0 \\ 0 & 0 & 0 & 0 \\ 1 & 0 & 0 & 0 \\ 0 & 0 & 0 & 0 \end{bmatrix}. \tag{17.31}$$

and the infinitesimal rotation generator is:

$$J_Z = -i \begin{bmatrix} 0 & 0 & 0 & 0 \\ 0 & 0 & 1 & 0 \\ 0 & -1 & 0 & 0 \\ 0 & 0 & 0 & 0 \end{bmatrix} \tag{17.32}$$

These generators give Eq. (17.27).
It is found that:

$$[B_X, B_Y] = i\left(\frac{v}{c}\right)^2 \left(1 - \frac{v^2}{c^2}\right)^{-1} J_Z. \tag{17.33}$$

Therefore the relativistically corrected infinitesimal rotation generator is:

$$J'_Z = \left(\frac{v}{c}\right)^2 \left(1 - \frac{v^2}{c^2}\right)^{-1} J_Z. \tag{17.34}$$

Therefore the relativistically corrected, infinitesimal torsion generator is:

$$T'_Z = \left(\frac{v}{c}\right)^2 \frac{\Omega}{c} J_Z \tag{17.35}$$

where:

$$\Omega = \omega \left(1 - \frac{v^2}{c^2}\right)^{-1} \qquad (17.36)$$

is the Thomas angular velocity (17.10). The latter has therefore been derived directly from the commutator of Lorentz boosts as in Eq. (17.33), and it has been proven that the Thomas angular velocity is part of the relativistic correction to the infinitesimal torsion generator.

Finally the tetrad postulate [1–11]:

$$\partial_\mu q^a_\sigma + \omega^a{}_{\mu b} q^b_\sigma - \Gamma^\lambda{}_{\mu\sigma} q^a_\lambda = 0 \qquad (17.37)$$

is used to show that there exists a precessional equation:

$$\partial_0 q^1_1 = \Gamma^0{}_{01} q^1_1 + \Gamma^2{}_{01} q^1_2 \qquad (17.38)$$

directly from Cartan geometry. A particular solution of Eq. (17.38) is:

$$\Gamma^2{}_{01} = -\frac{\omega}{c}, \quad \Gamma^1{}_{01} = 0. \qquad (17.39)$$

Note carefully that this is not a Christoffel connection, because in this example:

$$\Gamma^2{}_{10} = 0 \qquad (17.40)$$

and the Christoffel connection must be symmetric in its lower two indices:

$$\Gamma^\kappa{}_{\mu\nu} = \Gamma^\kappa{}_{\nu\mu}. \qquad (17.41)$$

The Christoffel torsion:

$$T^\kappa{}_{\mu\nu} = \Gamma^\kappa{}_{\mu\nu} - \Gamma^\kappa{}_{\nu\mu} \qquad (17.42)$$

is always zero, and the Cartan torsion defined by the Christoffel torsion:

$$T^a{}_{\mu\nu} = q^a{}_\kappa T^\kappa{}_{\mu\nu} \qquad (17.43)$$

is always zero. Therefore the gamma connection of type (17.39) is a rotational gamma connection defined by a non-zero Cartan torsion introduced by rotating the Minkowski frame. It is not a Christoffel connection.

This relatively straightforward analysis of a rotating Minkowski frame can be applied to rotate classes of line elements that are exact solutions of the

Einstein Hilbert (EH) field equation. It was discovered in 2007 (paper 93 of www.aias.us) that the geometry of the EH field equation is incorrect due to its neglect of torsion. In subsequent papers the torsion was used self-consistently, and this paper (paper 110 of the ECE series) it has been used to describe the simplest rotational phenomenon of special relativity - the Thomas precession. Other well known precessional effects include the Lense Thirring rotational frame dragging, and in subsequent papers it will be shown that the Lense Thirring effect is also based on the Cartan torsion, it is effectively a rotation of a line element that is a solution of the EH field equation. The conventional or standard model explanation of the Lense Thirring effect is now known to be incorrect, because it still uses a Christoffel connection in which torsion is zero.

Acknowledgments

The British Government is thanked for a Civil List Pension for pre-eminent contributions to Britain in science. The staff of AIAS and the Telesio-Galilei Association are thanked for many interesting discussions.

References

[1] S. P. Carroll, "Spacetime and Geometry: an Introduction to General Relativity" (Addison Wesley, New York, 2004, and 1997 online notes).

[2] M. W. Evans, "Generally Covariant Unified Field Theory: the Geometrization of Physics" (Abramis Academic, Suffolk, 2005 onwards), in four volumes to date (further volumes in preparation, papers 71 onwards on www.aias.us).

[3] L. Felker, "The Evans Equations of Unified Field Theory" (Abramis 2007).

[4] K. Pendergast, "Crystal Spheres" (Abramis in prep., preprint on www.aias.us).

[5] M. W. Evans, Acta Phys. Polonica, **38B**, 2211 (2007).

[6] M. W. Evans and H. Eckardt, Physica B, **400**, 175 (2007).

[7] M. W. Evans, Physica B, **403**, 517 (2008).

[8] M. W. Evans, (ed.), Adv.Chem. Phys., **119** (2001); M. W. Evans and S. Kielich (eds.), Adv. Chem. Phys. **85** (1992, 1993, 1997).

[9] M. W. Evans and L. B. Crowell, "Classical and Quantum Electrodynamics and the B(3) Field" (World Scientific 2001).

[10] M. W. Evans and J.-P Vigier, "The Enigmatic Photon" (Kluwer, 1994 to 2002, hardback and softback), in five volumes.

[11] M. W. Evans, Omnia Opera section of www.aias.us (1992 to present, hyperlinked original papers).

[12] P. W. Atkins, "Molecular Quantum Mechanics" (Oxford University Press, 1983, 2nd ed. and subsequent editions).

[13] L. H. Ryder, "Quantum Field Theory" (Cambridge Univeristy Press, 1996, 2nd ed.).

[14] J. B. Marion and S. T. Thornton, "Classical Dynamics of Particles and Systems" (HBC College Publishing, New York, 1988, 3rd ed.).

18

The Origin of Orbits in Spherically Symmetric Space-Time

by

Myron W. Evans,
Alpha Institute for Advanced Study, Civil List Scientist.
(emyrone@aol.com and www.aias.us)

and

Horst Eckardt,
Alpha Institute for Advanced Study/S.G.
(www.telesio-galilei.com)

Abstract

It is shown that the origin of all known orbital trajectories is the spherical symmetry of space-time, and not the Einstein field equation. A consideration of simple integral equations leads to the mathematical structure of the line element needed to describe all known orbits. The orbital equation is deduced as usual from the line element, which results directly from the spherical symmetry of space-time. All known orbits are described therefore by the spherical symmetry of space-time, irrespective of any field equation. The Einstein field equation is shown to be geometrically incorrect, and replaced by the ECE field equation of dynamics, based directly on the Bianchi identity.

Keywords: Origin of orbits, spherical symmetry of space-time, ECE equations of dynamics.

18.1 Introduction

Contemporary high precision satellite data are available to re-assess the validity of the Einstein field equation. It is known [1–12] that this equation is based directly on a type of Riemann geometry that is intrinsically self-inconsistent.

This type of Riemann geometry is based in turn on the well known Christoffel connection, which is symmetric in its lower two indices. Such a connection means that the torsion tensor is zero. However, this is inconsistent with the fact that the torsion tensor is produced by the commutator of covariant derivatives [11] acting on the vector in any dimension and irrespective of any assumed connection. The torsion tensor is thereby an ineluctable counterpart of the curvature tensor and it cannot be assumed that the torsion tensor vanishes. It was pointed out by Alley [13], in the 2006 Wheeeler Fest at Princeton, that as early as 1918, Bauer and Schroedinger demonstrated independently that there are severe internal inconsistencies in the approach to relativity taken by Einstein and by Hilbert in 1915. Einstein agreed with Schroedinger that his field equation is self-inconsistent [13], but this fact has been forgotten by history. Bauer pointed out that the Minkowski metric, which is constructed in the absence of gravitation, nevertheless produces a non zero type of stress energy momentum tensor for gravitation, if Einstein's method is applied. Schroedinger pointed out that the Einstein field equation does not self-consistently produce this type of tensor at all. Yilmaz [13] has attempted to remedy this self-inconsistency, as has Wheeler [13]. However, the attempts by Yilmaz and Wheeler still assume a zero torsion tensor. In the work of Rapoport, Santilli and others [14], the original approach by Einstein has been greatly extended. Rapoport [14] has successfully applied the torsion tensor to several areas of physics, including Brownian motion, stochastic vacuum physics, fractal physics and thermodynamics.

From 2003 onwards a generally covariant unified field theory [1–10] has been developed based on standard Cartan geometry [11], namely the two Cartan structure equations, the Bianchi/Cartan identity, and the tetrad postulate. The first and second structure equations respectively define the torsion as the covariant exterior derivative of the tetrad, and define the curvature in terms of the spin connection. The Bianchi identity has been proven [1–10] to be invariant under the Hodge duality transformation in four dimensions. In general both the Bianchi identity and its Hodge dual must be considered. They are two parts of the Bianchi identity and it is not admissible to consider one without the other. As pointed out by Santilli [14] there are several logical flaws in the Einstein field equation and in its geometry. Another one of these was discovered in 2007 in paper 93 of the ECE series on www.aias.us. Computer algebra was applied to the complete, duality invariant, Bianchi identity. It was found by direct computation that exact solutions of the Einstein field equation do not in general obey the complete Bianchi identity. The reason is the neglect of torsion, i.e. the use of the Christoffel connection is not consistent with the Bianchi identity. The so called first and second Bianchi identities that appear in the gravitational literature are incomplete, because they neglect torsion. The so-called first Bianchi identity was in fact inferred by Ricci and Levi-Civita, and is not in fact an identity. It is an equation that is true if an only if the metric and connection are symmetric, i.e. if and only if the torsion vanishes. The so called second Bianchi identity is in fact not an identity, it is an equation which is again true if and only if the torsion vanishes. The Einstein approach to relativity is based directly on this torsion-less second Bianchi identity. Paper 93 of www.aias.us shows that the Einstein approach is not just incomplete, but incorrect. Such a conclusion has been reached repeatedly, this is the ninetieth anniversary of the criticisms by Bauer and Schroedinger, criticisms with which Einstein agreed [13] and which are reviewed in a book by Santilli on www.telesio-galilei.com and www.aias.us.

The contemporary assertion that there is a "standard model" of physics therefore collapses entirely, because standard cosmologies are based on the incorrect Einstein field equation. The ECE field equations of dynamics, electrodynamics and cosmology [1–10] on the other hand are based on the self-consistent use of torsion in the duality invariant Bianchi identity of Cartan geometry. Clearly, ECE theory can be extended with the use of other geometries, such as those advocated by Rapoport [14], Santilli [14], Schadeck and Connes [14].

In Section 18.2 it is demonstrated straightforwardly that all known orbits are due to the spherical symmetry of space-time, irrespective of any assumed field equation or connection. The general line element for a spherically symmetric space-time is well known to have a general mathematical structure, and in Section 18.2 it is demonstrated in a very simple way that a particular solution this general structure is responsible for all known orbits to known experimental precision.

In Section 18.3 the line element derived in Section 18.2 is developed with the ECE equations of dynamics to produce a self consistent result. Computer algebra is used in both Sections 18.2 and 18.3 to eliminate the labor of calculation and to eliminate all possibility of purely calculational human error.

18.2 Line Element for a Spherically Symmetric Space-Time

Consider the simple equation:

$$r = \int dr \tag{18.1}$$

where r is the radial vector. This equation is developed in a spherically symmetric space-time which can be represented with coordinate systems such as the spherical polar (r, θ, ϕ) or Cartesian (X, Y, Z). In Eq. (18.1) the constant of integration, denoted μ, has been assumed to be zero:

$$\mu = 0. \tag{18.2}$$

18 The Origin of Orbits in Spherically Symmetric Space-Time

The following line element [11] is a representation of a spherically symmetric space-time:

$$ds^2 = -mc^2 dt^2 + n dr^2 + r^2 d\Omega^2 \tag{18.3}$$

where $d\Omega^2$ is the infinitesimal volume element in spherical polar coordinates. Eq. (18.3) is not the most general line element possible mathematically for a spherically symmetric space-time, but as the experimental satellite data show, will be sufficient to describe the experiments.

Assume that the function m is defined by:

$$mr = \int dr \tag{18.4}$$

and that the function n is defined by:

$$\frac{r}{n} = \int dr \tag{18.5}$$

and assume that the constant of integration μ is not zero, and is the same in Eqs. (18.4) and (18.5). It follows that:

$$m = \frac{1}{n} = 1 + \frac{\mu}{r} \tag{18.6}$$

and that a spherically symmetric metric can be described by the structure:

$$ds^2 = -\left(1 + \frac{\mu}{r}\right) c^2 dt^2 + \left(1 + \frac{\mu}{r}\right)^{-1} + r^2 d\Omega^2. \tag{18.7}$$

It is known from experimental satellite data that in almost all known orbits:

$$\mu = -\frac{2MG}{c^2} \tag{18.8}$$

where M is the mass of the object about which orbits an object of mass m. Here G is Newton's constant and c is the vacuum speed of light.

Therefore nearly all known orbits are manifestations of the spherically symmetry of space-time, because the orbital equation is obtained from the line element as is well known [1–12]. The only known deviations from Eq. (18.8) are the orbits of binary pulsars, in which the orbits slowly spiral inwards by an extremely tiny amount per revolution. It was shown in paper 108 of the

18.2 Line Element for a Spherically Symmetric Space-Time

ECE series (www.aias.us) that binary pulsar orbits are described in all detail by:

$$\mu = -\left(\frac{2MG}{c^2} + \frac{a}{r}\right) \tag{18.9}$$

where a is a constant perturbation of very tiny magnitude. There are well known systematic anomalies recorded by the Cassini and Pioneer spacecraft independently, these may indicate a deviation from Eq. (18.8), but may be due to mundane artifacts such as leaking gas in the propellant system. If these anomalies are indeed systematic within the uncertainty, as seems likely, they are described by Eq. (18.9), again with a spherically symmetric space-time.

Therefore equations of type (18.1), (18.4) and (18.5) are the basic reason for all known orbits. It may be that this type of simple integral structure has a fundamental significance in topology.

Using computer algebra (Maxima, paper 93 of www.aias.us) a line element of type (18.8) was evaluated with an assumed Christoffel connection. It was found that this assumption produces a correct solution of the Bianchi identity:

$$D \wedge T^a = R^a{}_b \wedge q^b \tag{18.10}$$

and also of the Hodge dual identity

$$D \wedge \tilde{T}^a = \tilde{R}^a{}_b \wedge q^b. \tag{18.11}$$

Here T is shorthand (index - less) notation for the Cartan torsion, $D\wedge$ denotes the covariant exterior derivative and R denotes the Cartan curvature. The tilde denotes Hodge dual. The reason for this is that if the Christoffel connection is used with the spherically symmetric line element of type (18.8), then it follows that:

$$T^a = 0, \ \tilde{T}^a = 0, \ R^a{}_b \wedge q^b = 0, \ \tilde{R}^a{}_b \wedge q^b = 0. \tag{18.12}$$

This result may be considered to be the fundamental reason why orbits are as observed in the vast majority of cases. The reason is that orbits are central phenomena, for which a symmetric connection suffices. If rotation is considered, a symmetric connection no longer suffices to describe the dynamics. Note carefully that the result (18.7) does not rely on any field equation. The explanation for orbits cannot be the incorrect Einstein field equation. The result (18.12) is fortuitously true for vacuum (Ricci flat) solutions of the Einstein field equation, but such vacuum solutions cannot have any physical significance.. As soon as a finite energy momentum density tensor is introduced, i.e. as soon as we consider:

$$R_{\mu\nu} - \frac{1}{2} R g_{\mu\nu} := G_{\mu\nu} = k T_{\mu\nu} \tag{18.13}$$

all exact solutions of the Einstein field equation VIOLATE the Bianchi identity [1–10], and so the Einstein field equation is clearly seen to be geometrically incorrect. Crothers has shown on www.aias.us and elsewhere that the Ricci flat assumption violates the Einstein principle of equivalence. The reason why central orbits are as observed is the spherical symmetry of space-time, not the vanishing of the Ricci tensor, and above all, not the Einstein field equation (18.13).

If the binary pulsar perturbation a/r is added, i.e. if we consider:

$$ds^2 = -\left(1 - \frac{2MG}{c^2 r} - \frac{a}{r^2}\right) c^2 dt^2 + \left(1 - \frac{2MG}{c^2 r} - \frac{a}{r^2}\right)^{-1} dr^2 + r^2 d\Omega^2 \tag{18.14}$$

computer algebra can be used to work out all the Christoffel connections from Eq. (18.14), assuming metric compatibility [11]. Knowing the Christoffel connections, the computer algebra is used to work out all the Riemann tensor elements from (18.14), the Ricci tensor elements, and the Ricci scalar. Finally the code was used to investigate whether Eq. (18.14) obeys metric compatibility and also both Eq. (18.10) and Eq. (18.11) under the assumption of a Christoffel connection. It was found that Eq. (18.14), used with an assumed Christoffel connection, obeys metric compatibility and Eq. (18.10), but does not obey Eq. (18.11). It does not produce a zero Ricci tensor in general, but produces a zero Ricci scalar:

$$R = g^{\mu\nu} R_{\mu\nu} = 0. \tag{18.15}$$

Therefore if a Christoffel connection is assumed, internal inconsistencies appear in the approach taken by Einstein to relativity, in that the complete, duality invariant, Bianchi identity is not obeyed, and in that:

$$G_{\mu\nu} = R_{\mu\nu} - \frac{1}{2} R g_{\mu\nu} = k T_{\mu\nu} \neq 0 \tag{18.16}$$

but, self-inconsistently:

$$R = -kT = 0. \tag{18.17}$$

It is known from paper 108 of www.aias.us that Eq. (18.14) produces all known features of the orbits of binary pulsars, so it is concluded that the Einstein field equation cannot describe such orbits. In the following section the ECE field equation is shown to provide a self-consistent explanation of all known orbits. This is achieved directly from the Bianchi identity and its Hodge dual, which give rise to the homogeneous and inhomogeneous ECE field equations of dynamics, cosmology and electrodynamics, and all fundamental interactions of classical fields.

18.3 General Equation for the Metric in ECE Theory

It has been shown in paper 108 of the ECE series (www.aias.us) that the general expression for the potential energy from a line element is:

$$V = \frac{m}{2} g_{00} \left(c^2 + \frac{L^2}{r^2} \right) \tag{18.18}$$

where m is the mass of an object orbiting an object of mass M, g_{00} is the time-like metric, L is the reduced angular momentum, the constant of motion:

$$L = r^2 \frac{\partial \phi}{\partial \lambda} \tag{18.19}$$

and λ a differential parameter. The gravitational potential Φ is obtained from the potential energy by:

$$\Phi = \frac{V}{m} \tag{18.20}$$

and in ECE theory [1–10] the acceleration due to gravity is related to the gravitational potential by:

$$\boldsymbol{g} = -(\boldsymbol{\nabla} + \boldsymbol{\omega})\Phi \tag{18.21}$$

where $\boldsymbol{\omega}$ is the spin connection vector. The Bianchi identity leads to the equation:

$$\boldsymbol{\nabla} \cdot \boldsymbol{g} = c^2(R - \omega T) \tag{18.22}$$

which is the gravitational equivalent of the Coulomb law. Therefore:

$$\boldsymbol{\nabla} \cdot (\boldsymbol{\nabla}\Phi + \boldsymbol{\omega}\Phi) = c^2(\omega T - R) \tag{18.23}$$

which is a differential equation for g_{00} in terms of ω, T, R and L in general. This equation makes no assumption other than those of Cartan geometry [1–10] (paper 105 of www.aias.us). In spherical polar coordinates:

$$\nabla \cdot (\nabla \Phi) = \frac{\partial^2 \Phi}{\partial r^2} + \frac{2}{r}\frac{\partial \Phi}{\partial r}, \tag{18.24}$$

$$\nabla \cdot (\omega \Phi) = \omega \cdot \nabla \Phi + (\nabla \cdot \omega)\Phi \tag{18.25}$$

so:

$$\frac{\partial^2 \Phi}{\partial r^2} + \left(\frac{2}{r} + \omega\right)\frac{\partial \Phi}{\partial r} + \frac{\partial \omega}{\partial r}\Phi = c^2(\omega T - R). \tag{18.26}$$

This equation has resonance solutions under well defined conditions.

From Eqs. (18.20) and (18.26):

$$\frac{1}{2}\left(c^2 + \frac{L^2}{r^2}\right)\frac{\partial^2 g_{00}}{\partial r^2} + \left(\frac{1}{2}\left(\frac{2}{r} + \omega\right)\left(c^2 + \frac{L^2}{r^2}\right) - \frac{L^2}{r^3}\right)\frac{\partial g_{00}}{\partial r}$$
$$+ \left(L^2\left(\frac{6}{r^4} - \left(\frac{2}{r} + \omega\right)\frac{1}{r^3}\right) + \frac{1}{2}\left(c^2 + \frac{L^2}{r^2}\right)\frac{\partial \omega}{\partial r}\right) g_{00}$$
$$= c^2(\omega T - R) \tag{18.27}$$

This equation may be integrated numerically to give g_{00}, or otherwise solved by computer algebra in given approximations. Precise satellite data show that:

$$g_{00} = 1 + \frac{\mu}{r} \tag{18.28}$$

where

$$\mu = -\frac{2MG}{c^2} \tag{18.29}$$

except for binary pulsars and the apparent Pioneer/Cassini anomalies. In the limit:

$$r \to \infty, \quad g_{00} \to 1 \tag{18.30}$$

so

$$\frac{\partial^2 g_{00}}{\partial r^2} \to 0, \quad \frac{\partial g_{00}}{\partial r} \to 0 \tag{18.31}$$

In this limit:

$$\omega \to 0, \quad \frac{\partial \omega}{\partial r} \to 0 \tag{18.32}$$

so:

$$\omega T - R \to \frac{4L^2}{c^2 R^4}\left(1 + \frac{\mu}{r}\right) \tag{18.33}$$

A particular solution of Eq. (18.33) is:

$$\omega T = \frac{4L^2}{c^2 r^4}\left(\frac{\mu}{r}\right) \tag{18.34}$$

so:

$$\nabla \cdot \boldsymbol{g} = c^2 R \left(1 - \frac{\mu}{r}\right). \tag{18.35}$$

Empirically, except for binary pulsars:

$$\nabla \cdot \boldsymbol{g} = c^2 k \rho_m \left(1 + \frac{2MG}{c^2 r}\right) \tag{18.36}$$

where ρ_m is the mass density (paper 105 of www.aias.us). The structure of ECE theory means that the Coulomb law is corrected in this weak field limit by:

$$\nabla \cdot \boldsymbol{E} = \frac{\rho_e}{\epsilon_0}\left(1 + \frac{2MG}{c^2 r}\right) \tag{18.37}$$

where ρ_e is the charge density. In this weak field limit:

$$\boldsymbol{E} \to -\nabla \phi \tag{18.38}$$

and the Poisson equation is corrected by mass as follows:

$$\nabla^2 \phi = -\frac{\rho_e}{\epsilon_0}\left(1 + \frac{2MG}{c^2 r}\right). \tag{18.39}$$

Therefore the scalar potential of electrodynamics is corrected by:

$$\phi = \frac{e}{4\pi\epsilon_0 r}\left(1 + \frac{MG}{rc^2}\right). \tag{18.40}$$

In the laboratory and in quantum chemistry this correction of the Coulomb law is entirely negligible, except when spin connection resonance occurs (for example papers 63 and 94 of www.aias.us). When the mass M becomes very large however, the correction becomes significant and may become experimentally observable.

In this analysis the torsion is self-consistently incorporated. The analysis starts from the Hodge dual of the Bianchi identity, which is:

$$D \wedge \tilde{T}^a := \tilde{R}^a{}_b \wedge q^b. \tag{18.41}$$

In tensor notation this becomes:

$$D_\mu T^{a\mu\nu} := R^a{}_\mu{}^{\mu\nu} \tag{18.42}$$

a particular solution of which is:

$$D_\mu T^{\kappa\mu\nu} = R^\kappa{}_\mu{}^{\mu\nu} \tag{18.43}$$

Eq. (18.43) may be written as:

$$\partial_\mu T^{\kappa\mu\nu} = R^\kappa{}_\mu{}^{\mu\nu} - \omega^\kappa{}_{\mu\lambda} T^{\lambda\mu\nu} \tag{18.44}$$

and this can be expressed as two vector equations, the inhomogeneous field equations of ECE theory. One of these may be written as:

$$\nabla \cdot \boldsymbol{T} = R - \omega T \tag{18.45}$$

and using the hypothesis in paper 105 of www.aias.us:

$$\boldsymbol{g} = c^2 \boldsymbol{T} \tag{18.46}$$

Eq. (18.22) is obtained. Here [1–10]:

$$\boldsymbol{T} := T^{010}\boldsymbol{i} + T^{020}\boldsymbol{j} + T^{030}\boldsymbol{k}, \tag{18.47}$$

$$R := R^0{}_1{}^{10} + R^0{}_2{}^{20} + R^0{}_3{}^{30}, \tag{18.48}$$

$$\omega T := (\omega^0{}_{1\lambda} T^{\lambda 10} + \omega^0{}_{2\lambda} T^{\lambda 20} + \omega^0{}_{3\lambda} T^{\lambda 30}) \tag{18.49}$$

The relation between $\boldsymbol{g}$ and Φ in Eq. (18.21) is derived from the first Cartan structure equation [1–10]. Therefore the equations of dynamics and cosmology, and of classical electrodynamics, are all based rigorously on Cartan geometry with torsion.

In his two original papers of 1916, Schwarzschild derived:

$$g_{00} = 1 + \frac{\mu}{r} \tag{18.50}$$

without consideration of torsion. Eq. (18.50) as derived by Schwarzschild is a purely geometrical solution (as pointed out by Crothers on www.aias.us). It is a solution of

$$R_{\mu\nu} - \frac{1}{2}Rg_{\mu\nu} = 0 \tag{18.51}$$

upon assuming that the connection is symmetric. As such it has no physical significance, because Eq. (18.51) eliminates all mass density from consideration, meaning that the object M cannot interact with the object m. Crothers has also shown that Eq. (18.51) violates the Einstein equivalence principle, and has also demonstrated basic errors in the theory of "big bang" and of "black holes". There are plentiful experimental data [1–10] that refute big bang comprehensively. In this paper it has been shown that the satellite data now available can be explained by considering only the spherical symmetry of space-time.

Acknowledgments

The British Government is thanked for a Civil List Pension to MWE and the staff of AIAS/SG for many interesting discussions.

References

[1] M. W. Evans, "Generally Covariant Unified Field Theory" (Abramis Academic, 2005 onwards), vols. 1–4, vol. 5 in prep. (preprints on www.aias.us).
[2] L .Felker, "The Evans Equations of Unified Field Theory" (Abramis, 2007).
[3] K. Pendergast, "Crystal Spheres" (preprint on www.aias.us).
[4] M. W. Evans, Acta Physica Polonica, **38B**, 2211 (2007).
[5] M. W. Evans and H. Eckardt, Physica B, **400**, 175 (2007).
[6] M. W. Evans, Physica B, **403**, 517 (2008).
[7] M. W. Evans (ed.), Adv. Chem. Phys., **119** (2001), ibid., M. W. Evans and S. Kielcih (eds.), Adv. Chem. Phys., **85** (1992, 1993, 1997).
[8] M. W. Evans and L. Crowell, "Classical and Quantum Electrodynamics and the B(3) Field" (World Scientific, 2001).
[9] M. W. Evans and J.-P. Vigier, "The Enigmatic Photon" (Kluwer, 1994 to 2002, hardback and softback), in five volumes.
[10] M. W. Evans, Omnia Opera section of www.aias.us (1992 onwards).
[11] S. P. Carroll, "Space-time and Geometry - an Introduction to General Relativity" (Addisaon-Wesley, New York, 2004, online notes of 1997).
[12] R. Wald, "General Relativity" (Univ. Chicago Press, 1986).
[13] C. Alley, tycho.usno.navy.mil/IAU31/alley.ppt (2006).
[14] AIAS/SG Discussions of May 2008.

19

On the Violation of the Bianchi Identity by the Einstein Field Equation and Big Bang Cosmologies

by

Myron W. Evans,
Alpha Institute for Advanced Study, Civil List Scientist.
(emyrone@aol.com and www.aias.us)

Abstract

It is shown that the Bianchi identity and its Hodge dual identity are cyclic sums of curvature tensors identically equal to the cyclic sums of the definitions of the same curvature tensors. In consequence, the torsion tensor cannot be neglected or assumed to be zero. This is a key result that shows the fundamental incorrectness of the Einstein field equation and all big bang cosmologies. The correct field equations of cosmology must be based directly on the Bianchi identity with finite torsion and curvature.

Keywords: Violation of the Bianchi identity by the Einstein field equation, incorrectness of big bang cosmologies, ECE equations of dynamics.

19.1 Introduction

The Bianchi identity of Cartan geometry [1–12] is:

$$D \wedge T^a := R^a{}_b \wedge q^b \qquad (19.1)$$

in the notation of differential geometry [1]. Here $D\wedge$ is the covariant exterior derivative, T^a is the torsion form, $R^a{}_b$ is the curvature form, $\wedge$ denoted wedge product, and q^b is the tetrad form. This identity is the homogeneous field equation of dynamics in ECE theory. Eq. (19.1) implies that there exists the identity:

$$D \wedge \tilde{T}^a := \tilde{R}^a{}_b \wedge q^b \tag{19.2}$$

where the tilde denotes Hodge duality in four dimensions. Eq. (19.2) is the inhomogeneous field equation of dynamics in ECE theory. It is seen that the equations of dynamics in ECE field theory are duality invariant, a fundamental symmetry property. In Section 2 it is proven that Eqs. (19.1) and (19.2) in tensor notation are both cyclic sums of curvature tensors identically equal to the same cyclic sum of definitions of the same curvature tensors. In order to arrive at this result the tetrad postulate is used:

$$D_\mu q^a_\nu = 0 \tag{19.3}$$

and the torsion tensor must be identically non-zero. The proof of Eq. (19.2) from Eq. (19.1) is therefore straightforward, because the Hodge dual of the curvature tensor in four dimensions is another curvature tensor. The cyclic sum of the latter is identically equal again to the sum of definitions, QED. Various other simple proofs of Eq. (19.2) are given in this section. All solutions of the Einstein field equation in the presence of finite canonical energy momentum density give the result:

$$D \wedge \tilde{T}^a = 0 \tag{19.4}$$

and the result:

$$\tilde{R}^a{}_b \wedge q^b \neq 0 \tag{19.5}$$

and so the Einstein field equation violates the Bianchi identity in its form (19.2). No physical inference of any kind can be drawn from the Einstein field equation because of its neglect of torsion. The only exception is the class of Ricci flat solutions of the Einstein field equation. Computer algebra showed that simple members of this class fortuitously obey both equations (19.1) and (19.2), but as shown by Crothers and independently elsewhere [1–12], vacuum solutions have no physical meaning. They are just equations of geometry that assume a priori that the Ricci tensor vanishes. The era of the great Einstein field equation has therefore drawn to a close. The correct field equations are derived directly from Eqs. (19.1) and (19.2) and are respectively the homogeneous and inhomogeneous dynamical equations of Einstein Cartan Evans (ECE) field theory [1–12]. The electro-dynamical field equations of ECE theory have the same structure as the dynamical field equations.

19.2 The Tensorial Formats of the Bianchi Identity

Expanding Eq. (19.1) using the well known [1–12] definitions of $D\wedge$ and of the wedge product, the following equation is obtained:

$$D_\mu T^a_{\nu\rho} + D_\rho T^a_{\mu\nu} + D_\nu T^a_{\rho\mu} := R^a_{\mu\nu\rho} + R^a_{\rho\mu\nu} + R^a_{\nu\rho\mu} \qquad (19.6)$$

in which appears the curvature and torsion tensors, and the spin connection. By considering the well known [1–12] action of the commutator of covariant derivatives on a four-vector the curvature and torsion tensors are obtained as follows:

$$[D_\mu, D_\nu]V^\rho = R^\rho{}_{\sigma\mu\nu}V^\sigma - T^\lambda_{\mu\nu}D_\lambda V^\rho \qquad (19.7)$$

irrespective of any metric and of any connection. There is no reason why the torsion tensor should be zero a priori, as incorrectly assumed in the standard model and big bang cosmologies. The two tensors are defined by the commutator, [1–12], and are:

$$R^\rho{}_{\sigma\mu\nu} = \partial_\mu \Gamma^\rho_{\nu\sigma} - \partial_\mu \Gamma^\rho_{\mu\sigma} + \Gamma^\rho_{\mu\lambda}\Gamma^\lambda_{\nu\sigma} - \Gamma^\rho_{\nu\lambda}\Gamma^\lambda_{\mu\sigma} \qquad (19.8)$$

and

$$T^\lambda_{\mu\nu} = \Gamma^\lambda_{\mu\nu} - \Gamma^\lambda_{\nu\mu} \qquad (19.9)$$

where $T^\lambda_{\mu\nu}$ is the connection defined by the covariant derivative:

$$D_\mu V^\rho = \partial_\mu V^\rho + \Gamma^\rho_{\mu\lambda}V^\lambda. \qquad (19.10)$$

The standard model of physics, and big bang cosmologies, incorrectly assume that the connection is symmetric:

$$\Gamma^\lambda_{\mu\nu} = \Gamma^\lambda_{\nu\mu} \qquad (19.11)$$

and that in consequence the torsion vanishes. As argued already, this assumption leads to a violation of the Bianchi identity, i.e. to an incorrect geometry. This basic flaw in the Einstein equation was discovered by computer algebra in papers 93 ff of www.aias.us. It was obviously not known to Einstein in 1915 and for ninety two years thereafter.

The connection defined in Eq. (19.10) is related to the spin connection by the tetrad postulate [1–12]:

$$D_\mu q^a_\nu = \partial_\mu q^a_\nu - q^a_\lambda \Gamma^\lambda_{\mu\nu} + q^b_\nu \omega^a_{\mu b} = 0. \qquad (19.12)$$

There is no situation in physics in which the tetrad postulate is not true, because the postulate is the fact that a complete vector field is independent of the coordinate system used to define the vector field. The curvature and torsion forms and their respective tensors are related by the tetrad [1–12] as follows:

$$T^\lambda_{\mu\nu} = q^\lambda_a T^a_{\mu\nu}, \tag{19.13}$$

$$R^\rho{}_{\sigma\mu\nu} = q^\rho_a q^b_\sigma R^a{}_{b\mu\nu}. \tag{19.14}$$

Using Eqs. (19.8) to (19.14) it follows [1–12] that the Bianchi identity (19.1) is:

$$\begin{aligned}
R^\lambda{}_{\rho\mu\nu} &+ R^\lambda{}_{\mu\nu\rho} + R^\lambda{}_{\nu\rho\mu} \\
:= &\partial_\mu \Gamma^\lambda_{\nu\rho} - \partial_\nu \Gamma^\lambda_{\mu\rho} + \Gamma^\lambda_{\mu\sigma}\Gamma^\sigma_{\nu\rho} - \Gamma^\lambda_{\nu\sigma}\Gamma^\sigma_{\mu\rho} \\
&+ \partial_\nu \Gamma^\lambda_{\rho\mu} - \partial_\rho \Gamma^\lambda_{\nu\mu} + \Gamma^\lambda_{\nu\sigma}\Gamma^\sigma_{\rho\mu} - \Gamma^\lambda_{\rho\sigma}\Gamma^\sigma_{\nu\mu} \\
&+ \partial_\rho \Gamma^\lambda_{\mu\nu} - \partial_\mu \Gamma^\lambda_{\rho\nu} + \Gamma^\lambda_{\rho\sigma}\Gamma^\sigma_{\mu\nu} - \Gamma^\lambda_{\mu\sigma}\Gamma^\sigma_{\rho\nu} \neq 0
\end{aligned} \tag{19.15}$$

where:

$$\begin{aligned}
R^\lambda{}_{\rho\mu\nu} &= \partial_\mu \Gamma^\lambda_{\nu\rho} - \partial_\nu \Gamma^\lambda_{\mu\rho} + \Gamma^\lambda_{\mu\sigma}\Gamma^\sigma_{\nu\rho} - \Gamma^\lambda_{\nu\sigma}\Gamma^\sigma_{\mu\rho}, \\
R^\lambda{}_{\mu\nu\rho} &= \partial_\nu \Gamma^\lambda_{\rho\mu} - \partial_\rho \Gamma^\lambda_{\nu\mu} + \Gamma^\lambda_{\nu\sigma}\Gamma^\sigma_{\rho\mu} - \Gamma^\lambda_{\rho\sigma}\Gamma^\sigma_{\nu\mu}, \\
R^\lambda{}_{\nu\rho\mu} &= \partial_\rho \Gamma^\lambda_{\mu\nu} - \partial_\mu \Gamma^\lambda_{\rho\nu} + \Gamma^\lambda_{\rho\sigma}\Gamma^\sigma_{\mu\nu} - \Gamma^\lambda_{\mu\sigma}\Gamma^\sigma_{\rho\nu}.
\end{aligned} \tag{19.16}$$

Therefore the Bianchi identity is a true identity, one side is identically the same as the other by definition. This result of Cartan geometry is true if and only if the torsion tensor is identically non-zero. Unfortunately, the standard model literature uses the equation:

$$R^\lambda{}_{\rho\mu\nu} + R^\lambda{}_{\mu\nu\rho} + R^\lambda{}_{\nu\rho\mu} = 0 \tag{19.17}$$

and names it the "first Bianchi identity". From the foregoing it is clear that Eq. (19.17) is true only if the connection is a priori assumed to be symmetric as in Eq. (19.11). There is no basis of this assumption. This error of the standard model is compounded by its use of the equation:

$$D_\mu R^\kappa{}_{\lambda\nu\rho} + D_\rho R^\kappa{}_{\lambda\mu\nu} + D_\nu R^\kappa{}_{\lambda\rho\mu} = 0 \tag{19.18}$$

which is referred to as "the second Bianchi identity". Unfortunately Eq. (19.18) again neglects torsion and again leads [1–12] to an incorrect geometry. There is therefore a catastrophic failure of the standard model because

19.2 The Tensorial Formats of the Bianchi Identity

Einstein based his field equation directly on Eq. (19.18) as is well known. The true Bianchi identity is Eq. (19.1), which in tensorial format is Eq. (19.15).

The easiest way to prove Eq. (19.2) from Eq. (19.1) is to consider the Hodge dual of the curvature form in four dimensions. This Hodge dual is well known [1–2] to be defined as:

$$\tilde{R}^\rho{}_\sigma{}^{\alpha\beta} = \frac{1}{2} \| g \|^{\frac{1}{2}} \epsilon^{\mu\nu\alpha\beta} R^\rho{}_{\sigma\mu\nu} \qquad (19.19)$$

where $\| g \|^{\frac{1}{2}}$ is the square root of the determinant of the metric and where $\epsilon^{\mu\nu\alpha\beta}$ is the totally anti-symmetric unit tensor. The latter is defined to be the Minkowski space-time tensor [1–12]. The Hodge dual of the curvature form is therefore another curvature form, the cyclic sum of which obeys the equation (19.15) again, Q.E.D. The catastrophic failure of the standard model of cosmology emerged in paper 93 (www.aias.us) when considering the Hodge dual Bianchi identity (19.2) in the tensor format:

$$D_\mu \tilde{T}^\kappa{}_{\nu\rho} + D_\rho \tilde{T}^\kappa{}_{\mu\nu} + D_\nu \tilde{T}^\kappa{}_{\rho\mu} := \tilde{R}^\kappa{}_{\mu\nu\rho} + \tilde{R}^\kappa{}_{\rho\mu\nu} + \tilde{R}^\kappa{}_{\nu\rho\mu}. \qquad (19.20)$$

This equation is the same as:

$$D_\mu T^{\kappa\mu\nu} := R^\kappa{}_\mu{}^{\mu\nu}. \qquad (19.21)$$

It was found by computer algebra in paper 93 ff. of www.aias.us that the Einstein field equation violates Eq. (19.21) because the Einstein field equation gives the result:

$$T^{\kappa\mu\nu} = 0,\ R^\kappa{}_\mu{}^{\mu\nu} \neq 0 \qquad (19.22)$$

in general. For example the result (19.22) was obtained for the Robertson Walker metric which is the basis for big bang. No big bang metric survived the test of Eq. (19.21), the inhomogeneous field equation of ECE cosmology.

This disaster for the standard model remained hidden for so many years because the Einstein field equation fortuitously obeys the homogeneous ECE field equation, whose tensorial format is:

$$D_\mu T^\kappa{}_{\nu\rho} + D_\rho T^\kappa{}_{\mu\nu} + D_\nu T^\kappa{}_{\rho\mu} := R^\kappa{}_{\mu\nu\rho} + R^\kappa{}_{\rho\mu\nu} + R^\kappa{}_{\nu\rho\mu} \qquad (19.23)$$

which is the same as:

$$D_\mu \tilde{T}^{\kappa\mu\nu} := \tilde{R}^\kappa{}_\mu{}^{\mu\nu}. \qquad (19.24)$$

The standard model has assumed for over ninety years that:

$$\widetilde{T}^{\kappa\mu\nu} = 0, \widetilde{R}^{\kappa}{}_{\mu}{}^{\mu\nu} = 0. \tag{19.25}$$

In the rest of this section we give simple proofs of Eq. (19.2) from Eq. (19.1). Given the tetrad postulate, the Bianchi identity as we have argued can be constructed from commutator relations such as:

$$[D_0, D_1]V^\rho = D_0(D_1 V^\rho) - D_1(D_0 V^\rho) \tag{19.26}$$
$$= R^\rho{}_{\sigma 01} V^\sigma - T^\lambda_{01} D_\lambda V^\rho.$$

The Hodge dual of Eq. (19.26) with lowered indices is:

$$[D_2, D_3]V^\rho = D_2(D_3 V^\rho) - D_3(D_2 V^\rho) = R^\rho{}_{\sigma 23} V^\sigma - T^\lambda_{23} D_\lambda V^\rho. \tag{19.27}$$

Equations (19.26) and (19.27) are both examples of:

$$[D_\mu, D_\nu]V^\rho = D_\mu(D_\nu V^\rho) - D_\nu(D_\mu V^\rho). \tag{19.28}$$

So it follows that:

$$D \wedge \widetilde{T}^a := \widetilde{R}^a{}_b \wedge q^b \tag{19.29}$$

is an example of

$$D \wedge T^a := R^a{}_b \wedge q^b \tag{19.30}$$

Q.E.D.

This is a simple proof given in outline. Some more details can be given as follows. Carrying out Hodge duals on both sides of Eq. (19.26):

$$\epsilon^{0123}[D_0, D_1]V^\rho = \epsilon^{0123} R^\rho{}_{\sigma 01} V^\sigma - \epsilon^{0123} T^\lambda_{01} D_\lambda V^\rho \tag{19.31}$$

i.e.:

$$[D^2, D^3]V^\rho = R^\rho{}_\sigma{}^{23} V^\sigma - T^{\lambda 23} D_\lambda V^\rho. \tag{19.32}$$

Indices are lowered as follows:

$$[D_2, D_3] = g_{22} g_{33} [D^2, D^3] \tag{19.33}$$

19.2 The Tensorial Formats of the Bianchi Identity

on both sides, giving Eq. (19.27). Therefore if

$$\mu = 0, \nu = 1 \tag{19.34}$$

in the original identity, then:

$$\mu = 2, \nu = 3 \tag{19.35}$$

in the Hodge dual identity. Both the original and Hodge dual identities are cyclic sums of curvature tensors as discussed already.

It is helpful to give examples of the results of paper 93. It is clear that these are technically correct because they were generated by computer. It was found that for exact solutions of the Einstein field equation with finite canonical energy-momentum density:

$$T_{\mu\nu} \neq 0 \tag{19.36}$$

the key curvature tensor $R^{\kappa}{}_{\mu}{}^{\mu\nu}$ was non-zero. For example:

$$R^0{}_1{}^{01} + R^0{}_2{}^{02} + R^0{}_3{}^{03} \neq 0. \tag{19.37}$$

These exact solutions use the Christoffel connection, so by definition the solutions assume that the torsion tensor vanishes. Thus for example:

$$T^{001} + T^{002} + T^{003} = 0. \tag{19.38}$$

Therefore:

$$D_1 T^{001} + D_2 T^{002} + D_3 T^{003} = 0. \tag{19.39}$$

Therefore the Hodge dual of the Bianchi identity is not obeyed:

$$D_1 T^{001} + D_2 T^{002} + D_3 T^{003} \neq R^0{}_1{}^{01} + R^0{}_2{}^{02} + R^0{}_3{}^{03} \tag{19.40}$$

a fundamental error of the Einstein field equation which is the basis of all current cosmologies.

Using the Hodge dual relations:

$$\widetilde{R}^0{}_{123} = \epsilon_{0123} R^0{}_1{}^{01},$$
$$\widetilde{R}^0{}_{312} = \epsilon_{0312} R^0{}_3{}^{03}, \tag{19.41}$$
$$\widetilde{R}^0{}_{231} = \epsilon_{0231} R^0{}_2{}^{02},$$

it is found that:

$$\tilde{R}^0{}_{123} + \tilde{R}^0{}_{312} + \tilde{R}^0{}_{231} \neq 0 \tag{19.42}$$

for all solutions of the Einstein field equation with finite $T_{\mu\nu}$. These are the type of solutions used in big bang cosmology. This result is technically irrefutable because it was generated by computer algebra. The latter also showed in paper 93 of www.aias.us that:

$$R^0{}_{123} + R^0{}_{312} + R^0{}_{231} = 0 \tag{19.43}$$

for all correct solutions of the Einstein field equation. Eq. (19.43) is an example of what is incorrectly referred to in the standard model as "the first Bianchi identity". The Hodge dual of Eq. (19.43) is:

$$\tilde{R}^0{}_1{}^{01} + \tilde{R}^0{}_2{}^{02} + \tilde{R}^0{}_3{}^{03} = 0. \tag{19.44}$$

So the Bianchi identity:

$$D_\mu \tilde{T}^{\kappa\mu\nu} := \tilde{R}^\kappa{}_\mu{}^{\mu\nu} \tag{19.45}$$

is obeyed fortuitously, but the Hodge dual:

$$D_\mu T^{\kappa\mu\nu} := R^\kappa{}_\mu{}^{\mu\nu} \tag{19.46}$$

is not obeyed. This result was first shown in paper 93 of www.aias.us. It is difficult to find because the calculations needed are far to complicated to be carried out by hand. They need the development of powerful Maxima based code.

These are examples of the general theorem:

$$\begin{aligned}[D_\mu, D_\nu] V^\rho &= D_\mu(D_\nu V^\rho) - D_\nu(D_\mu V^\rho) \\ &= R^\rho{}_{\sigma\mu\nu} V^\sigma - T^\lambda{}_{\mu\nu} D_\lambda V^\rho\end{aligned} \tag{19.47}$$

which is the result of the anti-symmetry of the commutator:

$$[D_\mu, D_\nu] = -[D_\nu, D_\mu] \tag{19.48}$$

of covariant derivatives. This anti-symmetry defines the anti-symmetry in the last two indices of the curvature tensor:

$$R^\rho{}_{\sigma\mu\nu} = -R^\rho{}_{\sigma\nu\mu} \tag{19.49}$$

19.2 The Tensorial Formats of the Bianchi Identity

and the torsion tensor:

$$T^\lambda_{\mu\nu} = -T^\lambda_{\nu\mu}. \tag{19.50}$$

The catastrophic situation has arisen that the incorrect assumption:

$$T^\lambda_{\mu\nu} = 0 \tag{19.51}$$

of Einstein has worked its way uncritically into the subject of cosmology for ninety years. The assumption of vanishing torsion was shown in paper 93 of www.aias.us to violate the complete Bianchi identity. In order to show this, the Hodge dual of the identity has to be taken into account. The resulting calculations are so complicated that computer algebra must be used throughout. Correct cosmology must be based on a computer solution of the Bianchi identity simultaneously with its Hodge dual. In ECE theory these are the equations of relativistic dynamics and cosmology. In so doing the Christoffel connection cannot be used, the connection is in general not symmetric in its lower two indices. So the well known equation that links the Christoffel connection to the symmetric metric:

$$\Gamma^\sigma_{\mu\nu} = \frac{1}{2} g^{\sigma\rho}(\partial_\mu g_{\nu\rho} + \partial_\nu g_{\rho\mu} - \partial_\rho g_{\mu\nu}) \tag{19.52}$$

must be discarded, because this leads to a violation of the Bianchi identity. The latter must always be written with non-zero torsion. As a result of this major change in cosmology it has been shown in ECE papers on www.aias.us that there exist relations such as:

$$\nabla \cdot g = c^2 T \tag{19.53}$$

where g is the acceleration due to gravity and where T is a well defined scalar torsion. It is found that the acceleration due to gravity g and the electric field strength E are both due to space-time torsion. The basic structure of spiral galaxies is also due to space-time torsion, so maps of the so called "dark matter" are maps of space-time torsion

The Hodge dual of the commutator operator is another commutator operator:

$$[D^\alpha, D^\beta]_{HD} = \frac{1}{2} \| g \|^{\frac{1}{2}} \epsilon^{\mu\nu\alpha\beta}[D_\mu, D_\nu] \tag{19.54}$$

and it follows that:

$$[D^\alpha, D^\beta]_{HD} V^\sigma = \tilde{R}^\rho{}_\sigma{}^{\alpha\beta} V^\sigma - \tilde{T}^{\lambda\alpha\beta} D_\lambda V^\rho \tag{19.55}$$

where:

$$\tilde{R}^{\rho}{}_{\sigma}{}^{\alpha\beta} = \frac{1}{2} \| g \|^{\frac{1}{2}} \epsilon^{\mu\nu\alpha\beta} R^{\rho}{}_{\sigma\mu\nu}. \tag{19.56}$$

and

$$\tilde{T}^{\lambda\alpha\beta} = \frac{1}{2} \| g \|^{\frac{1}{2}} \epsilon^{\mu\nu\alpha\beta} T^{\lambda}{}_{\mu\nu}. \tag{19.57}$$

In general, indices are lowered by:

$$\begin{aligned}
[D_\alpha, D_\beta]_{HD} &= g_{\alpha\mu} g_{\beta\nu} [D^\mu, D^\nu]_{HD} \\
\tilde{R}^{\rho}{}_{\sigma\alpha\beta} &= g_{\alpha\mu} g_{\beta\nu} \tilde{R}^{\rho}{}_{\sigma}{}^{\mu\nu} \\
\tilde{T}^{\lambda}{}_{\alpha\beta} &= g_{\alpha\mu} g_{\beta\nu} \tilde{T}^{\lambda\alpha\beta}
\end{aligned} \tag{19.58}$$

so it is found that:

$$[D_\alpha, D_\beta]_{HD} V^\sigma = \tilde{R}^{\rho}{}_{\alpha\beta} V^\sigma - \tilde{T}^{\lambda}{}_{\alpha\beta} D_\lambda V^\rho. \tag{19.59}$$

This is the Hodge dual with lowered indices of Eq. (19.47). The Hodge duals of the curvature and torsion tensors are curvature and torsion tensors with different indices. So it follows that they obey the Hodge dual identity:

$$D \wedge \tilde{T}^a := R^a{}_b \wedge q^b. \tag{19.60}$$

This is the identity:

$$D_\mu T^{\kappa\mu\nu} := R^{\kappa}{}_{\mu}{}^{\mu\nu} \tag{19.61}$$

which is the following cyclic sum identity:

$$\begin{aligned}
\tilde{R}^{\lambda}{}_{\rho\mu\nu} + \tilde{R}^{\lambda}{}_{\mu\nu\rho} + \tilde{R}^{\lambda}{}_{\nu\rho\mu} &\\
:= (\partial_\mu \Gamma^{\lambda}_{\nu\rho} - \partial_\nu \Gamma^{\lambda}_{\mu\rho} + \Gamma^{\lambda}_{\mu\sigma}\Gamma^{\sigma}_{\nu\rho} - \Gamma^{\lambda}_{\nu\sigma}\Gamma^{\sigma}_{\mu\rho})_{HD} &\\
+ (\partial_\nu \Gamma^{\lambda}_{\rho\mu} - \partial_\rho \Gamma^{\lambda}_{\nu\mu} + \Gamma^{\lambda}_{\nu\sigma}\Gamma^{\sigma}_{\rho\mu} - \Gamma^{\lambda}_{\rho\sigma}\Gamma^{\sigma}_{\nu\mu})_{HD} &\\
+ (\partial_\rho \Gamma^{\lambda}_{\mu\nu} - \partial_\mu \Gamma^{\lambda}_{\rho\nu} + \Gamma^{\lambda}_{\rho\sigma}\Gamma^{\sigma}_{\mu\nu} - \Gamma^{\lambda}_{\mu\sigma}\Gamma^{\sigma}_{\rho\nu})_{HD} &\neq 0
\end{aligned} \tag{19.62}$$

where the individual definitions are as follows:

$$\begin{aligned}
\tilde{R}^{\lambda}{}_{\rho\mu\nu} &= (\partial_\mu \Gamma^{\lambda}_{\nu\rho} - \partial_\nu \Gamma^{\lambda}_{\mu\rho} + \Gamma^{\lambda}_{\mu\sigma}\Gamma^{\sigma}_{\nu\rho} - \Gamma^{\lambda}_{\nu\sigma}\Gamma^{\sigma}_{\nu\mu})_{HD}, \\
\tilde{R}^{\lambda}{}_{\mu\nu\rho} &= (\partial_\nu \Gamma^{\lambda}_{\rho}\mu - \partial_\rho \Gamma^{\lambda}_{\nu\mu} + \Gamma^{\lambda}_{\nu\sigma}\Gamma^{\sigma}_{\rho\mu} - \Gamma^{\lambda}_{\rho\sigma}\Gamma^{\sigma}_{\nu\mu})_{HD}, \\
\tilde{R}^{\lambda}{}_{\nu\rho\mu} &= (\partial_\rho \Gamma^{\lambda}_{\mu\nu} - \partial_\mu \Gamma^{\lambda}_{\rho\nu} + \Gamma^{\lambda}_{\rho\sigma}\Gamma^{\sigma}_{\mu\nu} - \Gamma^{\lambda}_{\mu\sigma}\Gamma^{\sigma}_{\rho\nu})_{HD}.
\end{aligned} \tag{19.63}$$

Acknowledgments

The British Government is thanked for a Civil List Pension for pre-eminent contributions to Britain in science, and the staffs of AIAS and SG for many interesting discussions.

References

[1] S. P. Carroll, "Space-time and Geometry, an Introduction to General Relativity" (Addison Wesley, New York, 2004, also 1997 online notes).

[2] M. W. Evans, "Generally Covariant Unified Field Theory: the Geometrization of Physics" (Abramis 2005 onwards), vols. One to four, vol. five in prep. (Collection of the ECE papers on the www.aias.us website).

[3] L. Felker, "The Evans Equations of Unified Field Theory" (Abramis 2007, virtual best seller in pre-print form on www.aias.us, circa 2005 to 2007).

[4] K. Pendergast, "Crystal Spheres" (preprint on www.aias.us).

[5] Educational articles on ECE theory on the www.aias.us website by several authors).

[6] M. W. Evans, Collected Papers (www.aias.us Omnia Opera), 1992 to present, hyperlinked to original papers on the precursor gauge theories of ECE.

[7] M. W. Evans, ed., "Modern Non-linear Optics", in "Advances in Chemical Physics", vol. 119 (Wiley Interscience, New York, 2001, second and updated edition of a collection of 35 review articles); ibid., M. W. Evans and S. Kielich (eds.), vol. 85 (first edition, published, 1992, 1993 and 1997).

[8] M. W. Evans and L. B. Crowell, "Classical and Quantum Electrodynamics and the B(3) Field" (World Scientific, 2001).

[9] M. W. Evans and J.-P. Vigier, "The Enigmatic Photon" (Kluwer, 1994 to 2002, hardback and softback) in five volumes.

[10] M. W. Evans, ECE papers in Found. Phys. Lett., 2003 to 2005.

[11] M. W. Evans, Acta Phys. Polonica, **38B**, 2211 (2007).

[12] M. W. Evans and H. Eckardt, Physica B, **400**, 175 (2007), M. W. Evans, Physica B, **403**, 517 (2008).

20

The Complete Equations of Classical Dynamics in ECE Theory

by

Myron W. Evans,
Alpha Institute for Advanced Study, Civil List Scientist.
(emyrone@aol.com and www.aias.us)

Abstract

The complete equations of classical dynamics in ECE theory are deduced directly from Cartan geometry, the first Cartan structure equation, the Bianchi identity, and the Hodge dual of the Bianchi identity. These equations are fully relativistic and reduce to the well known non-relativistic limits when the spin connection becomes very small. The equations of ECE dynamics have the same structure as the equations of electrodynamics in ECE unified field theory, and the ECE equations of dynamics generalize the gravito-magnetic equations of the standard model to fully relativistic situations without any assumption on linearization.

Keywords: ECE equations of relativistic classical dynamics, reduction to non-relativistic dynamics.

20.1 Introduction

The fundamental idea of relativity is that physics is an objective subject, the observation of nature and its explanation without anthropomorphic distortion. Geometry is one method of representing objectivity: it is well known that special relativity uses the Minkowski space-time, and that Einstein's approach to objectivity (general relativity) used Riemann geometry. The use of Riemann geometry by Einstein and Hilbert in 1915 led to the well known Einstein-Hilbert (EH) field equation. The latter reduces to Newtonian dynamics in the weak field limit. The latter can be thought of as the approach to

Minkowski space-time. Recently, during the development of Einstein Cartan Evans (ECE) field theory [1–10], it has been realized that the Einsteinian geometry is severely self-inconsistent because of its neglect of space-time torsion. This omission means that the Einsteinian geometry does not obey the Bianchi identity of Cartan geometry and also its Hodge dual identity. This means that the Einstein field equation is obsolete, no physical inference can be drawn from an incorrect geometry in relativity theory because objectivity has been incorrectly represented.

In Section 20.2 the geometrically correct equations of relativistic dynamics are inferred directly from the correct Cartan geometry, space-time torsion playing its proper central role. There are two basic equation structures: based on orbital torsion and spin torsion. The equation based on orbital torsion for example is the dynamical analogue of the Coulomb inverse square law of ECE electrodynamics [1–10], and in the non-relativistic limit gives the Newton inverse square law. This equation of orbital torsion has its Hodge dual equation, and the two equations together control the dynamics. Similarly there is a dynamical equation of spin torsion which is the direct analogue of the Ampere Maxwell law of ECE electrodynamics. This spin torsion equation again has a Hodge dual spin torsion equation, the dynamical analogue of the Faraday law of induction of ECE electrodynamics. Finally in Section 20.2 the generally relativistic Euler type equation fo motion is obtained from the first Cartan structure equation, which defines the Cartan torsion as the covariant exterior derivative of the Cartan tetrad. The Euler equation is regained in the limit of very small, but non-zero, spin connection. The concept of force field in classical dynamics is inferred to be orbital space-time torsion multiplied by rest energy. The concept of torque field in classical dynamics is inferred to be the integral over distance of the spin torsion multiplied by rest energy. These equations of relativistic dynamics are fully compatible with the Bianchi identity and the dual identity as required by objectivity. Orbits are not due to the EH field equation, that obsolete inference was an anthropomorphic distortion introduced by the neglect of the central feature of both dynamics and electrodynamics-space-time torsion.

20.2 The Equations of Relativistic Dynamics

The basic structure of the equations is given in the standard notation [11] of differential geometry by the first Bianchi identity:

$$D \wedge T^a := R^a{}_b \wedge q^b \qquad (20.1)$$

and its Hodge dual identity:

$$D \wedge \tilde{T}^a := \tilde{R}^a{}_b \wedge q^b \qquad (20.2)$$

20.2 The Equations of Relativistic Dynamics

where $D\wedge$ denotes the covariant exterior derivative [1–11], T^a the space-time torsion form, R^a_b the space-time curvature form, and q^b the space-time tetrad form. The symbol $\wedge$ denotes the wedge product of differential geometry [1–11] and the tilde symbol denotes the Hodge dual transform in four dimensions. In tensorial notation Eq. (20.1) can be reduced to [1–10]:

$$D_\mu \widetilde{T}^{\kappa\mu\nu} = \widetilde{R}^\kappa{}_\mu{}^{\mu\nu} \tag{20.3}$$

and Eq. (20.2) can be reduced to:

$$D_\mu T^{\kappa\mu\nu} = R^\kappa{}_\mu{}^{\mu\nu}. \tag{20.4}$$

Eq. (20.3) is the homogeneous equation of Cartan geometry, and Eq. (20.4) is the inhomogeneous equation of Cartan geometry. Each tensorial equation can be developed as two vectorial equations. It is convenient to rewrite the equations as:

$$\partial_\mu \widetilde{T}^{\kappa\mu\nu} = \widetilde{J}^{\kappa\nu} \tag{20.5}$$

and

$$\partial_\mu T^{\kappa\mu\nu} = J^{\kappa\nu} \tag{20.6}$$

where the terms on the right hand side subsume the spin connection and are current terms in analogy with ECE electrodynamics [1–11]. Eqs. (20.5) and (20.6) are similar to the well known Maxwell Heaviside (MH) field equations:

$$\partial_\mu \widetilde{F}^{\mu\nu} = 0 \tag{20.7}$$

and

$$\partial_\mu F^{\mu\nu} = J^\nu/\epsilon_0 \tag{20.8}$$

where $F^{\mu\nu}$ is the electromagnetic field tensor, $\widetilde{F}^{\mu\nu}$ is its Hodge dual, J^ν is charge current density and ϵ_0 is the vacuum permittivity in S.I. units. However, Eqs. (20.5) and (20.6) are ones of general relativity, written in a four dimensional, dynamic, space-time with torsion and curvature both present. The MH equations are written in the Minkowski space-time of special relativity, the flat and static space-time with no torsion and no curvature.

The laws of orbital torsion are obtained with the index choice:

$$\kappa = 0 \tag{20.9}$$

giving:

$$\partial_\mu \widetilde{T}^{0\mu\nu} = \widetilde{J}^{0\nu} \tag{20.10}$$

and

$$\partial_\mu T^{0\mu\nu} = J^{0\nu}. \tag{20.11}$$

In vector notation these are respectively:

$$\boldsymbol{\nabla} \cdot \widetilde{\boldsymbol{T}} = \widetilde{J}^0 \tag{20.12}$$

and

$$\boldsymbol{\nabla} \cdot \boldsymbol{T} = J^0. \tag{20.13}$$

These are respectively the dynamical analogues of the ECE Gauss law:

$$\boldsymbol{\nabla} \cdot \boldsymbol{B} = 0 \tag{20.14}$$

and the ECE Coulomb law

$$\boldsymbol{\nabla} \cdot \boldsymbol{E} = \rho/\epsilon_0 \tag{20.15}$$

where $\boldsymbol{B}$ is magnetic flux density, $\boldsymbol{E}$ is electric field strength, and ρ is charge density.

In Cartesian coordinates the orbital torsion vector is [1–10]:

$$\boldsymbol{T} = T^{010}\boldsymbol{i} + T^{020}\boldsymbol{j} + T^{030}\boldsymbol{k} \tag{20.16}$$

and its Hodge dual is:

$$\widetilde{\boldsymbol{T}} = \widetilde{T}^{010}\boldsymbol{i} + \widetilde{T}^{020}\boldsymbol{j} + \widetilde{T}^{030}\boldsymbol{k}. \tag{20.17}$$

From the well known experimental fact that both the Coulomb and Newton laws are inverse square laws, Eq. (20.13) is identified as giving the Newton inverse square law in the non-relativistic limit. The latter is defined as the limit in which the spin connection becomes very small, i.e. the limit in which Minkowski space-time is approached. In this limit the torsion and curvature approach zero, but are still finite. By analogy between electric field strength $\boldsymbol{E}$ and acceleration due to gravity $\boldsymbol{g}$ it may be inferred that:

$$\boldsymbol{g} = c^2 \boldsymbol{T}. \tag{20.18}$$

20.2 The Equations of Relativistic Dynamics

This inference is an example of the basic ECE hypothesis [1–10]:

$$A^a = A^{(0)} q^a \tag{20.19}$$

where A^a is the electromagnetic potential form and where $cA^{(0)}$ is a primordial or fundamental voltage proportional to the charge e on the proton. The ECE hypothesis turns geometry into physics.

Therefore the fundamental relativistic equations of motion that reduce to Newtonian dynamics are:

$$\nabla \cdot \boldsymbol{g} = c^2 J^0 \tag{20.20}$$

and its Hodge dual:

$$\nabla \cdot \tilde{\boldsymbol{g}} = c^2 \tilde{J}^0. \tag{20.21}$$

From analogy with electrodynamics the Hodge dual current is zero for all practical purposes, and this analogy with electrodynamics comes from the fundamental hypothesis (20.19). Therefore $\tilde{\boldsymbol{g}}$ is the dynamical analogue of the magnetic flux density in the ECE Gauss law of magnetism.

In the non-relativistic limit the spin connection becomes very small, so:

$$J^0 \to R^0{}_1{}^{10} + R^0{}_2{}^{20} + R^0{}_3{}^{30} \tag{20.22}$$

where the right hand side is a sum of curvature elements. The mass density (kilograms per cubic meter) is identified from Eqs. (20.20) and (20.22) as:

$$\rho_m = \frac{1}{k}(R^0{}_1{}^{10} + R^0{}_2{}^{20} + R^0{}_3{}^{30}) \tag{20.23}$$

where k is the Einstein constant in meters per kilogram. Therefore in the non-relativistic limit:

$$\nabla \cdot \boldsymbol{g} = c^2 k \rho_m \tag{20.24}$$

which is the direct analogy of the ECE Coulomb law. More generally mass density is defined by:

$$\rho_m = J^0/k. \tag{20.25}$$

The Newtonian force is:

$$\boldsymbol{F} = m\boldsymbol{g} \tag{20.26}$$

where m is the mass of an object in a gravitational acceleration g. Therefore the concept of field of force in Newtonian dynamics is inferred to be:

$$\boldsymbol{F} = E_0 \boldsymbol{T} \qquad (20.27)$$

where:

$$E_0 = mc^2 \qquad (20.28)$$

is rest energy. The Newtonian force is defined by the orbital torsion:

$$\boldsymbol{F} = E_0(T^{010}\boldsymbol{i} + T^{020}\boldsymbol{j} + T^{030}\boldsymbol{k}) \qquad (20.29)$$

and the inverse square law is described by:

$$\nabla \cdot \boldsymbol{F} = E_0 k \rho_m. \qquad (20.30)$$

However, the Newtonian dynamics are controlled not only by Eqs. (20.29) and (20.30) but also by their Hodge duals:

$$\widetilde{\boldsymbol{F}} = E_0(\widetilde{T}^{010}\boldsymbol{i} + \widetilde{T}^{020}\boldsymbol{j} + \widetilde{T}^{030}\boldsymbol{k}) \qquad (20.31)$$

and

$$\nabla \cdot \widetilde{\boldsymbol{F}} = 0 \qquad (20.32)$$

and this is a new feature of Newtonian dynamics given by ECE theory.

It has also been inferred [1–10] that all orbital dynamics are controlled not by the obsolete EH field equation but by the much simpler and more powerful Theorem of Orbits:

$$n(r)r = \frac{r}{m(r)} = \int dr = r + \mu. \qquad (20.33)$$

In a spherically symmetric space-time this theorem gives the line element:

$$ds^2 = -\left(1 + \frac{\mu}{r}\right)c^2 dt^2 + \left(1 + \frac{\mu}{r}\right)^{-1} dr^2 + r^2 d\Omega^2 \qquad (20.34)$$

in spherical polar co-ordinates. If the symmetric or Christoffel connection [1–11]:

$$\Gamma^\kappa_{\mu\nu} = \Gamma^\kappa_{\nu\mu} \qquad (20.35)$$

is assumed, the following results are obtained from the line element (20.34) by computer algebra:

$$\left.\begin{aligned} R^0{}_1{}^{10} + R^0{}_2{}^{20} + R^0{}_3{}^{30} &= 0 \\ \tilde{R}^0{}_1{}^{10} + \tilde{R}^0{}_2{}^{20} + \tilde{R}^0{}_3{}^{30} &= 0 \\ T^{010} + T^{020} + T^{030} &= 0 \\ \tilde{T}^{010} + \tilde{T}^{020} + \tilde{T}^{030} &= 0. \end{aligned}\right\} \quad (20.36)$$

Therefore the Christoffel connection of Riemann geometry without torsion cannot be used to describe orbits or dynamics in general, because this connection produces zero force and zero mass density, i.e. a universe without matter and without energy- momentum density. That is the basic problem of the EH field equation. The correct description of relativistic dynamics requires an asymmetric connection, a major advance in understanding given by ECE theory.

The spin torsion laws of relativistic dynamics are obtained by the index choice:

$$\kappa = 1, 2, 3 \quad (20.37)$$

and these laws are the direct analogues of the Ampère Maxwell law of ECE theory:

$$\nabla \times \boldsymbol{B} - \frac{1}{c^2} \frac{\partial \boldsymbol{E}}{\partial t} = \mu_0 \boldsymbol{J} \quad (20.38)$$

and its Hodge dual law, the Faraday law of induction of ECE theory:

$$\nabla \times \boldsymbol{E} + \frac{\partial \boldsymbol{B}}{\partial t} = \boldsymbol{0}. \quad (20.39)$$

Here μ_0 is vacuum permeability and $\boldsymbol{J}$ is current density. In ECE theory the components appearing in these laws are as in Table 1.

So by Eq. (20.19) there are two spin torsion laws of relativistic dynamics, one being the Hodge dual of the other. The spin torsion law corresponding to the Faraday law of induction is:

$$\nabla \times \widetilde{\boldsymbol{T}}_1 + \frac{\partial \widetilde{\boldsymbol{T}}_2}{\partial t} = \boldsymbol{0} \quad (20.40)$$

where:

$$\widetilde{\boldsymbol{T}}_1 = \widetilde{T}^{332}\boldsymbol{i} + \widetilde{T}^{113}\boldsymbol{j} + \widetilde{T}^{221}\boldsymbol{k}. \quad (20.41)$$

Table 20.1 Components in the ECE Laws of Electrodynamics

Law	Electric Field Strength	Magnetic Flux Density	Type
Gauss		$B^{001}, B^{002}, B^{003}$	Orbital
Coulomb	$E^{010}, E^{020}, E^{030}$		Orbital
Faraday	$E^{332}, E^{113}, E^{221}$	$B^{101}, B^{202}, B^{303}$	Spin
Ampère Maxwell	$E^{110}, E^{220}, E^{330}$	$B^{332}, B^{113}, B^{221}$	Spin

and

$$\tilde{\boldsymbol{T}}_2 = \tilde{T}^{101}\boldsymbol{i} + \tilde{T}^{202}\boldsymbol{j} + \tilde{T}^{303}\boldsymbol{k}. \tag{20.42}$$

The spin law corresponding to the Ampère Maxwell law is

$$\nabla \times \boldsymbol{T}_1 - \frac{\partial \boldsymbol{T}_2}{\partial t} = \boldsymbol{J}_m \tag{20.43}$$

where

$$\boldsymbol{T}_1 = T^{332}\boldsymbol{i} + T^{113}\boldsymbol{j} + T^{221}\boldsymbol{k} \tag{20.44}$$

and

$$\boldsymbol{T}_2 = T^{101}\boldsymbol{i} + T^{202}\boldsymbol{j} + T^{303}\boldsymbol{k}. \tag{20.45}$$

The current density $\boldsymbol{J}_m$ is defined in Cartesian coordinates by:

$$\boldsymbol{J}_m = J_X\boldsymbol{i} + J_Y\boldsymbol{j} + J_Z\boldsymbol{k} \tag{20.46}$$

where

$$J_X = R^0{}_1{}^{01} + R^1{}_2{}^{22} + R^1{}_3{}^{31} \tag{20.47}$$

$$J_Y = R^2{}_0{}^{02} + R^2{}_1{}^{12} + R^2{}_3{}^{32} \tag{20.48}$$

20.2 The Equations of Relativistic Dynamics

$$J_Z = R^3{}_0{}^{03} + R^3{}_1{}^{13} + R^3{}_2{}^{23}. \tag{20.49}$$

Eq. (20.40) is the gravitational equivalent of the Faraday law of induction observed experimentally [1–10] in a recent European Space Agency cooperative experiment with the Austrian group of Tajmar et al.

The Euler equation of motion is given a relativistic meaning in ECE theory by considering the first Cartan structure equation:

$$T^a = d \wedge q^a + \omega^a{}_b \wedge q^b \tag{20.50}$$

where $\omega^a{}_b$ is the spin connection form [1–11]. The torsion form in Eq. (20.50) is defined by a frame of reference which is itself moving. If the spin connection were not considered the torsion would be due entirely to:

$$T^a(\text{static}) = d \wedge q^a \tag{20.51}$$

in a static frame. Therefore the first Cartan structure equation can be re-written as:

$$T^a = T^a(\text{static}) + \omega^a{}_b \wedge q^b. \tag{20.52}$$

The Euler equation of classical dynamics can be written [12] as:

$$\boldsymbol{T}q = \boldsymbol{T}q(\text{static}) - \boldsymbol{\Omega} \times \boldsymbol{L} \tag{20.53}$$

where $\boldsymbol{T}q$ denotes torque, $\boldsymbol{\Omega}$ denotes angular velocity and $\boldsymbol{L}$ denotes angular momentum. Thus $\boldsymbol{\Omega} \times \boldsymbol{L}$ causes the precession of for example a gyroscope. The Newtonian or inertial frame definition of torque is:

$$\boldsymbol{T}q(\text{static}) = \frac{\partial \boldsymbol{L}}{\partial t} \tag{20.54}$$

so gyroscope type precession is not present in Newtonian dynamics and was first inferred by Euler. In ECE theory it is inferred in analogy with force (Eq. (20.27)) that torque is due to spin torsion, (force as argued being due to orbital torsion). Thus by unit analysis:

$$\boldsymbol{T}q = E_0 \int \boldsymbol{T}(\text{spin}) dr. \tag{20.55}$$

In tensor notation, the first Cartan structure equation is:

$$T^a_{\mu\nu} = \partial_\mu q^a_\nu - \partial_\nu q^a_\mu + \omega^a_{\mu b} q^b_\nu - \omega^a_{\nu b} q^b_\mu. \tag{20.56}$$

Considering the spin torsion component T_1 it is seen that in vector notation that it is defined by:

$$T_1 = \nabla \times q - \omega \times q \qquad (20.57)$$

which is the direct analogy of the ECE relation:

$$B = \nabla \times A - \omega \times A \qquad (20.58)$$

where ω is the spin connection vector [1–10]. Thus it is inferred that:

$$Tq(\text{moving}) = E_0 \int T_1 \, dr \qquad (20.59)$$

$$Tq(\text{fixed}) = E_0 \int \nabla \times q \, dr \qquad (20.60)$$

and

$$\Omega \times L = E_0 \int \omega \times q \, dr. \qquad (20.61)$$

Therefore the Euler equation can be derived from the first Cartan structure equation in the limit of small but non-zero spin connection. It is also inferred from EEC theory that the structure of the Euler equation is correctly covariant, it retains its form under the general coordinate transformation. Gyroscope precession is due to spinning space-time.

Therefore ECE theory has been used to infer the correctly objective equations of translational and rotational motion in relativistic dynamics. The Theorem of Orbits and the spherical symmetry of space-time describe all known orbits.

Acknowledgments

The British Government is thanked for the award of a Civil List Pension and Armorial Ensigns for pre-eminent contributions to Britain in science, and the staffs of AIAS and SG and many others for interesting internet discussions.

References

[1] M. W. Evans, "Generally Covariant Unified Field Theory" (Abramis, 2005 onwards), four volumes, fifth and sixth in preparation (papers 71 onwards on www.aias.us).

[2] L. Felker, "The Evans Equations of Unified Field Theory" (Abramis 2007).

[3] K. Pendergast, "Crystal Spheres" (preprint on www.aias.us).

[4] Educational Articles on www.aias.us, notably by Horst Eckardt.

[5] M. W. Evans, Omnia Opera section of www.aias.us from 1992 onwards, hyperlinked to original articles.

[6] M. W. Evans, Adv. Chem. Phys., vol. 119 (2001); M. W Evans and S. Kielich, Adv. Chem. Phys., vol. 85 (1992, 1993, 1997).

[7] M. W. Evans and L. B. Crowell, "Classical and Quantum Electrodynamics and the B(3) Field" (World Scientific, 2001).

[8] M. W .Evans and J.-P. Vigier, "The Enigmatic Photon" (Kluwer, 1994 to 2002, hardback and softback), in five volumes.

[9] M. W. Evans, fifteen ECE papers in Found. Phys. Lett. (2003 to 2005).

[10] M. W. Evans, Acta Phys. Polon., **38B**, 2211 (2007); M. W. Evans and H. Eckardt, Physica B, **400**, 175 (2007), M. W. Evans, Physica B, **403**, 517 (2008).

[11] S. P. Carroll, "Space-time and Geometry: an Introduction to General Relativity" (Addison Wesley, New York, 2004, and 1997 downloadable lecture notes).

[12] J. B. Marion and S. T. Thornton, "Classical Dynamics" (HBJ, New York, 1988, 3rd. Ed.)

Derivation of the Gravitational Red Shift from the Theorem of Orbits

by

Myron W. Evans,
Alpha Institute for Advanced Study, Civil List Scientist.
(emyrone@aol.com and www.aias.us)

Abstract

The experimentally observable gravitational red shift is derived by rotating the line element derived from the Theorem of Orbits. The latter is a simple special case of the Frobenius Theorem for a spherically symmetric space-time. All known orbits are described by the geometry of the Theorem of Orbits, and the gravitational red shift is shown to be the precession or phase shift caused by rotating the line element of the Theorem of Orbits.

Keywords: ECE Theory, gravitational red shift, Theorem of Orbits, line element rotation.

21.1 Introduction

Recently in the ECE series of papers [1–10] it has been shown that all known orbits can be described directly by the spherical symmetry of space-time with torsion and curvature without having to use any field equation a priori. The Theorem of Orbits (paper 111) has been derived from the well known [11] Frobenius Theorem applied to a spherically symmetric space-time. From the Theorem of Orbits the line element is derived, giving the orbital equation. Therefore the field of force (which becomes the Newtonian field of force in the appropriate limit) is derived directly from spherical space-time symmetry. This procedure is summarized in Section 21.2, and in Section 21.3 the well known gravitational red shift is given a new meaning by deriving it from rotation of the line element of Section 21.2. It is found that the gravitational red shift is a precession or phase shift - essentially a property purely of spherical space-time and not of any field equation. In the standard model the gravitational red shift is thought to be a wavelength change and incorrectly derived from a space-time that has no torsion.

21.2 Line Element and Orbital Equation from Theorem of Orbits

The Theorem of Orbits is a simple example of the Frobenius Theorem [11] which defines the most general line element. The Theorem of Orbits is:

$$nr = \frac{r}{m} = \int dr = r + \mu \tag{21.1}$$

where n and m are functions of r, the radial coordinate in spherical polar coordinates. The constant of integration is in general non-zero, and goes to zero in a Minkowski space-time. If the Frobenius Theorem is applied [11] to a spherically symmetric space-time the line element is:

$$ds^2 = -n(r)c^2 dt^2 + m(r)dr^2 + r^2 d\Omega^2. \tag{21.2}$$

From the Theorem of Orbits it is found that:

$$n = 1 + \frac{\mu}{r}, \tag{21.3}$$

$$m = \left(1 + \frac{\mu}{r}\right)^{-1}, \tag{21.4}$$

so that the line element becomes:

$$ds^2 = -\left(1 + \frac{\mu}{r}\right)c^2 dt^2 + \left(1 + \frac{\mu}{r}\right)^{-1} dr^2 + r^2 d\Omega^2 \tag{21.5}$$

in spherical polar co-ordinates.

The orbital equation is obtained by considering the special case of orbits in a plane, so the line element (21.5) reduces to:

$$ds^2 = -\left(1 + \frac{\mu}{r}\right)c^2 dt^2 + \left(1 + \frac{\mu}{r}\right)^{-1} dr^2 + r^2 d\phi^2. \tag{21.6}$$

21.2 Line Element and Orbital Equation from Theorem of Orbits

Define [1–11] the constant of motion:

$$-\epsilon = -\left(\frac{ds}{d\lambda}\right)^2 = -c^2\left(\frac{d\tau}{d\lambda}\right)^2$$

$$= -\left(1+\frac{\mu}{r}\right)c^2\left(\frac{dt}{d\lambda}\right)^2 + \left(1+\frac{\mu}{r}\right)^{-1}\left(\frac{dr}{d\lambda}\right)^2 + r^2\left(\frac{d\phi}{d\lambda}\right)^2 \quad (21.7)$$

where $d\tau$ is the infinitesimal element of proper time. Now make the choice:

$$\lambda = \tau \quad (21.8)$$

to find:

$$-c^2 = -\left(1+\frac{\mu}{r}\right)c^2\left(\frac{dt}{d\tau}\right)^2 + \left(1+\frac{\mu}{r}\right)^{-1}\left(\frac{dr}{d\tau}\right)^2 + r^2\left(\frac{d\phi}{d\tau}\right)^2. \quad (21.9)$$

To convert to S.I. units multiply throughout by $\frac{1}{2}m$, where m is to be determined:

$$\frac{1}{2}mr^2\left(\frac{d\phi}{d\tau}\right)^2 - \frac{1}{2}m\left(1+\frac{\mu}{r}\right)c^2\left(\frac{dt}{d\tau}\right)^2 + \frac{1}{2}m\left(1+\frac{\mu}{r}\right)^{-1}\left(\frac{dr}{d\tau}\right)^2$$

$$= -\frac{1}{2}mc^2. \quad (21.10)$$

Multiply through by $\left(1+\frac{\mu}{r}\right)$: to find that:

$$\frac{1}{2}mr^2\left(\frac{d\phi}{d\tau}\right)^2\left(1+\frac{\mu}{r}\right) - \frac{1}{2}m\left(1+\frac{\mu}{r}\right)^2 c^2\left(\frac{dt}{d\tau}\right)^2 + \frac{1}{2}m\left(\frac{dr}{d\tau}\right)^2$$

$$= -\frac{1}{2}mc^2\left(1+\frac{\mu}{r}\right). \quad (21.11)$$

This is the orbital equation:

$$\frac{1}{2}m\left(\frac{dr}{d\tau}\right)^2 + V = E. \quad (21.12)$$

The total energy in S.I. units is:

$$E = \frac{1}{2}mc^2\left(1+\frac{\mu}{r}\right)^2\left(\frac{dt}{d\tau}\right)^2. \quad (21.13)$$

The potential energy in S.I. units is:

$$V = \frac{1}{2}m\left(1 + \frac{\mu}{r}\right)\left(c^2 + \frac{L^2}{r^2}\right) \tag{21.14}$$

where:

$$L = r^2 \frac{d\phi}{d\tau} \tag{21.15}$$

is a constant of motion having the units of angular momentum per unit mass. The factor $\frac{1}{2}$ is introduced [11] to write the equation in standard dynamical form. The potential energy is therefore:

$$V = \frac{1}{2}mc^2 + \frac{1}{2}mc^2\frac{\mu}{r} + \frac{1}{2}m\frac{L^2}{r^2} + \frac{1}{2}\frac{mL^2\mu}{r^3} \tag{21.16}$$

and is made up of four terms which are identified below. For all orbits excluding binary pulsars and the Cassini/Pioneer anomaly it is found by experimental observation that:

$$\mu = -\frac{2mG}{c^2}. \tag{21.17}$$

Therefore the potential energy becomes:

$$V = \frac{1}{2}mc^2 - m\frac{mG}{r} + \frac{1}{2}\frac{mL^2}{r^2} - \frac{L^2 mMG}{c^2 r^3}. \tag{21.18}$$

Therefore it becomes possible to identify the four terms as follows.

1) A constant term proportional to rest energy, $\frac{1}{2}mc^2$.

2) The Newtonian potential of attraction, $-mMG/r$.

3) The centripetal repulsion, $mL^2/(2r^2)$.

4) The relativistic correction to the Newtonian attraction, $-L^2 mMG/(c^2 r^3)$.

Therefore the factor m is the mass of an object attracted by an object of mass M. The Theorem of Orbits (21.1) is the "geometrical control" over the way m and M interact. The introduction of m, M and G introduces physics into pure geometry.

The Newtonian limit is defined by:

$$r \to \infty \tag{21.19}$$

21.2 Line Element and Orbital Equation from Theorem of Orbits

when the familiar Newtonian terms (21.2) and (21.3) dominate. The Newtonain force of attraction is:

$$F = -\frac{\partial V}{\partial r} = -\frac{mMG}{r^2} \tag{21.20}$$

which is the inverse square law of Newton. From Eqs (21.18) and (21.20) the total force between m and M is:

$$F = -\frac{mMG}{r^2} + \frac{mL^2}{r^3} - \frac{3L^2 mMG}{c^2 r^4}. \tag{21.21}$$

This force law describes the vast majority of known orbits with great accuracy. It describes perihelion advance, deflection of light by gravity, frame dragging, Shapiro time delay and all the phenomena incorrectly attributed in the standard model to the now obsolete [1–10] Einstein field equation. As argued, these phenomena are due purely to the spherical symmetry of space-time. The masses m and M are introduced following experimental observation. The Newtonian force is Eq. (21.20). Newton did not realize the existence of the centripetal force, and of course did not realize the existence of the relativistic correction.

In binary pulsars (paper 108) the orbits decrease by a few millimeters per revolution. This effect is described by the addition of a very small perturbation as follows:

$$\mu = -\left(\frac{2mG}{c^2} + \frac{a}{r}\right) \tag{21.22}$$

which generates an additional attraction potential:

$$\Delta V = -\frac{1}{2}ma\left(\frac{c^2}{r^2} + \frac{L^2}{c^2 r^4}\right) \tag{21.23}$$

and an additional force of attraction:

$$\Delta F = -ma\left(\frac{c^2}{r^3} + \frac{2L^2}{c^2 r^5}\right) \tag{21.24}$$

which causes the two objects of a binary pulsar to spiral in towards each other from a relativistic orbit whose perihelion advance is a few degrees per revolution. This is a very small effect, but reproducible and repeatable. The same type of phenomenon is found in the solar system in the well known Pioneer/Cassini anomalies. Both spacecraft see a tiny additional force of attraction not present in Eq. (21.21). The complete force law for binary pulsars and the Pioneer Cassini orbits is therefore:

$$F = -\frac{mMG}{r^2} + \frac{m}{r^3}(L^2 - ac^2) - \frac{3LmMG}{c^2 r^4} - \frac{2amL^2}{c^2 r^5}, \quad (21.25)$$

the Newtonian force in this case being only one of five terms.

21.3 The Gravitational Red Shift

Consider the line element (21.5) in cylindrical polar co-ordinates:

$$-ds^2 = \left(1 + \frac{\mu}{r}\right)c^2 dt^2 - \left(1 + \frac{\mu}{r}\right)^{-1} dr^2 - r^2 d\phi^2 - dZ^2. \quad (21.26)$$

Now rotate it (see paper 110) at an angular velocity ω as follows:

$$\phi' = \phi + \omega t. \quad (21.27)$$

The rotated line element is therefore:

$$-ds'^2 = \left(1 + \frac{\mu}{r}\right)c^2 dt^2 - \left(1 + \frac{\mu}{r}\right)^{-1} dr^2 - r^2 d\phi'^2 - dZ^2 \quad (21.28)$$

where:

$$d\phi' = d\phi + \omega dt \quad (21.29)$$

and

$$d\phi'^2 = d\phi^2 + 2\omega d\phi dt + \omega^2 dt^2. \quad (21.30)$$

It is found that:

$$-ds'^2 = \left(1 + \frac{\mu}{r} - \frac{v^2}{c^2}\right)(c^2 dt^2 - 2r^2 \Omega d\phi dt) \\ - \left(1 + \frac{\mu}{r}\right)^{-1} dr^2 - r^2 d\phi^2 - dZ^2 \quad (21.31)$$

where the orbital linear velocity of rotation is defined by

$$v = r\omega. \quad (21.32)$$

21.3 The Gravitational Red Shift

Identify the relativistic angular velocity (compare paper 110 on www.aias.us) as:

$$\Omega = \omega \left(1 + \frac{\mu}{r} - \frac{v^2}{c^2}\right)^{-1} \tag{21.33}$$

and the infinitesimal of proper time by:

$$d\tau = \left(1 + \frac{\mu}{r} - \frac{v^2}{c^2}\right)^{\frac{1}{2}} dt. \tag{21.34}$$

The change of phase, or precession, upon rotating by 2π radians is:

$$\alpha = \Omega d\tau - \omega dt = 2\pi \left(\left(1 + \frac{\mu}{r} - \frac{v^2}{c^2}\right)^{-\frac{1}{2}} - 1\right). \tag{21.35}$$

The limit

$$r \to \infty \tag{21.36}$$

defines the Thomas precession (paper 110):

$$\alpha(\text{Thomas}) = 2\pi \left(\left(1 - \frac{v^2}{c^2}\right)^{-\frac{1}{2}} - 1\right) \tag{21.37}$$

and the limit:

$$v \to 0 \tag{21.38}$$

defines the gravitational red shift:

$$\alpha(\text{grav}) = 2\pi \left(\left(1 + \frac{\mu}{r}\right)^{-\frac{1}{2}} - 1\right). \tag{21.39}$$

For almost all orbits, as argued in Section 21.2, it is found by experimental observation that:

$$\mu = -\frac{2mG}{c^2} \tag{21.40}$$

so the gravitational red shift is:

$$\alpha(\text{grav}) = 2\pi \left(\left(1 - \frac{2mG}{c^2 r}\right)^{-\frac{1}{2}} - 1 \right) \sim \frac{2\pi mG}{c^2 r}. \tag{21.41}$$

as observed experimentally as is well known. The Thomas precession is well observed experimentally in atomic and molecular spectra in spin orbit coupling.

It is concluded that both the gravitational red shift and the Thomas precession are due purely to the spherical symmetry of space-time. The standard model's Einstein field equation is known to be geometrically incorrect because of its neglect of torsion, so the standard explanation of the gravitational red shift cannot be correct. Similarly, the standard model's cosmological red shift (which is different from the gravitational red shift) is an artifact based on the Roberston Walker metric, which in paper 93 on www.aias.us was shown to violate basic geometry (the Hodge dual of the Bianchi identity).

Acknowledgments

The British Government is thanked for the award of a Civil List Pension and armorial ensigns for pre-eminent contributions to Britain and the Commonwealth in science and the staffs of AIAS and many other colleagues worldwide thanked for interesting discussions.

References

[1] M. W. Evans, "Generally Covariant Unified Field Theory" (Abramis 2005 onwards), volumes one to four, volumes five and six in prep. (papers 71 onwards on www.aias.us).

[2] L. Felker, "The Evans Equations of Unified Field Theory" (Abramis, 2007).

[3] K. Pendergast, "Crystal Spheres" (preprint on www.aias.us).

[4] M. W. Evans, Omnia Opera section of www.aias.us from 1992 onwards.

[5] M. W. Evans (ed.), Adv. Chem. Phys, vol. 119 (2001); M. W. Evans and S. Kielich (eds.), Adv. Chem. Phys., vol. 85 (1992, 1993, 1997).

[6] M. W. Evans and L. B. Crowell, "Classical and Quantum Electrodynamics and the B(21.3) Field" (World Scientific, 2001).

[7] M. W. Evans and J.-P. Vigier, "The Enigmatic Photon" (Kluwer, 1994 to 2002, hardback and softback), in five volumes.

[8] M. W. Evans, fifteen papers in Found. Phys. Lett., 2003 to 2005.

[9] M. W. Evans, Acta Phys. Polon., **38B**, 2211 (2007); Physica B, **403**, 517 2008).

[10] M. W. Evans and H. Eckardt, Physica B, **400**, 175 (2007).

[11] S. P. Carroll, "Space-time and Geometry: an Introduction to General Relativity" (Addison Wesley, New York, 2004, and downloadable lecture notes), chapter 7.

Invariance, Covariance and Duality Properties of the ECE Laws of Dynamics and Electrodynamics

by

Myron W. Evans,
Alpha Institute for Advanced Study, Civil List Scientist.
(emyrone@aol.com and www.aias.us)

Abstract

The ECE laws of classical dynamics and electrodynamics are based on the Bianchi identity and their duality, covariance and invariance properties in general relativity are also based directly on Cartan geometry. It is shown that the vacuum equations are generally invariant, and that the field matter equations are generally covariant. The structure of the components of the electric and magnetic fields is also determined by the Bianchi identity, and there exist Hodge duals of both the vacuum and field matter equations. In general the equations transform according to the rules of Cartan geometry. The general coordinate transformation becomes the Lorentz transformation in the limit of Minkowski space-time.

Keywords: ECE laws of dynamics and electrodynamics, invariance, covariance, duality.

22.1 Introduction

Recently [1–10] the generally covariant laws of dynamics and electrodynamics have been developed with Cartan geometry [11]. It has been found that they are based on the Bianchi identity of Cartan geometry and the Hodge dual identity. The laws of classical electrodynamics are the same in overall format as the familiar Maxwell Heaviside (MH) field equations of special relativity, but the Einstein Cartan Evans (ECE) laws are written in a space-time with torsion and curvature, and give more information, notably they include the spin connection of space-time. The ECE laws of dynamics are the same in structure as those of electrodynamics, so there are four laws of generally covariant dynamics, akin to the Gauss, Faraday, Coulomb and Ampère-Maxwell laws of classical ECE electrodynamics. In both subject areas there are two orbital laws (Gauss and Coulomb) and two spin laws (Faraday and Ampère Maxwell). The Gauss and Faraday laws refer to the free field or vacuum field and the Coulomb and Ampère Maxwell laws refer to field matter interaction. In Section 22.2, the properties of these laws under coordinate transformation are developed using the methods [1–11] of geometry. It is found that the free field or vacuum laws are invariant under coordinate transformation, and that the field - matter laws are covariant under coordinate transformation. In Section 22.3 the Hodge duality properties of the laws are given in the vacuum and in the presence of field matter interaction.

22.2 Invariance and Covariance

In previous work [1–10] it has been shown that the homogeneous laws of dynamics and electrodynamics are based on the geometrical structure:

$$D_\mu \widetilde{T}^{\kappa\mu\nu} = \widetilde{R}^\kappa{}_\mu{}^{\mu\nu} \tag{22.1}$$

where $\widetilde{T}^{\kappa\mu\nu}$ is a rank three torsion tensor and where $\widetilde{R}^\kappa{}_\mu{}^{\mu\nu}$ is a rank four curvature tensor. The D_μ denotes covariant derivative in a space-time with torsion and curvature [11]. The inhomogeneous laws are based on the Hodge dual of Eq. (22.1):

$$D_\mu T^{\kappa\mu\nu} = R^\kappa{}_\mu{}^{\mu\nu}. \tag{22.2}$$

These equations can be rearranged to give:

$$\partial_\mu \widetilde{T}^{\kappa\mu\nu} = \widetilde{j}^{\kappa\nu} \tag{22.3}$$

and

$$\partial_\mu T^{\kappa\mu\nu} = j^{\kappa\nu} \tag{22.4}$$

where the right hand side terms include the spin connection in their structure. It has been shown that Eqs. (22.3) and (22.4) give the equations of dynamics in generally covariant unified field theory [1–10]. If a primordial voltage is defined by:

$$\phi = cA^{(0)} \tag{22.5}$$

where c is the vacuum speed of light, the equations of classical electrodynamics are given by the fundamental hypothesis:

$$F^{\kappa\mu\nu} = A^{(0)} T^{\kappa\mu\nu} \qquad (22.6)$$

which defines the electromagnetic field tensor in terms of the torsion tensor. The ECE equations of electrodynamics are therefore:

$$\partial_\mu \widetilde{F}^{\kappa\mu\nu} = A^{(0)} j^{\kappa\nu}, \qquad (22.7)$$
$$\partial_\mu F^{\kappa\mu\nu} = A^{(0)} \widetilde{j}^{\kappa\nu'}. \qquad (22.8)$$

Experimentally it is found that:

$$\partial_\mu \widetilde{F}^{\kappa\mu\nu} = 0, \qquad (22.9)$$
$$\partial_\mu F^{\kappa\mu\nu} = A^{(0)} j^{\kappa\nu}, \qquad (22.10)$$

because a magnetic monopole (part of $j^{\kappa\nu}$) has never been observed. There have been some claims to magnetic monopole observation, but these claims appear not to be reproducible. For all practical purposes therefore the ECE laws of classical electrodynamics are:

$$\partial_\mu \widetilde{F}^{\kappa\mu\nu} = 0, \qquad (22.11)$$
$$\partial_\mu F^{\kappa\mu\nu} = J^{\kappa\nu}/\epsilon_0. \qquad (22.12)$$

For comparison, the MH laws are well known to be:

$$\partial_\mu \widetilde{F}^{\mu\nu} = 0, \qquad (22.13)$$
$$\partial_\mu F^{\mu\nu} = J^\nu/\epsilon_0. \qquad (22.14)$$

The main differences between the ECE and MH laws include the fact that the latter are restricted to Minkowski space-time, whereas the former are written in a space-time with torsion and curvature present and are part of a generally covariant unified field. In ECE theory the electromagnetic field is the correct rank three tensor related [1–10] to canonical angular momentum/energy density as required in general relativity. In MH theory the electromagnetic field is a rank two tensor which is an integral [12–14] over a rank three density.

The concept of vacuum electromagnetic field is used routinely in classical electrodynamics and is defined as the field propagating infinitely far from its source. This concept is a mathematical limit, not a physical reality, because without a source there is no field. However, if we transfer this concept of the received view to ECE theory the vacuum field is defined by:

$$\partial_\mu \widetilde{F}^{\kappa\mu\nu} = 0 \qquad (22.15)$$

$$\partial_\mu F^{\kappa\mu\nu} = 0 \qquad (22.16)$$

and in vector notation the ECE vacuum equations of electrodynamics are:

$$\nabla \cdot \boldsymbol{B} = 0. \qquad (22.17)$$

$$\nabla \times \boldsymbol{E} + \frac{\partial \boldsymbol{B}}{\partial t} = \boldsymbol{0}, \qquad (22.18)$$

$$\nabla \cdot \boldsymbol{E} = 0, \qquad (22.19)$$

$$\nabla \times \boldsymbol{B} - \frac{1}{c^2}\frac{\partial \boldsymbol{E}}{\partial t} = \boldsymbol{0}, \qquad (22.20)$$

in which $\boldsymbol{B}$ is the magnetic flux density in tesla and $\boldsymbol{E}$ is the electric field strength in volts per meter. Eqs. (22.17) and (22.19) are laws of orbital torsion [1–10] in which the field components are defined by:

$$\boldsymbol{B} = B^{001}\boldsymbol{i} + B^{002}\boldsymbol{j} + B^{003}\boldsymbol{k} \qquad (22.21)$$

and

$$\boldsymbol{E} = E^{010}\boldsymbol{i} + E^{020}\boldsymbol{j} + E^{030}\boldsymbol{k} \qquad (22.22)$$

respectively. Eq. (22.18) is a law of spin torsion in which:

$$\boldsymbol{E} = E^{332}\boldsymbol{i} + E^{113}\boldsymbol{j} + E^{221}\boldsymbol{k} \qquad (22.23)$$

and:

$$\boldsymbol{B} = B^{101}\boldsymbol{i} + B^{202}\boldsymbol{j} + B^{303}\boldsymbol{k} \qquad (22.24)$$

and Eq. (22.20) is a law of spin torsion in which:

$$\boldsymbol{E} = E^{110}\boldsymbol{i} + E^{220}\boldsymbol{j} + E^{330}\boldsymbol{k} \qquad (22.25)$$

and

$$\boldsymbol{B} = B^{332}\boldsymbol{i} + B^{113}\boldsymbol{j} + B^{221}\boldsymbol{k}. \qquad (22.26)$$

The right hand sides of Eqs. (22.15) and (22.16) are null rank two tensors. If the latter is denoted by:

$$X^{\mu\nu} = 0 \qquad (22.27)$$

22.2 Invariance and Covariance

it transforms as another null tensor:

$$X'^{\mu\nu} = \frac{\partial x^{\mu'}}{\partial x^{\mu}} \frac{\partial x^{\nu'}}{\partial x^{\nu}} X^{\mu\nu} = 0 \qquad (22.28)$$

where x^{μ} is the coordinate four-vector [1–11]. Therefore this property is necessary and sufficient to show that the ECE vacuum equations are generally invariant. This means that the equations are the same under arbitrary coordinate transform from a frame K to K'. In contrast the vacuum MH equations are invariant only under the Lorentz transform from frame K to K'. In other words the ECE equations are those of a generally covariant unified field, and the MH equations are those of Lorentz covariant and un-unified field. Note carefully that the electromagnetic field tensor itself transforms as a rank three tensor as follows [1–11]:

$$T^{\kappa\mu\nu'} = \left(\frac{\partial x^{\kappa'}}{\partial x^{\kappa}}\right)\left(\frac{\partial x^{\mu'}}{\partial x^{\nu}}\right)\left(\frac{\partial x^{\nu'}}{\partial x^{\nu}}\right) T^{\kappa\mu\nu} \qquad (22.29)$$

and is not invariant. It is well known that the MH field tensor transform as a rank two tensor using two Lorentz transform matrices. They are transformed from frame K to a frame K' moving at a constant velocity v with respect to K. When:

$$v/c \ll 1 \qquad (22.30)$$

this type of transform produces:

$$\boldsymbol{E}' = \boldsymbol{E} + \boldsymbol{v} \times \boldsymbol{B} \qquad (22.31)$$

and

$$\boldsymbol{B}' = \boldsymbol{B} - \frac{1}{c^2} \boldsymbol{v} \times \boldsymbol{E}, \qquad (22.32)$$

and the Lorentz force law is:

$$\boldsymbol{F} = e(\boldsymbol{E} + \boldsymbol{v} \times \boldsymbol{B}) \qquad (22.33)$$

where e is a charge. So the transform of frames has a physical effect. Similarly in ECE theory the general transform (22.29) will have a physical effect. Therefore coordinate transform of the vacuum ECE field equations leaves them invariant, but the fields themselves are changed.

In order to develop the concept of coordinate transform the general rule for any tensor is [11]:

$$T^{\mu'_1\ldots\mu'_\kappa}_{\nu'_1\ldots\nu'_\ell} = \left(\frac{\partial x^{\mu'_1}}{\partial x^{\mu_1}}\ldots\frac{\partial x^{\mu'_\kappa}}{\partial x^{\mu_\kappa}}\right)\left(\frac{\partial x^{\nu_1}}{\partial x^{\nu'_1}}\ldots\frac{\partial x^{\nu_\ell}}{\partial x^{\nu'_\ell}}\right) T^{\mu_1\ldots\mu_\kappa}_{\nu_1\ldots\nu_\ell} \tag{22.34}$$

For a mixed index tensor in Cartan geometry [1–10]:

$$T^{a'\mu'}_{b'\nu'} = \Lambda^{a'}_a \frac{\partial x^{\mu'}}{\partial x^\mu} \Lambda^b_{b'} \frac{\partial x^\nu}{\partial x^{\nu'}} T^{a\mu}_{b\nu} \tag{22.35}$$

where $\Lambda^{a'}_a$ denotes Lorentz transform of the Minkowski tangent space-time at point P to the base manifold. The partial derivative transforms as:

$$\partial_{\mu'} = \frac{\partial x^\mu}{\partial x^{\mu'}} \partial_\mu. \tag{22.36}$$

The covariant derivative of a vector transforms as:

$$(D_\mu V^\nu)' = \frac{\partial x^\mu}{\partial x^{\mu'}} \frac{\partial x^{\nu'}}{\partial x^\nu} D_\mu V^\nu \tag{22.37}$$

because by definition D_μ is covariant, whereas ∂_μ acting on a tensor produces extra terms. For example the homogeneous MH equation transforms using the Leibniz theorem as:

$$(\partial_\mu \tilde{F}^{\mu\nu})' = \Lambda^{\nu'}_\nu \partial_\mu \tilde{F}^{\mu\nu} + \Lambda^\mu_{\mu'} \tilde{F}^{\mu\nu} \partial_\mu (\Lambda^{\mu'}_\mu \Lambda^{\nu'}_\nu) \tag{22.38}$$

and because of the second term on the right hand side the homogeneous MH equation would appear not to transform covariantly. However, using Eq. (22.28) it is known that:

$$(\partial_\mu \tilde{F}^{\mu\nu})' = \partial_\mu \tilde{F}^{\mu\nu} = 0 \tag{22.39}$$

so the homogeneous MH equation is frame invariant.

Similarly the homogeneous ECE equation transforms as:

$$(\partial_\mu \tilde{F}^{\kappa\mu\nu})' = \frac{\partial x^{\nu'}}{\partial x^\nu} \frac{\partial x^{\kappa'}}{\partial x^\kappa} \partial_\mu \tilde{F}^{\kappa\mu\nu} + \tilde{F}^{\kappa\mu\nu} \partial_\mu \left(\frac{\partial x^{\mu'}}{\partial x^\mu} \frac{\partial x^{\nu'}}{\partial x^\nu} \frac{\partial x^{\kappa'}}{\partial x^\kappa}\right) \tag{22.40}$$

and from Eq. (22.15):

$$(\partial_\mu \tilde{F}^{\kappa\mu\nu})' = \partial_\mu \tilde{F}^{\kappa\mu\nu} = 0. \tag{22.41}$$

22.2 Invariance and Covariance

For example, consider a rotation of the ECE Gauss law about the Z axis and without loss of generality assume that:

$$\kappa' = \kappa = 0 \tag{22.42}$$

so

$$\frac{\partial x^{\kappa'}}{\partial x^{\kappa}} = 1. \tag{22.43}$$

For rotation about the Z axis:

$$\frac{\partial x^{\nu'}}{\partial x^{\nu}} = \begin{bmatrix} 1 & 0 & 0 & 0 \\ 0 & \cos\theta & \sin\theta & 0 \\ 0 & -\sin\theta & \cos\theta & 0 \\ 0 & 0 & 0 & 0 \end{bmatrix}. \tag{22.44}$$

Therefore:

$$\begin{bmatrix} V^{0'} \\ V^{1'} \\ V^{2'} \\ V^{3'} \end{bmatrix} = \begin{bmatrix} 1 & 0 & 0 & 0 \\ 0 & \cos\theta & \sin\theta & 0 \\ 0 & -\sin\theta & \cos\theta & 0 \\ 0 & 0 & 0 & 0 \end{bmatrix} \begin{bmatrix} V^{0} \\ V^{1} \\ V^{2} \\ V^{3} \end{bmatrix} \tag{22.45}$$

which is equivalent to the rotation of a four-vector:

$$V^{0'} = V^0 \tag{22.46}$$
$$V^{1'} = V^0 \cos\theta + V^1 \sin\theta$$
$$V^{2'} = V^1 \sin\theta + V^2 \cos\theta$$
$$V^{3'} = V^3.$$

The time-like part of this vector is:

$$V^0 = \boldsymbol{\nabla} \cdot \boldsymbol{B}, \tag{22.47}$$

and the space-like part is:

$$\boldsymbol{V} = \boldsymbol{\nabla} \times \boldsymbol{E} + \frac{\partial \boldsymbol{B}}{\partial t}, \tag{22.48}$$

where the components are:

$$V^1 = V_X = \left(\boldsymbol{\nabla} \times \boldsymbol{E} + \frac{\partial \boldsymbol{B}}{\partial t}\right)_X \tag{22.49}$$

and so on. The structure of this vector is derived from the vector equivalent of Eq. (22.15):

$$\partial_\mu \widetilde{F}^{0\mu\nu} = 0 := V^\nu \qquad (22.50a)$$
$$\nabla \cdot \boldsymbol{B} = 0$$
$$\nabla \times \boldsymbol{E} + \frac{\partial \boldsymbol{B}}{\partial t} = 0 \qquad (22.50b)$$

and the vector is defined by:

$$V^\nu = \left(c\nabla \cdot \boldsymbol{B}, \quad \nabla \times \boldsymbol{E} + \frac{\partial \boldsymbol{B}}{\partial t} \right) = 0. \qquad (22.51)$$

The rotation of the column vector about the Z axis is therefore:

$$V^{\nu'} = \Lambda^{\nu'}{}_\nu V^\nu \qquad (22.52)$$

which gives the four equations:

$$(\nabla \cdot \boldsymbol{B})' = \nabla \cdot \boldsymbol{B} = 0 \qquad (22.53)$$
$$\left(\nabla \times \boldsymbol{E} + \frac{\partial \boldsymbol{B}}{\partial t} \right)'_X = \left(\nabla \times \boldsymbol{E} + \frac{\partial \boldsymbol{B}}{\partial t} \right)_X = 0 \qquad (22.54)$$

The rotation of the ECE Gauss law about the Z axis is given by Eq. (22.46a), from which:

$$(\nabla' \cdot \boldsymbol{B}' = 0) = (\nabla \cdot \boldsymbol{B} = 0). \qquad (22.55)$$

This result means that the ECE Gauss law is invariant under Z axis rotation, Q.E.D. This is an example of the fact that the ECE Gauss law is invariant under any type of transform from frame K to K'.

Similarly, Eq. (22.46b) gives the transform under Z axis rotation of the X component of the ECE Faraday law:

$$\left(\nabla \times \boldsymbol{E} + \frac{\partial \boldsymbol{B}}{\partial t} \right)'_X = c\nabla \cdot \boldsymbol{B} \cos\theta + \left(\nabla \times \boldsymbol{E} + \frac{\partial \boldsymbol{B}}{\partial t} \right)_X \sin\theta = 0. \qquad (22.56)$$

Using the K frame results:

$$c\nabla \cdot \boldsymbol{B} = \left(\nabla \times \boldsymbol{E} + \frac{\partial \boldsymbol{B}}{\partial t} \right)_X = 0 \qquad (22.57)$$

it is found that:

$$\left(\nabla \times \boldsymbol{E} + \frac{\partial \boldsymbol{B}}{\partial t}\right)'_X = 0 \qquad (22.58)$$

and Z axis rotation gives us the original equation again, QED. Similarly for the Y and Z components of the ECE Faraday law. The same result is true for the MH Gauss and Faraday laws, which are also invariant under Z axis rotation. As argued, the basic reason for the invariance is that a null tensor in frame K is a null tensor in frame K' (Eq. (22.28)).

Adopting differential form notation [1–11] the free space or vacuum ECE equations of classical electrodynamics are defined by:

$$d \wedge \widetilde{F}^a = 0 \qquad (22.59)$$

and

$$d \wedge F^a = 0. \qquad (22.60)$$

Therefore in the vacuum:

$$\widetilde{R}^a{}_b \wedge q^b = \omega^a{}_b \wedge \widetilde{T}^b \qquad (22.61)$$

and

$$R^a{}_b \wedge q^b = \omega^a{}_b \wedge T^b. \qquad (22.62)$$

As shown in precious ECE papers this vacuum geometry can be interpreted to mean that the spin connection form $\omega^a{}_b$ is the tensor dual of the tetrad form:

$$\omega^a{}_b = -\frac{\kappa}{2}\epsilon^a{}_{bc}q^c \qquad (22.63)$$

and that the curvature form is the tensor dual of the torsion form:

$$R^a{}_b = -\frac{\kappa}{2}\epsilon^a{}_{bc}T^c. \qquad (22.64)$$

Here $\frac{k}{2}$ is a proportionality coefficient with the units of wave-number. Such dualities define the vacuum electromagnetic field in ECE theory, and also the vacuum dynamical field. They are analogous to, and generalize, the well known duality in Euclidean space between an axial vector V_k and a anti-symmetric tensor V_{ij}:

$$V_{ij} = \frac{1}{2}\epsilon_{ijk}V_k \qquad (22.65)$$

where ϵ_{ijk} is the rank three totally anti-symmetric unit tensor. For example the rotation generator is, within a factor $-i$, an anti-symmetric unit tensor dual to an axial vector. Therefore Eq (22.58) and its Hodge dual (22.59) define the motion for example of a vacuum plane wave, both in ECE electrodynamical radiation and in ECE gravitational radiation. For a plane wave propagating in the Z axis, the motion is a rotation about the Z axis superimposed on translation. The electric and magnetic components of the plane wave are:

$$\boldsymbol{E} = \frac{E^{(0)}}{\sqrt{2}}(\boldsymbol{i} - i\boldsymbol{j})\ \exp(i(\omega t - \kappa Z)) \qquad (22.66)$$

and

$$\boldsymbol{B} = \frac{B^{(0)}}{\sqrt{2}}(i\boldsymbol{i} + \boldsymbol{j})\ \exp(i(\omega t - \kappa Z)) \qquad (22.67)$$

where ω is the angular frequency of the wave at instant t, and κ is its wavenumber at point Z. The mathematical law (22.58) indicates that $\boldsymbol{E}$ and $\boldsymbol{B}$ of the plane wave will transform covariantly in general. However the vacuum plane wave is already propagating at c, so the addition of v to c in special relativity produces c again by the velocity addition law of special relativity [1–12]. This law is derived from the Lorentz transform as is well known. So vacuum plane waves propagating at c are invariant under the Lorentz transform. They are already propagating at c and cannot travel faster than c. In ECE theory the constancy of c is also a fundamental hypothesis and plane waves such as (22.65) and (22.66) are also solutions in ECE theory. Note carefully that plane waves are mathematical idealizations, not physical entities.

In vector notation the inhomogeneous field equation (22.10) can be written as the two vector equations that constitute the Coulomb law:

$$\boldsymbol{\nabla} \cdot \boldsymbol{D} = \rho \qquad (22.68)$$

and the Ampère Maxwell law:

$$\boldsymbol{\nabla} \times \boldsymbol{H} - \frac{\partial \boldsymbol{D}}{\partial t} = \boldsymbol{J} \qquad (22.69)$$

where $\mathbf{D}$ is the electric displacement and where $\mathbf{H}$ is the magnetic field strength. Here $\mathbf{J}$ is the current density and ρ is the charge density. The electric displacement in the Coulomb law is defined by an orbital torsion as follows:

$$\mathbf{D} = D^{010}\mathbf{i} + D^{020}\mathbf{j} + D^{030}\mathbf{k} \tag{22.70}$$

The magnetic field strength in the Ampère Maxwell law is defined in terms of spin torsion as:

$$\mathbf{H} = H^{332}\mathbf{i} + H^{113}\mathbf{j} + H^{221}\mathbf{k}. \tag{22.71}$$

and the electric displacement in the Ampère Maxwell law is defined in terms of spin torsion by:

$$\mathbf{D} = D^{110}\mathbf{i} + D^{220}\mathbf{j} + D^{330}\mathbf{k}. \tag{22.72}$$

Using the methods given above for the vacuum laws, it is found that under a Z axis rotation the ECE Coulomb and Ampère Maxwell laws are generally covariant. Under the arbitrary transformation from frame K to K' the laws in frame K' become:

$$(\boldsymbol{\nabla} \cdot \mathbf{D} = \rho) \to (\boldsymbol{\nabla} \cdot \mathbf{D} = \rho)' \tag{22.73}$$

and

$$\left(\boldsymbol{\nabla} \times \mathbf{H} - \frac{\partial \mathbf{D}}{\partial t} = \mathbf{J}\right) \to \left(\boldsymbol{\nabla} \times \mathbf{H} - \frac{\partial \mathbf{D}}{\partial t} = \mathbf{J}\right)'. \tag{22.74}$$

They retain their vector format but

$$\mathbf{D} \to \mathbf{D}', \quad \rho \to \rho', \quad \mathbf{H} \to \mathbf{H}', \quad \mathbf{J} \to \mathbf{J}' \tag{22.75}$$

and:

$$\boldsymbol{\nabla} \to \boldsymbol{\nabla}', \quad \frac{\partial}{\partial t} \to \frac{\partial}{\partial t}, \tag{22.76}$$

So the general transformation in this case produces new physical effects, the essential reason being that the charge current density is changed.

As argued, the vacuum field equations of both ECE and the standard model are invariant under respectively the general coordinate transformation and the Lorentz transformation, while the vacuum fields $\mathbf{E}$ and $\mathbf{B}$ themselves change. One consequence of this property is that the standard model is unable to describe the Faraday disk generator - the well known Faraday paradox. In the K frame the Faraday law of induction in the standard model is:

$$\nabla \times E + \frac{\partial B}{\partial t} = 0 \qquad (22.77)$$

and in order for induction to take place of a field E by a field B the experimental condition needed is:

$$\frac{\partial B}{\partial t} \neq 0. \qquad (22.78)$$

In the Faraday disk generator this condition is not fulfilled. The generator consists of a disk of uncharged metal placed on a magnet. The condition for induction is that the disk rotates relative to the observing apparatus in frame K. Whether or not the magnet is static or spinning about its own Z axis:

$$\frac{\partial B}{\partial t} = 0 \qquad (22.79)$$

and no induction of E occurs according to Eq. (22.76). This paradox is not resolved by Lorentz transformation, under which Eq. (22.76) stays the same and under which the E and B fields change as in Eqs. (22.31) and (22.32). After Lorentz transformation therefore:

$$\nabla \times (E + V \times B) + \frac{\partial}{\partial t}\left(B - \frac{1}{c^2} V \times E\right) = 0. \qquad (22.80)$$

Using Eq. (22.76):

$$\nabla \times (v \times B) - \frac{1}{c^2} \frac{\partial}{\partial t}(v \times E) = 0. \qquad (22.81)$$

Now use the vector identities [15]:

$$\nabla \times (a \times b) = a \nabla \cdot b - b \nabla \cdot a + (b \cdot \nabla)a - (a \cdot \nabla)b = 0 \qquad (22.82)$$

and

$$a \times (\nabla \times a) = \frac{1}{2} \nabla a^2 - (a \cdot \nabla)a \qquad (22.83)$$

to find that:

$$\nabla \times (v \times B) = v \nabla \cdot B - B \nabla \cdot v + (B \cdot \nabla)v - (v \cdot \nabla)B. \qquad (22.84)$$

22.2 Invariance and Covariance

For a constant v:

$$\nabla \times (v \times B) = -(v \cdot \nabla)B. \qquad (22.85)$$

From Eq. (22.82):

$$v \times (\nabla \times B) = \frac{1}{2}\nabla(v \cdot B) - (v \cdot \nabla)B = -(v \cdot \nabla)B \qquad (22.86)$$

for constant v. Therefore:

$$\nabla \times (v \times B) = v \times (\nabla \times B) \qquad (22.87)$$

and for constant v:

$$\frac{\partial}{\partial t}(v \times E) = v \times \frac{\partial E}{\partial t}. \qquad (22.88)$$

So Eq. (22.79) is:

$$v \times \left(\nabla \times B - \frac{1}{c^2}\frac{\partial E}{\partial t}\right) = 0 \qquad (22.89)$$

which is the vacuum Ampère Maxwell law of the standard model cross multiplied by v. Therefore the Lorentz transformation has produced the Hodge dual of the Faraday law, which is the vacuum Ampère Maxwell law:

$$\nabla \times B - \frac{1}{c^2}\frac{\partial E}{\partial t} = 0. \qquad (22.90)$$

Since:

$$\nabla \times B = 0 \qquad (22.91)$$

experimentally the B field cannot again induce an E field in the standard model, and the Faraday paradox remains in the standard model. There have been some claims in the literature that Lorentz induction explains the Faraday disk, but these are based on the transform of E, neglecting the transform of B. When both transforms are accounted for correctly it is seen that

$$\left(\nabla \times E + \frac{\partial B}{\partial t}\right)' = 0 \qquad (22.92)$$

and there is no induction of E by B in any frame of reference, contrary to observation. This result is shown in another way by noting that the linear velocity at the rim of a disk rotating at an angular frequency Ω is the real part of:

$$\boldsymbol{v} = \frac{v^{(0)}}{\sqrt{2}}(\boldsymbol{i} - i\boldsymbol{j}) \, \exp(i\Omega t) \qquad (22.93)$$

which is:

$$\text{Real}\,(\boldsymbol{v}) = \frac{v^{(0)}}{\sqrt{2}}(\boldsymbol{j}\cos(\Omega t) + \boldsymbol{j}\sin(\Omega t)). \qquad (22.94)$$

The Lorentz transform of the electric field:

$$\boldsymbol{E'} = \boldsymbol{E} + \boldsymbol{v} \times \boldsymbol{B} \qquad (22.95)$$

produces:

$$\boldsymbol{v} \times \boldsymbol{B} = v^{(0)} \frac{B^{(0)}}{\sqrt{2}}(\boldsymbol{i}\sin(\Omega t) - \boldsymbol{j}\cos(\Omega t)) \qquad (22.96)$$

where:

$$\boldsymbol{B} = B^{(0)}\boldsymbol{k}, \qquad (22.97)$$

which is the product of $v^{(0)}$ with a rotating magnetic field:

$$\boldsymbol{B} = -\frac{B^{(0)}}{\sqrt{2}}(i\boldsymbol{i} + \boldsymbol{j}) \, \exp(i\Omega t) \qquad (22.98)$$

which has the real part of:

$$\text{Real}\,(\boldsymbol{B}) = \frac{B^{(0)}}{\sqrt{2}}(\boldsymbol{i}\sin(\Omega t) - \boldsymbol{j}\cos(\Omega t)). \qquad (22.99)$$

The Lorentz transform does not produce a rotating electric field as claimed for example by Feynman [16]. It produces a rotating magnetic field. Also, when rotation is present, the inertial Lorentz transform is not applicable, and the standard model cannot explain the Faraday paradox. For this ECE theory is needed [1–10].

In contrast to the standard model the ECE explanation of the Faraday disk generator is based directly on the fundamental hypothesis:

$$A^a_\mu = A^{(0)} q^a_\mu \qquad (22.100)$$

that a vector potential is generated by the Cartan tetrad. The disk rotates at an angular frequency Ω and generates the potential [1–10]:

$$\boldsymbol{A}^{(2)} = \boldsymbol{A}^{(2)*} = \frac{A^{(0)}}{\sqrt{2}}(\boldsymbol{i} - i\boldsymbol{j})e^{i\Omega t} \tag{22.101}$$

where the C negative A is the magnitude of the vector potential of the magnet. From the ECE equations linking the field and potential, the following rotating electric field is generated:

$$\boldsymbol{E}^{(2)} = \boldsymbol{E}^{(1)*} = -\left(\frac{\partial}{\partial t} + i\Omega\right)\boldsymbol{A}^{(2)} \tag{22.102}$$

where the spin connection in units of inverse meters is:

$$\omega = \frac{\Omega}{c}. \tag{22.103}$$

The real part of this electric field rotates around the rim of the Faraday disk and is detected by apparatus in the observer frame K. As observed experimentally, $\boldsymbol{E}^{(1)}$ is proportional to Ω multiplied by $A^{(0)}$, the magnitude of the vector potential of the magnetic flux density $\boldsymbol{B}$. As noted, the standard model has no explanation for the Faraday disk generator.

Similarly the standard model has no explanation for the Sagnac effect, which is a phase shift induced by rotation. As argued, the free space Maxwell Heaviside (MH) equations are invariant under frame rotation, so they cannot explain the Sagnac effect because their solutions are the same in any frame of reference. This has been a well known cause of difficulty since the effect was discovered in 1913, and in the standard physics there continues to be no satisfactory explanation for the effect. In ECE theory [1–10] the effect is explained in the same way as the Faraday disk. A vector potential rotates around the platform of the Sagnac interferometer (ring laser gyroscope). Rotation to the left is described by:

$$\boldsymbol{A}_L^{(2)} = \frac{A^{(0)}}{\sqrt{2}}(\boldsymbol{i} - i\boldsymbol{j})\exp(i\omega_1 t) \tag{22.104}$$

and to the right by:

$$\boldsymbol{A}_R^{(2)} = \frac{A^{(0)}}{\sqrt{2}}(\boldsymbol{i} + i\boldsymbol{j})\exp(i\omega_1 t). \tag{22.105}$$

When the platform is at rest the time delay is:

$$\Delta t = 2\pi \left(\frac{1}{\omega_1} - \frac{1}{\omega_1} \right) = 0. \tag{22.106}$$

Eqs. (22.103) and (22.104) are tetrad equations of spinning space-time, a concept that does not exist in standard electrodynamics. In ECE theory the electromagnetic field is the frame of reference itself, so a beam of light traveling in a circular path is equivalent to a rotating tetrad multiplied by $A^{(0)}$, giving Eq. (22.103). When the platform is spun left at an angular frequency Ω:

$$\omega_1 \to \omega_1 + \Omega \tag{22.107}$$

for the left rotating beam and:

$$\omega_1 \to \omega_1 - \Omega \tag{22.108}$$

for the right rotating beam. This is because the rotating platform causes an additional or subtractive frame rotation, i.e. of space-time itself. There is a time delay for light going around a right or left spinning platform:

$$\Delta t = 2\pi \left(\frac{1}{\omega_1 - \Omega} - \frac{1}{\omega_1 + \Omega} \right) \tag{22.109}$$

which is the well known Sagnac effect.

Experimentally it is found that:

$$\Delta t = \frac{4}{c^2} Ar \Omega \tag{22.110}$$

where Ar is the area of the platform. For a circular platform:

$$Ar = \pi r^2 \tag{22.111}$$

and experimentally:

$$\Omega \ll \omega_1 \tag{22.112}$$

so

$$\omega_1 = \frac{c}{r} := \kappa_1 c. \tag{22.113}$$

The Sagnac effect is therefore an effect of spinning space-time, and Eq. (22.113) defines the frequency ω_1 and wave-number k_1 for a platform radius r.

22.3 Hodge Duality

In this section Hodge dual transformations are defined for free and interacting fields in ECE theory. In index-less shorthand notation [1–10] the Bianchi identity is:

$$D \wedge T := R \wedge q \qquad (22.114)$$

and its Hodge dual is:

$$D \wedge \tilde{T} := \tilde{R} \wedge q. \qquad (22.115)$$

These equations are written as:

$$d \wedge T = j = R \wedge q - \omega \wedge T \qquad (22.116)$$
$$d \wedge \tilde{T} = \tilde{j} = \tilde{R} \wedge q - \omega \wedge \tilde{T}. \qquad (22.117)$$

Free fields are defined by the geometry:

$$j = \tilde{j} = 0 \qquad (22.118)$$

i.e.

$$R \wedge q = \omega \wedge T, \qquad (22.119)$$
$$\tilde{R} \wedge q = \omega \wedge \tilde{T}. \qquad (22.120)$$

The free field is defined as the field infinitely distant from its source, a mathematical limit defined by:

$$j \to 0, \qquad (22.121)$$
$$\tilde{j} \to 0. \qquad (22.122)$$

Therefore the free field geometry is:

$$d \wedge T \to 0, \qquad (22.123)$$
$$d \wedge \tilde{T} \to 0. \qquad (22.124)$$

a) Free Electromagnetic Field
   Use the ECE hypothesis:

$$F = A^{(0)} T \qquad (22.125)$$

to find the ECE equations of the free electromagnetic field:

$$d \wedge F = 0, \qquad (22.126)$$

$$d \wedge \tilde{F} = 0. \qquad (22.127)$$

These equations imply:

$$R \wedge A = \omega \wedge F, \qquad (22.128)$$

$$\tilde{R} \wedge A = \omega \wedge \tilde{F}. \qquad (22.129)$$

For the free field, propagating in vacuo at c, the spin connection is of the same order as the torsion, tetrad and curvature. Therefore this is not Minkowski space-time because the free field is due to space-time torsion. In a Minkowski space-time, the torsion, curvature and spin connection all vanish.

Translating into vector notation, Eqs. (22.125) and (22.126) become the familiar:

$$\boldsymbol{\nabla} \cdot \boldsymbol{B} = 0, \qquad (22.130)$$

$$\boldsymbol{\nabla} \times \boldsymbol{E} + \frac{\partial \boldsymbol{B}}{\partial t} = \boldsymbol{0}, \qquad (22.131)$$

$$\boldsymbol{\nabla} \cdot \boldsymbol{E} = 0, \qquad (22.132)$$

$$\boldsymbol{\nabla} \times \boldsymbol{B} - \frac{1}{c^2} \frac{\partial \boldsymbol{E}}{\partial t} = \boldsymbol{0}. \qquad (22.133)$$

For example, the Coulomb law for the free field is Eq. (22.131), with the solution:

$$\boldsymbol{E} \to \boldsymbol{0}. \qquad (22.134)$$

This result means that in electro-statics, the static electric field tends to zero if the distance between two charges approaches infinity. This result can be seen from the Coulomb law:

$$\boldsymbol{F} = \frac{e_1 e_2}{4 \pi \epsilon_0 r^3} \boldsymbol{r} \qquad (22.135)$$

where:

$$\boldsymbol{E} = \frac{e_2}{4 \pi \epsilon_0 r^3} \boldsymbol{r}. \qquad (22.136)$$

Here e_1 is a charge interacting with e_2, ϵ_0 is the S.I. vacuum permittivity, $\boldsymbol{r}$ is the radial coordinate, and $\boldsymbol{F}$ is the Coulomb force of repulsion. In electro-dynamics there are plane wave solutions to Eqs. (22.129) to (22.132), plane waves which propagate through the vacuum infinitely distant from the source. This is a mathematical concept. The plane waves are:

$$\mathbf{E} = \frac{E^{(0)}}{\sqrt{2}}(\mathbf{i} - i\mathbf{j})\ \exp(i(\omega t - \boldsymbol{\kappa}\cdot\mathbf{r})) \tag{22.137}$$

and

$$\mathbf{B} = \frac{B^{(0)}}{\sqrt{2}}(i\mathbf{i} + \mathbf{j})\ \exp(i(\omega t - \boldsymbol{\kappa}\cdot\mathbf{r})) \tag{22.138}$$

and there are other types of dynamical solution such as spherical waves based on spherical harmonics. The plane waves have angular frequency ω at an instant t, and wave-number $\mathbf{k}$ at position $\mathbf{r}$.

b) Free Gravitational Fields

The free gravitational fields are defined in ECE theory by equations which have the same structure as (22.129) to (22.132) [1–10]:

$$\boldsymbol{\nabla}\cdot\mathbf{h} = 0, \tag{22.139}$$

$$\boldsymbol{\nabla}\times\mathbf{g} + \frac{1}{c}\frac{\partial\mathbf{h}}{\partial t} = \mathbf{0}, \tag{22.140}$$

$$\boldsymbol{\nabla}\cdot\mathbf{g} = 0, \tag{22.141}$$

$$\boldsymbol{\nabla}\times\mathbf{h} - \frac{1}{c}\frac{\partial\mathbf{g}}{\partial t} = \mathbf{0}, \tag{22.142}$$

where $\mathbf{g}$ is the acceleration due to gravity (an orbital component of torsion). Eq. (22.140) can be interpreted in the same way as Eq. (22.131) for the static electric field. The Newton inverse square law is a well defined limit of ECE theory and is a (negative valued) force of attraction between two masses m and M:

$$\mathbf{F} = -\frac{mMG}{r^3}\mathbf{r} \tag{22.143}$$

in which:

$$\mathbf{g} = -\frac{mG}{r^3}\mathbf{r} \tag{22.144}$$

from the weak equivalence principle. Therefore for infinitely distant masses the acceleration due to gravity approaches zero, which is the physical interpretation of Eq. (22.140). Eqs. (22.138) to (22.141) show that besides this well known law, there are three other laws of classical gravitation whose structure is the same as the laws of classical electrodynamics. This is a major result of the EEC unified field theory. The classical vacuum is defined as the absence of mass and charge, so:

$$\mathbf{E} \to 0, \ \mathbf{g} \to 0 \tag{22.145}$$

in the vacuum.

The radiated gravitational plane waves are evidently:

$$\mathbf{g} = \frac{g^{(0)}}{\sqrt{2}}(i\mathbf{i} - \mathbf{j}) \ \exp(i(\omega t - \boldsymbol{\kappa} \cdot \mathbf{r})) \tag{22.146}$$

and

$$\mathbf{h} = \frac{h^{(0)}}{\sqrt{2}}(i\mathbf{i} + \mathbf{j}) \ \exp(i(\omega t - \boldsymbol{\kappa} \cdot \mathbf{r})) \tag{22.147}$$

and are about twenty one orders of magnitude weaker in the laboratory than the radiated plane waves of classical electromagnetism. Note carefully that gravitational radiation cannot be deduced from the Einstein field equation, because of the latter's neglect of torsion. As shown in papers 93 and following on www.aias.us, the neglect of torsion leads to an incorrect geometry, so nothing can be deduced form the Einstein field equation. The latter is regarded in ECE theory as obsolete.

When fields interact with matter the basic geometry is:

$$d \wedge T = (R \wedge q - \omega \wedge T)_{\text{int}} \tag{22.148}$$

and its Hodge dual:

$$d \wedge \widetilde{T} = (\widetilde{R} \wedge q - \omega \wedge \widetilde{T})_{\text{int}}. \tag{22.149}$$

a) In electrodynamics in the laboratory, the magnetic monopole is unmeasurably small experimentally, so

$$d \wedge F \to 0, \tag{22.150}$$

$$d \wedge \widetilde{F} = \widetilde{j}/\epsilon_0. \tag{22.151}$$

Eqs. (22.149) and (22.150) translate into the free field homogeneous equations:

$$\boldsymbol{\nabla} \cdot \mathbf{B} = 0, \tag{22.152}$$

$$\boldsymbol{\nabla} \times \mathbf{E} + \frac{\partial \mathbf{B}}{\partial t} = \mathbf{0}, \tag{22.153}$$

and the field matter inhomogeneous equations:

$$\nabla \cdot \boldsymbol{D} = \rho, \tag{22.154}$$

$$\nabla \times \boldsymbol{H} - \frac{\partial \boldsymbol{D}}{\partial t} = \boldsymbol{J}, \tag{22.155}$$

where $\boldsymbol{E}$ is the electric field strength, $\boldsymbol{B}$ is the magnetic flux density, $\boldsymbol{D}$ is the electric displacement, ρ is the electric charge density, $\boldsymbol{H}$ is the magnetic field strength, and $\boldsymbol{J}$ is the electric current density. The displacement is defined as:

$$\boldsymbol{D} = \epsilon_0 \boldsymbol{E} + \boldsymbol{P} \tag{22.156}$$

where $\boldsymbol{P}$ is the electric polarization, and the magnetic field strength is defined by:

$$\boldsymbol{B} = \mu_0(\boldsymbol{H} + \boldsymbol{M}) \tag{22.157}$$

where $\boldsymbol{M}$ is the magnetization and μ_0 is the magnetic vacuum permeability [1–10].

In general, an asymmetric connection must be used to find j in Eq. (22.150). The Hodge dual of the interaction current:

$$\widetilde{j} := (\widetilde{R} \wedge q - \omega \wedge \widetilde{T})_{\text{int}} \tag{22.158}$$

is another interaction current:

$$j := (R \wedge q - \omega \wedge T)_{\text{int}}. \tag{22.159}$$

Therefore the Hodge dual of the pair of vector equations:

$$\nabla \cdot \boldsymbol{D} = \rho, \tag{22.160}$$

$$\nabla \times \boldsymbol{H} + \frac{\partial \boldsymbol{D}}{\partial t} = \boldsymbol{J} \tag{22.161}$$

is

$$\nabla \cdot \boldsymbol{H} = -\nabla \cdot \boldsymbol{M} := \widetilde{\rho}, \tag{22.162}$$

$$\nabla \times \boldsymbol{D} + \frac{1}{c^2}\frac{\partial \boldsymbol{H}}{\partial t} = \nabla \times \boldsymbol{P} - \frac{1}{c^2}\frac{\partial \boldsymbol{M}}{\partial t} := \boldsymbol{J} \tag{22.163}$$

where we have used Eqs. (22.151) and (22.152). The Hodge dual of these latter equations are Eqs. (22.131) and (22.132).

b) The interaction of the gravitational field with matter is given by:

$$d \wedge \widetilde{T} = \widetilde{j}_{\text{int}} = (\widetilde{R} \wedge q - \omega \wedge \widetilde{T})_{\text{int}} \tag{22.164}$$

so that Eq. (141) for example, becomes:

$$\nabla \cdot g = 4\pi G \rho_m \tag{22.165}$$

where G is the Newton constant and ρ_m is the mass density [1–10]. This is the gravitational analogue of the Coulomb law (22.159). In ECE theory there is also a gravitational analogue of Eq. (22.160), a law that should be investigated experimentally. The gravitational law (22.139) has already been observed by Tajmar, de Matos et al. [16].

The homogeneous gravitational current:

$$j = (R \wedge q - \omega \wedge T)_{\text{free}} \tag{22.166}$$

must be interpreted carefully as being due to the interaction of gravitational and electromagnetic fields infinitely far from their sources. This type of interaction in classical electrodynamics would mean that the Gauss law of magnetism and Faraday law of induction become minutely different due to the existence of a magnetic charge density (monopole) and magnetic current density. The magnetic charge density or monopole, and the magnetic current density are both due to the geometry of Eq. (22.1). This is a concept of general relativity and not of special relativity. The ECE magnetic monopole is not a Dirac monopole, and not a topological monopole of gauge theory in special relativity. The ECE monopole is due to the geometry of space-time.

Acknowledgments

The British Government is thanked for the award of a Civil List pension and the staff of A.I.A.S. and many others for interesting discussions.

References

[1] M. W. Evans, "Generally Covariant Unified Field Theory" (Abramis 2005 onwards), volumes one to four, volumes five and six in prep. (papers 71 onwards on www.aias.us).

[2] L. Felker, "The Evans Equations of Unified Field Theory" (Abramis 2007).

[3] K. Pendergast, "Crystal Spheres" (preprint on www.aias.us).

[4] M. W. Evans, Omnia Opera section of www.aias.us from 1992 onwards.

[5] M. W. Evans (ed.), Adv. Chem. Phys., vol 119 (2001); M. W. Evans and S. Kielich (eds.), Adv. Chem. Phys., vol. 85 (1992, 1993, 1997).

[6] M. W. Evans and L. B. Crowell, "Classical and Quantum Electrodynamics and the B(22.3) Field" (World Scientific, 2001).

[7] M. W. Evans and J.-P. Vigier, "The Enigmatic Photon" (Kluwer, 1994 to 2002, hardback and softback).

[8] M. W. Evans, fifteen papers on ECE in Found. Phys. Lett., 2003 to 2005.

[9] M. W. Evans, Acta Phys. Polon., **38B**, 2211 (2007); Physica B, **403**, 517 (2008).

[10] M. W. Evans and H. Eckardt, Physia B, **400**, 175 (2007).

[11] S. P. Carroll, "Space-time and Geometry: an Introduction to General Relativity" (Addison-Wesley, New York, 2004, and downloadable lecture notes).

[12] L .H. Ryder, "Quantum Field Theory" (Cambridge University Press, 1996, 2nd. Edition).

[13] J. D. Jackson, "Classical Electrodynamics" (Wiley, 1999, 3rd Edition).

[14] P. W. Atkins, "Molecular Quantum Mechanics" (Oxford University Press, 1983, 2nd. Edition).

[15] E. Milewski (ed.), "Vector Analysis Problem Solver" (REA, New York, 1987).

[16] R. Feynman's "Lectures in Physics".

[17] Austrian and European Space Agency experiments.

The Continuity Equation in ECE Theory

by

Myron W. Evans,
Alpha Institute for Advanced Study, Civil List Scientist.
(emyrone@aol.com and www.aias.us)

Abstract

In Einstein Cartan Evans (ECE) field theory the charge on the electron, -e, is a fundamental constant, and therefore, electric charge/current density is conserved in a fundamental continuity equation. In this paper the latter is derived in a space-time with torsion and curvature within the context of Cartan geometry, thus proving that the continuity equation is valid in a generally covariant unified field theory.

Keywords: ECE theory, unified field theory, continuity equation, conservation of charge/current density.

23.1 Introduction

The ECE theory has been accepted as being a valid unified field theory within the context of general relativity. It is therefore a generally covariant unified field theory of physics, or natural philosophy. The equations of dynamics and electrodynamics are expressed in a space-time with torsion and curvature present in general. In both subject areas the vector equations have the same format, and this is also the same as the Maxwell Heaviside and gravitomagnetic equations of standard physics, but expressed in ECE self consistently in a generally covariant mathematical framework based on Cartan geometry [1–12]. In ECE theory the charge on the electron, -e, is a universal constant, which implies that charge/current density must be conserved in a continuity equation that must be developed in a space-time with curvature and torsion both present. In standard physics the continuity equation [13, 14] is part of Noether's Theorem expressed in a Minkowski space-time with no torsion and no curvature. In Section 23.2 the ECE equations of classical electrodynamics and dynamics are reviewed and the electric charge and current density defined. In Section 23.3 the ECE continuity equation is derived self-consistently with the generally covariant Proca equation of ECE theory [1–12] by making the charge current density proportional to the vector potential and using an identically non-zero photon mass. In so doing the ECE continuity equation is identified as an example of the tetrad postulate of Cartan geometry and the continuity equation is derived from geometry as required in the philosophy of relativity.

23.2 The ECE Equations of Classical Dynamics and Electrodynamics

The ECE equations of dynamics are found from the following equation of geometry:

$$D_\mu T^{\kappa\mu\nu} = R^\kappa{}_\mu{}^{\mu\nu} \tag{23.1}$$

where $T^{\kappa\mu\nu}$ is the Cartan torsion tensor and $R^\kappa{}_\mu{}^{\mu\nu}$ is the Cartan curvature tensor. This is a tensor equation in a space-time with curvature and torsion [1–12] and is a tensorial expression of the Hodge dual of the Bianchi identity. The covariant derivative of the torsion is the curvature. In vector format this tensor equation becomes two generally covariant gravitomagnetic equations valid for all field strengths. The first one is the covariant generalization of the Newton inverse square law:

$$\nabla \cdot \boldsymbol{g} = 4\pi G \rho_m. \tag{23.2}$$

Here G is Newton's gravitational constant. The acceleration due to gravity is:

$$\boldsymbol{g} = c^2(T^{010}\boldsymbol{i} + T^{020}\boldsymbol{j} + T^{030}\boldsymbol{k}) \tag{23.3}$$

and the mass density is:

$$\rho = J^0{}_1{}^{10} + J^0{}_2{}^{20} + J^0{}_3{}^{30} \tag{23.4}$$

where c is the vacuum speed of light, a universal constant in the theory of relativity, of which ECE is the most developed form to date. The current terms defining the mass density are made up of curvature, torsion and spin connection elements as in the following general formula [1–12]:

$$J^\kappa{}_\mu{}^{\mu\nu} = \frac{c^2}{4\pi G}\left(R^\kappa{}_\mu{}^{\mu\nu} - \omega^\kappa{}_{\mu\lambda}T^{\lambda\mu\nu}\right) \tag{23.5}$$

23.2 The ECE Equations of Classical Dynamics and Electrodynamics

where $\omega^\kappa{}_{\mu\lambda}$ denotes the spin connection. The second vector equation is the gravitomagnetic analogue of the Ampère Maxwell law of electrodynamics written in a space-time with torsion and curvature both present in general, and is:

$$\nabla \times \boldsymbol{h} - \frac{1}{c}\frac{\partial \boldsymbol{g}}{\partial t} = 4\pi G \boldsymbol{J}_m \tag{23.6}$$

where $\boldsymbol{h}$ is the gravitomagnetic analogue of the magnetic field strength $\boldsymbol{H}$ of electrodynamics, $\boldsymbol{g}$ being the gravitomagnetic analogue of the electric displacement $\boldsymbol{D}$. In Eq. (23.6):

$$\boldsymbol{h} = c^2(T^{332}\boldsymbol{i} + T^{113}\boldsymbol{j} + T^{221}\boldsymbol{k}) \tag{23.7}$$

and

$$\boldsymbol{g} = c^2(T^{110}\boldsymbol{i} + T^{220}\boldsymbol{j} + T^{330}\boldsymbol{k}). \tag{23.8}$$

The current density term of the gravitomagnetic equation (23.6) is defined by:

$$\boldsymbol{J}_m = J_X \boldsymbol{i} + J_Y \boldsymbol{j} + J_Z \boldsymbol{k} \tag{23.9}$$

where:

$$J_X = J^1{}_0{}^{01} + J^1{}_2{}^{21} + J^1{}_3{}^{31}, \tag{23.10}$$
$$J_Y = J^2{}_0{}^{02} + J^2{}_1{}^{12} + J^2{}_3{}^{32}, \tag{23.11}$$
$$J_Z = J^3{}_0{}^{03} + J^3{}_1{}^{13} + J^3{}_2{}^{23}. \tag{23.12}$$

Eq. (23.6) is a spin equation whereas Eq. (23.2) is an orbital equation. Note carefully that the torsion components defining $\boldsymbol{g}$ are in general different in the two equations.

There are two more gravitomagnetic field equations which are found by Hodge dual transformation of Eqs. (23.2) and (23.6). The Hodge dual of the orbital equation (23.2) is:

$$\nabla \cdot \boldsymbol{h} = 4\pi G \widetilde{\rho}_m \tag{23.13}$$

where:

$$\boldsymbol{h} = c^2(\widetilde{T}^{010}\boldsymbol{i} + \widetilde{T}^{020}\boldsymbol{j} + \widetilde{T}^{030}\boldsymbol{k}) \tag{23.14}$$

and

$$\tilde{\rho} = \tilde{J}^0{}_1{}^{01} + \tilde{J}^0{}_2{}^{02} + \tilde{J}^0{}_3{}^{03}. \tag{23.15}$$

The tilde denoting Hodge dual transformation. The Hodge dual of the spin equation (23.6) is:

$$\nabla \times \boldsymbol{g} + \frac{1}{c}\frac{\partial \boldsymbol{h}}{\partial t} = 4\pi G \tilde{\boldsymbol{J}}_m \tag{23.16}$$

where:

$$\boldsymbol{g} = c^2(\tilde{T}^{332}\boldsymbol{i} + \tilde{T}^{113}\boldsymbol{j} + \tilde{T}^{221}\boldsymbol{k}) \tag{23.17}$$

and

$$\boldsymbol{h} = c^2(\tilde{T}^{101}\boldsymbol{i} + \tilde{T}^{202}\boldsymbol{j} + \tilde{T}^{303}\boldsymbol{k}) \tag{23.18}$$

and where the Hodge dual current is:

$$\tilde{\boldsymbol{J}}_m = \tilde{J}_X \boldsymbol{i} + \tilde{J}_Y \boldsymbol{j} + \tilde{J}_Z \boldsymbol{k}, \tag{23.19}$$

$$\tilde{J}_X = \tilde{J}^1{}_0{}^{01} + \tilde{J}^1{}_2{}^{21} + \tilde{J}^1{}_3{}^{31} \tag{23.20}$$

$$\tilde{J}_Y = \tilde{J}^2{}_0{}^{02} + \tilde{J}^2{}_1{}^{12} + \tilde{J}^2{}_3{}^{32} \tag{23.21}$$

$$\tilde{J}_Z = \tilde{J}^3{}_0{}^{03} + \tilde{J}^3{}_1{}^{13} + \tilde{J}^3{}_2{}^{23}. \tag{23.22}$$

These four equations of gravitomagnetism are valid for any field strength and are generally covariant. They describe the interaction of field and matter. If the field is propagating infinitely distant from its source the four equations reduce to the free field or vacuum gravitomagnetic equations:

$$\nabla \cdot \boldsymbol{h}_0 = 0, \tag{23.23}$$

$$\nabla \times \boldsymbol{g}_0 + \frac{1}{c}\frac{\partial \boldsymbol{h}_0}{\partial t} = \boldsymbol{0}, \tag{23.24}$$

$$\nabla \cdot \boldsymbol{g}_0 = 0, \tag{23.25}$$

$$\nabla \times \boldsymbol{h}_0 - \frac{1}{c}\frac{\partial \boldsymbol{g}_0}{\partial t} = \boldsymbol{0}, \tag{23.26}$$

with plane wave solutions indicating gravitational radiation. In the laboratory the gravitational radiation is about twenty one orders of magnitude weaker

23.2 The ECE Equations of Classical Dynamics and Electrodynamics

than electromagnetic radiation. The gravitomagnetic field/potential relations are:

$$g_0 = -\nabla \Phi - c\frac{\partial q}{\partial t} + \Phi \omega - c\omega^0 q, \qquad (23.27)$$

$$h_0 = c^2(\nabla \times q - \omega \times q), \qquad (23.28)$$

where q is the vector potential of the gravitomagnetic equations and where Φ is the scalar potential of the gravitomagnetic equations. Here ω° is the spin connection scalar and ω is the spin connection vector. In the Newtonian limit:

$$g_0 \rightarrow -\nabla \Phi. \qquad (23.29)$$

The generally covariant equations of classical electromagnetism in ECE theory have the same vector structure as the standard Maxwell Heaviside equations but are written as follows in a space-time with torsion and curvature. The Coulomb law of ECE theory is an orbital torsion law:

$$\nabla \cdot D = \rho \qquad (23.30)$$

where the electric displacement is:

$$D = D^{010}i + D^{020}j + D^{030}k \qquad (23.31)$$

and where the electric charge density is:

$$\rho = J^0{}_1{}^{10} + J^0{}_2{}^{20} + J^0{}_3{}^{30}. \qquad (23.32)$$

The Ampère Maxwell law is a spin torsion law:

$$\nabla \times H - \frac{\partial D}{\partial t} = J \qquad (23.33)$$

in which the magnetic field strength is:

$$H = H^{332}i + H^{113}j + H^{221}k \qquad (23.34)$$

and the electric displacement is:

$$D = D^{110}i + D^{220}j + D^{330}k. \qquad (23.35)$$

The electric current density in Eq. (23.33) has the same form as in Eq. (23.19)–(23.22):

$$\boldsymbol{J} = J_X \boldsymbol{i} + J_Y \boldsymbol{j} + J_Z \boldsymbol{k}. \tag{23.36}$$

The other two laws of ECE classical electrodynamics are the Gauss law of magnetism, an orbital law:

$$\boldsymbol{\nabla} \cdot \boldsymbol{B} = 0 \tag{23.37}$$

in which the magnetic flux density is:

$$\boldsymbol{B} = B^{010} \boldsymbol{i} + B^{020} \boldsymbol{j} + B^{030} \boldsymbol{k} \tag{23.38}$$

and the Faraday law of induction, a spin law where:

$$\boldsymbol{\nabla} \times \boldsymbol{E} + \frac{\partial \boldsymbol{B}}{\partial t} = 0 \tag{23.39}$$

and:

$$\boldsymbol{E} = E^{332} \boldsymbol{i} + E^{113} \boldsymbol{j} + E^{221} \boldsymbol{k}, \tag{23.40}$$
$$\boldsymbol{B} = B^{101} \boldsymbol{i} + B^{202} \boldsymbol{j} + B^{303} \boldsymbol{k}, \tag{23.41}$$

The field potential relations are:

$$\boldsymbol{E} = -\boldsymbol{\nabla}\phi - \frac{\partial \boldsymbol{A}}{\partial t} + \phi\boldsymbol{\omega} - \omega^0 \boldsymbol{A} \tag{23.42}$$

and

$$\boldsymbol{B} = \boldsymbol{\nabla} \times \boldsymbol{A} - \boldsymbol{\omega} \times \boldsymbol{A} \tag{23.43}$$

in which ϕ is the scalar potential and $\boldsymbol{A}$ is the vector potential.

23.3 Derivation of the Generally Covariant Continuity Equation

In this section it is shown that the continuity equation is:

$$D_\mu j^a_\nu = 0 \tag{23.44}$$

23.3 Derivation of the Generally Covariant Continuity Equation

where j_ν^a is a vector valued differential one-form defined by:

$$j_\nu^a = -\epsilon_0 k T A_\nu^a. \tag{23.45}$$

Here, the generally covariant Proca equation of ECE theory is [1–12]:

$$(\Box + kT)A_\nu^a = 0. \tag{23.46}$$

where k is the Einstein equation, T is the index reduced canonical energy-momentum density, and where the potential is defined by the fundamental ECE postulate:

$$A_\nu^a = A^{(0)} q_\nu^a \tag{23.47}$$

where q_ν^a is the Cartan tetrad. In Eq. (23.45) ϵ_0 is the vacuum permittivity. In the limit of Minkowski space-time:

$$kT \rightarrow \left(\frac{mc}{\hbar}\right)^2 = \frac{1}{\lambda_c^2} \tag{23.48}$$

where m is the photon mass, $\hbar$ is the reduced Planck constant and where λ_c is the Comton wavelength.

Therefore Eq. (23.44) is an example of the tetrad postulate [1–12]:

$$D_\nu q_\mu^a = 0. \tag{23.49}$$

The structure of the inhomogeneous ECE field equation is:

$$\partial_\mu F^{\kappa\mu\nu} = j^\kappa{}_\mu{}^{\mu\nu}/\epsilon_0 \tag{23.50}$$

therefore by index contraction the charge-current density is a rank two tensor in ECE theory:

$$j^{\kappa\nu} = j^\kappa{}_\mu{}^{\mu\nu}. \tag{23.51}$$

Lowering an index:

$$j^\kappa{}_\nu = j^\kappa{}_{\mu\nu}{}^\mu \tag{23.52}$$

and by definition [1–12]:

$$j_\nu^a = q_\kappa^a j_\nu^\kappa. \tag{23.53}$$

Therefore in general, Eq. (23.50) is:

$$\partial_\mu F^{a\mu}{}_\nu = j^a{}_\nu/\epsilon_0 \tag{23.54}$$

Proceed now with reference to the Proca equation in standard physics. The Proca equation in standard physics is defined in a Minkowski space-time by:

$$\partial_\mu F^{\mu\nu} = j^\nu/\epsilon_0 = -\left(\frac{mc}{\hbar}\right)^2 A^\nu \tag{23.55}$$

where the field tensor is:

$$F^{\mu\nu} = \partial^\nu A^\nu - \partial^\nu A^\mu. \tag{23.56}$$

Using Eq. (23.56) in Eq. (23.55) we obtain the Lorentz covariant Proca equation of standard physics:

$$\left(\Box + \left(\frac{mc}{\hbar}\right)^2\right) A^\nu = 0 \tag{23.57}$$

provided that:

$$\partial_\mu A^\mu = 0. \tag{23.58}$$

The latter "Lorenz gauge" result follows from the continuity equation of standard physics:

$$\partial_\mu j^\mu = 0 \tag{23.59}$$

and Eq. (23.55). It is well known that if the photon mass is not zero:

$$m \neq 0 \tag{23.60}$$

the Proca equation is not gauge invariant, and the "Lorenz gauge" is not arbitrary. This leads to the collapse of gauge theory if the photon mass is not identically zero. In general relativity on the other hand the photon mass is identically non-zero, as seen in the bending of light by gravity for example. Therefore in ECE theory the gauge principle is rejected and the potential is considered to be physically meaningful, as observed in such phenomena as ESR and NMR, and in the Aharonov Bohm effects [1–12].

The generally covariant Proca equation (23.46) is based on

$$D_\nu A^a_\mu = 0 \tag{23.61}$$

23.3 Derivation of the Generally Covariant Continuity Equation

from which:

$$\partial^\mu \left(\partial_\mu A^a_\nu + \omega^a_{\mu b} A^b_\nu - \Gamma^\lambda_{\mu\nu} A^a_\lambda \right) = 0 \tag{23.62}$$

i.e.:

$$\Box A^a_\mu = \frac{j\nu^a}{\epsilon_0} = \partial^\mu (\Gamma^\lambda_{\mu b} A^a_\lambda - \omega^a_{\mu b} A^b_\nu). \tag{23.63}$$

This equation is the correctly covariant form of the standard physics equation:

$$\Box A^\mu = \frac{j^\mu}{\epsilon_0} \tag{23.64}$$

whose solutions are the Lienard Wiechert potentials [13]. In ECE theory:

$$F^{a\mu\nu} = \partial^\mu A^{a\nu} - \partial^\nu A^{a\mu} + \omega^{a\mu}_b A^{b\nu} - \omega^{a\nu}_b A^{b\mu} \tag{23.65}$$

Using this equation with Eq. (23.54) it is found that:

$$(\Box + kT) A^{a\nu} = \partial_\mu (\partial^\nu A^{a\mu} - \omega^{a\mu}_b A^{b\nu} + \omega^{a\nu}_b A^{b\mu}) = 0 \tag{23.66}$$

which is the generally covariant "Lorenz gauge" condition, but is now a rigorous geometrical requirement and not an arbitrary choice of gauge. If Eq. (23.66) is compared with the tetrad postulate:

$$D^\nu A^{a\mu} = \partial^\mu A^{a\nu} + \omega^{a\nu}_b A^{b\mu} - \Gamma^{\lambda\nu\mu} A^a_\lambda = 0 \tag{23.67}$$

it is found that Eq. (23.66) is true if:

$$\omega^{a\mu}_b A^{b\nu} = \Gamma^{\lambda\nu\mu} A^a_\lambda \tag{23.68}$$

which may be taken as the condition for Eq. (23.45).

The charge current density from Eq. (23.45) may be written as:

$$j^a_\mu = j^{(0)} q^a_\mu \tag{23.69}$$

where:

$$j^{(0)} = -\epsilon_0 k T A^{(0)}. \tag{23.70}$$

An example of Eq. (23.69) is:

$$j^{\kappa\nu} = j^{(0)} q^{\kappa\nu} \tag{23.71}$$

with:

$$D_\mu j^{\kappa\nu} = 0. \tag{23.72}$$

This is the continuity equation associated with Eq. (23.50). The covariant derivative in Eq. (23.44) may be replaced by the ordinary derivative if:

$$\omega^a{}_{\mu b} j^b_\nu = \Gamma^\lambda_{\mu\nu} j^a_\lambda \tag{23.73}$$

and under this condition, also that of Eq. (23.68), the continuity equation is:

$$\partial_\mu j^a_\nu = 0. \tag{23.74}$$

A special case of Eq. (23.74) is:

$$\partial_\mu j^{\kappa\mu} = 0 \tag{23.75}$$

and in ECE theory [1–12]:

$$\rho = \frac{1}{c} j^{00}, \quad \mathbf{J} = j^{11}\mathbf{i} + j^{22}\mathbf{j} + j^{33}\mathbf{k} \tag{23.76}$$

so:

$$\frac{1}{c}\frac{\partial j^{00}}{\partial t} + \frac{\partial j^{11}}{\partial X} + \frac{\partial j^{22}}{\partial Y} + \frac{\partial j^{33}}{\partial Z} = 0 \tag{23.77}$$

which in vector notation is the continuity equation:

$$\frac{\partial \rho}{\partial t} + \boldsymbol{\nabla} \cdot \mathbf{J} = 0 \tag{23.78}$$

but now written in a space-time with torsion and curvature as required, and not in the flat or Minkowski equation of standard electrodynamics, which is Lorentz covariant but not generally covariant as required.

Acknowledgments

The British Government is thanked for a Civil List Pension and the staffs of AIAS and TGA for many interesting discussions.

References

[1] M. W. Evans, "Generally Covariant Unified Field Theory" (Abramis 2005 onwards), in six volumes to date, volumes 1- 4 published and volumes 5 and 6 in press.

[2] L .Felker, "The Evans Equations of Unified Field Theory" (Abramis 2007).

[3] K. Pendergast, "Crystal Spheres" (Abramis in prep., preprint on www.aias.us).

[4] M. W. Evans, Omnia Opera section of www.aias.us (1992 to present).

[5] M. W. Evans (ed.), "Modern Non-linear Optics", volume 119 of "Advances in Chemical Physics" (Wiley 2001); ibid. first edition edited by M. W .Evans and S. Kielich, vol. 85.

[6] M. W. Evans and L. B. Crowell, "Classical and Quantum Electrodynamics and the B(23.3) Field" (World Scientific 2001).

[7] M. W. Evans and J.-P. Vigier, "The Enigmatic Photon" (Kluwer 1994 to 2002, hardback and softback), in five volumes.

[8] M. W. Evans et al., fifteen papers in Found. Phys. Lett., 2003 onwards.

[9] M. W. Evans, Acta Phys. Polon., **33B**, 2211 (2007).

[10] M. W. Evans, Physica B, **403**, 517 (2008).

[11] M. W. Evans and H. Eckardt, Physica B, **400**, 175 (2007).

[12] S. P. Carroll, "Space-time and Geometry: an Introduction to General Relativity" (Addison-Wesley, New York, 2004).

[13] J. D. Jackson, "Classical Electrodynamics" (Wiley, New York, 1999, 3rd Ed.).

[14] L .H. Ryder, "Quantum Field Theory" (Cambridge Univ. Press, 1996, 2nd Ed.).

ECE Theory of the Earth's Gravitomagnetic Precession

by

Myron W. Evans,
Alpha Institute for Advanced Study, Civil List Scientist.
(emyrone@aol.com and www.aias.us)

and

Horst Eckardt,
Alpha Institute for Advanced Studies (A.I.A.S) and Telesio Galilei Association (T.G.A).

Abstract

The ECE equation of static gravitomagnetism is used to calculate the angle of precession due to the earth of a gyroscope carried in an orbiting satellite. The origin of the earth's gravitomagnetic angular frequency is thereby identified as space-time torsion. The ECE result is fortuitously in exact agreement with the first term of a dipole approximation in a recent result of H. Pfister from a revised theory of the standard physics' Lense Thirring effect. However, the latter is obtained from the incorrect Einstein field equation using the incorrect Kerr metric. Metrics used to calculate the so called Lense-Thirring effect from the Einstein field equation are shown to violate the fundamental dual identity of Cartan geometry and ECE theory.

Keywords: ECE theory, gravitomagnetic precession of the earth.

24.1 Introduction

It has been shown recently [1–12] that the Einstein field equation is incorrect due to its neglect of torsion. This is a fundamental error which shows up in the Hodge dual of the well known Bianchi identity as given by Cartan [13, 14]. Exact solutions of the Einstein field equation are given in terms of line elements [15], and in paper 93 of this series several such line elements were shown to violate the dual identity:

$$D_\mu T^{\kappa\mu\nu} = R^\kappa{}_\mu{}^{\mu\nu} \qquad (24.1)$$

of Cartan geometry. Here $T^{\kappa\mu\nu}$ is the torsion tensor and $R^\kappa{}_\mu{}^{\mu\nu}$ is the curvature tensor. The dual identity states that the covariant derivative of the torsion tensor is the curvature tensor. In paper 93 it was shown that the type of curvature tensor that appears in Eq. (24.1) is non-zero in general from the Einstein field equation, while the torsion tensor in that equation is by definition zero because of its use of the symmetric connection. Therefore the result is obtained that the Einstein field equation is geometrically incorrect, a severely negative result for modern physics of the standard type. In hindsight such a result was bound to occur, because the neglect of torsion by Einstein and his contemporaries was arbitrary. If the torsion is eliminated by choice of connection, the subject of general relativity becomes incorrectly constrained to curvature only. It has been known for ninety years that the Einstein equation had severe flaws in it but this criticism has gone unanswered. The result is a fiasco for the subject of natural philosophy, because the well known predictions attributed to this deeply flawed equation, over no less than ninety years, are entirely meaningless. Among these is the Lense Thirring effect [16] which is the attempted standard explanation for the earth's gravitomagnetic precession.

By use of the correct geometry due to Cartan [1–13] the Einstein Cartan Evans (ECE) theory has re-instated spacetime torsion in its rightful place in physics and has developed equations of dynamics based on the correct consideration of both torsion and curvature. One of these is the static gravitomagnetic equation, whose analogue in classical electrodynamics is the ECE Ampère law [1–12]. In Section 24.2 the simplest type of solution of this equation in the weak field approximation is found in order to give a first approximation to the angular frequency of the earth's gravitomagnetic precession. The latter gives an angle of precession when observed over a year's time - the aim of the well known Gravity Probe B experiment. The ECE theory gives the expected precession in a much simpler and much more direct way than the standard physics. In an article such as that by Pfister [16] it is seen that the history of the so-called Lense Thirring effect is convoluted, and even within the context of the standard model there are several errors in its development. It is not clear that these errors have been corrected and it is not even clear that Gravity Probe B has produced anything new experimentally. In Section 24.3, the Kerr metric and similar metrics used in the description of the Lense Thirring effect are shown to violate the dual identity (24.1) of Cartan geometry, and complete details of the computation are given.

24.2 Calculation of the Gravitomagnetic Angular Frequency

The ECE theory of gravitomagnetic precession is part of a generally covariant unified field theory [1–12] in which the equations of classical electrodynamics and dynamics have precisely the same structure based on the Bianchi identity of geometry. The ECE dynamical equation that gives gravitomagnetic precession is the precise analogue of the ECE Ampère law of classical electrodynamics, one of the law of magnetostatics. The ECE Ampère law is:

$$\nabla \times \mathbf{B} = \mu_0 \mathbf{J} \tag{24.2}$$

where $\mathbf{B}$ is magnetic flux density (defined by elements of spacetime torsion) and where $\mathbf{J}$ is part of the charge current four-density:

$$J^\mu = (c\rho, \mathbf{J}) \tag{24.3}$$

where ρ is charge density in coulombs per metre cubed and where c is the speed of light in vacuo. Here μ_0 is the permeability in vacuo in S.I. units. The other ECE law of magnetostatics is, for all practical purposes:

$$\nabla \cdot \mathbf{B} = 0. \tag{24.4}$$

In the limit of:

$$v \ll c \tag{24.5}$$

the inverse Lorentz transform gives the well known Biot Savart law [17] in the form:

$$\mathbf{B} = -\frac{1}{c^2}\mathbf{v} \times \mathbf{E} \tag{24.6}$$

where $\mathbf{E}$ is electric field strength in volts per metre.

In precise analogy, the mass four density is defined as:

$$j^\mu = (c\rho_m, \mathbf{j}) \tag{24.7}$$

where ρ_m is the mass density in kilograms per cubic metre and where $\mathbf{j}$ is the mass current density. The analogue of $\mathbf{E}$ in the ECE dynamical equations is the usual acceleration due to gravity $\mathbf{g}$ in metres per second squared. The dynamical analogue of $\mathbf{B}$ is defined as the quantity:

$$\mathbf{\Omega} = \frac{\mathbf{h}}{c} \tag{24.8}$$

which has the units of radians per second and which comes from the precise dynamical analogue of Eq. (24.6), i.e. from:

$$\mathbf{\Omega} = -\frac{1}{c^2} \mathbf{v} \times \mathbf{g}. \tag{24.9}$$

The dynamical analogue of the ECE Ampère law is therefore:

$$\nabla \times \mathbf{\Omega} = \left(\frac{4\pi G}{c^2}\right) \mathbf{j} \tag{24.10}$$

and the precise dynamical analogue of Eq. (24.4) is:

$$\nabla \cdot \mathbf{\Omega} = 0. \tag{24.11}$$

The ECE Coulomb law is:

$$\nabla \cdot \mathbf{E} = \frac{\rho}{\epsilon_0} \tag{24.12}$$

where ϵ_0 is the permittivity in vacuo, and its dynamical analogue is:

$$\nabla \cdot \mathbf{g} = 4\pi G \rho_m \tag{24.13}$$

which in the weak field approximation gives the Newton inverse square law where G is Newton's gravitational constant. Gravity Probe B [16] is a satellite that orbits at 650 kilometres over the poles, i.e. is a low orbit satellite whose mean distance above the earth's surface is small compared with the earth's radius. The aim of the experiment is to measure the angle defined by:

$$\theta = \Omega t \tag{24.14}$$

where the interval of time t is one year. The angle θ is given straightforwardly from the ECE equation (24.10), which is the Biot Savart type solution of

24.2 Calculation of the Gravitomagnetic Angular Frequency

Eq. (24.9). It is well known that if the earth is considered as a uniform sphere of mass M, then the acceleration due to gravity is:

$$\mathbf{g} = -\frac{MG}{r^3}\mathbf{r} \tag{24.15}$$

where r is the radial coordinate. In the present context this is the distance between the earth's centre of mass and the Gravity Probe B satellite. The earth's radius is denoted R, and its mass is denoted M. Its angular momentum [4] is therefore the well known angular momentum of a sphere of uniform mass:

$$L = \frac{2}{5}MR^2\omega \tag{24.16}$$

where ω is the angular frequency of diurnal rotation of the earth, a well measured quantity. It is also well known that the attraction between a mass m and the earth, a sphere of mass of radius R, can be represented by the Newtonian inverse square law:

$$\mathbf{F} = m\mathbf{g} = -mMG\frac{\mathbf{r}}{r^3} \tag{24.17}$$

for all r. Now apply Eq. (24.9) to find:

$$\mathbf{\Omega} = \frac{mG}{c^2r^3}\mathbf{L} \tag{24.18}$$

in which the integrated angular momentum of the earth is defined as:

$$\mathbf{L} = \sum_i m\mathbf{r}_i \times \mathbf{v}_i. \tag{24.19}$$

Therefore the angular frequency in radians per second of the earth's gravitomagnetic precession is:

$$\Omega = \frac{2}{5}\frac{mG}{c^2r^3}R^2\omega. \tag{24.20}$$

The data relevant to Gravity Probe B are as follows, in S.I. units:

$$\left.\begin{aligned} R &= 6.37 \times 10^6 \text{ m}, \\ r &= 7.02 \times 10^6 \text{ m}, \\ M &= 5.98 \times 10^{24} \text{ kg}, \\ c &= 2.998 \times 10^8 \text{ ms}^{-1}, \\ G &= 6.67 \times 10^{-11} \text{ m}^3\text{kg}^{-1}\text{s}^{-2}, \\ \omega &= 7.29 \times 10^{-5} \text{ rad s}^{-1}. \end{aligned}\right\} \quad (24.21)$$

Therefore the earth's angular momentum is:

$$L = 7.076 \times 10^{33} \text{ kg m}^2\text{s}^{-1} \quad (24.22)$$

and the earth's gravitomagnetic precession in an orbit 650 kilometres above the surface is:

$$\Omega = 1.52 \times 10^{-14} \text{ rad s}^{-1}. \quad (24.23)$$

One year is $3600 \times 24 \times 365.25 = 3.156 \times 10^7$ seconds, and in one year:

$$\theta = 4.80 \times 10^{-7} \text{ radians}. \quad (24.24)$$

Finally use:

$$1 \text{ radian} = 2.06265 \times 10^5 \text{ arcseconds} \quad (24.25)$$

so the angular change is:

$$\theta = 9.9 \times 10^{-2} \text{ arcseconds}. \quad (24.26)$$

It is not clear whether the Gravity Probe B experiment is free of artifact as claimed [16], but the observed angular change is expected to be of the order of the very simple first approximation given in this paper. No more is claimed of Eq. (24.26), because it is to be regarded as a first approximation. However, it is clear that the standard approach to the so called Lense Thirring effect is attributed [16] not to Lense and Thirring but to Einstein, whose field equation has been known to be incorrect in several ways for ninety years. The problems with this well known field equation began to emerge in 1918 [4], when Bauer and Schroedinger independently and severely criticised its energy momentum structure. Using Eq. (24.1) it becomes clear as in Section 24.3 that the equation is irretrievably self-inconsistent because of its use of a symmetric connection or zero torsion [1–12]. This should have been

24.2 Calculation of the Gravitomagnetic Angular Frequency

clear at the outset of general relativity, when the torsion was discarded in an entirely arbitrary manner. It appears that the standard theoretical approach to the Lense Thirring effect is based on metrics that involve rotation. One of these is the Kerr metric, which is shown in Section 24.3 to violate the Bianchi identity (24.1), and therefore to be incorrect geometrically. It is obvious that no physics can emerge from basically incorrect mathematics. Even worse (Section 24.3) is the failure of the Kerr Newman metric to obey the fundamental Ricci cyclic equation known in standard physics as the "first Bianchi identity". Proceeding in this way it becomes clear that no line element based on a symmetric connection can be correct mathematically, meaning that gravitational physics of the last ninety years is meaningless.

There are clear additional problems in the standard treatment [16] of the Lense Thirring effect. The usual theory of the effect is developed in a mass dipole approximation. However, a sphere of uniform mass has no multipoles except for the familar monopole - the Newtonian potential:

$$\mathbf{g} = -\boldsymbol{\nabla}\Phi, \quad \Phi = -\frac{GM}{r}. \tag{24.27}$$

In molecular physics, it is well known that the existence of multipoles is determined by the group theory of the molecule. Cyanogen (NCCN) for example [4] has no electric dipole moment, but has a large electric quadrupole moment. The more symmetric a molecule, the higher the multipole it possesses, so sulphur hexafluoride for example only has a hexadecapole moment. A perfectly spherical molecule would not have any multipole at all. Similarly therefore a sphere of uniform mass (the earth) does not have any multipole except for the Newtonian monopole. This means that the dipole approximation used in the standard physics [16] to describe the so called Lense Thirring effect is not correct. Even if it were correct it is valid only when:

$$d \ll r \tag{24.28}$$

where d is the length of the dipole. It is entirely unclear whether this is the case for Gravity Probe B, and using higher order multipole terms will not cure this problem, as Pfister apparently claims. In a sphere of mass, none of these multipole terms exist, including the dipole term itself. Pfister's Eq. (24.1) [16] is therefore the result of a hypothetical dipole approximation for a sphere, a contradiction, and his result applies if and only if Eq. (24.28) of this paper is true, and if and only if the hypothetical dipole is aligned in the Z axis. Fortuitously, the first term of Pfister's result is the same as our Eq. (24.20), but this is a coincidence only. Pfister however does give a detailed history of the so called Lense - Thirring effect, and reveals several basic errors in the standard approach. The latter approach should now be regarded as entirely obsolete, and the Einstein field equation similarly abandoned.

24.3 Testing Metrics of the Einstein Equation with Eq. (24.1)

The methods used to test metrics of the Einstein field equation are a development of our previous work in papers 93 and 95 of this series (www.aias.us). In this section Eq. (24.1) is evaluated directly for the well known Kerr, Kerr Newman and Reissner Nordstrom metrics, which are all solutions of the Einstein field equation. For reference, our results for the misnamed Schwarzschild metric solution are also given. As discussed in paper 93 and elsewhere on www.aias.us, Schwarzschild in 1916 did not derive this metric. The source mass M does not appear in the two original 1916 papers by Schwarzschild, which give purely geometrical solutions of the equation:

$$G_{\mu\nu} = 0 \qquad (24.29)$$

where $G_{\mu\nu}$ is the well known Einstein tensor:

$$G_{\mu\nu} = R_{\mu\nu} - \frac{1}{2} R g_{\mu\nu}. \qquad (24.30)$$

Here $R_{\mu\nu}$ is the well known Ricci tensor, $g_{\mu\nu}$ is the symmetric metric and R is the Ricci scalar. In the standard terminology, Eq. (24.29) is known as a Ricci flat solution. Crothers has argued correctly (www.aias.us for example) that the Ricci flat solution can have no physical meaning. This should be immediately clear from the fact that the Einstein field equation is:

$$G_{\mu\nu} = k T_{\mu\nu} \qquad (24.31)$$

where k is the Einstein constant and where $T_{\mu\nu}$ is the canonical energy momentum density of the covariant Noether theorem. A Ricci flat solution therefore has no physics in it, it assumes a priori that there is no energy density present in the calculation. Schwarzschild in 1916 produced Ricci flat solutions, a purely geometrical exercise. In particular, a Ricci flat solution is devoid of mass M by definition, because mass M is part of $T_{\mu\nu}$. Others incorrectly reinstated mass M into the geometry in order to obtain Newtonian mechanics as a limit. This is an entirely meaningless procedure. In the present context, Ricci flat solutions produce the trivial result

$$T^{\kappa\mu\nu} = 0, \quad R^{\kappa}{}_{\mu}{}^{\mu\nu} = 0 \qquad (24.32)$$

simply because $R^{\kappa}{}_{\mu}{}^{\mu\nu}$ is initially set to zero and because $T^{\kappa\mu\nu}$ is also set to zero initially by use of a symmetric connection.

Table 24.1 Definition of charged/rotational metrics (Q = charge, J = angular momentum).

	Non-rotating ($J = 0$)	Rotating ($J \neq 0$)
Uncharged ($Q = 0$)	Schwarzschild	Kerr
Charged ($Q \neq 0$)	Reissner-Nordstrom	Kerr-Newman

The Kerr, Kerr-Newman and Reissner Nordstrom metrics produce the geometrically incorrect result:

$$T^{\kappa\mu\nu} = 0, \quad R^{\kappa}{}_{\mu}{}^{\mu\nu} \neq 0, \tag{24.33}$$

and so cannot give meaningful physics. Our results for these four metrics are summarized in Table (24.1) and Figs. (24.1) to (24.3). These are developments of our previous results in papers 93 and 95 on www.aias.us where it was found that the Robertson Walker metric of big bang also gives the incorrect result (24.33), and indeed all solutions of the Einstein field equation. Subsequently the generally covariant equations of ECE dynamics and cosmology were developed in parallel with the ECE equations of electrodynamics, and named the ECE engineering model on www.aias.us. This is a precisely determined system of eight equations in eight unknowns and can be applied to any situation in classical physics.

24.4 Discussion of Experimental Results of Gravity Probe B

The Gravity Probe B website (http://einstein.stanford.edu) reports the well known geodetic effect to be 6.6 plus or minus 0.097 arcseconds a year. The satellite failed to measure the gravitomagnetic effect, which can therefore only be reported to be within the noise of the experiment. There is another disputed measurement [16] of the gravitomagnetic effect at 0.043 arcseconds a year from LAGEOS 1976 and 1992. Therefore the ECE result is satisfactory, because it is a first approximation that can be greatly refined. The theoretical result by Pfister [16] in his Eq. (24.1) is given in terms of a quantity which he denotes **H**. The first term of Pfister's eq. (24.1) is the same as the result in this paper, but as argued here, Pfister uses what appears to be either an incorrect or inapplicable dipole approximation. A sphere of mass density (the earth) doe not have a mass dipole or any multipole higher than the Newtonian monopole (the usual Newtonian potential), but Pfister may be using an experimentally measured mass dipole due to the well known irregularities in the earth's structure. Other sites attribute the Lense Thirring effect to the Kerr metric, which as shown in this paper is geometrically incorrect. In a first equatorial approximation the angular frequency of the Lense Thirring effect from the Kerr metric is five times greater than the result (24.26) of ECE theory.

To add to the confusion, the Gravity probe B website claims that the Lense Thirring effect is 170 times smaller than 6.6 arcseconds per year (i.e. 39 milliarcseconds a year) but does not define the gravitomagnetic effect mathematically. Pfister claims that his Eq. (24.1) is the Lense Thirring effect, but this claim can be discarded both on thereotical and experimental grounds as argued in this paper. NASA has decided to cease funding of Gravity probe B, which developed an artifact. In view of these confused and incorrect claims, and in view of the fact that the Kerr metric and Einstein field equation are geometrically incorrect, the close agreement between ECE theory and the available data is conclusive evidence in favour of ECE theory, i.e. in favour of the fact that ECE theory predicts the correct order of the earth's gravitomagnetic precession in the first approximation.

24.5 Detailed metrics

24.5.1 Schwarzschild metric

The so-called Schwarzschild metric is a pure vacuum metric. The interpretation of the parameters (M: mass, G: Newton's constant of gravitation, c: velocity of light) was added later. The Ricci tensor and Einstein tensor vanish as do the cosmological charge and current densities which are a measure for the violation of the dual Bianchi identity (24.1).

24.5.1.1 Coordinates

$$\mathbf{x} = \begin{pmatrix} t \\ r \\ \vartheta \\ \varphi \end{pmatrix}$$

24.5.1.2 Metric

$$g_{\mu\nu} = \begin{pmatrix} \frac{2GM}{c^2 r} - 1 & 0 & 0 & 0 \\ 0 & \frac{1}{1 - \frac{2GM}{c^2 r}} & 0 & 0 \\ 0 & 0 & r^2 & 0 \\ 0 & 0 & 0 & r^2 \sin^2 \vartheta \end{pmatrix}$$

24.5.1.3 Contravariant Metric

$$g^{\mu\nu} = \begin{pmatrix} \frac{c^2 r}{2GM - c^2 r} & 0 & 0 & 0 \\ 0 & -\frac{2GM - c^2 r}{c^2 r} & 0 & 0 \\ 0 & 0 & \frac{1}{r^2} & 0 \\ 0 & 0 & 0 & \frac{1}{r^2 \sin^2 \vartheta} \end{pmatrix}$$

24.5.1.4 Christoffel Connection

$$\Gamma^0{}_{01} = -\frac{GM}{2rGM - c^2 r^2}$$

$$\Gamma^0{}_{10} = -\frac{GM}{2rGM - c^2 r^2}$$

$$\Gamma^1{}_{00} = -\frac{2G^2 M^2 - c^2 r GM}{c^4 r^3}$$

$$\Gamma^1{}_{11} = \frac{GM}{2rGM - c^2 r^2}$$

$$\Gamma^1{}_{22} = \frac{2GM - c^2 r}{c^2}$$

$$\Gamma^1{}_{33} = \frac{2 \sin^2 \vartheta\, GM - c^2 r \sin^2 \vartheta}{c^2}$$

$$\Gamma^2{}_{12} = \frac{1}{r}$$

$$\Gamma^2{}_{21} = \frac{1}{r}$$

$$\Gamma^2{}_{33} = -\cos\vartheta \sin\vartheta$$

$$\Gamma^3{}_{13} = \frac{1}{r}$$

$$\Gamma^3{}_{23} = \frac{\cos\vartheta}{\sin\vartheta}$$

$$\Gamma^3{}_{31} = \frac{1}{r}$$

$$\Gamma^3{}_{32} = \frac{\cos\vartheta}{\sin\vartheta}$$

24.5.1.5 Metric Compatibility

———— o.k.

24.5.1.6 Riemann Tensor

$$R^0{}_{101} = -\frac{2GM}{2r^2 GM - c^2 r^3}$$

$$R^0{}_{110} = \frac{2GM}{2r^2 GM - c^2 r^3}$$

$$R^0{}_{202} = -\frac{GM}{c^2 r}$$

$$R^0{}_{220} = \frac{GM}{c^2 r}$$

$$R^0{}_{303} = -\frac{\sin^2\vartheta \, GM}{c^2 r}$$

24.5 Detailed metrics

$$R^0{}_{330} = \frac{\sin^2 \vartheta\, G\, M}{c^2\, r}$$

$$R^1{}_{001} = -\frac{4\, G^2\, M^2 - 2\, c^2\, r\, G\, M}{c^4\, r^4}$$

$$R^1{}_{010} = \frac{4\, G^2\, M^2 - 2\, c^2\, r\, G\, M}{c^4\, r^4}$$

$$R^1{}_{212} = -\frac{G\, M}{c^2\, r}$$

$$R^1{}_{221} = \frac{G\, M}{c^2\, r}$$

$$R^1{}_{313} = -\frac{\sin^2 \vartheta\, G\, M}{c^2\, r}$$

$$R^1{}_{331} = \frac{\sin^2 \vartheta\, G\, M}{c^2\, r}$$

$$R^2{}_{002} = \frac{2\, G^2\, M^2 - c^2\, r\, G\, M}{c^4\, r^4}$$

$$R^2{}_{020} = -\frac{2\, G^2\, M^2 - c^2\, r\, G\, M}{c^4\, r^4}$$

$$R^2{}_{112} = -\frac{G\, M}{2\, r^2\, G\, M - c^2\, r^3}$$

$$R^2{}_{121} = \frac{G\, M}{2\, r^2\, G\, M - c^2\, r^3}$$

$$R^2{}_{323} = \frac{2\sin^2\vartheta\, G\, M}{c^2\, r}$$

$$R^2{}_{332} = -\frac{2\sin^2\vartheta\, G\, M}{c^2\, r}$$

$$R^3{}_{003} = \frac{2\, G^2\, M^2 - c^2\, r\, G\, M}{c^4\, r^4}$$

$$R^3{}_{030} = -\frac{2\, G^2\, M^2 - c^2\, r\, G\, M}{c^4\, r^4}$$

$$R^3{}_{113} = -\frac{G\, M}{2\, r^2\, G\, M - c^2\, r^3}$$

$$R^3{}_{131} = \frac{G\, M}{2\, r^2\, G\, M - c^2\, r^3}$$

$$R^3{}_{223} = -\frac{2\, G\, M}{c^2\, r}$$

$$R^3{}_{232} = \frac{2\, G\, M}{c^2\, r}$$

24.5.1.7 Ricci Tensor

———— all elements zero

24.5.1.8 Ricci Scalar

$$R_{sc} = 0$$

24.5.1.9 Bianchi identity (Ricci cyclic equation $R^\kappa{}_{[\mu\nu\sigma]} = 0$)

———— o.k.

24.5 Detailed metrics

24.5.1.10 Einstein Tensor

———— all elements zero

24.5.1.11 Hodge Dual of Bianchi Identity

———— (see charge and current densities)

24.5.1.12 Scalar Charge Density $(-R^0{}_i{}^{i0})$

$$\rho = 0$$

24.5.1.13 Current Density Class 1 $(-R^i{}_\mu{}^{\mu j})$

$$J_1 = 0$$
$$J_2 = 0$$
$$J_3 = 0$$

24.5.1.14 Current Density Class 2 $(-R^i{}_\mu{}^{\mu j})$

$$J_1 = 0$$
$$J_2 = 0$$
$$J_3 = 0$$

24.5.1.15 Current Density Class 3 $(-R^i{}_\mu{}^{\mu j})$

$$J_1 = 0$$
$$J_2 = 0$$
$$J_3 = 0$$

24.5.2 Reissner-Nordstrom metric

This is a metric of a charged mass. M is a mass parameter, Q a charge parameter. Cosmological charge and current densities do exist.

24.5.2.1 Coordinates

$$\mathbf{x} = \begin{pmatrix} t \\ r \\ \vartheta \\ \varphi \end{pmatrix}$$

24.5.2.2 Metric

$$g_{\mu\nu} = \begin{pmatrix} -\frac{Q^2}{r^2} + \frac{2M}{r} - 1 & 0 & 0 & 0 \\ 0 & \frac{1}{\frac{Q^2}{r^2} - \frac{2M}{r} + 1} & 0 & 0 \\ 0 & 0 & r^2 & 0 \\ 0 & 0 & 0 & r^2 \sin^2 \vartheta \end{pmatrix}$$

24.5.2.3 Contravariant Metric

$$g^{\mu\nu} = \begin{pmatrix} -\frac{r^2}{Q^2 - 2rM + r^2} & 0 & 0 & 0 \\ 0 & \frac{Q^2 - 2rM + r^2}{r^2} & 0 & 0 \\ 0 & 0 & \frac{1}{r^2} & 0 \\ 0 & 0 & 0 & \frac{1}{r^2 \sin^2 \vartheta} \end{pmatrix}$$

24.5.2.4 Christoffel Connection

$$\Gamma^0{}_{01} = -\frac{Q^2 - rM}{rQ^2 - 2r^2 M + r^3}$$

$$\Gamma^0{}_{10} = -\frac{Q^2 - rM}{rQ^2 - 2r^2 M + r^3}$$

$$\Gamma^1{}_{00} = -\frac{Q^4 + (r^2 - 3rM)Q^2 + 2r^2 M^2 - r^3 M}{r^5}$$

$$\Gamma^1{}_{11} = \frac{Q^2 - rM}{rQ^2 - 2r^2 M + r^3}$$

$$\Gamma^1{}_{22} = -\frac{Q^2 - 2rM + r^2}{r}$$

$$\Gamma^1{}_{33} = -\frac{\sin^2\vartheta\, Q^2 - 2r\sin^2\vartheta\, M + r^2\sin^2\vartheta}{r}$$

$$\Gamma^2{}_{12} = \frac{1}{r}$$

$$\Gamma^2{}_{21} = \frac{1}{r}$$

$$\Gamma^2{}_{33} = -\cos\vartheta\,\sin\vartheta$$

$$\Gamma^3{}_{13} = \frac{1}{r}$$

$$\Gamma^3{}_{23} = \frac{\cos\vartheta}{\sin\vartheta}$$

$$\Gamma^3{}_{31} = \frac{1}{r}$$

$$\Gamma^3{}_{32} = \frac{\cos\vartheta}{\sin\vartheta}$$

24.5.2.5 Metric Compatibility

———— o.k.

24.5.2.6 Riemann Tensor

$$R^0{}_{101} = -\frac{3Q^2 - 2rM}{r^2 Q^2 - 2r^3 M + r^4}$$

$$R^0{}_{110} = \frac{3Q^2 - 2rM}{r^2 Q^2 - 2r^3 M + r^4}$$

$$R^0{}_{202} = \frac{Q^2 - rM}{r^2}$$

$$R^0{}_{220} = -\frac{Q^2 - rM}{r^2}$$

$$R^0{}_{303} = \frac{\sin^2\vartheta\, Q^2 - r\sin^2\vartheta\, M}{r^2}$$

$$R^0{}_{330} = -\frac{\sin^2\vartheta\, Q^2 - r\sin^2\vartheta\, M}{r^2}$$

$$R^1{}_{001} = -\frac{3Q^4 + (3r^2 - 8rM)\,Q^2 + 4r^2 M^2 - 2r^3 M}{r^6}$$

$$R^1{}_{010} = \frac{3Q^4 + (3r^2 - 8rM)\,Q^2 + 4r^2 M^2 - 2r^3 M}{r^6}$$

$$R^1{}_{212} = \frac{Q^2 - rM}{r^2}$$

$$R^1{}_{221} = -\frac{Q^2 - rM}{r^2}$$

$$R^1{}_{313} = \frac{\sin^2\vartheta\, Q^2 - r\sin^2\vartheta\, M}{r^2}$$

$$R^1{}_{331} = -\frac{\sin^2\vartheta\, Q^2 - r\sin^2\vartheta\, M}{r^2}$$

$$R^2{}_{002} = \frac{Q^4 + (r^2 - 3rM)\,Q^2 + 2r^2 M^2 - r^3 M}{r^6}$$

$$R^2{}_{020} = -\frac{Q^4 + (r^2 - 3rM)\,Q^2 + 2r^2 M^2 - r^3 M}{r^6}$$

24.5 Detailed metrics

$$R^2{}_{112} = -\frac{Q^2 - rM}{r^2 Q^2 - 2r^3 M + r^4}$$

$$R^2{}_{121} = \frac{Q^2 - rM}{r^2 Q^2 - 2r^3 M + r^4}$$

$$R^2{}_{323} = -\frac{\sin^2 \vartheta \, Q^2 - 2r \sin^2 \vartheta \, M}{r^2}$$

$$R^2{}_{332} = \frac{\sin^2 \vartheta \, Q^2 - 2r \sin^2 \vartheta \, M}{r^2}$$

$$R^3{}_{003} = \frac{Q^4 + (r^2 - 3rM) Q^2 + 2r^2 M^2 - r^3 M}{r^6}$$

$$R^3{}_{030} = -\frac{Q^4 + (r^2 - 3rM) Q^2 + 2r^2 M^2 - r^3 M}{r^6}$$

$$R^3{}_{113} = -\frac{Q^2 - rM}{r^2 Q^2 - 2r^3 M + r^4}$$

$$R^3{}_{131} = \frac{Q^2 - rM}{r^2 Q^2 - 2r^3 M + r^4}$$

$$R^3{}_{223} = \frac{Q^2 - 2rM}{r^2}$$

$$R^3{}_{232} = -\frac{Q^2 - 2rM}{r^2}$$

24.5.2.7 Ricci Tensor

$$\text{Ric}_{00} = \frac{Q^2 \left(Q^2 - 2rM + r^2\right)}{r^6}$$

$$\text{Ric}_{11} = -\frac{Q^2}{r^2 \left(Q^2 - 2rM + r^2\right)}$$

$$\text{Ric}_{22} = \frac{Q^2}{r^2}$$

$$\text{Ric}_{33} = \frac{\sin^2 \vartheta \, Q^2}{r^2}$$

24.5.2.8 Ricci Scalar

$$R_{sc} = 0$$

24.5.2.9 Bianchi identity (Ricci cyclic equation $R^\kappa{}_{[\mu\nu\sigma]} = 0$)

———— o.k.

24.5.2.10 Einstein Tensor

———— not zero:

$$G_{00} = \frac{Q^4 + \left(r^2 - 2rM\right) Q^2}{r^6}$$

$$G_{11} = -\frac{Q^2}{r^2 Q^2 - 2r^3 M + r^4}$$

$$G_{22} = \frac{Q^2}{r^2}$$

$$G_{33} = \frac{\sin^2 \vartheta \, Q^2}{r^2}$$

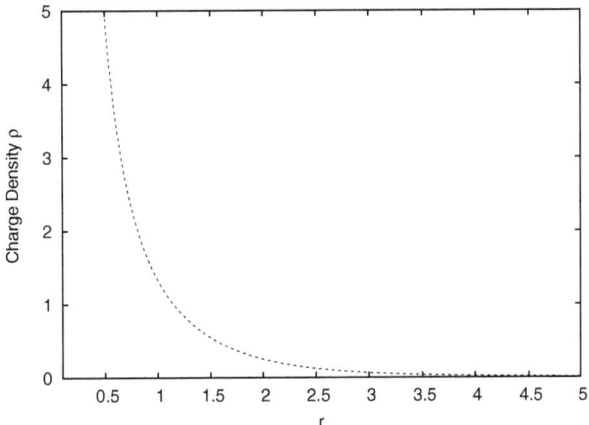

Fig. 24.1. Reissner-Nordstrom metric, cosmological charge density ρ for M = 1, Q = 2.

24.5.2.11 Hodge Dual of Bianchi Identity

———— (see charge and current densities)

24.5.2.12 Scalar Charge Density $(-R^0{}_i{}^{i0})$

$$\rho = \frac{Q^2}{r^2 Q^2 - 2 r^3 M + r^4}$$

24.5.2.13 Current Density Class 1 $(-R^i{}_\mu{}^{\mu j})$

$$J_1 = \frac{Q^4 + (r^2 - 2 r M) Q^2}{r^6}$$

$$J_2 = -\frac{Q^2}{r^6}$$

$$J_3 = -\frac{Q^2}{r^6 \sin^2 \vartheta}$$

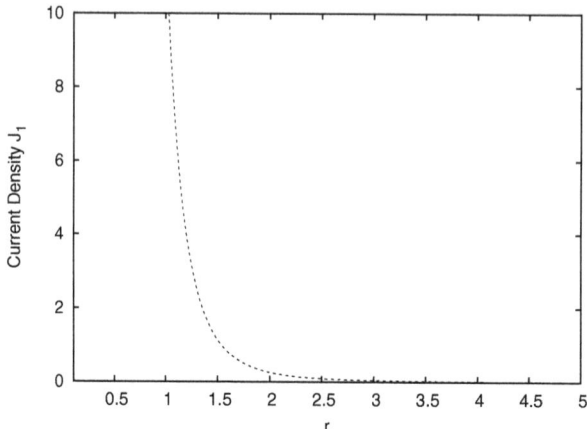

Fig. 24.2. Reissner-Nordstrom metric, cosmological current density J_r for M = 1, Q = 2.

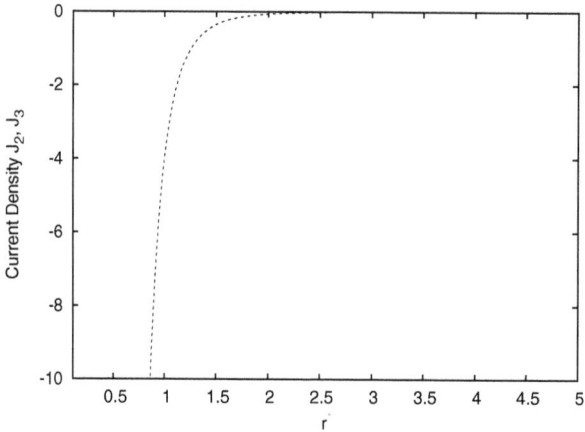

Fig. 24.3. Reissner-Nordstrom metric, cosmological current density J_θ, J_φ for M = 1, Q = 2.

24.5.2.14 Current Density Class 2 ($-R^i{}_\mu{}^{\mu j}$)

$$J_1 = 0$$
$$J_2 = 0$$
$$J_3 = 0$$

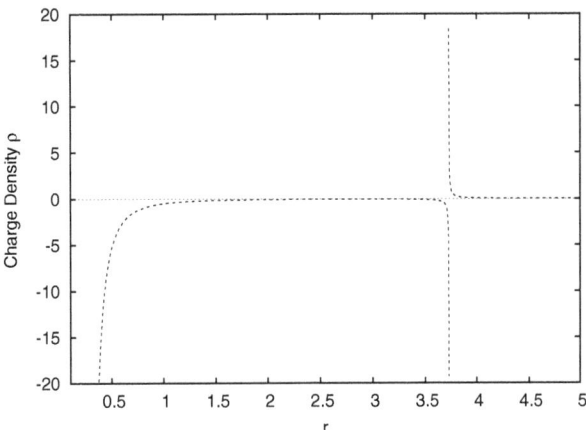

Fig. 24.4. Reissner-Nordstrom metric, cosmological charge density ρ for M=2, Q=1.

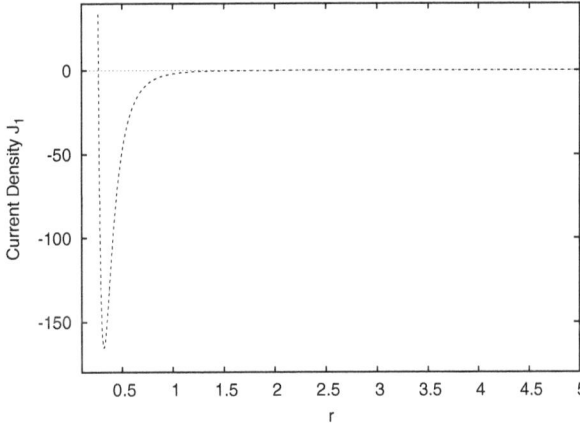

Fig. 24.5. Reissner-Nordstrom metric, cosmological current density J_r, for J_r M=2, Q=1.

24.5.2.15 Current Density Class 3 $(-R^i{}_\mu{}^{\mu j})$

$$J_1 = 0$$
$$J_2 = 0$$
$$J_3 = 0$$

442   24 ECE Theory of the Earth's Gravitomagnetic Precession

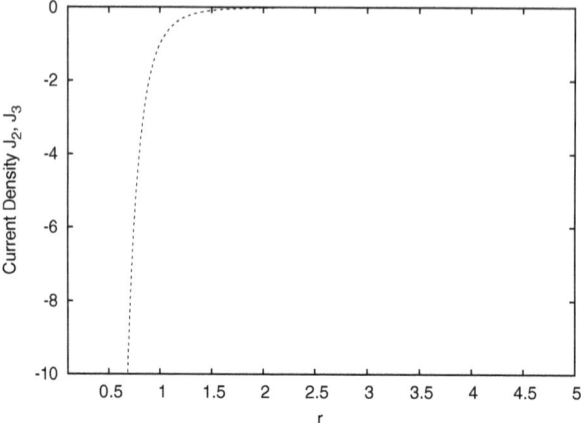

Fig. 24.6. Reissner-Nordstrom metric, cosmological current density J_θ, J_φ for M=2, Q=1.

24.5.3 Kerr metric

This metric describes a rotating mass without charge. M is the mass parameter, J the parameter of angular momentum. Cosmological charge and current densities do exist. There are horizons (pole locations) in these quantities which give hint to an irregular behaviour of this metric.

24.5.3.1 Coordinates

$$\mathbf{x} = \begin{pmatrix} t \\ r \\ \vartheta \\ \varphi \end{pmatrix}$$

24.5.3.2 Metric

$$g_{\mu\nu} = \begin{pmatrix} \frac{2M}{r} - 1 & 0 & 0 & -\frac{4\sin^2\vartheta\, J}{r} \\ 0 & \frac{1}{1-\frac{2M}{r}} & 0 & 0 \\ 0 & 0 & r^2 & 0 \\ -\frac{4\sin^2\vartheta\, J}{r} & 0 & 0 & r^2\sin^2\vartheta \end{pmatrix}$$

24.5.3.3 Contravariant Metric

$$g^{\mu\nu} = \begin{pmatrix} \frac{r^4}{2r^3 M - 16 \sin^2 \vartheta \, J^2 - r^4} & 0 & 0 & \frac{4rJ}{2r^3 M - 16 \sin^2 \vartheta \, J^2 - r^4} \\ 0 & -\frac{2M-r}{r} & 0 & 0 \\ 0 & 0 & \frac{1}{r^2} & 0 \\ \frac{4rJ}{2r^3 M - 16 \sin^2 \vartheta \, J^2 - r^4} & 0 & 0 & \frac{r(2M-r)}{\sin^2 \vartheta \, (2r^3 M - 16 \sin^2 \vartheta \, J^2 - r^4)} \end{pmatrix}$$

24.5.3.4 Christoffel Connection

$$\Gamma^0{}_{01} = -\frac{r^3 M - 8 \sin^2 \vartheta \, J^2}{2r^4 M - 16r \sin^2 \vartheta \, J^2 - r^5}$$

$$\Gamma^0{}_{02} = -\frac{16 \cos \vartheta \, \sin \vartheta \, J^2}{2r^3 M - 16 \sin^2 \vartheta \, J^2 - r^4}$$

$$\Gamma^0{}_{10} = -\frac{r^3 M - 8 \sin^2 \vartheta \, J^2}{2r^4 M - 16r \sin^2 \vartheta \, J^2 - r^5}$$

$$\Gamma^0{}_{13} = \frac{6r^2 \sin^2 \vartheta \, J}{2r^3 M - 16 \sin^2 \vartheta \, J^2 - r^4}$$

$$\Gamma^0{}_{20} = -\frac{16 \cos \vartheta \, \sin \vartheta \, J^2}{2r^3 M - 16 \sin^2 \vartheta \, J^2 - r^4}$$

$$\Gamma^0{}_{31} = \frac{6r^2 \sin^2 \vartheta \, J}{2r^3 M - 16 \sin^2 \vartheta \, J^2 - r^4}$$

$$\Gamma^1{}_{00} = -\frac{2M^2 - rM}{r^3}$$

$$\Gamma^1{}_{03} = \frac{4 \sin^2 \vartheta \, J M - 2r \sin^2 \vartheta \, J}{r^3}$$

$$\Gamma^1{}_{11} = \frac{M}{2rM - r^2}$$

$$\Gamma^1{}_{22} = 2M - r$$

$$\Gamma^1{}_{30} = \frac{4\sin^2\vartheta\, JM - 2r\sin^2\vartheta\, J}{r^3}$$

$$\Gamma^1{}_{33} = 2\sin^2\vartheta\, M - r\sin^2\vartheta$$

$$\Gamma^2{}_{03} = \frac{4\cos\vartheta\,\sin\vartheta\, J}{r^3}$$

$$\Gamma^2{}_{12} = \frac{1}{r}$$

$$\Gamma^2{}_{21} = \frac{1}{r}$$

$$\Gamma^2{}_{30} = \frac{4\cos\vartheta\,\sin\vartheta\, J}{r^3}$$

$$\Gamma^2{}_{33} = -\cos\vartheta\,\sin\vartheta$$

$$\Gamma^3{}_{01} = -\frac{2J}{2r^3 M - 16\sin^2\vartheta\, J^2 - r^4}$$

$$\Gamma^3{}_{02} = -\frac{8\cos\vartheta\, JM - 4r\cos\vartheta\, J}{2r^3\sin\vartheta\, M - 16\sin^3\vartheta\, J^2 - r^4\sin\vartheta}$$

$$\Gamma^3{}_{10} = -\frac{2J}{2r^3 M - 16\sin^2\vartheta\, J^2 - r^4}$$

24.5 Detailed metrics

$$\Gamma^3{}_{13} = \frac{2r^3 M + 8\sin^2\vartheta J^2 - r^4}{2r^4 M - 16r\sin^2\vartheta J^2 - r^5}$$

$$\Gamma^3{}_{20} = -\frac{8\cos\vartheta J M - 4r\cos\vartheta J}{2r^3\sin\vartheta M - 16\sin^3\vartheta J^2 - r^4\sin\vartheta}$$

$$\Gamma^3{}_{23} = \frac{\cos\vartheta}{\sin\vartheta}$$

$$\Gamma^3{}_{31} = \frac{2r^3 M + 8\sin^2\vartheta J^2 - r^4}{2r^4 M - 16r\sin^2\vartheta J^2 - r^5}$$

$$\Gamma^3{}_{32} = \frac{\cos\vartheta}{\sin\vartheta}$$

24.5.3.5 Metric Compatibility

———— o.k.

24.5.3.6 Riemann Tensor

$$R^0{}_{003} = \frac{8r^3\sin^2\vartheta J M^2 + (32\sin^4\vartheta J^3 - 4r^4\sin^2\vartheta J) M + (48r\sin^4\vartheta - 64r\sin^2\vartheta) J^3}{2r^7 M - 16r^4\sin^2\vartheta J^2 - r^8}$$

$$R^0{}_{012} = \frac{48r^2\cos\vartheta\sin\vartheta J^2 M - 24r^3\cos\vartheta\sin\vartheta J^2}{4r^6 M^2 + (-64r^3\sin^2\vartheta J^2 - 4r^7) M + 256\sin^4\vartheta J^4 + 32r^4\sin^2\vartheta J^2 + r^8}$$

$$R^0{}_{021} = -\frac{48r^2\cos\vartheta\sin\vartheta J^2 M - 24r^3\cos\vartheta\sin\vartheta J^2}{4r^6 M^2 + (-64r^3\sin^2\vartheta J^2 - 4r^7) M + 256\sin^4\vartheta J^4 + 32r^4\sin^2\vartheta J^2 + r^8}$$

$$R^0{}_{030} = -\frac{8r^3\sin^2\vartheta J M^2 + (32\sin^4\vartheta J^3 - 4r^4\sin^2\vartheta J) M + (48r\sin^4\vartheta - 64r\sin^2\vartheta) J^3}{2r^7 M - 16r^4\sin^2\vartheta J^2 - r^8}$$

$$R^0{}_{101} = -R^0{}_{110}$$

$$R^0{}_{102} = -\frac{48r^3\cos\vartheta\sin\vartheta J^2 M - 384\cos\vartheta\sin^3\vartheta J^4 - 32r^4\cos\vartheta\sin\vartheta J^2}{4r^7 M^2 + (-64r^4\sin^2\vartheta J^2 - 4r^8) M + 256r\sin^4\vartheta J^4 + 32r^5\sin^2\vartheta J^2 + r^9}$$

$$R^0{}_{110} = \frac{8r^6 M^3 + (-128r^3\sin^2\vartheta J^2 - 8r^7) M^2 + (512\sin^4\vartheta J^4 + 112r^4\sin^2\vartheta J^2 + 2r^8) M - 192r\sin^4\vartheta J^4 - 28r^5\sin^2\vartheta J^2}{8r^8 M^3 + (-128r^5\sin^2\vartheta J^2 - 12r^9) M^2 + (512r^2\sin^4\vartheta J^4 + 128r^6\sin^2\vartheta J^2 + 6r^{10}) M - 256r^3\sin^4\vartheta J^4 - 32r^7\sin^2\vartheta J^2 - r^{11}}$$

$$R^0{}_{113} = -\frac{24r^4 \sin^2 \vartheta J M^2 + (96r \sin^4 \vartheta J^3 - 24r^5 \sin^2 \vartheta J) M - 96 r^2 \sin^4 \vartheta J^3 + 6r^6 \sin^2 \vartheta J}{8r^6 M^3 + (-128 r^3 \sin^2 \vartheta J^2 - 12 r^7) M^2 + (512 \sin^4 \vartheta J^4 + 128 r^4 \sin^2 \vartheta J^2 + 6 r^8) M - 256 r \sin^4 \vartheta J^4 - 32 r^5 \sin^2 \vartheta J^2 - r^9}$$

$$R^0{}_{120} = \frac{48 r^3 \cos \vartheta \sin \vartheta J^2 M - 384 \cos \vartheta \sin^3 \vartheta J^4 - 32 r^4 \cos \vartheta \sin \vartheta J^2}{4 r^7 M^2 + (-64 r^4 \sin^2 \vartheta J^2 - 4 r^8) M + 256 r \sin^4 \vartheta J^4 + 32 r^5 \sin^2 \vartheta J^2 + r^9}$$

$$R^0{}_{123} = \frac{24 r^5 \cos \vartheta \sin \vartheta J M - 96 r^2 \cos \vartheta \sin^3 \vartheta J^3 - 12 r^6 \cos \vartheta \sin \vartheta J}{4 r^6 M^2 + (-64 r^3 \sin^2 \vartheta J^2 - 4 r^7) M + 256 \sin^4 \vartheta J^4 + 32 r^4 \sin^2 \vartheta J^2 + r^8}$$

$$R^0{}_{131} = \frac{24 r^4 \sin^2 \vartheta J M^2 + (96 r \sin^4 \vartheta J^3 - 24 r^5 \sin^2 \vartheta J) M - 96 r^2 \sin^4 \vartheta J^3 + 6 r^6 \sin^2 \vartheta J}{8 r^6 M^3 + (-128 r^3 \sin^2 \vartheta J^2 - 12 r^7) M^2 + (512 \sin^4 \vartheta J^4 + 128 r^4 \sin^2 \vartheta J^2 + 6 r^8) M - 256 r \sin^4 \vartheta J^4 - 32 r^5 \sin^2 \vartheta J^2 - r^9}$$

$$R^0{}_{132} = -\frac{24 r^5 \cos \vartheta \sin \vartheta J M - 96 r^2 \cos \vartheta \sin^3 \vartheta J^3 - 12 r^6 \cos \vartheta \sin \vartheta J}{4 r^6 M^2 + (-64 r^3 \sin^2 \vartheta J^2 - 4 r^7) M + 256 \sin^4 \vartheta J^4 + 32 r^4 \sin^2 \vartheta J^2 + r^8}$$

$$R^0{}_{201} = -\frac{96 r^3 \cos \vartheta \sin \vartheta J^2 M - 384 \cos \vartheta \sin^3 \vartheta J^4 - 56 r^4 \cos \vartheta \sin \vartheta J^2}{4 r^7 M^2 + (-64 r^4 \sin^2 \vartheta J^2 - 4 r^8) M + 256 r \sin^4 \vartheta J^4 + 32 r^5 \sin^2 \vartheta J^2 + r^9}$$

$$N = 4 r^7 M^2 - 64 r^4 \sin^2 \vartheta J^2 M - 4 r^8 M + 256 r \sin^4 \vartheta J^4 + 32 r^5 \sin^2 \vartheta J^2 + r^9$$

$$R^0{}_{202} = \frac{-4 r^6 M^3 + 64 r^3 \sin^2 \vartheta J^2 M^2 + 4 r^7 M^2 - 256 \sin^4 \vartheta J^4 M - 112 r^4 \sin^2 \vartheta J^2 M}{N}$$
$$+ \frac{32 r^4 J^2 M - r^8 M + 384 r \sin^4 \vartheta J^4 + 40 r^5 \sin^2 \vartheta J^2 - 16 r^5 J^2}{N}$$

$$R^0{}_{210} = \frac{96 r^3 \cos \vartheta \sin \vartheta J^2 M - 384 \cos \vartheta \sin^3 \vartheta J^4 - 56 r^4 \cos \vartheta \sin \vartheta J^2}{4 r^7 M^2 + (-64 r^4 \sin^2 \vartheta J^2 - 4 r^8) M + 256 r \sin^4 \vartheta J^4 + 32 r^5 \sin^2 \vartheta J^2 + r^9}$$

$$R^0{}_{213} = \frac{6 r^2 \cos \vartheta \sin \vartheta J}{2 r^3 M - 16 \sin^2 \vartheta J^2 - r^4}$$

$$R^0{}_{220} = -R^0{}_{202}$$

$$R^0{}_{223} = -\frac{12 r^2 \sin^2 \vartheta J M - 6 r^3 \sin^2 \vartheta J}{2 r^3 M - 16 \sin^2 \vartheta J^2 - r^4}$$

24.5 Detailed metrics

$$R^0{}_{231} = -\frac{6\,r^2\,\cos\vartheta\,\sin\vartheta\,J}{2\,r^3\,M - 16\,\sin^2\vartheta\,J^2 - r^4}$$

$$R^0{}_{232} = \frac{12\,r^2\,\sin^2\vartheta\,J\,M - 6\,r^3\,\sin^2\vartheta\,J}{2\,r^3\,M - 16\,\sin^2\vartheta\,J^2 - r^4}$$

$$R^0{}_{303} = -\frac{2\,r^3\,\sin^2\vartheta\,M^2 + \left(8\,\sin^4\vartheta\,J^2 - r^4\,\sin^2\vartheta\right)M + \left(12\,r\,\sin^4\vartheta - 16\,r\,\sin^2\vartheta\right)J^2}{2\,r^4\,M - 16\,r\,\sin^2\vartheta\,J^2 - r^5}$$

$$R^0{}_{312} = -\frac{12\,r^5\,\cos\vartheta\,\sin\vartheta\,J\,M - 6\,r^6\,\cos\vartheta\,\sin\vartheta\,J}{4\,r^6\,M^2 + \left(-64\,r^3\,\sin^2\vartheta\,J^2 - 4\,r^7\right)M + 256\,\sin^4\vartheta\,J^4 + 32\,r^4\,\sin^2\vartheta\,J^2 + r^8}$$

$$R^0{}_{321} = \frac{12\,r^5\,\cos\vartheta\,\sin\vartheta\,J\,M - 6\,r^6\,\cos\vartheta\,\sin\vartheta\,J}{4\,r^6\,M^2 + \left(-64\,r^3\,\sin^2\vartheta\,J^2 - 4\,r^7\right)M + 256\,\sin^4\vartheta\,J^4 + 32\,r^4\,\sin^2\vartheta\,J^2 + r^8}$$

$$R^0{}_{330} = \frac{2\,r^3\,\sin^2\vartheta\,M^2 + \left(8\,\sin^4\vartheta\,J^2 - r^4\,\sin^2\vartheta\right)M + \left(12\,r\,\sin^4\vartheta - 16\,r\,\sin^2\vartheta\right)J^2}{2\,r^4\,M - 16\,r\,\sin^2\vartheta\,J^2 - r^5}$$

$$R^1{}_{001} = -\frac{8\,r^3\,M^3 + \left(-64\,\sin^2\vartheta\,J^2 - 8\,r^4\right)M^2 + \left(32\,r\,\sin^2\vartheta\,J^2 + 2\,r^5\right)M - 4\,r^2\,\sin^2\vartheta\,J^2}{2\,r^7\,M - 16\,r^4\,\sin^2\vartheta\,J^2 - r^8}$$

$$R^1{}_{002} = \frac{16\,\cos\vartheta\,\sin\vartheta\,J^2\,M - 8\,r\,\cos\vartheta\,\sin\vartheta\,J^2}{2\,r^5\,M - 16\,r^2\,\sin^2\vartheta\,J^2 - r^6}$$

$$R^1{}_{010} = \frac{8\,r^3\,M^3 + \left(-64\,\sin^2\vartheta\,J^2 - 8\,r^4\right)M^2 + \left(32\,r\,\sin^2\vartheta\,J^2 + 2\,r^5\right)M - 4\,r^2\,\sin^2\vartheta\,J^2}{2\,r^7\,M - 16\,r^4\,\sin^2\vartheta\,J^2 - r^8}$$

$$R^1{}_{013} = -\frac{16\,r^3\,\sin^2\vartheta\,J\,M^2 + \left(-128\,\sin^4\vartheta\,J^3 - 20\,r^4\,\sin^2\vartheta\,J\right)M + 48\,r\,\sin^4\vartheta\,J^3 + 6\,r^5\,\sin^2\vartheta\,J}{2\,r^7\,M - 16\,r^4\,\sin^2\vartheta\,J^2 - r^8}$$

$$R^1{}_{020} = -\frac{16\,\cos\vartheta\,\sin\vartheta\,J^2\,M - 8\,r\,\cos\vartheta\,\sin\vartheta\,J^2}{2\,r^5\,M - 16\,r^2\,\sin^2\vartheta\,J^2 - r^6}$$

$$R^1{}_{023} = \frac{48 r^3 \cos\vartheta \sin\vartheta J M^2 + \left(-192 \cos\vartheta \sin^3\vartheta J^3 - 48 r^4 \cos\vartheta \sin\vartheta J\right) M + 96 r \cos\vartheta \sin^3\vartheta J^3 + 12 r^5 \cos\vartheta \sin\vartheta J}{2 r^6 M - 16 r^3 \sin^2\vartheta J^2 - r^7}$$

$$R^1{}_{031} = \frac{16\,r^3\,\sin^2\vartheta\,J\,M^2 + \left(-128\,\sin^4\vartheta\,J^3 - 20\,r^4\,\sin^2\vartheta\,J\right)M + 48\,r\,\sin^4\vartheta\,J^3 + 6\,r^5\,\sin^2\vartheta\,J}{2\,r^7\,M - 16\,r^4\,\sin^2\vartheta\,J^2 - r^8}$$

$$R^1{}_{032} = -\frac{48 r^3 \cos\vartheta \sin\vartheta J M^2 + \left(-192 \cos\vartheta \sin^3\vartheta J^3 - 48 r^4 \cos\vartheta \sin\vartheta J\right) M + 96 r \cos\vartheta \sin^3\vartheta J^3 + 12 r^5 \cos\vartheta \sin\vartheta J}{2 r^6 M - 16 r^3 \sin^2\vartheta J^2 - r^7}$$

$$R^1{}_{203} = \frac{24\,\cos\vartheta\,\sin\vartheta\,J\,M^2 - 24\,r\,\cos\vartheta\,\sin\vartheta\,J\,M + 6\,r^2\,\cos\vartheta\,\sin\vartheta\,J}{2\,r^3\,M - 16\,\sin^2\vartheta\,J^2 - r^4}$$

$$R^1{}_{212} = -\frac{M}{r}$$

$$R^1{}_{221} = \frac{M}{r}$$

$$R^1{}_{230} = -\frac{24\cos\vartheta\,\sin\vartheta\,J\,M^2 - 24\,r\cos\vartheta\,\sin\vartheta\,J\,M + 6\,r^2\cos\vartheta\,\sin\vartheta\,J}{2\,r^3\,M - 16\sin^2\vartheta\,J^2 - r^4}$$

$$R^1{}_{301} = \frac{16\,r^3\sin^2\vartheta\,J\,M^2 + \left(-128\sin^4\vartheta\,J^3 - 20\,r^4\sin^2\vartheta\,J\right)M + 48\,r\sin^4\vartheta\,J^3 + 6\,r^5\sin^2\vartheta\,J}{2\,r^7\,M - 16\,r^4\sin^2\vartheta\,J^2 - r^8}$$

$$R^1{}_{302} = -\frac{12\cos\vartheta\,\sin\vartheta\,J\,M - 6\,r\cos\vartheta\,\sin\vartheta\,J}{r^3}$$

$$R^1{}_{310} = -\frac{16\,r^3\sin^2\vartheta\,J\,M^2 + \left(-128\sin^4\vartheta\,J^3 - 20\,r^4\sin^2\vartheta\,J\right)M + 48\,r\sin^4\vartheta\,J^3 + 6\,r^5\sin^2\vartheta\,J}{2\,r^7\,M - 16\,r^4\sin^2\vartheta\,J^2 - r^8}$$

$$R^1{}_{313} = -\frac{2\,r^3\sin^2\vartheta\,M^2 + \left(56\sin^4\vartheta\,J^2 - r^4\sin^2\vartheta\right)M - 36\,r\sin^4\vartheta\,J^2}{2\,r^4\,M - 16\,r\sin^2\vartheta\,J^2 - r^5}$$

$$R^1{}_{320} = \frac{12\cos\vartheta\,\sin\vartheta\,J\,M - 6\,r\cos\vartheta\,\sin\vartheta\,J}{r^3}$$

$$R^1{}_{331} = \frac{2\,r^3\sin^2\vartheta\,M^2 + \left(56\sin^4\vartheta\,J^2 - r^4\sin^2\vartheta\right)M - 36\,r\sin^4\vartheta\,J^2}{2\,r^4\,M - 16\,r\sin^2\vartheta\,J^2 - r^5}$$

$$R^2{}_{001} = -\frac{8\cos\vartheta\,\sin\vartheta\,J^2}{2\,r^6\,M - 16\,r^3\sin^2\vartheta\,J^2 - r^7}$$

$$R^2{}_{002} = \frac{4r^3 M^3 + \left(-32\sin^2\vartheta\,J^2 - 4r^4\right)M^2 + \left(\left(48r\sin^2\vartheta - 32r\right)J^2 + r^5\right)M + \left(16r^2 - 16r^2\sin^2\vartheta\right)J^2}{2r^7 M - 16r^4\sin^2\vartheta\,J^2 - r^8}$$

$$R^2{}_{010} = \frac{8\cos\vartheta\,\sin\vartheta\,J^2}{2\,r^6\,M - 16\,r^3\sin^2\vartheta\,J^2 - r^7}$$

$$R^2{}_{013} = -\frac{6\cos\vartheta\,\sin\vartheta\,J}{r^4}$$

$$R^2{}_{020} = -\frac{4r^3 M^3 + \left(-32\sin^2\vartheta\,J^2 - 4r^4\right)M^2 + \left(\left(48r\sin^2\vartheta - 32r\right)J^2 + r^5\right)M + \left(16r^2 - 16r^2\sin^2\vartheta\right)J^2}{2r^7 M - 16r^4\sin^2\vartheta\,J^2 - r^8}$$

$$R^2{}_{023} = \frac{4\sin^2\vartheta\,J\,M - 6\,r\sin^2\vartheta\,J}{r^4}$$

24.5 Detailed metrics

$$R^2{}_{031} = \frac{6\cos\vartheta\sin\vartheta\, J}{r^4}$$

$$R^2{}_{032} = -\frac{4\sin^2\vartheta\, J\, M - 6r\sin^2\vartheta\, J}{r^4}$$

$$R^2{}_{103} = \frac{12\cos\vartheta\sin\vartheta\, J\, M - 6r\cos\vartheta\sin\vartheta\, J}{2r^4 M - 16r\sin^2\vartheta\, J^2 - r^5}$$

$$R^2{}_{112} = -\frac{M}{2r^2 M - r^3}$$

$$R^2{}_{121} = \frac{M}{2r^2 M - r^3}$$

$$R^2{}_{130} = -\frac{12\cos\vartheta\sin\vartheta\, J\, M - 6r\cos\vartheta\sin\vartheta\, J}{2r^4 M - 16r\sin^2\vartheta\, J^2 - r^5}$$

$$R^2{}_{301} = \frac{24r^3\cos\vartheta\sin\vartheta\, J\, M - 96\cos\vartheta\sin^3\vartheta\, J^3 - 12r^4\cos\vartheta\sin\vartheta\, J}{2r^7 M - 16r^4\sin^2\vartheta\, J^2 - r^8}$$

$$R^2{}_{302} = -\frac{4\sin^2\vartheta\, J\, M - 6r\sin^2\vartheta\, J}{r^4}$$

$$R^2{}_{310} = -\frac{24r^3\cos\vartheta\sin\vartheta\, J\, M - 96\cos\vartheta\sin^3\vartheta\, J^3 - 12r^4\cos\vartheta\sin\vartheta\, J}{2r^7 M - 16r^4\sin^2\vartheta\, J^2 - r^8}$$

$$R^2{}_{320} = \frac{4\sin^2\vartheta\, J\, M - 6r\sin^2\vartheta\, J}{r^4}$$

$$R^2{}_{323} = \frac{2\sin^2\vartheta\, M}{r}$$

$$R^2{}_{332} = -\frac{2\sin^2\vartheta\, M}{r}$$

$$R^3{}_{003} = \frac{4r^3 M^3 + \left(16\sin^2\vartheta\, J^2 - 4r^4\right)M^2 + \left(\left(16r\sin^2\vartheta - 32r\right)J^2 + r^5\right)M + \left(16r^2 - 12r^2\sin^2\vartheta\right)J^2}{2r^7 M - 16r^4\sin^2\vartheta\, J^2 - r^8}$$

$$R^3{}_{012} = \frac{24r^2\cos\vartheta\, J M^2 - 24r^3\cos\vartheta\, J M + 6r^4\cos\vartheta\, J}{4r^6\sin\vartheta\, M^2 + \left(-64r^3\sin^3\vartheta\, J^2 - 4r^7\sin\vartheta\right)M + 256\sin^5\vartheta\, J^4 + 32r^4\sin^3\vartheta\, J^2 + r^8\sin\vartheta}$$

$$R^3{}_{021} = -\frac{24r^2\cos\vartheta\, J M^2 - 24r^3\cos\vartheta\, J M + 6r^4\cos\vartheta\, J}{4r^6\sin\vartheta\, M^2 + \left(-64r^3\sin^3\vartheta\, J^2 - 4r^7\sin\vartheta\right)M + 256\sin^5\vartheta\, J^4 + 32r^4\sin^3\vartheta\, J^2 + r^8\sin\vartheta}$$

$$R^3{}_{030} = -\frac{4^3 M^3 + \left(16\sin^2\vartheta J^2 - 4r^4\right)M^2 + \left(\left(16r\sin^2\vartheta - 32r\right)J^2 + r^5\right)M + \left(16r^2 - 12r^2\sin^2\vartheta\right)J^2}{2r^7 M - 16r^4 \sin^2\vartheta J^2 - r^8}$$

$$R^3{}_{101} = -\frac{24r^3 J M^2 + \left(-96\sin^2\vartheta J^3 - 24r^4 J\right)M + 32r\sin^2\vartheta J^3 + 6r^5 J}{8r^7 M^3 + \left(-128r^4 \sin^2\vartheta J^2 - 12r^8\right)M^2 + \left(512r\sin^4\vartheta J^4 + 128r^5\sin^2\vartheta J^2 + 6r^9\right)M - 256r^2 \sin^4\vartheta J^4 - 32r^6 \sin^2\vartheta J^2 - r^{10}}$$

$$R^3{}_{102} = -\frac{24r^3 \cos\vartheta J M^2 + \left(-192\cos\vartheta \sin^2\vartheta J^3 - 24r^4 \cos\vartheta J\right)M + 64r\cos\vartheta \sin^2\vartheta J^3 + 6r^5 \cos\vartheta J}{4r^7 \sin\vartheta M^2 + \left(-64r^4 \sin^3\vartheta J^2 - 4r^8 \sin\vartheta\right)M + 256r\sin^5\vartheta J^4 + 32r^5 \sin^3\vartheta J^2 + r^9 \sin\vartheta}$$

$$R^3{}_{110} = \frac{24r^3 J M^2 + \left(-96\sin^2\vartheta J^3 - 24r^4 J\right)M + 32r\sin^2\vartheta J^3 + 6r^5 J}{8r^7 M^3 + \left(-128r^4 \sin^2\vartheta J^2 - 12r^8\right)M^2 + \left(512r\sin^4\vartheta J^4 + 128r^5\sin^2\vartheta J^2 + 6r^9\right)M - 256r^2 \sin^4\vartheta J^4 - 32r^6 \sin^2\vartheta J^2 - r^{10}}$$

$$R^3{}_{113} = -R^3{}_{131}$$

$$R^3{}_{120} = \frac{24r^3 \cos\vartheta J M^2 + \left(-192\cos\vartheta \sin^2\vartheta J^3 - 24r^4 \cos\vartheta J\right)M + 64r\cos\vartheta \sin^2\vartheta J^3 + 6r^5 \cos\vartheta J}{4r^7 \sin\vartheta M^2 + \left(-64r^4 \sin^3\vartheta J^2 - 4r^8 \sin\vartheta\right)M + 256r\sin^5\vartheta J^4 + 32r^5 \sin^3\vartheta J^2 + r^9 \sin\vartheta}$$

$$R^3{}_{123} = \frac{96 r^3 \cos\vartheta \sin\vartheta J^2 M - 384\cos\vartheta \sin^3\vartheta J^4 - 48 r^4 \cos\vartheta \sin\vartheta J^2}{4 r^7 M^2 + \left(-64 r^4 \sin^2\vartheta J^2 - 4 r^8\right) M + 256 r \sin^4\vartheta J^4 + 32 r^5 \sin^2\vartheta J^2 + r^9}$$

$$R^3{}_{131} = \frac{4r^6 M^3 + \left(176r^3 \sin^2\vartheta J^2 - 4r^7\right)M^2 + \left(-512\sin^4\vartheta J^4 - 208r^4\sin^2\vartheta J^2 + r^8\right)M + 192r\sin^4\vartheta J^4 + 60r^5 \sin^2\vartheta J^2}{8r^8 M^3 + \left(-128r^5 \sin^2\vartheta J^2 - 12r^9\right)M^2 + \left(512r^2 \sin^4\vartheta J^4 + 128r^6 \sin^2\vartheta J^2 + 6r^{10}\right)M - 256r^3 \sin^4\vartheta J^4 - 32r^7 \sin^2\vartheta J^2 - r^{11}}$$

$$R^3{}_{132} = -\frac{96 r^3 \cos\vartheta \sin\vartheta J^2 M - 384\cos\vartheta \sin^3\vartheta J^4 - 48 r^4 \cos\vartheta \sin\vartheta J^2}{4 r^7 M^2 + \left(-64 r^4 \sin^2\vartheta J^2 - 4 r^8\right) M + 256 r \sin^4\vartheta J^4 + 32 r^5 \sin^2\vartheta J^2 + r^9}$$

$$R^3{}_{201} = -\frac{48r^3 \cos\vartheta J M^2 + \left(-192\cos\vartheta \sin^2\vartheta J^3 - 48r^4 \cos\vartheta J\right)M + 64r\cos\vartheta \sin^2\vartheta J^3 + 12r^5 \cos\vartheta J}{4r^7 \sin\vartheta M^2 + \left(-64r^4 \sin^3\vartheta J^2 - 4r^8 \sin\vartheta\right)M + 256r\sin^5\vartheta J^4 + 32r^5 \sin^3\vartheta J^2 + r^9 \sin\vartheta}$$

$$R^3{}_{202} = -\frac{24 r^3 J M^2 + \left(\left(-64\sin^2\vartheta - 128\right)J^3 - 24 r^4 J\right)M + \left(32 r \sin^2\vartheta + 64 r\right)J^3 + 6 r^5 J}{4 r^6 M^2 + \left(-64 r^3 \sin^2\vartheta J^2 - 4 r^7\right) M + 256 \sin^4\vartheta J^4 + 32 r^4 \sin^2\vartheta J^2 + r^8}$$

$$R^3{}_{210} = \frac{48 r^3 \cos\vartheta J M^2 + \left(-192\cos\vartheta \sin^2\vartheta J^3 - 48 r^4 \cos\vartheta J\right) M + 64 r \cos\vartheta \sin^2\vartheta J^3 + 12 r^5 \cos\vartheta J}{4 r^7 \sin\vartheta M^2 + \left(-64 r^4 \sin^3\vartheta J^2 - 4 r^8 \sin\vartheta\right) M + 256 r \sin^5\vartheta J^4 + 32 r^5 \sin^3\vartheta J^2 + r^9 \sin\vartheta}$$

$$R^3{}_{213} = \frac{24 \cos\vartheta \sin\vartheta J^2}{2 r^4 M - 16 r \sin^2\vartheta J^2 - r^5}$$

$$R^3{}_{220} = \frac{24 r^3 J M^2 + \left(\left(-64\sin^2\vartheta - 128\right)J^3 - 24 r^4 J\right)M + \left(32 r \sin^2\vartheta + 64 r\right)J^3 + 6 r^5 J}{4 r^6 M^2 + \left(-64 r^3 \sin^2\vartheta J^2 - 4 r^7\right) M + 256 \sin^4\vartheta J^4 + 32 r^4 \sin^2\vartheta J^2 + r^8}$$

$$R^3{}_{223} = -\frac{4r^3 M^2 + (16 \sin^2 \vartheta J^2 - 2r^4) M - 24r \sin^2 \vartheta J^2}{2r^4 M - 16r \sin^2 \vartheta J^2 - r^5}$$

$$R^3{}_{231} = -\frac{24 \cos \vartheta \sin \vartheta J^2}{2r^4 M - 16r \sin^2 \vartheta J^2 - r^5}$$

$$R^3{}_{232} = \frac{4r^3 M^2 + (16 \sin^2 \vartheta J^2 - 2r^4) M - 24r \sin^2 \vartheta J^2}{2r^4 M - 16r \sin^2 \vartheta J^2 - r^5}$$

$$R^3{}_{303} = -\frac{8r^3 \sin^2 \vartheta J M^2 + (32 \sin^4 \vartheta J^3 - 4r^4 \sin^2 \vartheta J) M + (48r \sin^4 \vartheta - 64r \sin^2 \vartheta) J^3}{2r^7 M - 16r^4 \sin^2 \vartheta J^2 - r^8}$$

$$R^3{}_{312} = -\frac{48 r^2 \cos \vartheta \sin \vartheta J^2 M - 24 r^3 \cos \vartheta \sin \vartheta J^2}{4r^6 M^2 + (-64 r^3 \sin^2 \vartheta J^2 - 4 r^7) M + 256 \sin^4 \vartheta J^4 + 32 r^4 \sin^2 \vartheta J^2 + r^8}$$

$$R^3{}_{321} = \frac{48 r^2 \cos \vartheta \sin \vartheta J^2 M - 24 r^3 \cos \vartheta \sin \vartheta J^2}{4r^6 M^2 + (-64 r^3 \sin^2 \vartheta J^2 - 4 r^7) M + 256 \sin^4 \vartheta J^4 + 32 r^4 \sin^2 \vartheta J^2 + r^8}$$

$$R^3{}_{330} = \frac{8r^3 \sin^2 \vartheta J M^2 + (32 \sin^4 \vartheta J^3 - 4r^4 \sin^2 \vartheta J) M + (48r \sin^4 \vartheta - 64r \sin^2 \vartheta) J^3}{2r^7 M - 16r^4 \sin^2 \vartheta J^2 - r^8}$$

24.5.3.7 Ricci Tensor

$$\mathrm{Ric}_{00} = \frac{8 J^2 \left(6 \cos^2 \vartheta M^2 - 6 M^2 + 4r \cos^2 \vartheta M + 4r M - 3r^2 \cos^2 \vartheta - r^2\right)}{r^4 \left(2r^3 M - 16 \sin^2 \vartheta J^2 - r^4\right)}$$

$$\mathrm{Ric}_{03} = \frac{32 \sin^2 \vartheta J^3 \left(3 \sin^2 \vartheta M + 3r \sin^2 \vartheta - 2r\right)}{r^4 \left(2r^3 M - 16 \sin^2 \vartheta J^2 - r^4\right)}$$

$$\mathrm{Ric}_{11} = \frac{8 \sin^2 \vartheta J^2 \left(30 r^3 M^2 - 96 \sin^2 \vartheta J^2 M - 36 r^4 M + 48r \sin^2 \vartheta J^2 + 11 r^5\right)}{r^2 \left(2M - r\right) \left(2r^3 M - 16 \sin^2 \vartheta J^2 - r^4\right)^2}$$

$$\mathrm{Ric}_{12} = -\frac{16 \cos \vartheta \sin \vartheta J^2 \left(9 r^3 M - 48 \sin^2 \vartheta J^2 - 5 r^4\right)}{r \left(2r^3 M - 16 \sin^2 \vartheta J^2 - r^4\right)^2}$$

$$\mathrm{Ric}_{21} = -\frac{16 \cos \vartheta \sin \vartheta J^2 \left(9 r^3 M - 48 \sin^2 \vartheta J^2 - 5 r^4\right)}{r \left(2r^3 M - 16 \sin^2 \vartheta J^2 - r^4\right)^2}$$

$$\mathrm{Ric}_{22} = \frac{16J^2 \left(6r^3 \sin^2\vartheta M^2 - 48\sin^4\vartheta J^2 M - 11r^4 \sin^2\vartheta M + 2r^4 M + 48r\sin^4\vartheta J^2 + 4r^5 \sin^2\vartheta - r^5\right)}{r\left(2r^3 M - 16\sin^2\vartheta J^2 - r^4\right)^2}$$

$$\mathrm{Ric}_{30} = \frac{32\sin^2\vartheta J^3 \left(3\sin^2\vartheta M + 3r\sin^2\vartheta - 2r\right)}{r^4 \left(2r^3 M - 16\sin^2\vartheta J^2 - r^4\right)}$$

$$\mathrm{Ric}_{33} = -\frac{8\sin^2\vartheta J^2 \left(12\sin^2\vartheta M - 3r\sin^2\vartheta - 2r\right)}{r\left(2r^3 M - 16\sin^2\vartheta J^2 - r^4\right)}$$

24.5.3.8 Ricci Scalar

$$N = 4r^9 M^2 - 64r^6 \sin^2\vartheta J^2 M - 4r^{10} M + 256r^3 \sin^4\vartheta J^4 + 32r^7 \sin^2\vartheta J^2 + r^{11}$$

$$R_{sc} = \frac{-384r^3 \sin^2\vartheta J^2 M^2 + 768\sin^4\vartheta J^4 M + 224r^4 \sin^2\vartheta J^2 M + 128r^4 J^2 M}{N}$$
$$+ \frac{1152r\sin^4\vartheta J^4 - 512r\sin^2\vartheta J^4 - 24r^5 \sin^2\vartheta J^2 - 64r^5 J^2}{N}$$

24.5.3.9 Bianchi identity (Ricci cyclic equation $R^\kappa{}_{[\mu\nu\sigma]} = 0$)

———— o.k.

24.5.3.10 Einstein Tensor

———— not zero:

$$N = 4r^9 M^2 - 64r^6 \sin^2\vartheta J^2 M - 4r^{10} M + 256r^3 \sin^4\vartheta J^4 + 32r^7 \sin^2\vartheta J^2 + r^{11}$$

$$G_{00} = \frac{288r^2 \sin^2\vartheta J^2 M^3 - 432r^3 \sin^2\vartheta J^2 M^2 - 512\sin^2\vartheta J^4 M + 216r^4 \sin^2\vartheta J^2 M + 192r\sin^4\vartheta J^4 + 256r\sin^2\vartheta J^4 - 36r^5 \sin^2\vartheta J^2}{N}$$

$$G_{03} = -\frac{576r^2 \sin^4\vartheta J^3 M^2 + \left(-544r^3 \sin^4\vartheta - 128r^3 \sin^2\vartheta\right) J^3 M - 768\sin^6\vartheta J^5 + \left(144r^4 \sin^4\vartheta + 64r^4 \sin^2\vartheta\right) J^3}{4r^9 M^2 + \left(-64r^6 \sin^2\vartheta J^2 - 4r^{10}\right) M + 256r^3 \sin^4\vartheta J^4 + 32r^7 \sin^2\vartheta J^2 + r^{11}}$$

$$G_{11} = \frac{48r^3 \sin^2\vartheta J^2 M^2 + \left(\left(64r^4 - 176r^4 \sin^2\vartheta\right) J^2 - 384\sin^4\vartheta J^4\right) M + \left(960r\sin^4\vartheta - 256r\sin^2\vartheta\right) J^4 + \left(76r^5 \sin^2\vartheta - 32r^5\right) J^2}{8r^8 M^3 + \left(-128r^5 \sin^2\vartheta J^2 - 12r^9\right) M^2 + \left(512r^2 \sin^4\vartheta J^4 + 128r^6 \sin^2\vartheta J^2 + 6r^{10}\right) M - 256r^3 \sin^4\vartheta J^4 - 32r^7 \sin^2\vartheta J^2 - r^{11}}$$

$$G_{12} = -\frac{144\,r^3\,\cos\vartheta\,\sin\vartheta\,J^2\,M - 768\,\cos\vartheta\,\sin^3\vartheta\,J^4 - 80\,r^4\,\cos\vartheta\,\sin\vartheta\,J^2}{4\,r^7\,M^2 + (-64\,r^4\,\sin^2\vartheta\,J^2 - 4\,r^8)\,M + 256\,r\,\sin^4\vartheta\,J^4 + 32\,r^5\,\sin^2\vartheta\,J^2 + r^9}$$

$$G_{21} = -\frac{144\,r^3\,\cos\vartheta\,\sin\vartheta\,J^2\,M - 768\,\cos\vartheta\,\sin^3\vartheta\,J^4 - 80\,r^4\,\cos\vartheta\,\sin\vartheta\,J^2}{4\,r^7\,M^2 + (-64\,r^4\,\sin^2\vartheta\,J^2 - 4\,r^8)\,M + 256\,r\,\sin^4\vartheta\,J^4 + 32\,r^5\,\sin^2\vartheta\,J^2 + r^9}$$

$$G_{22} = \frac{288r^3\sin^2\vartheta J^2 M^2 + ((-288r^4\sin^2\vartheta - 32r^4)J^2 - 1152\sin^4\vartheta J^4)M + (192r\sin^4\vartheta + 256r\sin^2\vartheta)J^4 + (76r^5\sin^2\vartheta + 16r^5)J^2}{4r^7 M^2 + (-64r^4\sin^2\vartheta J^2 - 4r^8)M + 256r\sin^4\vartheta J^4 + 32r^5\sin^2\vartheta J^2 + r^9}$$

$$G_{30} = -\frac{576r^2\sin^4\vartheta J^3 M^2 + (-544r^3\sin^4\vartheta - 128r^3\sin^2\vartheta)J^3 M - 768\sin^6\vartheta J^5 + (144r^4\sin^4\vartheta + 64r^4\sin^2\vartheta)J^3}{4r^9 M^2 + (-64r^6\sin^2\vartheta J^2 - 4r^{10})M + 256r^3\sin^4\vartheta J^4 + 32r^7\sin^2\vartheta J^2 + r^{11}}$$

$$G_{33} = \frac{(1152\sin^6\vartheta J^4 + (32r^4\sin^4\vartheta - 32r^4\sin^2\vartheta)J^2)M - 960r\sin^6\vartheta J^4 + (16r^5\sin^2\vartheta - 12r^5\sin^4\vartheta)J^2}{4r^7 M^2 + (-64r^4\sin^2\vartheta J^2 - 4r^8)M + 256r\sin^4\vartheta J^4 + 32r^5\sin^2\vartheta J^2 + r^9}$$

24.5.3.11 Hodge Dual of Bianchi Identity

———— (see charge and current densities)

24.5.3.12 Scalar Charge Density $(-R^0{}_i{}^{i0})$

$$N = 8r^9 M^3 + \left(-192r^6\sin^2\vartheta J^2 - 12r^{10}\right)M^2 + \left(1536r^3\sin^4\vartheta J^4 + 192r^7\sin^2\vartheta J^2 + 6r^{11}\right)M$$
$$- 4096\sin^6\vartheta J^6 - 768r^4\sin^4\vartheta J^4 - 48r^8\sin^2\vartheta J^2 - r^{12}$$

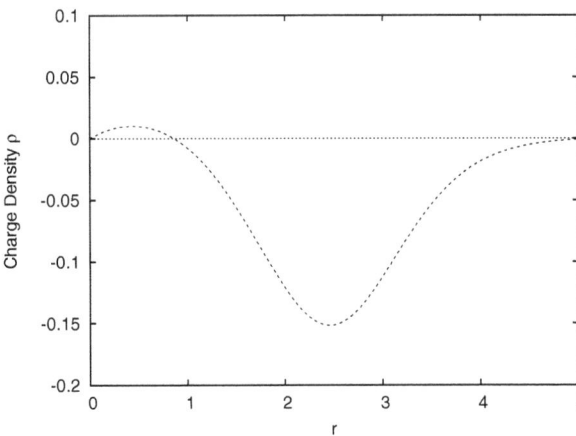

Fig. 24.7. Kerr metric, cosmological charge density ρ for M=1, J=2.

454   24 ECE Theory of the Earth's Gravitomagnetic Precession

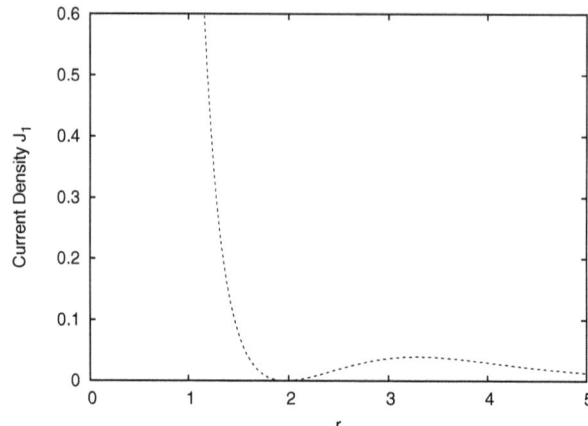

Fig. 24.8. Kerr metric, cosmological current density J_r for M=1, J=2.

$$\rho = \frac{\left(48r^4 \cos^2 \vartheta - 48r^4\right) J^2 M^2 + \left(\left(-768r \cos^4 \vartheta + 1536r \cos^2 \vartheta - 768r\right) J^4 + \left(32r^5 \cos^2 \vartheta + 32r^5\right) J^2\right) M}{N}$$
$$+ \frac{\left(1152r^2 \cos^4 \vartheta - 2048r^2 \cos^2 \vartheta + 896r^2\right) J^4 + \left(-24r^6 \cos^2 \vartheta - 8r^6\right) J^2}{N}$$

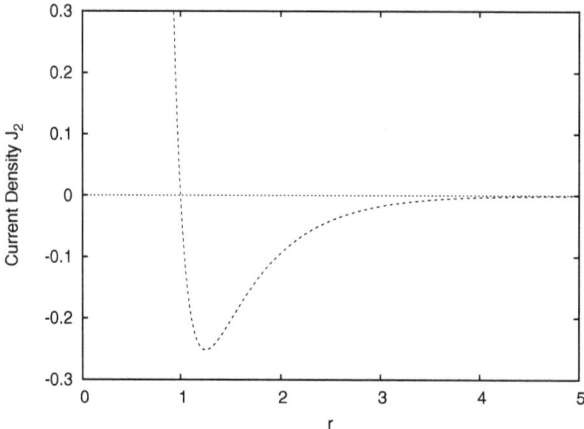

Fig. 24.9. Kerr metric, cosmological current density J_ϑ for M=1, J=2.

24.5 Detailed metrics   455

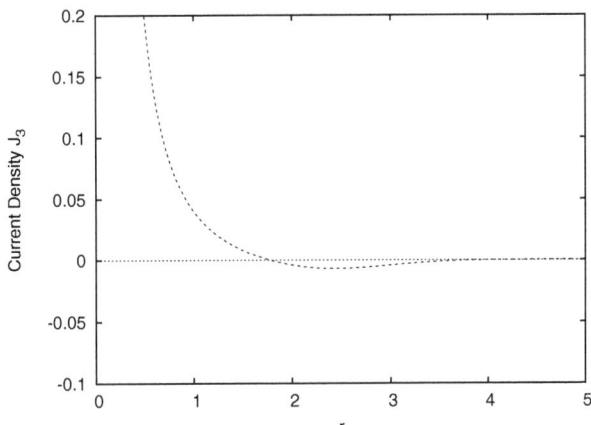

Fig. 24.10. Kerr metric, cosmological current density J_φ for M=1, J=2.

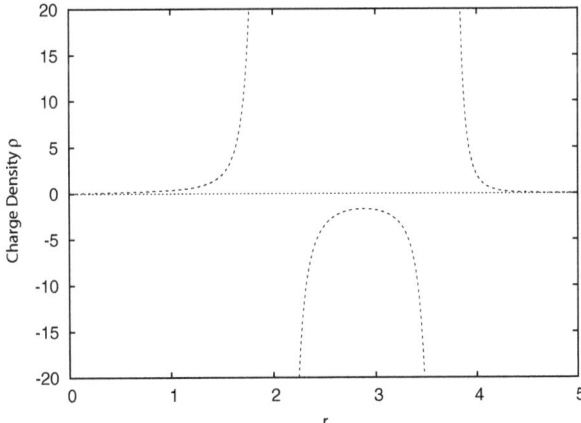

Fig. 24.11. Kerr metric, cosmological charge density ρ for M=2, J=1.

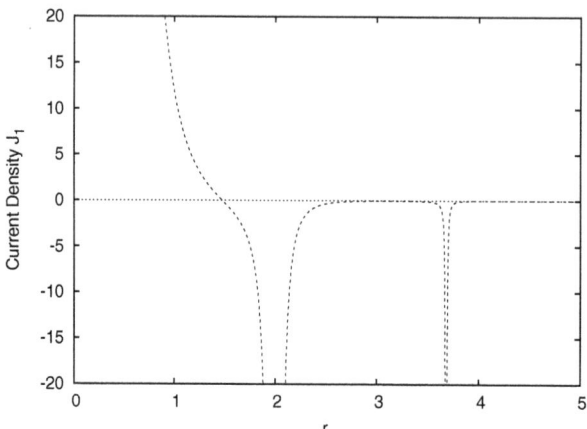

Fig. 24.12. Kerr metric, cosmological current density J_r for M=2, J=1.

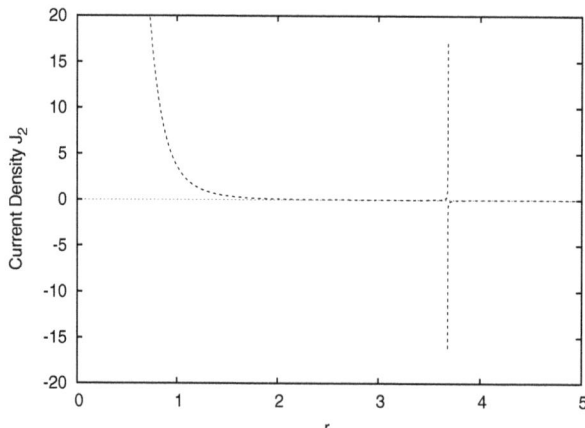

Fig. 24.13. Kerr metric, cosmological current density J_θ for M=2, J=1.

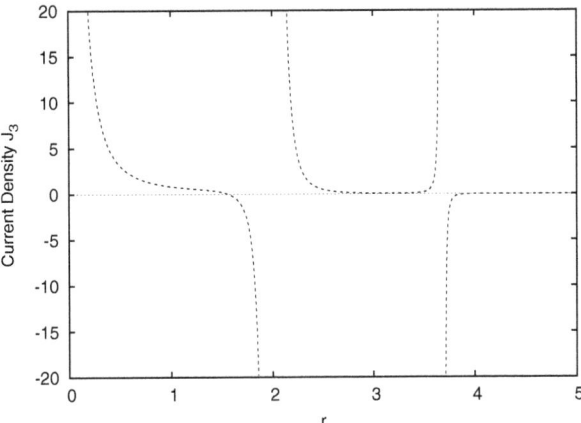

Fig. 24.14. Kerr metric, cosmological current density J_φ for M=2, J=1.

24.5.3.13 Current Density Class 1 $(-R^i{}_\mu{}^{\mu j})$

$$N = 4r^{10}M^2 + \left(-64r^7 \sin^2 \vartheta J^2 - 4r^{11}\right)M + 256r^4 \sin^4 \vartheta J^4 + 32r^8 \sin^2 \vartheta J^2 + r^{12}$$

$$J_1 = -\frac{480r^3 \sin^2 \vartheta J^2 M^3 + \left(-1536 \sin^4 \vartheta J^4 - 816r^4 \sin^2 \vartheta J^2\right)M^2 + \left(1536r \sin^4 \vartheta J^4 + 464r^5 \sin^2 \vartheta J^2\right)M}{N} - \frac{384r^2 \sin^4 \vartheta J^4 - 88r^6 \sin^2 \vartheta J^2}{N}$$

$$J_2 = -\frac{96r^3 \sin^2 \vartheta J^2 M^2 + \left(\left(32r^4 - 176r^4 \sin^2 \vartheta\right)J^2 - 768 \sin^4 \vartheta J^4\right)M + 768r \sin^4 \vartheta J^4 + \left(64r^5 \sin^2 \vartheta - 16r^5\right)J^2}{4r^{11}M^2 + \left(-64r^8 \sin^2 \vartheta J^2 - 4r^{12}\right)M + 256r^5 \sin^4 \vartheta J^4 + 32r^9 \sin^2 \vartheta J^2 + r^{13}}$$

$$N = 8r^{11} \sin^2 \vartheta\, M^3 + \left(-192r^8 \sin^4 \vartheta\, J^2 - 12r^{12} \sin^2 \vartheta\right)M^2$$
$$+ \left(1536r^5 \sin^6 \vartheta\, J^4 + 192r^9 \sin^4 \vartheta\, J^2 + 6r^{13} \sin^2 \vartheta\right)M$$
$$- 4096r^2 \sin^8 \vartheta\, J^6 - 768r^6 \sin^6 \vartheta\, J^4 - 48r^{10} \sin^4 \vartheta\, J^2 - r^{14} \sin^2 \vartheta$$

$$J_3 = \frac{384r^3 \sin^2 \vartheta\, J^2 M^3 + \left(\left(-480r^4 \sin^2 \vartheta - 64r^4\right)J^2 - 768 \sin^4 \vartheta\, J^4\right)M^2}{N}$$
$$+ \frac{\left(\left(192r^5 \sin^2 \vartheta + 64r^5\right)J^2 - 256r \sin^4 \vartheta\, J^4\right)M + 384r^2 \sin^4 \vartheta\, J^4 + \left(-24r^6 \sin^2 \vartheta - 16r^6\right)J^2}{N}$$

24.5.3.14 Current Density Class 2 ($-R^i{}_\mu{}^{\mu j}$)

$$J_1 = 0$$

$$J_2 = -\frac{288r^3 \cos\vartheta \sin\vartheta J^2 M^2 + \left(-1536\cos\vartheta\sin^3\vartheta J^4 - 304r^4\cos\vartheta\sin\vartheta J^2\right)M + 768r\cos\vartheta\sin^3\vartheta J^4 + 80r^5\cos\vartheta\sin\vartheta J^2}{4r^{10}M^2 + \left(-64r^7\sin^2\vartheta J^2 - 4r^{11}\right)M + 256r^4\sin^4\vartheta J^4 + 32r^8\sin^2\vartheta J^2 + r^{12}}$$

$$J_3 = 0$$

24.5.3.15 Current Density Class 3 ($-R^i{}_\mu{}^{\mu j}$)

$$J_1 = -\frac{288r^3 \cos\vartheta \sin\vartheta J^2 M^2 + \left(-1536\cos\vartheta\sin^3\vartheta J^4 - 304r^4\cos\vartheta\sin\vartheta J^2\right)M + 768r\cos\vartheta\sin^3\vartheta J^4 + 80r^5\cos\vartheta\sin\vartheta J^2}{4r^{10}M^2 + \left(-64r^7\sin^2\vartheta J^2 - 4r^{11}\right)M + 256r^4\sin^4\vartheta J^4 + 32r^8\sin^2\vartheta J^2 + r^{12}}$$

$$J_2 = 0$$
$$J_3 = 0$$

24.5.4 Kerr-Newman metric

This is the most complex metric of this group for a charged mass with rotation. The functions occuring in the metric are defined as follows:

$$\rho^2 = r^2 + a^2 \cos^2\theta$$
$$\Delta = r^2 - 2Mr + a^2 + Q^2$$
$$a = \frac{J}{M}$$

where a is the angular momentum per unit mass. The inverse metric is highly complex and not shown. The same holds for most of the derived quantities like Christoffel symbols, Riemann, Ricci and Einstein tensors and cosmological charge and current density. In particular the charge and current density are not zero.

24.5.4.1 Coordinates

$$\mathbf{x} = \begin{pmatrix} t \\ r \\ \vartheta \\ \varphi \end{pmatrix}$$

24.5.4.2 Metric

$$g_{\mu\nu} = \begin{pmatrix} \frac{2rM-Q^2}{a^2\cos^2\vartheta+r^2}-1 & 0 & 0 & -\frac{a\sin^2\vartheta(4rM-2Q^2)}{a^2\cos^2\vartheta+r^2} \\ 0 & \frac{a^2\cos^2\vartheta+r^2}{Q^2-2rM+r^2+a^2} & 0 & 0 \\ 0 & 0 & a^2\cos^2\vartheta+r^2 & 0 \\ -\frac{a\sin^2\vartheta(4rM-2Q^2)}{a^2\cos^2\vartheta+r^2} & 0 & 0 & \sin^2\vartheta\left(\frac{a^2\sin^2\vartheta(2rM-Q^2)}{a^2\cos^2\vartheta+r^2}+r^2+a^2\right) \end{pmatrix}$$

24.5.4.3 Christoffel Connection

$$\Gamma^0{}_{01} \neq 0$$

$$\Gamma^0{}_{02} \neq 0$$

$$\Gamma^0{}_{10} \neq 0$$

$$\Gamma^0{}_{13} \neq 0$$

$$\Gamma^0{}_{20} \neq 0$$

$$\Gamma^0{}_{23} \neq 0$$

$$\Gamma^0{}_{31} \neq 0$$

$$\Gamma^0{}_{32} \neq 0$$

$$\Gamma^1{}_{00} = -\frac{rQ^4+\left(\left(a^2\cos^2\vartheta-3r^2\right)M+r^3+a^2r\right)Q^2+\left(2r^3-2a^2r\cos^2\vartheta\right)M^2+\left(\left(a^2r^2+a^4\right)\cos^2\vartheta-r^4-a^2r^2\right)M}{a^6\cos^6\vartheta+3a^4r^2\cos^4\vartheta+3a^2r^4\cos^2\vartheta+r^6}$$

$$\Gamma^1{}_{03} \neq 0$$

$$\Gamma^1{}_{11} = \frac{rQ^2+\left(-a^2\sin^2\vartheta-r^2+a^2\right)M+a^2r\sin^2\vartheta}{(a^2\cos^2\vartheta+r^2)Q^2+\left(-2a^2r\cos^2\vartheta-2r^3\right)M+(a^2r^2+a^4)\cos^2\vartheta+r^4+a^2r^2}$$

$$\Gamma^1{}_{12} = -\frac{a^2\cos\vartheta\sin\vartheta}{a^2\cos^2\vartheta+r^2}$$

$$\Gamma^1{}_{21} = -\frac{a^2 \cos\vartheta \sin\vartheta}{a^2 \cos^2\vartheta + r^2}$$

$$\Gamma^1{}_{22} = -\frac{rQ^2 - 2r^2 M + r^3 + a^2 r}{a^2 \cos^2\vartheta + r^2}$$

$$\Gamma^1{}_{30} \neq 0$$

$$\Gamma^1{}_{33} \neq 0$$

$$\Gamma^2{}_{00} = \frac{a^2 \cos\vartheta \sin\vartheta Q^2 - 2a^2 r \cos\vartheta \sin\vartheta M}{a^6 \cos^6\vartheta + 3a^4 r^2 \cos^4\vartheta + 3a^2 r^4 \cos^2\vartheta + r^6}$$

$$\Gamma^2{}_{03} = -\frac{\left(2ar^2 + 2a^3\right) \cos\vartheta \sin\vartheta Q^2 + \left(-4ar^3 - 4a^3 r\right) \cos\vartheta \sin\vartheta M}{a^6 \cos^6\vartheta + 3a^4 r^2 \cos^4\vartheta + 3a^2 r^4 \cos^2\vartheta + r^6}$$

$$\Gamma^2{}_{11} = \frac{a^2 \cos\vartheta \sin\vartheta}{\left(a^2 \cos^2\vartheta + r^2\right) Q^2 + \left(-2a^2 r \cos^2\vartheta - 2r^3\right) M + \left(a^2 r^2 + a^4\right) \cos^2\vartheta + r^4 + a^2 r^2}$$

$$\Gamma^2{}_{12} = \frac{r}{a^2 \cos^2\vartheta + r^2}$$

$$\Gamma^2{}_{21} = \frac{r}{a^2 \cos^2\vartheta + r^2}$$

$$\Gamma^2{}_{22} = -\frac{a^2 \cos\vartheta \sin\vartheta}{a^2 \cos^2\vartheta + r^2}$$

$$\Gamma^2{}_{30} = -\frac{\left(2ar^2 + 2a^3\right) \cos\vartheta \sin\vartheta Q^2 + \left(-4ar^3 - 4a^3 r\right) \cos\vartheta \sin\vartheta M}{a^6 \cos^6\vartheta + 3a^4 r^2 \cos^4\vartheta + 3a^2 r^4 \cos^2\vartheta + r^6}$$

$$\Gamma^2{}_{33} \neq 0$$

$$\Gamma^3{}_{01} \neq 0$$

$$\Gamma^3{}_{02} \neq 0$$

$$\Gamma^3{}_{10} \neq 0$$

$$\Gamma^3{}_{13} \neq 0$$

$$\Gamma^3{}_{20} \neq 0$$

$$\Gamma^3{}_{23} \neq 0$$

$$\Gamma^3{}_{31} \neq 0$$

$$\Gamma^3{}_{32} \neq 0$$

24.5.4.4 Metric Compatibility

────── o.k.

24.5.4.5 Bianchi identity (Ricci cyclic equation $R^\kappa{}_{[\mu\nu\sigma]} = 0$)

────── o.k.

24.5.4.6 Einstein Tensor

────── not zero:

$$G_{00} \neq 0$$

$$G_{03} \neq 0$$

$$G_{11} \neq 0$$

$$G_{12} \neq 0$$

$$G_{21} \neq 0$$

$$G_{22} \neq 0$$

. . .

Acknowledgments

The British Government is thanked for a Civil List Pension to MWE for distinguished contributions to Britain in science and colleagues of AIAS and TGA are thanked for many interesting discussions.

A

Appendix 1: Validity of the Dipole Approximation

The ECE theory of gravitomagnetism is based on the ECE theory of magnetostatics, which can be built from electrostatics as is well known [17]. It is sufficient for the purposes of this appendix to consider the static electric field without spin connection:

$$\mathbf{E} = -\boldsymbol{\nabla}\Phi. \tag{A.1}$$

The multipole expansion used in electrostatics [17] is:

$$\Phi(\mathbf{x}) = \frac{1}{4\pi\epsilon_0} \int \frac{\rho(\mathbf{x}')}{|\mathbf{x}-\mathbf{x}'|} d^3x' = \frac{1}{4\pi\epsilon_0} \sum_{l=0}^{\infty} \sum_{m=-l}^{l} \frac{4\pi}{2l+1} q_{lm} \frac{Y_{lm}(\theta,\phi)}{r^{l+1}}. \tag{A.2}$$

The choice of constant coefficients is a convention. A localized distribution of charge is described by the charge density $\rho(\mathbf{x}')$, which is non-vanishing only inside a sphere of radius R, defined around an origin. The sphere is a concept used only to divide space into regions with and without charge. The multipole expansion is valid if and only if the charge density falls off with distance faster than any power of r.

Under these assumptions:

$$\Phi(\mathbf{x}) = \frac{1}{4\pi\epsilon_0}\left(\frac{q}{r} + \frac{\mathbf{p}\cdot\mathbf{x}}{r^3} + ...\right) \tag{A.3}$$

where the integrated charge or monopole moment is:

$$q = \int \rho(\mathbf{x}') d^3x' \tag{A.4}$$

and where the electric dipole moment is:

$$\mathbf{p} = \int \mathbf{x}\,\rho(\mathbf{x'})d^3x'. \tag{A.5}$$

In spherical polar coordinates, the electric field strength components for a dipole aligned in the Z axis are:

$$E_r = \frac{2p\cos\theta}{4\pi\epsilon_0\,r^3}, \quad E_\theta = \frac{p\sin\theta}{4\pi\epsilon_0\,r^3}, \quad E_\phi = 0 \tag{A.6}$$

and the total field vector is:

$$\mathbf{E} = E_r\mathbf{e}_r + E_\theta\mathbf{e}_\theta + E_\phi\mathbf{e}_\phi \tag{A.7}$$

where:

$$\mathbf{e}_r = \sin\theta\cos\phi\,\mathbf{i} + \sin\theta\sin\phi\,\mathbf{j} + \cos\theta\,\mathbf{k},$$
$$\mathbf{e}_\theta = \cos\theta\cos\phi\,\mathbf{i} + \cos\theta\sin\phi\,\mathbf{j} - \sin\theta\,\mathbf{k},$$
$$\mathbf{e}_\phi = -\sin\phi\,\mathbf{i} + \cos\phi\,\mathbf{j}.$$

Therefore:

$$\mathbf{E} = \frac{p}{4\pi\epsilon_0\,r^3}\left(3\sin\theta\cos\theta\cos\phi\,\mathbf{i} + 3\sin\theta\cos\theta\sin\phi\,\mathbf{j} + \left(2\cos^2\theta - \sin^2\theta\right)\mathbf{k}\right) \tag{A.8}$$

where:

$$\sin\theta\cos\phi = \frac{x}{r},$$
$$\sin\theta\sin\phi = \frac{y}{r},$$
$$\cos\theta = \frac{z}{r},$$
$$\sin\theta = \left(1 - \frac{z^2}{r^2}\right)^{1/2}.$$

The dipole field is therefore:

$$\mathbf{E} = \frac{p}{4\pi\epsilon_0\,r^3}\left(\frac{3z}{r^2}\left(x\,\mathbf{i} + y\,\mathbf{j} + z\,\mathbf{k}\right) - \mathbf{k}\right). \tag{A.9}$$

This result is denoted in ref. [17], eq. (4.13), as:

$$\mathbf{E}(\mathbf{x}) = \frac{3\mathbf{n}(\mathbf{p} \cdot \mathbf{n}) - \mathbf{p}}{4\pi\epsilon_0 |\mathbf{x} - \mathbf{x}_0|^3} \tag{A.10}$$

where $\mathbf{n}$ is a unit vector directed from $\mathbf{x}_0$ to $\mathbf{x}$. Thus:

$$\mathbf{n} = \frac{\mathbf{r}}{|\mathbf{r}|}, \tag{A.11}$$

$$r = |\mathbf{x} - \mathbf{x}_0|, \tag{A.12}$$

$$\mathbf{n}\,(\mathbf{p} \cdot \mathbf{n}) = \frac{p\,z}{r^2}\,(x\,\mathbf{i} + y\,\mathbf{j} + z\,\mathbf{k}). \tag{A.13}$$

The electric dipole moment (A.5) is defined in the range of validity:

$$|\mathbf{x}'| \ll |\mathbf{x} - \mathbf{x}_0| \tag{A.14}$$

where $|\mathbf{x}'|$ is the distance between the two charges of the dipole. By reference to Fig. A1, the exact solution:

$$\Phi = \frac{1}{4\pi\epsilon_0} \left(\frac{q}{\left(\left(z - \frac{d}{2}\right)^2 + x^2 + y^2\right)^{1/2}} - \frac{q}{\left(\left(z + \frac{d}{2}\right)^2 + x^2 + y^2\right)^{1/2}} \right) \tag{A.15}$$

should be used. Eq. (A.9) is obtained only if:

$$d \ll |\mathbf{r}| \tag{A.16}$$

when:

$$\left(z - \frac{d}{2}\right)^2 \approx z^2 - z\,d \tag{A.17}$$

and

$$\left(1 - \frac{zd}{r^2}\right)^{-1/2} \approx 1 + \frac{1}{2}\frac{zd}{r^2}. \tag{A.18}$$

Appendix 1: Validity of the Dipole Approximation

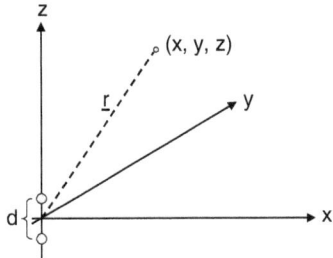

Fig. A.1. Position of dipole charges in the coordinate system.

If

$$p = q\,d \tag{A.19}$$

then

$$\Phi \approx \frac{1}{4\pi\epsilon_0} \frac{zqd}{r^3}. \tag{A.20}$$

Now use:

$$\cos\theta = \frac{z}{r} \tag{A.21}$$

so

$$\Phi = \frac{1}{4\pi\epsilon_0} \frac{p\cos\theta}{r^2}. \tag{A.22}$$

The vector **p** is defined as along the Z axis from −q to q, with

$$|\mathbf{p}| = q\,d. \tag{A.23}$$

Therefore:

$$\mathbf{p}\cdot\mathbf{r} = r\,p\,\cos\theta \tag{A.24}$$

i.e.

$$p\cos\theta = \frac{\mathbf{p}\cdot\mathbf{r}}{r} \tag{A.25}$$

and

$$\Phi = \frac{1}{4\pi\epsilon_0 r^3} \mathbf{p} \cdot \mathbf{r}. \tag{A.26}$$

The magnetic flux density from the electric dipole field can be found in the first approximation from Eq. (24.6) of the text, i.e. from:

$$\mathbf{B}(\text{dipole}) = -\frac{1}{c^2} \mathbf{v} \times \mathbf{E}(\text{dipole}) \tag{A.27}$$

and the rigorously correct method to use is from Eq. (A.15).

In Gravity Probe B the satellite orbits a distance r from the centre of the earth of radius R, where r is of the order of R. Newton's law holds accurately for this situation because the mass M of the earth can be considered as being at its centre of mass, which is the centre of the earth. So:

$$\mathbf{F} = m\mathbf{g} = -\frac{mMG}{r^2}\mathbf{e}_r. \tag{A.28}$$

The acceleration due to gravity is:

$$\mathbf{g} = -\boldsymbol{\nabla}\Phi = -\frac{MG}{r^2}\mathbf{e}_r \tag{A.29}$$

where only one (monopole) potential is needed:

$$\Phi = -\frac{MG}{r}. \tag{A.30}$$

The mass is the integral of the mass density:

$$M = \int \rho_m(\mathbf{r}')d^3r' \tag{A.31}$$

and:

$$\boldsymbol{\nabla} \cdot \mathbf{g} = 4\pi G \rho_m. \tag{A.32}$$

Comparing Eq. (A.30) to (A.3) it is seen that only the first term of the multipole expansion is needed for an accurate description of the Newtonian attraction between the satellite and the earth if the latter were a perfect sphere. The gravitational potential at any point outside a spherical distribution of matter, a solid or a shell, is independent of the size of the distribution as is well known. However the earth is not a perfect sphere, and gravitational multipoles of the earth are experimentally observable. The expression given by Pfister is the dipole approximation:

$$\mathbf{\Omega} = \frac{2}{5}\frac{MGR^2}{c^2 r^3}\left(\boldsymbol{\omega} - 3\mathbf{n}(\boldsymbol{\omega}\cdot\mathbf{n})\right) \tag{A.33}$$

which if valid, corrects our Eq. (24.20). Its analogue in magnetostatics is the magnetic flux density in the dipole approximation [17]:

$$\mathbf{B} = \frac{\mu_0}{4\pi r^3}\left(\mathbf{m} - 3\mathbf{n}(\mathbf{m}\cdot\mathbf{n})\right) \tag{A.34}$$

where **m** is the magnetic dipole moment. Eq. (A.33) is a solution of Eq. (24.10) in the far field approximation (A.15). Ref. [16] claims to have verified Eq. (A.33) experimentally, and if this claim is accepted, our Eq. (24.10) is also verified experimentally.

References

[1] M. W. Evans, "Generally Covariant Unified Field Theory" (Abramis 2005 onwards), volumes one to four, volumes five and six in preparation.
[2] L. Felker, "The Evans Equations of Unified Field Theory" (Abramis 2007).
[3] K. Pendergast, "Crystal Spheres" (Abramis, in preparation, preprint on www.aias.us).
[4] M. W. Evans, Omnia Opera section of the British Civil List Scientist's website www.aias.us., hyperlinked to original papers.
[5] M. W. Evans, ed., "Modern Nonlinear Optics", in Adv. Chem. Phys. Vol. 119 (Wiley 2001, second edition); ibid. M. W. Evans and S. Kielich (eds.), first edition volume 85 (Wiley 1992, 1993 and 1997).
[6] M. W. Evans and L. B. Crowell, "Classical and Quantum Electrodynamics and the B(3) Field" (World Scientific 2001).
[7] M. W. Evans and J.-P. Vigier, "The Enigmatic Photon" (Kluwer, Dordrecht, 1994 to 2002, hardback and softback), in five volumes.
[8] M. W. Evans et al., fifteen papers in "Foundations of Physics Letters" on ECE theory, 2003 to present.
[9] M. W. Evans, Acta Phys. Polon., **33B**, 2211 (2007).
[10] M. W. Evans and H. Eckardt, Physica B, **400**, 175 (2007).
[11] M. W. Evans, Physica B, **403**, 517 (2008).
[12] Documentary Film, "The Universe of Myron Evans" (in prep.) Directed by Jack Iandoli and produced by Francesco Fucilla; Documentary Film, "All about Tesla", directed by Michael Krause (premiered in Berlin 2007, Cannes Film Festival, Dylan Thomas Centre).
[13] S. P. Carroll, "Space-time and Geometry: an Introduction to General Relativity" (Addison Wesley, New York, 2004).
[14] R. Wald, "General Relativity" (Chicago, 1984).
[15] H. Stephani, D. Kramer, M. MacCallum, C. Hoenselaers and E. Hertl, "Exact Solutions to Einstein's Field Equation" (Cambridge, 2nd. Ed., 2003).
[16] H. Pfister, http://philsci-archive.pitt.edu/archive/00002681/01/lense.pdf.
[17] J. D. Jackson, "Classical Electrodynamics" (Wiley, 1999, 3rd ed.).

Explanation of the Cosmological Red Shift

by

Myron W. Evans,
Alpha Institute for Advanced Study, Civil List Scientist.
(emyrone@aol.com and www.aias.us)

Abstract

A simple explanation is given of the cosmological red shift using the ECE equations of the classical electrodynamics of a non-conducting medium and the Planck law. The resulting equation shows that the cosmological red shift is due to the nature of inter-galactic space-time, and not due to the incorrect metric of big bang, the Friedmann Lemaitre Robertson Walker (FLRW) metric. The FLRW metric is incorrect because of its neglect of space-time torsion, a fundamental error which was demonstrated in paper 93 of this series.

Keywords: Cosmological red shift, ECE equations, big bang.

25.1 Introduction

It is well known that the conventional idea of an expanding universe ("big bang") is based on the Friedmann Lemaitre Robertson Walker (FLRW) metric [1]. In this model the observable red shift of objects is explained by the Einstein equation of gravitational general relativity, of which the FLRW metric is a solution in the presence of canonical energy-momentum density. During the course of development of the Einstein Cartan Evans (ECE) unified field theory [2–12], the Einstein field equation has been shown conclusively to be incorrect because of its arbitrary neglect of space-time torsion. In tensorial notation, the dual identity of geometry states that:

$$D_\mu T^{\kappa\mu\nu} = R^\kappa{}_\mu{}^{\mu\nu} \qquad (25.1)$$

where $T^{\kappa\mu\nu}$ is the torsion tensor and $R^\kappa{}_\mu{}^{\mu\nu}$ is a well defined type of curvature tensor. Summation occurs as usual over repeated indices. It was found in paper 93 and following papers that Eq. (25.1) is not obeyed by the Einstein equation in the presence of energy momentum density because the equation omits torsion by using a symmetric connection:

$$T^\kappa_{\mu\nu} = \Gamma^\kappa_{\mu\nu} - \Gamma^\kappa_{\nu\mu} = 0. \qquad (25.2)$$

Unfortunately this omission by Einstein has been repeated uncritically, leading to gross errors in standard physics and cosmology. ECE theory has not repeated this error and has developed [1–12] a torsion based physics and cosmology.

Therefore nothing can be concluded about gravitational physics and cosmology from the Einstein equation, in which the omission of torsion is a basic error which came about from Einstein's limited knowledge of geometry and tensor analysis. It appears that the torsion was omitted by mathematicians prior to Einstein in order to simplify the problem. This procedure has simply been repeated throughout the twentieth century, but at the same time, careful scholarship has repeatedly criticised the Einstein field equation during the same century. These criticisms were initiated [13] in 1918 by Bauer and Schroedinger independently, but were apparently ignored by Eddington et al., who incorrectly claimed to have verified the flawed Einstein equation – an early example of media hype. By repeatedly ignoring valid criticism, standard physics has been reduced to unscientific dogma, dogma which is regularly propagated by unscientific and unscholarly methods. The evaluation of the tensor on the right hand side of Eq. (25.1) was carried out in paper 93 by computer, however, the equation itself is simple in structure, it shows immediately that the covariant derivative of torsion is the non-zero curvature. Therefore to assert that torsion vanishes will lead to a gross error (i.e. to zero torsion, but non-zero curvature). Essentially no textbooks in standard gravitational relativity develop torsion, but ECE has shown conclusively that torsion is the central idea of physics on all scales. Additionally, the so called "precision tests" of the Einstein equation have been revised and the data explained to high accuracy with the orbital theorem of ECE paper 111. Crothers [14] has shown that the so-called Schwarzschild metric was not obtained by Schwarzschild in 1916, whose procedure was to solve a geometrical problem in which the Ricci tensor was identically zero by construction. Therefore energy-momentum density and mass M were eliminated by construction by Schwarzschild and cannot appear therefore in the final metric. Schwarzschild was aware of this and mass M indeed does not appear in his two 1916 papers. The mass was inserted into the metric by others as a means of forcing the Schwarzschild solution to fit orbital data via the Newtonian limit. Therefore the Einstein equation does not predict data at all, it follows data by adapting geometry phenomenologically to the Newtonian limit. This geometry is now known to be basically incorrect because it directly violates Eq. (25.1). Numerous flaws in black hole mathematics have been pointed out in clear and unambiguous terms by Crothers [14] and others [15].

In Section 25.2 a review is given of the experimental data which refute big bang independently of any other theory – the Baconian principle. There are many instances known for example of objects or clusters being far older than big bang (the instant at which the universe is asserted to have "begun"). These data are well known but are ignored by the dogmatists who masquerade as scientists in the standard cosmology. Ignoring experimental data is by definition unscientific, and ignoring mathematics such as Eq. (25.1) is equally culpable. It is shown in Section 25.2 that much of big bang is empty speculation, it is merely an incorrect mathematical contrivance, not a theory of physics. In Section 25.3 a simple outline explanation is suggested for the cosmological red shift by using the ECE equations of classical electrodynamics in a non-conducting medium combined with a simple use of the Planck law. These procedures give the main properties of the cosmological red shift, and also allows for blue shifts, for which big bang can have no explanation.

25.2 A Summary of Experimental Data that Refute Big Bang

There are observable objects or clusters of galaxies that are far older than big bang [15], for example long chains of galaxies requiring hundreds of billions of years to have formed, while big bang is about ten billion years old theoretically. In other words there are formations of galaxies that are ten TIMES older than the "start" of the universe. Globular clusters in our galaxy are older than big bang, and the uranium content of stars is about twelve billion years old, again older than big bang. The most ancient spiral galaxies have already developed two or three arms, meaning that they are evolved and are older than big bang. If the universe started ten billion years ago, as asserted by big bang, then there could be no objects older than this. The most distant observable objects would ALL be defined by ten billion years multiplied by the speed of light in metres per year, and we would expect these most distant objects to be densely packed together in the part of the universe in which big bang "started". The observational truth is that the most ancient spiral galaxies are not clustered together at all, and have two or three arms, meaning that at that point in time (supposedly the start point of the entire universe) they were already evolved and therefore much older than big bang. If the latter were true they should be densely packed together in a given point, because the initial event of big bang is asserted to be a state of effectively infinite density and no volume. There is no sign of this mythical genesis in any data. The most ancient and distant objects are as far apart as near objects, implying an unbounded universe with no beginning or end. The unbounded universe was actually advocated by Einstein himself, also by Hubble and by many others, notably Hoyle. Another conclusive piece of evidence against big bang is that galaxies collide, they are not flying away from each other at an ever expanding rate as asserted by big bang.

As developed in paper 49 of the ECE series (www.aias.us), the 2.7 K background radiation temperature is easily calculated by elementary thermodynamics from an unbounded universe. This was the procedure adapted by Regener, Nernst, Herzberg, Finlay-Freundlich, Born, Assis and many others. The existence of the background radiation does not imply an expanding universe. Crothers [14] has cited work that suggest that the background radiation may merely be an artifact of observation. If the background radiation is artifact free as claimed in the standard physics, it is almost perfectly homogeneous, has only slight inhomogeneities, photons from opposite regions of the sky were never in contact with each other, contrary to big bang. This means that the background radiation is black body radiation which has always existed. The second law of thermodynamics requires entropy to increase following big bang, so that the universe would be disordered and very inhomogeneous contrary to observation in the almost completely homogeneous background radiation. Therefore there is no observational support for big bang, the background radiation is in fact strong evidence AGAINST big bang. This is a major and well known flaw of big bang, one of many. The other obvious conflict is with the first law of thermodynamics, because total energy in the universe must be conserved, the total energy is never observationally infinite, and therefore could not have been infinite at a speculative initial event of zero volume and infinite energy. Another major problem for big bang is that the universe is composed overwhelmingly of matter, indeed anti-matter can only be produced artificially in particle colliders. This observation has to be explained by speculation, the unsupported assertion of baryon asymmetry. This necessity leads in turn to more speculation, notably the speculation of cosmic inflation. The latter is asserted quite arbitrarily to be a phase transition, a simplistic speculation that after 10^{-35} secs the universe suddenly and without cause expanded exponentially to give a quark gluon plasma. It is then speculated without data that conservation of baryon number was violated, leading to the great predominance of matter over anti-matter in the current universe. In big bang a series of symmetry breaking phase transitions is speculated without data. A few minutes after the speculated initial event we are told that neutrons combined with protons to give deuterium and helium in big bang nucleosynthesis. However, Hoyle is well known to have had developed a successful and well known theory of nucleosynthesis prior to the empty speculation of big bang, with many arguments of his own against big bang. Pinter [16] in a scholarly multidisciplinary treatise, has argued that nearly all aspects of big bang nucleosynthesis are contested currently by scientists of various disciplines. Another severe weak point of big bang is the speculation that rest mass energy density gravitationally dominated over photon radiation. There is no clear mechanism for this, and some scientists such as Alfven argued for a universe that evolved from plasma. At 379,000 years after big bang it is speculated that radiation somehow "decoupled" from matter to give the background radiation. Another major weak point of big bang is that the homogeneity of the background radiation is speculated to be homogeneity of some kind prior to the inflation, and in violation of the second law of thermodynamics, this primordial homogeneity somehow persisted without entropy increase for ten billion years following exponential expansion at 10^{-35} secs. This is wildly unscientific and contrary to thermodynamics. Commonplace experience shows that an explosion scatters matter in an inhomogeneous manner. So the big bang argument starts to degenerate into speculation piled upon empty speculation, data to the contrary being ignored, and now, mathematics to the contrary (Eq. (25.1)) also being ignored.

Another major weakness of big bang is that it is unable to describe the structure of spiral galaxies without the introduction of yet more speculation, known as cold and hot dark matter and dark energy. The composition and mechanism of dark matter is unknown, and it is irrationally speculated that it causes the universe to "accelerate". In ECE theory [2–12] the structure of spiral galaxies is a direct consequence of geometry as required by relativity, the spiral galaxy vividly shows the underlying torsion, and the theory of this effect is simple and therefore preferred by Okham's Razor and by the philosophy of relativity. Above all, the lambda CDM model of big bang is based on basically incorrect mathematics, the FLRW metric that violates Eq. (25.1), i. e. basic geometry. One cannot violate Eq. (25.1) any more than one can violate the Pythagoras Theorem.

Other explanations for the cosmological red shift are available in the literature [17], notably explanations based on the Compton effect, and explanations based on optical theory as in paper 49. Hubble himself rejected the idea of the FLRW metric, as did Einstein, Vigier, Hoyle and many others. The sun's red shift for example is a Compton effect of the order of one part in a million. The sun is not receding from the earth, so this property is not a cosmological red shift and not a relativistic Doppler shift or a gravitational red shift. The sun's red shift can become as high as one part in a hundred in gamma rays emitted by a solar flare [17]. This suggests that there may be shifts of wavelength due to the Compton effect in inter-galactic space, which is by no means devoid of matter such as electron plasma, hydrogen molecules and so on. There would not be much scattering because the inter-galactic matter is very dilute, quite obviously.

We can therefore entirely discard big bang as obsolete and incorrect dogma. In the next section a simple optical explanation is suggested for the outline properties of the observable red shifts of cosmology.

25.3 ECE Explanation of the Cosmological Red Shift

In this section the main features of the cosmological red shift are calculated from the ECE equations [2–12] of plane waves propagating in a nonconducting, ponderable medium with polarizability and magnetizability. The starting point is the ECE Ampère Maxwell law written as:

$$\nabla \times \boldsymbol{B} - \frac{1}{c^2}\frac{\partial \boldsymbol{E}}{\partial t} = \mu_0 \boldsymbol{J} \tag{25.3}$$

where $\boldsymbol{B}$ is magnetic flux density, $\boldsymbol{E}$ is electric field strength, c is the vacuum speed of light, μ_0 is the vacuum permeability in SI units, and $\boldsymbol{J}$ is the interaction current of light propagating through inter galactic space. The electric displacement $\boldsymbol{D}$ and the magnetic field strength $\boldsymbol{H}$ are defined in general [18, 19] by the polarization $\boldsymbol{P}$ and the magnetization $\boldsymbol{M}$ where ϵ_0 is the permittivity of the vacuum in SI units. In general:

$$\begin{aligned}\boldsymbol{D} &= \epsilon_0 \boldsymbol{E} + \boldsymbol{P}, \quad \boldsymbol{B} = \mu_0(\boldsymbol{H} + \boldsymbol{M}), \\ &= \epsilon \boldsymbol{E} \qquad\qquad = \mu \boldsymbol{H}\end{aligned} \tag{25.4}$$

and in SI units:

$$\mu_0 \epsilon_0 = \frac{1}{c^2}. \tag{25.5}$$

Therefore in Eq. (25.3):

$$\boldsymbol{E} = \frac{1}{\epsilon_0}(\boldsymbol{D} - \boldsymbol{P}), \boldsymbol{B} = \mu_0(\boldsymbol{H} + \boldsymbol{M}) \tag{25.6}$$

and eq. (25.3) becomes:

$$\nabla \times \boldsymbol{H} - \frac{\partial \boldsymbol{D}}{\partial t} = \boldsymbol{J} - \left(\nabla \times \boldsymbol{M} - \frac{\partial \boldsymbol{P}}{\partial t}\right). \tag{25.7}$$

It is enough for our present purposes to consider the case

$$\boldsymbol{J} = \nabla \times \boldsymbol{M} - \frac{\partial \boldsymbol{P}}{\partial t} \tag{25.8}$$

where the current is defined by the polarization and magnetization. In this case:

$$\nabla \times \boldsymbol{H} - \frac{\partial \boldsymbol{D}}{\partial t} = \boldsymbol{0} \tag{25.9}$$

25.3 ECE Explanation of the Cosmological Red Shift

where D and H are expressed in terms of E and H by the permittivity ϵ of the inter-galactic ponderable medium, and by its permeability μ. From eqs (25.4) and (25.9):

$$\nabla \times B + i\omega\mu\epsilon E = 0 \qquad (25.10)$$

if a harmonic time dependence of type $e^{-i\omega t}$ is assumed [18, 19] in the solution.

Consider a plane wave with phase [18, 19]:

$$\phi = \omega t - \kappa Z \qquad (25.11)$$

where ω is its angular frequency at instant t and κ is its wavenumber at point Z for propagation along the Z axis. From Eq. (25.10) the wavenumber and frequency are related by:

$$\kappa = (\mu\epsilon)^{\frac{1}{2}}\omega. \qquad (25.12)$$

The phase velocity of the wave is [19]:

$$v = \frac{\omega}{\kappa} = \frac{c}{n} = \frac{1}{(\mu\epsilon)^{\frac{1}{2}}} \qquad (25.13)$$

where the refractive index is:

$$n = \left(\frac{\mu\epsilon}{\mu_0\epsilon_0}\right)^{\frac{1}{2}} = (\epsilon_r\mu_r)^{\frac{1}{2}} \qquad (25.14)$$

and where the relative permittivity and permeability are:

$$\epsilon_r = \frac{\epsilon}{\epsilon_0}, \mu_r = \frac{\mu}{\mu_0}. \qquad (25.15)$$

In the presence of absorption [19] the wavenumber is in general a complex number, conventionally denoted:

$$\kappa = \beta + i\frac{\alpha}{2}. \qquad (25.16)$$

Here α is the power absorption coefficient defined by the Beer Lambert law:

$$\alpha = \frac{1}{z}\log_e \frac{I_0}{I} \qquad (25.17)$$

where I is the power density and I_0 the initial power density. Therefore in the presence of absorption the angular frequency may be developed as a complex number:

$$\omega = \omega' + i\omega'' = \left(\frac{c}{n}\right)\left(\beta + i\frac{\alpha}{2}\right). \tag{25.18}$$

In some texts the angular frequency is kept constant and the wavenumber developed as a complex number. In a medium in which the phase velocity of the wave is v, the relation between angular frequency and wavenumber is:

$$v = \frac{\omega}{\kappa}. \tag{25.19}$$

Consider for simplicity of argument a relative permeability of unity:

$$\mu_r = 1 \tag{25.20}$$

then:

$$\omega = \frac{c}{\epsilon_r^{\frac{1}{2}}}\left(\beta + i\frac{\alpha}{2}\right) \tag{25.21}$$

where ϵ_r is the complex relative permittivity:

$$\epsilon_r = \epsilon_r' + i\epsilon_r'' \tag{25.22}$$

made up of dielectric dispersion ϵ_r' and dielectric loss ϵ_r''. Therefore:

$$\begin{aligned}\omega^2 &= \frac{c^2}{\epsilon_r' + i\epsilon_r''}\left(\beta + i\frac{\alpha}{2}\right)^2 \\ &= \frac{c^2(\epsilon_r' - i\epsilon_r'')}{\epsilon_r'^2 - \epsilon_r''^2}\left(\beta + i\frac{\alpha}{2}\right)^2 \\ &= (\omega' + i\omega'')^2 = \omega'^2 + 2i\omega'\omega'' - \omega''^2.\end{aligned} \tag{25.23}$$

and:

$$\begin{aligned}\omega'^2 - \omega''^2 &= \frac{c^2}{\epsilon_r'^2 - \epsilon_r''^2}\left(\epsilon_r'\left(\beta^2 - \frac{\alpha^2}{4}\right) + \alpha\beta\epsilon_r''\right), \\ 2\omega'\omega'' &= \frac{c^2}{\epsilon_r'^2 - \epsilon_r''^2}\left(\epsilon_r'\frac{\alpha}{2} - \epsilon_r''\beta\right).\end{aligned} \tag{25.24}$$

In general, in the presence of absorption, the frequency is by no means constant, so light travelling through the inter galactic medium is governed by

25.3 ECE Explanation of the Cosmological Red Shift

these equations. We already see that there are optical explanations for the cosmological red shift.

If the problem is developed in terms of fixed ω and varying wavenumber then:

$$\epsilon_r^2 = n = \kappa \left(\frac{c}{\omega}\right) \tag{25.25}$$

i.e.:

$$(\epsilon_r' + i\epsilon_r'')^2 = n' + in'' = \frac{c}{\omega}(\kappa' + i\kappa''). \tag{25.26}$$

This means that the relative permittivity changes the wavenumber for constant ϵ_r or given ω. In the case of no absorption:

$$\kappa = \left(\frac{\omega}{c}\right) \epsilon_r^2. \tag{25.27}$$

The standard SI unit of wavenumber is $\bar{\nu}$ (Neper cm^{-1}), and is defined [20]:

$$\omega = 2\pi\bar{\nu}c = \kappa v. \tag{25.28}$$

Therefore:

$$\bar{\nu} = \frac{\kappa}{2\pi} \tag{25.29}$$

and

$$\bar{\nu} = \frac{\omega}{2\pi c}\epsilon_r^2 = \frac{f}{c}\epsilon_r^2 \tag{25.30}$$

where f is the frequency in hertz. Sometimes this is assumed to be the fixed frequency of the source, so $\bar{\nu}$ is the observed wavenumber of light after it has travelled through a nonconducting medium. In the absence of absorption α:

$$\bar{\nu} = \left(\frac{f}{c}\right) \epsilon_r^2 \tag{25.31}$$

where ϵ_r is frequency independent, but in the presence of absorption there is dielectric dispersion and dielectric loss as is well known [18–20].

In the presence of absorption the dependence of $\bar{\nu}$ on ϵ_r is given by:

$$\bar{\nu}' + i\bar{\nu}'' = \frac{f}{c}(\epsilon_r' + i\epsilon_r'')^2 \tag{25.32}$$

i.e.:

$$\bar{\nu}' = \frac{f}{c}(\epsilon_r'^2 - \epsilon_r''^2) \tag{25.33}$$

and:

$$\bar{\nu}'' = \frac{2f}{c}\epsilon_r'\epsilon_r''. \tag{25.34}$$

The power absorption coefficient and dielectric loss are related by [20]:

$$\alpha = \frac{\omega \epsilon_r''}{n'c} \tag{25.35}$$

so:

$$\bar{\nu}'' = \frac{n'\epsilon_r'}{\pi}\alpha. \tag{25.36}$$

Therefore the imaginary part of the complex wavenumber is proportional to the power absorption coefficient.

From Eqs. (25.33) and (25.36) it is seen that the medium regarded as nonconducting ponderable matter, changes the observable wavenumber of the light. This shift depends only on the medium (inter galactic space) so is such that:

$$\frac{\bar{\nu}'}{f} = \frac{1}{c}(\epsilon_r'^2 - \epsilon_r''^2) \tag{25.37}$$

is relatively the same for each spectral line, as observed in the cosmological red shift because

$$\Delta\epsilon^2 := \epsilon_r'^2 - \epsilon_r''^2 \tag{25.38}$$

is a property of the inter galactic matter only. By observation it is seen that $\Delta\epsilon^2$ is on average a constant property. The quantity:

$$\frac{\bar{\nu}'}{f} = \frac{1}{c}\Delta\epsilon^2 \tag{25.39}$$

must be greater than one, otherwise the observed wavenumber would be negative. This is observed in dielectric spectroscopy of a non-conducting medium.

The second main feature of the cosmological red shift is that it is observed to be proportional to distance, or sample length of the Beer Lambert law, denoted Z in Eq. (25.17). This feature is explained in the simplest way by

25.3 ECE Explanation of the Cosmological Red Shift

considering a monochromatic beam made up of one photon. The Planck law for the photon is:

$$E = hf \tag{25.40}$$

where E is its quantum of energy and h is the Planck constant [18]. The energy density of the photon is:

$$U = \frac{E}{V} \tag{25.41}$$

where V is the volume it occupies. Its intensity or power density in watts per metre squared is:

$$I = cU = \left(\frac{hc}{V}\right) f. \tag{25.42}$$

From eqs. (25.17) and (25.42):

$$f = f_0 \exp(-\alpha Z) \tag{25.43}$$

so the energy or frequency of the photon decreases with distance. If the light is completely absorbed no energy emerges at the detector and there is no measurable frequency at all. This is the ultimate red shift. The average energy of n oscillators of a monochromatic beam of light of frequency f is given by [18]:

$$E = \sum_n p_n E_n \tag{25.44}$$

where p_n is the probability of finding it in a state with energy E_n. Using the Boltzmann distribution [18]:

$$p_n = \exp(-E_n/(kT))/\sum_n \exp(-E_n/(kT)) \tag{25.45}$$

and this choice leads to thermodynamic equilibrium as is well known. The mean energy of an oscillator of frequency f may then be calculated [18]:

$$\langle E \rangle = hf \left(\frac{x}{1-x}\right), \quad x = \exp\left(-\frac{hf}{kT}\right). \tag{25.46}$$

This is the mean energy of a monochromatic beam at frequency f containing n photons. It is calculated in the limit [18]:

$$hf \ll kT. \tag{25.47}$$

When this quantity is much less than unity Eq. (25.46) reduces to Eq. (25.40).
Combining Eqs. (25.37) and (25.43):

$$\frac{\bar{\nu}'}{f} = \frac{f_0}{c}\exp\left(-\frac{\omega\epsilon'_r Z}{n'c}\right)(\epsilon'^2_r - \epsilon''^2_2), \tag{25.48}$$

so the way in which the real part of the observed wavenumber is shifted depends on the relative values of the dielectric permittivity and dielectric loss of intergalactic space, or deep space. In general this is not a simplistic red shift as in Big Bang. There may be blue shifts as well as red shifts. Finally if there is an electron plasma in deep space the medium develops a conductivity, and the optical properties change. In general all the optical properties of light may be changed on its long inter galactic journey from source to observer. As in any spectrum there may be several absorption and dispersion features, and for plasma, the highly developed theory of plasma [19] is needed.

As argued in Section 25.2, Compton shifts also occur as the photons interact with inter galactic electrons. In this development it is seen from eq. (25.29) that it is one in terms of wavenumber, which may be related to wavelength by:

$$\bar{\nu} = 2\pi\omega c, \quad \lambda = 2\pi\frac{V}{\omega}. \tag{25.49}$$

In the absence of absorption the refractive index is a constant greater than unity, so the phase velocity v is lower than c. In the presence of absorption the refractive index is complex as argued.

Acknowledgments

The British Government is thanked for a Civil List Pension and the staffs of AIAS and TGA for many interesting discussions.

References

[1] S. P. Carroll, "Spacetime and Geometry: an Introduction to General Relativity" (Addison Wesley, New York, 2004).

[2] M. W. Evans, "Generally Covariant Unified Field Theory" (Abramis, 2005 onwards), volumes 1–4, volumes 5 and 6 in prep.

[3] L. Felker, "The Evans Equations of Unified Field Theory" (Abramis 2007).

[4] K .Pendergast, "Crystal Spheres" (Abramis in prep., preprint on www.aias.us).

[5] M. W. Evans, Omnia Opera section of the British Civil List Scientist's website www.aias.us.

[6] M. W. Evans et al., Adv. Chem. Phys. Vol. 119 (Wiley 2001); ibid vol. 85 (Wiley 1992, 1993, 1997).

[7] M. W. Evans and J.-P. Vigier, "The Enigmatic Photon" (Kluwer, Dordrecht, 1994 to 2002, hardback and softback), in five volumes.

[8] M. W. Evans and L. B. Crowell, "Classical and Quantum Electrodynamics and the $B(3)$ Field" (World Scientific, 2001).

[9] M. W. Evans et al., fifteen papers on ECE theory in Found. Phys. Lett., 2003 to 2005.

[10] M. W. Evans, Acta Phys. Polonica, **33B**, 2211 (2007).

[11] M. W. Evans and H. Eckardt, Physica B, **400**, 175 (2007).

[12] M. W. Evans, Physica B, **403**, 517 (2008).

[13] C. Alley, Wheeler Conference at Princeton, 2006.

[14] S. Crothers, papers on www.aias.us and paper 93 of the ECE series.

[15] J. Dunning-Davies, "Exploding a Myth", (Horwood, 2007)

[16] P. Pinter, www.originoflife.org.uk and book published by Abramis.

[17] http://www.angelfire.com/az/Bigbangiswrong/index.html

[18] P. W. Atkins, "Molecular Quantum Mechanics (Oxford, 2nd ed. 1983 and subsequent editions).

[19] J. D. Jackson, "Classical Electrodynamics" (Wiley, 1999, 3rd. Ed.).

[20] M. W. Evans, G. J. Evans, W. T Coffey and P. Grigolini, "Moelcualr Dyanmics (Wiley, 1982, no 108 on the Omnia Opera of www.aias.us)

ECE Theory of the Equinoctial Precession and Galactic Dynamics

by

Myron W. Evans,
Alpha Institute for Advanced Study, Civil List Scientist.
(emyrone@aol.com and www.aias.us)

Abstract

The equinoctial precession of the earth is shown to be due to the gravitomagnetic equation of ECE dynamics. The gravitomagnetic precession is caused by the almost constant orbital velocity of the sun around the galactic centre. The sun's orbit is non-Newtonian, as is explained by another ECE law of dynamics without the assumption of dark matter. These are straightforward relativistic explanations of phenomena that cannot be explained in standard physics without ad hoc assumptions such as dark matter.

Keywords: ECE theory, equinoctial precession, galactic dynamics.

26.1 Introduction

Recently it has been shown that there are four generally covariant equations of dynamics whose structure is the same as the generally covariant equations of electrodynamics of Einstein Cartan Evans (ECE) unified field theory [1–10]. The basic hypothesis of unification requires the existence gravitomagnetic equations of dynamics in addition to the familiar Newtonian structure. In paper 117 of this series (www.aias.us), the precession due to the earth's daily spin was explained with one of the ECE gravitomagnetic equations, giving plausible results of order of magnitude about 0.1 arcseconds a year. In this paper the simplest type of gravitomagnetic equation explains straightforwardly the earth's equinoctial precession, which is currently 50.29 arcseconds a year. In section 26.2, a summary of Cruttenden's arguments [11] is given. This paper [11] clearly shows that the equinoctial precession is that of the solar system with respect to a distant star, a precession caused by the motion of the sun. The earth does not show an equinoctial precession with respect to objects within the solar system. Cruttenden [11] argues for the existence of a binary star system of which the sun is one star. This is a partially correct argument, because the precession according to ECE gravitomagnetic theory is due to the velocity of the sun with respect to the galactic centre. This is a non-Newtonian orbit with almost constant velocity because the sun is positioned on that part of the Milky Way's galactic velocity curve. The ECE theory gives a straightforward explanation for galactic dynamics in terms of general relativity with torsion. In this theory no concepts extraneous to relativity (notably dark matter) are required and so the ECE theory is preferred by Ockham's Razor.

In Section 26.2 a summary of Cruttenden's arguments are given and concepts defined. In Section 26.3 the equinoctial precession is explained straightforwardly with the simplest type of gravitomagnetic equation, and in Section 26.4 galactic dynamics are explained starightforwardly with ECE theory (general relativity with torsion correctly incorporated).

26.2 The Equinoctial Precession of the Earth

Cruttenden argues convincingly [11] that the equinoctial precession is observed to be a precession with respect to distant objects, not with respect to objects within the solar system. The equinox is defined as the point when the earth's axis is at right angles to a line drawn from the centre of the sun to the centre of the earth. The solar year is defined as the time it takes for the earth to complete one rotation from equinox to equinox, and the sidereal year is defined as the time taken by the earth to realign with a fixed distant star. The solar year is 365.2422 earth spins, and the sidereal year is 365.2563 earth spins. The entire solar system is well known to be moving through space around the galactic centre, and it is this movement that causes the precession of the equinox. There is no precession of the earth relative to the sun, or planets, or any object in the solar system [11]. The precession is a term used to describe a reorientation of the earth's axis, the earth is held in a synchronous position and this causes the reorientation. The time form equinox to equinox is a 360 degree rotation of the earth around the sun, but due to to the velocity v of the sun around the galactic centre, the earth appears to fall short of 360 degrees by 50.29 arcseconds every sidereal year. In other words the entire solar system has moved with respect to a distant star by this amount. The earth has not moved by this amount with respect to Perseid showers for example [11]. These occur inside the solar system. The equinox moves by 50.29 arcseconds per sidereal year but does not move with respect to the ecliptic, i.e. with respect to the sun. Another way of stating

26.2 The Equinoctial Precession of the Earth

this result is that the earth does not align with the fixed or distant stars at the vernal equinox, it shifts by 50.29 arcseconds clockwise every sidereal year. This means that the entire solar system is moving clockwise because of the sun's well known velocity with respect to the galactic centre of the Milky Way. In addition to this, the entire Milky Way Galaxy is now known to be moving towards the Hydra constellation. The earth moves 360 degrees relative to the fixed stars in a sidereal year, but the sun also moves at an almost constant 210 kilometres per second with respect to the galactic centre, which is itself moving with respect to fixed stars and constellations such as Hydra. These are the causes of the equinoctial precession, not gravitation of the sun and moon as postulated originally by Newton in a time where nothing was known about the sun's motion or the structure of galaxies.

The current standard model of the earth's equinoctial precession has not changed since the time of Newton, and is a complicated concoction of ideas based on the gravitational pull of objects inside the solar system. This is contrary to data [11] and also violates basic theory. This has been argued by Santagata [12], who has pointed out several problems [11] in the development of the gravitational theory of equinoctial precession by Newton, d'Alembert and many others. Negut [11] has pointed out that the earth cannot precess or nutate in the absence of a support point. The dynamics of a spinning top with one point fixed were first worked out by Lagrange [13] as is well known, but the earth obviously has no support point. The usual explanation of the equinoctial precession is that a net torque is present on the earth due to the moon and sun, but as ref. (26.13), problem 10.10 shows, this torque can only result in precession of a symmetrical top if the origins of the laboratory and rigid body fixed frames are the same, i.e. if the point of the top is fixed in space and coincides with the origin of the symmetric top frame. This is not the case for the earth, even if it were a symmetric top and not a spherical top. It is well known that the earth is indeed a slightly oblate symmetric top due to a bulge at the equator. The dynamics of the spinning top with one point fixed and of mass m in a gravitational field of acceleration g are due to the torque:

$$|\boldsymbol{Tq}| = |\boldsymbol{r} \times \boldsymbol{F}| = mgr \sin \phi \qquad (26.1)$$

created by the force of gravity:

$$\boldsymbol{F} = m\boldsymbol{g} \qquad (26.2)$$

on the top's centre of mass. Here r is the distance between the top's fixed support point and the centre of mass, and ϕ is the angle between the vectors F and r. If the point is not fixed, the force F would cause the arm r simply to spin. No such spin occurs for the earth because its centre of mass is approximately that of a sphere located at the origin of the sphere. The origin of the earth's daily spin is not the sun's gravity, and is not the moon's gravity. The moon itself for example does not spin in the earth's gravity or sun's gravity. The torque (26.1) is also:

$$Tq = \omega_p \times L \qquad (26.3)$$

where the spin angular momentum vector L is parallel to r and ω_p is the angular precession vector. The precession of the spinning top is therefore the angular frequency defined by eq. (26.1). Its magnitude is:

$$\omega_p = \frac{mgr}{L} = \frac{g}{v} = \frac{d\theta}{dt}. \qquad (26.4)$$

However, in order for the torque (26.2) to exist, the point of the top must be fixed, so that there is a leverage and lever arm $r\sin\theta$. Otherwise the top would be put into a spinning motion only, meaning that the lever arm would not be stable. The angular momentum L is spin angular momentum defined by the moment of inertia I and the spin angular velocity vector ω. When the top slows down it begins to nutate, as first shown by Lagrange. This nutation is what is known as "wobbling". However, neither precession nor nutation occur if the point of the top is not fixed, as first shown by Lagrange in the eighteenth century. Numerous other criticisms of this standard model are summarized by Cruttenden [11], both on experimental and theoretical grounds.

26.3 Gravitomagnetic Explanation of the Equinoctial Precession

The equinoctial precession is described straightforwardly by the simplest ECE gravitomagnetic equation (paper 117 of www.aias.us):

$$\Omega = -\frac{1}{c^2} v \times g \qquad (26.5)$$

where g is the earth's acceleration due to gravity and v is the velocity of the sun around the galactic centre. Eq. (26.5) is the precise analogue of the magnetic equation:

$$B = -\frac{1}{c^2} v \times E \qquad (26.6)$$

obtained from the inverse Lorentz transform in the non-relativistic limit $v \ll c$. Therefore from eq. (26.5):

$$\Omega = \frac{vg}{c^2} \sin\theta \qquad (26.7)$$

where θ is the angle between v and g. For an observed precession of 50.29 arcseconds per year, $v \sin\theta$ is 70.8 kilometres per second. The sun's orbital velocity around the galactic centre is 220 kilometres per second, so the angle between v and g is 18.8°. The equinoctial precession magnitude is:

$$\Omega = 7.10 \times 10^{-4} \ v \sin\theta \tag{26.8}$$

in arcseconds a year. It is known that the sun's velocity v is almost constant, and that the sun orbits the galactic centre towards Cygnus, taking 220 million years to complete one orbit. The local standard of rest (LSR) is a reference frame in circular orbit around the galactic centre. The gravity of nearby stars causes the sun to move at 20 kilometres a second with respect to the LSR. The sun's orbit is 10% off from circular and it is falling slowly towards the centre of the Milky Way galaxy. At the centre of the Milky Way there is a mass of about 1 to 2.3 million sun masses. The sun is situated in the outer part of the galaxy, between the third and fourth arms, about 20 light years above the galactic plane and 28,000 light years from the galactic centre. It is on the non-Newtonian part of the galactic velocity curve (paper 76 of this series on www.aias.us) because its orbital velocity is nearly constant. It is not therefore governed by Kepler's laws but by the relevant ECE equation of motion as developed in the next section.

26.4 ECE Equation of Motion of Galaxies

Galactic dynamics are explained straightforwardly in ECE theory by the generally covariant law of dynamics:

$$\boldsymbol{\nabla} \cdot \boldsymbol{g} = 4\pi G \rho \tag{26.9}$$

where G is Newton's constant and ρ is mass density. Eq. (26.9) is a direct consequence of general relativity with torsion, and of the Bianchi identity as developed by Cartan [1–10] If spacetime torsion is represented by the shorthand symbol T (without indices for clarity) and if R represents curvature, then:

$$\boldsymbol{\nabla} \cdot \boldsymbol{g} = c^2 \boldsymbol{\nabla} \cdot \boldsymbol{T} = c^2 (R - \omega T) \tag{26.10}$$

where ω represents the connection of spacetime [1–10], in this case the spin connection. In the Newtonian limit, the spin connection approaches zero so that:

$$\boldsymbol{\nabla} \cdot \boldsymbol{g} = c^2 R. \tag{26.11}$$

This equation is the familiar Newtonian equation for continuous mass contained within a volume, and from it the Newtonian inverse square law emerges as a limit. In the opposite limit:

$$R = \omega T \tag{26.12}$$

of pure rotational dynamics [1–10]:

$$\boldsymbol{\nabla} \cdot \boldsymbol{g} = \boldsymbol{\nabla} \cdot \boldsymbol{T} = 0 \tag{26.13}$$

and

$$g = Tc^2 = \text{constant} \tag{26.14}$$

meaning that g and T are constants independent of distance. These are the non-Newtonian dynamics that govern the constant v region of a galactic velocity curve. For stars close to the galactic centre, Newtonian dynamics apply as in eq. (26.11), and in intermediate regions the dynamics are governed by a balance of T, R and ω. These are direct results of standard four dimensional general relativity with torsion [1–10], the ECE theory. There is no need to postulate dark matter and it is well known that big bang, black holes and dark matter are the erroneous results of the flawed Einstein equation [1–10]. String theory is merely mathematical hyper-complexity due to lack of understanding of torsion in general relativity. The mass at the centre of galaxies such as the Milky Way can be calculated with Kepler's third law applied to stars near the centre that obey Newtonian dynamics. It is not a black hole, which is an erroneous mathematical notion.

The canonical angular energy momentum density tensor is well known to be defined by [14]:

$$J^{\kappa\mu\nu} = -\frac{1}{2}\left(T^{\kappa\mu}x^\nu - T^{\kappa\nu}x^\nu\right) \tag{26.15}$$

where $T^{\kappa\mu}$ is the symmetric canonical energy momentum density tensor:

$$T^{\kappa\mu} = T^{\mu\kappa} \tag{26.16}$$

(not to be confused with the three index torsion tensor). Here x^μ is the four-coordinate:

$$x^\mu = (ct, X, Y, Z). \tag{26.17}$$

26.4 ECE Equation of Motion of Galaxies

The angular momentum tensor [14] is defined by integration:

$$J^{\mu\nu} = \int J^{\kappa\mu\nu} d^3 x_\kappa \qquad (26.18)$$

over an infinitesimal hypersurface in four dimensions ($d^3 x_\kappa$). In paper 103, Eq. (16) (www.aias.us) it was shown that:

$$T^{\kappa\mu\nu} = \frac{k}{c} J^{\kappa\mu\nu} \qquad (26.19)$$

where is Einstein's constant and where $T^{\kappa\mu\nu}$ is the three index torsion tensor of geometry. In paper 98, Eq. (62), it was shown that:

$$T^{\kappa\mu\nu} = \frac{1}{\hbar \kappa^2} J^{\kappa\mu\nu} \qquad (26.20)$$

where $\hbar$ is Planck's constant κ and a fundamental wave-number. Comparing Eqns. (26.19) and (26.20):

$$\kappa = \left(\frac{c}{\hbar k}\right)^{\frac{1}{2}} \qquad (26.21)$$

giving a wavelength akin to the Planck length. Therefore the torsion of spacetime is proportional to the canonical angular energy momentum density of matter. This is the correct version of the Einstein postulate between the Bianchi identity and Noether's Theorem. The Einstein postulate is well known to be incorrect because of neglect of torsion. Therefore the integral over torsion give angular momentum, and the constant v region of the galactic velocity curve is, self consistently, a region of constant angular momentum of stars and matter generated by the underlying constant torsion of spacetime. This is the correct relativistic explanation of galactic dynamics, and not dark matter.

The quantity [1–10] that defines g in Eq. (26.9) is an orbital torsion:

$$\boldsymbol{T} = T^{010}\boldsymbol{i} + T^{020}\boldsymbol{j} + T^{030}\boldsymbol{k} \qquad (26.22)$$

so:

$$\boldsymbol{g} = c^2 \boldsymbol{T} = ck\boldsymbol{J} \qquad (26.23)$$

where

$$\boldsymbol{J} = J^{010}\boldsymbol{i} + J^{020}\boldsymbol{j} + J^{030}\boldsymbol{k} \qquad (26.24)$$

is an orbital canonical angular energy momentum density vector formed from tensor elements. From paper 100, eq. (116) (www.aias.us) :

$$J^{\mu\nu} = \int J^{0\mu\nu} d^3x \qquad (26.25)$$

i.e.:

$$\left.\begin{aligned} J^{01} &= \int J^{010} d^3x \\ J^{02} &= \int J^{020} d^3x \\ J^{03} &= \int J^{030} d^3x. \end{aligned}\right\} \qquad (26.26)$$

In the galactic region where the sun is located, J^{01}, J^{02}, and J^{03} are constant:

$$\boldsymbol{J}_1 = J^{01}\boldsymbol{i} + J^{02}\boldsymbol{j} + J^{03}\boldsymbol{k}, \quad |\boldsymbol{J}_1| = mrv = \text{constant}. \qquad (26.27)$$

and this is a region of constant spacetime torsion that gives rise to the earth's equinoctial precession. In the Newtonian region of the Milky Way the torsion is r dependent, because its divergence is not zero, i.e. g is given by the Newtonian $-GM/r^2$ where M is the mass at the centre of the galaxy, and r is the distance from an inner star to the galactic centre. This is as observed experimentally, the inner stars of a galaxy obey Newtonian dynamics in the non-relativistic approximation.

Acknowledgments

The British Government is thanked for a Civil List Pension and many colleagues for interesting discussions. Axel Westrenius is thanked for bringing my attention to the paper by Walter Cruttenden.

References

[1] M. W. Evans, "Generally Covariant Unified Field Theory" (Abramis, Suffolk, 2005 onwards), in four volumes to date, volumes five and six in prep. (See also www.aias.us).

[2] L Felker, "The Evans Equations of Unified Field Theory" (Abramis, 2007).

[3] K. Pendergast, "Crystal Spheres" (Abramis in prep., see also www.aias.us).

[4] S. P. Carroll, "Spacetime and Geometry: an Introduction to General Relativity" (Addison-Wesley, New York, 2004).

[5] "The Universe of Myron Evans", a film directed by Jack Iandoli and produced by Francesco Fucilla, (2008).

[6] M. W. Evans et al., Omnia Opera section of www.aias.us, the website of the British Civil List Scientist, 1992 to present.

[7] M. W Evans (ed.), Advances in Chemical Physics, vol. 119 (2001); ibid., M. W. Evans and S. Kielich, ibid., vol. 85 (first edition, 1992, 1993, 1997).

[8] M. W. Evans, fifteen papers on ECE theory in Foundations of Physics Letters, 2003 to present, M. W. Evans, Acta Phys. Polonica, **33B**, 2211 (2007); M. W .Evans, Physica B, **403**, 517 (2008); M. W. Evans and H. Eckardt, Physica B, **400**, 175 (2007).

[9] M. W. Evans and L. B. Crowell, "Classical and Quantum Electrodynamics and the B(26.3) Field" (World Scientific, 2001).

[10] M. W. Evans and J.-P, Vigier, "The Enigmatic Photon" (Kluwer, Dordrecht, 1994 to 2002, hardback and softback), in five volumes.

[11] W. Cruttenden, communication of 2007 forwarded by A. Westrenius.

[12] C. Santagata, GRMA, 2007.

[13] J. B. Marion and S. T. Thornton, "Classical Dynamics" (HB, New York, 1988, 3rd. ed).

[14] L. H. Ryder, "Quantum Field Theory" (Cambridge, 1996, 2nd ed.)

Criticisms of Black Hole Theory

by

Myron W. Evans,
Alpha Institute for Advanced Study, Civil List Scientist.
(www.aias.us and www.atomicprecision.com),

and

S. Crothers, J. Dunning-Davies and Horst Eckardt,
Alpha Institute for Advanced Studies (A.I.A.S/T.G.A)
(www.aias.us, www.atomicprecision.com and www.telesio-galilei.com)

Abstract

The claims to existence of black holes in nature are refuted in several ways, and several fundamental errors of the theory are pointed out. The most basic error is that the Einstein field equation violates the Cartan/Evans dual identity of geometry, as shown in previous papers of this series. Several commonly used metrics of black hole theory are shown by computer algebra to violate basic geometry (the Cartan/Evans dual identity). These include the Kruskal metric and the metric used to claim the existence of Hawking radiation. The class of vacuum solutions refer only to a Ricci flat spacetime in which there is no energy momentum density and no physics. The so called Schwarzschild metric of black hole theory is incorrectly attributed to him and the conventional theory of singularities in this metric is fundamentally incorrect, as shown independently in this paper by Dunning-Davies and Crothers. The most fundamental error in black hole theory is the arbitrary neglect of space-time torsion. In the ECE equations of motion, the torsion is correctly re-instated, and plays a fundamental role in physics on all scales.

Keywords: criticism of black holes, torsion, ECE theory.

27.1 Introduction

The Einstein field equation is the archetypical construct of twentieth century physics, and general relativity is the best known idea of twentieth century science and thought in general. This was the great paradigm change in natural philosophy which gradually manifested itself in the years from 1887 (Michelson Morley experiment) to 1915 (when Einstein and Hilbert arrived at the famous field equation using different methods). This was the evolution of the theory of relativity, an attempt to describe nature objectively. The fundamental idea behind general relativity is that the laws or equations of physics must be generally covariant. This means that their tensor structure is governed by coordinate transformation in geometry. Therefore physics becomes geometry. It is of great importance to use the right geometry, and it has always been assumed that relativity has been based on the right geometry. In a series of papers [1–12] this claim has been refuted, the geometry used in 1915 omits spacetime torsion, and this leads to a violation of a development of the Cartan identity of 1922 [12] in which torsion is ineluctably linked to curvature. This became clear by using the Hodge transform in four dimensions to give the Cartan/Evans dual identity [1–12]. In tensor format this is:

$$D_\mu T^{\kappa\mu\nu} = R^\kappa{}_\mu{}^{\mu\nu} \tag{27.1}$$

where $T^{\kappa\mu\nu}$ is the torsion tensor and $R^\kappa{}_\mu{}^{\mu\nu}$ is the curvature tensor. Here D_μ denotes the covariant derivative. The torsion and curvature tensors are defined in Riemann geometry through the action of the commutator of covariant derivatives [1–12] on any tensor. It is incorrect to assert that the torsion tensor is zero. The torsion tensor is defined by:

$$T^\kappa_{\mu\nu} = \Gamma^\kappa_{\mu\nu} - \Gamma^\kappa_{\nu\mu} \tag{27.2}$$

where $\Gamma^\kappa_{\mu\nu}$ is the connection, and in general the torsion is not zero. In 1900, at the dawn of tensor theory and the twentieth century, Ricci and Levi-Civita [13] introduced the symmetric connection:

$$\Gamma^\kappa_{\mu\nu} = \Gamma^\kappa_{\nu\mu} \tag{27.3}$$

for which the torsion tensor is made zero in an arbitrary way, merely an assumption used apparently to simplify the calculations in 1900. The symmetric connection is commonly called the Christoffel, Levi-Civita, or Riemann connection, but was not introduced by Riemann or Christoffel. It was introduced by Ricci and Levi-Civita in 1900. Thereafter it was used uncritically by Einstein, who was not aware of the existence of torsion in 1915, and who based his field equation on a Bianchi identity without torsion. This is the so called "second Bianchi identity" in standard physics [14], now thoroughly obsolete. Einstein made this torsion-less identity proportional to the covariant Noether Theorem through k, the Einstein constant.

The fundamental importance of torsion was not realized clearly until about 1922, when Cartan developed his two structure equations in his elegant differential geometry. The first structure equation defines the torsion differential form (T^a) as the covariant derivative of the Cartan tetrad differential form (q^a), using the spin connection $\omega^a{}_b$ of Cartan:

$$T^a = d \wedge q^a + \omega^a{}_b \wedge q^b \tag{27.4}$$

where $\wedge$ denotes the wedge product [1–12]. The second Cartan structure equation defines the curvature differential form of Cartan ($R^a{}_b$) in terms of the spin connection:

$$R^a{}_b = d \wedge \omega^a{}_b + \omega^a{}_c \wedge \omega^a{}_b. \tag{27.5}$$

The spin connection and Riemann connection are related by the tetrad postulate of Cartan:

$$D_\mu q^a_\lambda = \partial_\mu q^a_\lambda + \omega^a{}_{\mu b} q^b_\lambda - \Gamma^\nu_{\mu\lambda} q^a_\nu = 0 \tag{27.6}$$

These equations imply [1–12] the Cartan identity:

$$d \wedge T^a + \omega^a{}_b \wedge T^b := R^a{}_b \wedge q^b \tag{27.7}$$

and the Cartan/Evans [1–12] dual identity

$$d \wedge \tilde{T}^a + \omega^a{}_b \wedge \tilde{T}^b := \tilde{R}^a{}_b \wedge q^b \tag{27.8}$$

where the tilde denotes the Hodge dual transform in four dimensions. Eq. (27.8) is an example of Eq. (27.7), and Eq. (27.7) is written in a spacetime with torsion and curvature non-zero. The Riemannian definition of torsion and curvature is:

$$[D_\mu, D_\nu]V^\rho = R^\rho{}_{\sigma\mu\nu}V^\sigma - T^\lambda_{\mu\nu}D_\lambda V^\rho \tag{27.9}$$

where:

$$[D_\mu, D_\nu] = -[D_\nu, D_\mu] \tag{27.10}$$

is the anti-symmetric commutator of covariant derivatives and V^ρ a vector (rank one tensor). The definition (27.9) can be extended to the commutator operator acting on any tensor [1–12] of any rank. The Riemannian equivalents of the Cartan and Cartan/Evans identities have been given in comprehensive detail [1–12], and are exact identities which state that the cyclic sums of three curvature tensors or Hodge duals thereof are identically equal to the same cyclic sum of definitions of the same tensors. The so called "first Bianchi identity" and "second Bianchi identity" of the obsolete twentieth century gravitational physics incorrectly omit torsion by using the symmetric connection (27.3) of Ricci and Levi-Civita.

In Section 27.2 the Cartan/Evans dual identity (27.1) is used to show by computer algebra that the Einstein field equation is incorrect geometrically because of its omission of torsion. In the context of this paper, all black hole metrics are either physically meaningless Ricci flat metrics, or, whenever the canonical energy momentum density tensor is non-zero, they are incorrect. The reason is that they produce a non-zero curvature $R^\kappa{}_\mu{}^{\mu\nu}$, but a zero torsion $T^{\kappa\mu\nu}$ by construction, so violate the Cartan/Evans dual identity. This situation has been remedied by basing the equations of motion of general relativity directly on the Cartan identity (homogeneous field equations of ECE theory) and the Cartan/Evans dual identity (inhomogeneous field equations of ECE theory). In Section 3, further criticisms of black hole theory are given by Dunning-Davies and Crothers, notably, the Schwarszchild metric has been incorrectly attributed to him, and the singularity analysis of the so called Schwarzschild metric is incorrect because it violates geometry. The introduction of mass M into any Ricci flat (vacuum) class of metric is self-contradictory. In these vacuum metrics mass is initially eliminated by construction, and cannot be assumed thereafter to be non-zero. A Ricci flat assumption means that the symmetric canonical energy momentum density tensor $T_{\mu\nu}$ is zero by the Einstein equation, so mass is zero by construction and cannot be arbitrarily asserted thereafter to be non-zero. Mass cannot appear in any vacuum metric, the vacuum extends throughout the whole universe, there is no source mass by construction. This is again a major flaw of black hole theory, which is all based on the incorrect and mis-named "Schwarzschild metric". The original 1916 paper of this author does not contain mass. There are no black holes in nature, no Hawking radiation, no dark matter. These are pseudo-scientific concepts of an incorrect and obsolete physics.

27.2 Computer Evaluation of Some Commonly Used Black Hole Metrics

In papers 93, 95 and 117 of the ECE series (www.aias.us) all solutions of the Einstein field equation were shown to violate the Cartan/Evans dual identity of geometry, eq. (27.1). The same methods are used in this section to test some commonly used metrics of black hole theory. The code is programmed to use the symmetric connection (27.3) of the Einstein field equation. So if the field equation were correct, it should produce a zero $R^\kappa{}_\mu{}^{\mu\nu}$. Such is not the case whenever the canonical energy momentum density $T_{\mu\nu}$ is finite. If

27.2 Computer Evaluation of Some Commonly Used Black Hole Metrics

$T_{\mu\nu}$ is zero there is no physics as argued in the introduction. The overall conclusion [1–12] is that the Einstein field equation is obsolete and has been replaced by the Einstein Cartan Evans (ECE) equations of general relativity and generally covariant classical and quantum unified field theory. The code (developed by Dr Eckardt and his group) also makes fundamental tests of each metric, notably tests its metric compatibility [1–12]:

$$D_\rho g_{\mu\nu} = D_\rho g^{\mu\nu} = 0 \qquad (27.11)$$

and its compatibility with the Ricci cyclic equation:

$$R_{\mu\nu\rho} + R_{\rho\mu\nu} + R_{\nu\rho\mu} = 0 \qquad (27.12)$$

which is mis-named "the first Bianchi identity" in the obsolete physics. If metrics fail either of these tests they are incorrect and self-inconsistent even within their assumptions. Surprisingly, it has been found that several obsolete metrics did fail these tests. One of these is the Kruskal metric of black hole theory as described in the following in more detail. Therefore not only is the standard physics obsolete, the standard mathematics are full of errors.

The metrics tested in this paper are as follows.

1) The wormhole metric [15]:

$$ds^2 = -c^2 dt^2 + d\ell^2 + (k^2 + \ell^2)d\ell^2. \qquad (27.13)$$

2) A wormhole metric with varying cosmological constant [15]:

$$ds^2 = -e^\gamma dt^2 + e^\mu dr^2 + r^2 d\Omega^2 \qquad (27.14)$$

where

$$e^{-\mu} = 1 - \frac{b(r)}{r}. \qquad (27.15)$$

Here $\gamma(r)$ is a red-shift function and $b(r)$ is a shape function.

3) The Morris Thorne wormhole [16]:

$$ds^2 = \left(1 - \frac{2m}{r}\right)\left(\frac{dr}{2\lambda r}\right)^2 - \left(1 - \frac{2m}{r}\right)dt^2. \qquad (27.16)$$

These authors also give a metric:

$$ds^2 = 2\left(1 - \frac{a}{r}\right)^{-1}\left(\frac{dr}{4\lambda r}\right)^2 - \frac{2}{r}dt^2 \qquad (27.17)$$

4) A flat wormhole metric form straight cosmic strings [17]:

$$ds^2 = dt^2 - d\sigma^2 - dZ^2 \qquad (27.18)$$

where

$$d\sigma^2 = \prod_i |\zeta - a_i|^{-8Gm_i} d\zeta d\zeta^*, \; \zeta = x + i\zeta, \; d\sigma^2 = du^2 + dv^2. \qquad (27.19)$$

This author also gives a metric for flat space-time with n wormholes and $2p$ ordinary cosmic strings:

$$d\sigma^2 = \frac{|\zeta^2 - c^2|^2}{|(\zeta^2 - a^2)^2 - b^4|} d\zeta d\zeta^*,$$
$$m_1 = m_2 = -\frac{1}{4G}, \qquad (27.20)$$
$$n = p = 2.$$

5) The Einstein Rosen Bridge [18]:

$$ds^2 = -a dt^2 + \frac{1}{a} dr^2 - r^2 d\Omega^2,$$
$$a = 1 - \frac{2m}{r} - \frac{\epsilon^2}{2r^2}. \qquad (27.21)$$

6) The massless Einstein Rosen Bridge [19].

$$ds^2 = a dt^2 - b(dr^2 + r^2 d\Omega^2),$$
$$a = \left(1 - \frac{m^2 + \beta^2}{4r^2}\right)^2 \left(1 + \frac{m}{r} + \frac{m^2 + \beta^2}{4r^2}\right)^{-2}, \qquad (27.22)$$
$$b = 1 + \frac{m}{r} + \frac{m^2 + \beta^2}{4r^2}.$$

These authors also give the general Morris Thorne wormhole:

$$ds^2 = e^{2\phi(r)} dt^2 - \frac{dr^2}{1 - b(r)/r} - r^2 d\Omega^2. \qquad (27.23)$$

7) Einstein Metric of 1936

$$ds^2 = \frac{\rho^2}{2m + \rho^2} dt^2 - 4(2m + \rho^2) d\rho^2 - (2m + \rho^2)^2 d\Omega^2 \qquad (27.24)$$

27.2 Computer Evaluation of Some Commonly Used Black Hole Metrics

where
$$\rho = r - 2m.$$

8) The Bekenstein Hawking Radiation Metric.
$$ds^2 = -\frac{u^2}{4m^2}dt^2 + du^2 + dX^2, \qquad (27.25)$$

where
$$r = 2m + \frac{u^2}{2m}$$

9) The Eddington Finkelstein Metric [20]
$$ds^2 = \left(1 - \frac{2m}{r}\right)dv^2 - 2dvdr - r^2 d\Omega^2 \qquad (27.26)$$

where
$$v = t + r + 2m \log_e \left(\frac{r}{2m} - 1\right)$$

10) The Kruskal Metric [20]
$$ds^2 = -32\frac{m^3}{r}\exp\left(-\frac{r}{2m}\right)(du^2 - dv^2),$$
$$u = \left(\frac{r}{2m} - 1\right)^{1/2} \exp\left(\frac{r}{4m}\right) \cosh\left(\frac{t}{4m}\right), \qquad (27.27)$$
$$v = \left(\frac{r}{2m} - 1\right)^{1/2} \exp\left(\frac{r}{4m}\right) \sinh\left(\frac{t}{4m}\right).$$

11) The Spherically Symmetric Metric in Four Dimensions
$$ds^2 = A dt^2 - 2B dt\, dr - C dr^2 - D d\Omega^2 \qquad (27.28)$$

12) Particular Example of a Spherically Symmetric Metric [1–12].
$$ds^2 = e^{2\alpha} dt^2 - e^{2\beta} dr^2 - r^2 d\Omega^2. \qquad (27.29)$$

The computer code found that the Cartan/Evans dual identity is not obeyed in general by spherically symmetric metrics in four dimensions if the connection is symmetric. This leads to the important inference that torsion must be non-zero for all spherically symmetric metrics in four dimensions, a general theorem. The connection cannot be symmetric in any spherically symmetric metric in four dimensions. Another general inference is that the Ricci tensor cannot be zero in physics, any metric that uses such an assumption is physically meaningless, i.e. the class of vacuum metrics is physically meaningless. This class includes the Eddington Finkelstein metric of black hole theory, and the central metric of black hole theory, the mis-named Schwarzschild metric. For these vacuum metrics the code showed that $R_\mu^{\kappa\mu\nu}$ is zero by construction. This is simply the result of assuming a Ricci flat condition initially. Both these metrics incorrectly include M, and as shown in the following section, the commonly used singularity analysis of both metrics is also incorrect. The Einstein Rosen bridge, the 1936 metric of Einstein, and the cosmic string metric are three further examples of physically meaningless vacuum metrics.

The worst error in black hole theory is the use of the Kruskal metric [1–12, 20]. The code found that this is mathematically erroneous because it produces a non-zero Ricci tensor and Einstein tensor, and also violates the Cartan/Evans dual identity. The Kruskal transformation, being a change of coordinates, must leave the physical energy momentum density unchanged, but it does not, it produces a non-zero Einstein tensor from an initially zero Einstein tensor. There exists no Bekenstein radiation or Hawking radiation in nature, because their metric (27.25) violates the Cartan/Evans dual identity and is geometrically incorrect and thus physically meaningless.

The Hayward Kim Lee metric (27.24) was found to fail the test of metric compatibility and Ricci cyclic compatibility, and also to violate the Cartan/Evans dual identity, so it is complete nonsense. This is also true of the general wormhole metric (27.23). Morris Thorne wormhole and similar.

It is therefore concluded that there are no black holes in nature. In paper 95 it was also shown that the Friedmann Lemaitre Robertson Walker (FLRW) metric of Big Bang theory is geometrically incorrect, again because of omission of torsion. In consequence there is no dark matter in nature, and none of the cosmologies based on the obsolete Einstein equation are correct. The Einstein Cartan Evans (ECE) version of general relativity is the correct one to use in physics and cosmology. Numerous other criticisms of black hole theory are given in section 27.3 by Dunning-Davies and Crothers. The details of the computations used in this section will be posted on www.aias.us as supplementary material for paper 120, and will be collected in a forthcoming publication [21].

Acknowledgments

The British Government is thanked for the award of a Civil List Pension to MWE and staffs of AIAS and TGA for many interesting discussions.

References

[1] M. W. Evans, "Generally Covariant Unified Field Theory" (Abramis Academic, 2005 onwards in softback), in six volumes to date, four published (see www.aias.us).
[2] L Felker, "The Evans Equations of Unified Field Theory" (Abramis 2007).
[3] K. Pendergast, "Crystal Spheres" (Abramis in press, www.aias.us).
[4] S. P. Carroll, "Spacetime and Geometry: an Introduction to General Relativity" (Addison Wesley, 2004, online notes).
[5] "The Universe of Myron Evans" (a film directed by Jack Iandoli and produced by Francesco Fucilla, to be release date Nov. 2008).
[6] M. W. Evans, Omnia Opera section of www.aias.us, the website of the British Civil List Scientist, 1992 to present.
[7] M. W. Evans (ed.), and M. W. Evans and S. Kielcih (eds.), Adv. Chem. Phys., volumes 85 and 119 (1992, 1993, 1997, 2001).
[8] M. W. Evans and L. B. Crowell, "Classical and Quantum Electrodynamics and the B(27.3) Field" (World Scientific, 2001).
[9] M. W. Evans and J.-P. Vigier, "The Enigmatic Photon" (Kluwer, Dordrecht, 1994 to 2002, hardback and softback), in five volumes.
[10] M. W. Evans, Found. Phys. Lett., fifteen papers on ECE theory (2003 to 2006).
[11] M. W. Evans, Acta Phys. Polonica, **33B**, 2211 (2007).
[12] M. W. Evans and H. Eckardt, Physica B, **400**, 175 (2007); M. W. Evans, Physica B, **403**, 517 (2008).
[13] This was the connection used by Einstein.
[14] Almost no twentieth century text in gravitational physics correctly recognizes torsion.
[15] F. Rahaman et al., gr-qc/0611133v1 (2006).
[16] S. A. Hayward, S.-W Kim and H. Lee, J. Korean Phys. Soc., **42**, 31 (2003).
[17] G. Clement, gr - gc/9607008v1 (1996).
[18] A. Einstein and N. Rosen, Phys. Rev., **48**, 73 (1935).
[19] K. K. Nandi and D. H. Xu, gr - qc/0410052v2 (2004).
[20] http://io.uwinnipeg.ca/~vincent/4500.6-001/Cosmology/Black_Holes.htm.
[21] M. W. Evans (ed.), "Criticims of the Einstein Field Equation" (Abramis in prep.), a collection of criticisms of the obsolete physics.

Conservation Theorem of Einstein Cartan Evans Field Theory

by

Myron W. Evans,
Alpha Institute for Advanced Study, Civil List Scientist.
(emyrone@aol.com and www.aias.us)

Abstract

The conservation theorems of physics are based on the tetrad postulate of differential geometry. It is shown that the tetrad postulate is invariant under the general coordinate transformation and that a frame invariant conservation theorem of physics can be based directly on the invariant tetrad postulate of geometry, as required by the philosophy of relativity. In special cases the conservation theorem reduces to the various conservation laws of physics, notably the conservation of canonical energy/momentum density. The conservation theorem and conservation laws apply to all the equations of physics derivable from ECE field theory, these include the wave equations of physics, also derivable from the tetrad postulate.

Keywords: Tetrad postulate, conservation theorem of ECE theory, conservation laws of physics.

28.1 Introduction

In gauge theory the conservation laws of physics are based on the Noether Theorem, which is derived from the invariance of action under various symmetry operations [1] in a Lagrangian formalism. These laws include the conservation of canonical energy momentum density, the covariant form of which is used in the Einstein field equation. The latter is well known [2–10] to be obsolete because of its neglect of space-time torsion, an essential part of differential geometry, and has been replaced by the well accepted ECE engineering model. The latter is based directly on Cartan's differential geometry both in respect of dynamics and electrodynamics, and the torsion plays a central role in both subjects. The ECE model allows new technologies to be developed using the concept of spin connection resonance (for example ECE papers 63, 92, 94 and 107 on www.aias.us). The physics of ECE theory is shown in Section 28.2 to be based on a general conservation theorem constructed in turn from the tetrad postulate [11] of differential geometry. The tetrad postulate is the most fundamental theorem of differential geometry, and states that the complete vector field is independent of its components and basis elements. In three dimensions for example a complete vector field V is the same if expressed in cartesian or circular polar coordinates. The same is true of a complete vector field in n dimensions. The tetrad postulate is used throughout differential geometry, and all the equations of Cartan geometry depend on it. The postulate may be seen as the link between Cartan and Riemann geometry. In view of its importance a proof of it is given in Section 28.2 with all details. In Section 28.3, it is shown that the tetrad postulate is invariant under the general coordinate transformation, i.e. is frame invariant. This property means that the postulate is the same for an observer moving arbitrarily with respect to another. In Section 28.4 the fundamental conservation theorem of ECE theory is based directly on the tetrad postulate, and is developed to give the conservation laws and wave equations of physics. In ECE theory the latter therefore obey the conservation laws by construction, because in ECE theory the fundamental wave equations are derived from the tetrad postulate by developing the latter into the ECE lemma and wave equation [2–10].

28.2 Proof of the Tetrad Postulate

Consider the complete vector field X in n dimensions and in a space-time with torsion and curvature. Denote the covariant derivative of the compete vector field by DX. In Riemannian geometry this quantity is expressed as [2–11]:

$$DX = D_\mu X^\nu dx^\mu \otimes \partial_\nu \tag{28.1}$$

where X^ν are the components of X, D_μ are the components of D, dx^μ are the basis elements of X, and ∂_ν are the basis elements of X^ν. The covariant derivative is defined as:

$$D_\mu X^\nu = \partial_\mu X^\nu + \Gamma^\nu_{\mu\lambda} X^\lambda \tag{28.2}$$

where $\Gamma^\nu_{\mu\lambda}$ is the connection. Therefore:

$$DX = (\partial_\mu X^\nu + \Gamma^\nu_{\mu\lambda} X^\lambda) dx^\mu \otimes \partial_\nu. \tag{28.3}$$

28.2 Proof of the Tetrad Postulate

In Cartan geometry a tangent Minkowski space-time is defined at point P to the base manifold and the covariant derivative is defined in terms of the spin connection $\omega^a_{\mu b}$:

$$D_\mu X^a = \partial_\mu X^a + \omega^a_{\mu b} X^b. \tag{28.4}$$

The quantity DX is the same in Riemannian and Cartan geometry, so:

$$DX = (\partial_\mu X^a + \omega^a_{\mu b} X^b) dx^\mu \otimes \hat{e}_a \tag{28.5}$$

where $\hat{e}_a$ is the basis element of the component X^a. By construction:

$$\hat{e}_a = q^\sigma_a \partial_\sigma \tag{28.6}$$

$$X^a = q^a_\nu X^\nu \tag{28.7}$$

where q^a_ν is the Cartan tetrad [2–11] and where q^σ_a is the inverse tetrad. These are related by:

$$q^\sigma_a q^a_\nu = \delta^\sigma_\nu \tag{28.8}$$

where:

$$\delta^\sigma_\nu = 1, \quad \sigma = \nu, \tag{28.9}$$

$$\delta^\sigma_\nu = 0, \quad \sigma \neq \nu. \tag{28.10}$$

Note carefully that by convention, there is no summation over repeated a indices in Eq. (28.8), the notation of which means that when:

$$\sigma = \nu \tag{28.11}$$

then:

$$q^\sigma_a q^a_\sigma = 1. \tag{28.12}$$

using Eqs. (28.6) and (28.7) in Eq. (28.5):

$$DX = (\partial_\mu(q^a_\nu X^\nu) + \omega^a_{\mu b} q^b_\nu X^\nu) dx^\mu \otimes (q^\sigma_a \partial_\sigma) \tag{28.13}$$

which may be re-expressed as:

$$DX = (q_a^\sigma q_\nu^a \partial_\mu X^\nu + q_a^\sigma X^\nu \partial_\mu q_\nu^a + q_a^\sigma \omega_{\mu b}^a q_\lambda^b X^\lambda) dx^\mu \otimes \partial_\sigma \quad (28.14)$$

Now compare Eqs. (28.3) and (28.14) when:

$$\sigma = \nu. \quad (28.15)$$

In this case, using Eq. (28.12), Eq. (28.14) becomes:

$$DX = (\partial_\mu X^\nu + q_a^\nu X^\lambda \partial_\mu q_\lambda^a + q_a^\nu \omega_{\mu b}^a q_\lambda^b X^\lambda) dx^\mu \otimes \partial_\nu. \quad (28.16)$$

Therefore we obtain:

$$\Gamma_{\mu\lambda}^\nu = q_a^\nu \partial_\mu q_\lambda^a + q_a^\nu \omega_{\mu b}^a q_\lambda^b. \quad (28.17)$$

Multiply both sides of Eq. (28.17) by q_λ^a to find:

$$\partial_\mu q_\lambda^a + \omega_{\mu b}^a q_\lambda^b = q_\nu^a \Gamma_{\mu\lambda}^\nu \quad (28.18)$$

i.e.:

$$\partial_\mu q_\lambda^a + \omega_{\mu b}^a q_\lambda^b - q_\nu^a \Gamma_{\mu\lambda}^\nu = 0. \quad (28.19)$$

Using the rule [2–11] for the covariant derivative of the tetrad, a rank two mixed-index tensor, Eq. (28.19) is:

$$D_\mu q_\lambda^a = 0 \quad (28.20)$$

which is the tetrad postulate Q.E.D.

28.3 Invariance of the Tetrad Postulate

In this section the tetrad postulate is subjected to a well defined general coordinate transformation [2–11]. The fundamental idea of relativity theory is that the equations of physics retain their format under the general coordinate transformation. They must be generally covariant. In Riemann geometry there are base manifold indices (labelled by Greek subscripts and superscripts), and in Cartan geometry there are additional Latin indices of the tangent space-time. So in Cartan geometry the general coordinate transformation consists in general of transformation matrices with base manifold and tangent indices [2–11]. The transformation matrix in the tangent space-time is the Lorentz transformation matrix $\Lambda_a^{a'}$. In the base manifold there occur transformation matrices such as $\partial x^\mu / \partial x^{\mu'}$. Therefore Eq. (28.20) transforms according to the rule for the coordinate transformation of a rank three mixed index tensor:

$$(D_\mu q_\nu^a)' = \left(\Lambda_a^{a'} \frac{\partial x^\mu}{\partial x^{\mu'}} \frac{\partial x^\nu}{\partial x^{\nu'}}\right) D_\mu q_\nu^a = 0 \qquad (28.21)$$

It follows that the tetrad postulate is true in any frame of reference, and is an invariant under the general coordinate transformation, Q.E.D.

28.4 Conservation Theorem of ECE Theory

For practical applications the tetrad postulate is developed in the base manifold using Eq. (28.8). Therefore:

$$q_\nu^\mu = \delta_\nu^\mu \qquad (28.22)$$

in the base manifold. The contravariant and covariant form of Eq. (28.22) are obtained using the inverse metric and metric respectively, to give:

$$q^{\mu\nu} = g^{\mu\sigma} \delta_\sigma^\nu = g^{\mu\nu}, \qquad (28.23)$$

$$q_{\mu\nu} = g_{\mu\nu}. \qquad (28.24)$$

The general conditions of metric compatibility [2–11] follow:

$$D^\sigma g_{\mu\nu} = 0, \qquad (28.25)$$

$$D_\sigma g^{\mu\nu} = 0, \qquad (28.26)$$

and these are forms of the ECE conservation theorem. The fundamental and most general conservation theorem of ECE theory is therefore the tetrad postulate itself.

The conservation of canonical energy momentum density used in the Einstein field equation is derived as a special case of Eq. (28.25) when:

$$\sigma = \mu. \qquad (28.27)$$

In general, the canonical energy momentum density is defined in ECE theory as:

$$T_\mu^a = T^{(0)} q_\mu^a \qquad (28.28)$$

so using Eq. (28.25):

$$D^\mu T_{\mu\nu} = 0. \tag{28.29}$$

This is the covariant law for conservation of energy momentum density which is obtainable from the invariance of action under space-time translation in the Noether Theorem formalism [1]. In ECE theory it is derived straightforwardly from geometry. During the course of development of ECE theory it has been proven rigorously [2–10] that the tetrad postulate may be developed into the ECE Lemma:

$$\Box q_\mu^a = R q_\mu^a \tag{28.30}$$

where R is a well defined scalar eigen-value with the units of curvature (inverse square metres). Therefore the general law for conservation of canonical energy - momentum density is also a wave equation:

$$\Box T_\mu^a = R T_\mu^a \tag{28.31}$$

This may be a wave equation of quantum mechanics or statistical mechanics. It follows that all process of quantum mechanics and statistical mechanics in ECE theory automatically obey all the conservation theorems of physics.

The law of conservation of canonical angular energy momentum density follows from the definition [1–10]:

$$J^{\kappa\mu\nu} = -\frac{1}{2}(T^{\kappa\mu} x^\nu - T^{\kappa\nu} x^\mu) \tag{28.32}$$

which is a rank three tensor density in the base manifold. The space-time torsion tensor in the base manifold [2–11] is also a rank three tensor related to a curvature tensor through the Cartan Evans dual identity:

$$D_\mu T^{\kappa\mu\nu} = R^\kappa{}_\mu{}^{\mu\nu}. \tag{28.33}$$

By index contraction (summation over internal indices), a rank two curvature tensor may be defined as follows:

$$R^{\kappa\nu} = R^\kappa{}_\mu{}^{\mu\nu} \tag{28.34}$$

and by hypothesis similar to that of Einstein this tensor is made proportional to the rank two canonical energy momentum density tensor through Einstein's constant k:

$$R^{\kappa\nu} = k T^{\kappa\nu} = D_\mu T^{\kappa\mu\nu}. \tag{28.35}$$

28.4 Conservation Theorem of ECE Theory

By similar hypothesis:

$$T^{\kappa\mu\nu} = kJ^{\kappa\mu\nu} \tag{28.36}$$

so we obtain:

$$D_\mu J^{\kappa\mu\nu} = T^{\kappa\nu}. \tag{28.37}$$

This is a field equation that automatically includes a non-zero torsion as required by the dual identity (28.33) [2–10]. It follows that the law for conservation of canonical angular energy momentum density is:

$$D^\mu T_{\mu\nu} = D^\mu(D^\kappa J_{\mu\kappa\nu}) = 0. \tag{28.38}$$

Similarly the charge current density in general is defined as:

$$J^a_\mu = J^{(0)} q^a_\mu \tag{28.39}$$

and leads to the covariant continuity equation:

$$D^\mu J_{\mu\nu} = 0 \tag{28.40}$$

(see also paper 116 of www.aias.us). Finally the fundamental hypothesis leading to the equations of classical dynamics in ECE tehory [2–10] is:

$$A^a_\mu = A^{(0)} q^a_\mu \tag{28.41}$$

where A^a_μ is the electromagnetic potential field. It follows that the electromagnetic potential in the base manifold is the metric within a factor $A^{(0)}$, where $cA^{(0)}$ is the primordial voltage of ECE theory [2–10]. The electromagnetic field tensor in ECE theory is a rank three tensor proportional to the space-time torsion, so from Cartan's structure equation:

$$F^{\kappa\mu\nu} = \partial^\mu A^{\kappa\mu} - \partial^\nu A^{\kappa\mu} + \omega^{\kappa\mu}{}_\lambda A^{\lambda\nu} - \omega^{\kappa\nu}{}_\lambda A^{\lambda\mu}. \tag{28.42}$$

For example, the electric field in Coulomb's law [2–10] is:

$$\mathbf{E} = E^{010}\mathbf{i} + E^{020}\mathbf{j} + E^{030}\mathbf{k} \tag{28.43}$$

in the Cartesian basis. Each electric field component is a component of orbital torsion, and in general:

$$F^{010} = \partial^1 A^{00} - \partial^0 A^{01} + \omega^{01}{}_\lambda A^{\lambda 0} - \omega^{00}{}_\lambda A^{\lambda 1} \tag{28.44}$$

where there is summation over repeated λ indices. Therefore the components of the potential tensor appearing in Eq. (28.44) are components of the inverse metric tensor in a space-time with both curvature an torsion present in general:

$$A^{\mu\nu} = A^{(0)} g^{\mu\nu}. \tag{28.45}$$

In vector format, Eq. (28.44) reduces to:

$$\boldsymbol{E} = -\boldsymbol{\nabla}\phi - \frac{\partial \boldsymbol{A}}{\partial t} + \phi\boldsymbol{\omega} - \omega\boldsymbol{A}. \tag{28.46}$$

where ω is the spin connection scalar and $\boldsymbol{\omega}$ is the spin connection vector. The Coulomb law itself reduces to:

$$\boldsymbol{\nabla} \cdot \boldsymbol{E} = \rho/\epsilon_0 \tag{28.47}$$

and Eqs. (28.46) and (28.47) give rise to spin connection resonance [2–10]. It is seen that this process obeys the conservation theorems of physics. The same is true for the other ECE equations of classical electrodynamics, and for all the ECE equations of classical dynamics. The same is also true for the ECE equations of wave mechanics and statistical mechanics. At spin connection resonance the primordial voltage $cA^{(0)}$ (which fills the vacuum in ECE theory and which is observable in the well known radiative corrections) is greatly amplified, giving rise to the possibility of electric power from space-time. This process obeys all the conservation theorems of physics.

Acknowledgments

The British Government is thanked for a Civil List Pension and the staffs of AIAS and TGA for many interesting discussions.

References

[1] L. H. Ryder, "Quantum Field Theory" (Cambridge Univ. Press, 1996, 2nd. Ed.), chapter 3.
[2] M. W. Evans, "Generally Covariant Unified Field Theory" (Abramis 2005 onwards), in six volumes to date, four published to date, see also www.aias.us.
[3] L. Felker, "The Evans Equations of Unified Field Theory" (Abramis, 2007).
[4] K. Pendergast, "Crystal Spheres" (Abramis in press, see also www.aias.us).
[5] M. W. Evans, Omnia Opera section of www.aias.us.
[6] F. Fucilla, producer and director, "The Universe of Myron Evans" (Nov. 2008).
[7] M. W. Evans et al., Found. Phys. Lett., fifteen papers on ECE theory, 2003 to 2005.
[8] M. W. Evans, Acta Phys. Polonica, **33B**, 2211 (2007).
[9] M. W. Evans and H. Eckardt, Physica B, **400**, 175 (2007).
[10] M. W. Evans, Physica B, **403**, 517 (2008).
[11] S. P. Carroll, "Space-time and Geometry: an Introduction to General Relativity" (Addison-Wesley, New York, 2005, and online lecture notes).

29

On the Symmetry of the Connection in Relativity and ECE Theory

by

Myron W. Evans,
Alpha Institute for Advanced Study, Civil List Scientist.
(emyrone@aol.com and www.aias.us)

Abstract

It is shown that the connection in relativity theory must always be anti-symmetric and that relativity theory must be based on a non-zero space-time torsion as in the Einstein Cartan Evans (ECE) field theory. These results follow straightforwardly from the action of the commutator of covariant derivatives on any tensor in any space-time. The commutator is anti-symmetric by construction and generates the torsion and curvature tensors with anti-symmetric connection. The assumption of a symmetric connection as used in the standard model is irretrievably incorrect, meaning that the Einsteinian era in gravitational theory and cosmology is over. The ECE equations are geometrically correct and provide new cosmologies and technologies.

Keywords: Anti-symmetric connection, ECE theory, space-time torsion.

29.1 Introduction

The philosophy of relativity is that of objectivity, without which there is no natural or life science. The central thesis of relativity is that physics, or natural philosophy, is geometry. Physics is also causal in nature. These ideas go back to ancient times, but the most well known example of relativity is the theory developed from the late eighteen eighties to 1915 by many scientists. General relativity is the term used to describe the type of relativity suggested by Einstein from about 1906 onwards, when he developed the metric diag

(−1, 1, 1, 1) of the Minkowski space-time into one which in general is space-time dependent. Tensor analysis had been proposed mainly by Ricci and Levi-Civita and published in about 1900. The idea of a space-time connection goes back via Christoffel and others to Riemann in the early nineteenth century, and the connection appears in the definition of the covariant derivative. An anti-symmetric commutator made up of covariant derivatives may act on any tensor [1] to produce torsion and curvature tensors. In ECE theory [2–10] the torsion has assumed central importance and its role in the natural and life sciences recognized. This is not the case in the Einsteinian era of about 1915 to 2003, when ECE theory first began to be developed. The reason is that the connection in the Einsteinian era was incorrectly assumed to be symmetric. In Section 2 it is shown straightforwardly that the connection must be anti-symmetric in both the torsion and curvature tensors, and that the torsion must always be non-zero in any theory of relativity. It follows that no physical inference may be drawn from the Einsteinian era in gravitation and cosmology. The proof of the anti-symmetry of the connection is simple, and it is not clear why a symmetric connection has been incorrectly used for over a century. The situation can be rescued by adopting the well known ECE equations of dynamics, electrodynamics and quantum mechanics [2–10]. In these equations a symmetric connection is not assumed, and the torsion is the centrally important concept.

Before proceeding to Section 2 a few historical remarks are given for ease of reference. It appears that the assumption of a symmetric connection was first made in about 1900 by Ricci and Levi-Civita for ease of calculation. This assumption should be researched further by historians of science, and it may emerge that Christoffel or Riemann assumed such a symmetry. It is well known that Einstein relied on advice by Grossman and that Einstein frequently read the 1900 paper, corresponding with Levi-Civita. The latter corrected some errors in Einstein's use of tensor analysis. In attempting to develop his 1915 field equation, Einstein made several errors and false turns before finally making a type of covariant Noether Theorem proportional to an identity of geometry known in the Einsteinian era as "the second Bianchi identity". The proportionality constant is the Einstein constant k as is well known. In the ECE era it has been recognized that this is not an identity because it omits space-time torsion. Similarly the "first Bianchi identity" of the Einsteinian era also omits torsion and is not correct. There is only one true identity, and it was first given by Cartan in the early twenties. In short hand notation (with indices left out for clarity), the Cartan Bianchi identity is:

$$D \wedge T := R \wedge q. \tag{29.1}$$

Here $D\wedge$ represents the covariant exterior derivative, defined by:

$$D\wedge := d\wedge + \omega\wedge \tag{29.2}$$

where $d\wedge$ is the exterior derivative and ω is the spin connection form of differential geometry. The symbol T denotes the torsion form, R denotes the curvature form, and q denotes the Cartan tetrad. In the ECE era it has also been recognized that there exists the Cartan Evans dual identity:

$$D \wedge \widetilde{T} := \widetilde{R} \wedge q \qquad (29.3)$$

where $\widetilde{T}$ is the Hodge dual of the torsion form, and $\widetilde{R}$ is the Hodge dual of the curvature form. Neglect of torsion leads to a violation of the dual identity [2–10], thus ending the Einsteinian era.

The fundamental reason for this error is the assumption by Einstein that the connection is symmetric. It was known as early as 1918, in independent criticisms by Bauer and Schroedinger, [11], that there was something amiss with the Einstein field equation of 1915. However, the concept of torsion was not fully developed until about 1922, when Cartan and Maurer gave the first structure equation:

$$T = D \wedge q = d \wedge q + \omega \wedge q. \qquad (29.4)$$

In tensor notation this is equivalent to [1–10]:

$$T^{\kappa}_{\mu\nu} = \Gamma^{\kappa}_{\mu\nu} - \Gamma^{\kappa}_{\nu\mu} \qquad (29.5)$$

where $T^{\kappa}_{\mu\nu}$ is the connection. If the latter is arbitrarily forced to be symmetric:

$$\Gamma^{\kappa}_{\mu\nu} = ? \; \Gamma^{\kappa}_{\nu\mu} \qquad (29.6)$$

the torsion vanishes. In the Einsteinian era therefore the torsion was seen as a complication which was arbitrarily removed. It was shown in paper 93 ff. on www.aias.us that this removal of torsion leads to a violation of geometry, i.e. a violation of the Cartan Evans dual identity. In Section 29.2 this conclusion is reinforced using a straightforward demonstration that the connection must be anti-symmetric, both in the torsion and curvature tensors. It appears that the erroneous use of a symmetric connection was perpetrated uncritically throughout the Einsteinian era, either by lack of scholarship, by peer pressure, or by ignoring new developments. The result is that pseudo-scientific concepts such as big bang, black hole theory, dark matter theory and associated paraphenalia have proliferated, and are being taught as if they were mathematically correct. The intense international interest in ECE theory however has brought this era to an end and ECE theory has been adapted in the industrial sector. Its equations are able to provide cosmologies and technologies based on correct geometry. The use of correct geometry is of course a fundamental requirement of relativity theory.

29.2 Anti-Symmetry of the Connection

Define the covariant derivative of a vector V^ν of any dimension in any space-time as

$$D_\mu V^\nu = \partial_\mu V^\nu + \Gamma^\nu_{\mu\lambda} V^\lambda. \qquad (29.7)$$

The commutator of the covariant derivative is anti-symmetric by construction:

$$[D_\mu, D_\nu] = -[D_\nu, D_\mu] \qquad (29.8)$$

and may operate on any tensor in any space-time of any dimension. Let the commutator (29.8) operate on the vector V^ρ. The result is well known [1–10] to be:

$$[D_\mu, D_\nu]V^\rho = R^\rho{}_{\sigma\mu\nu} V^\sigma - T^\lambda_{\mu\nu} D_\lambda V^\rho \qquad (29.9)$$

where $T^\lambda_{\mu\nu}$ is the torsion tensor:

$$T^\lambda_{\mu\nu} := \Gamma^\lambda_{\mu\nu} - \Gamma^\lambda_{\nu\mu} \qquad (29.10)$$

and

$$R^\rho{}_{\sigma\mu\nu} := \partial_\mu \Gamma^\rho_{\nu\sigma} - \partial_\nu \Gamma^\rho_{\mu\sigma} + \Gamma^\rho_{\mu\lambda}\Gamma^\lambda_{\nu\sigma} - \Gamma^\rho_{\nu\lambda}\Gamma^\lambda_{\mu\sigma} \qquad (29.11)$$

is the curvature tensor. These tensors are anti-symmetric in their last two indices by construction:

$$T^\lambda_{\mu\nu} = -T^\lambda_{\nu\mu} \qquad (29.12)$$
$$R^\rho{}_{\sigma\mu\nu} = -R^\rho{}_{\sigma\nu\mu}. \qquad (29.13)$$

If it is assumed that:

$$\mu = \nu \qquad (29.14)$$

the commutator operator becomes the null operator, and the curvature and torsion tensors BOTH vanish. It is incorrect to assume that the torsion vanishes and that the curvature does not vanish when the assumption (29.14) is made. This is an error that was perpetrated throughout the Einsteinian era for over a century. The torsion is defined as:

$$[D_\mu, D_\nu]V^\rho = -T^\lambda_{\mu\nu} D_\lambda V^\rho + \ldots := -\left(\Gamma^\lambda_{\mu\nu} - \Gamma^\lambda_{\nu\mu}\right) D_\lambda V^\rho + \ldots \qquad (29.15)$$

and is anti-symmetric, and it follows that:

$$\Gamma^\lambda_{\mu\nu} = -\Gamma^\lambda_{\nu\mu}. \tag{29.16}$$

The connection in the theory of relativity must be anti-symmetric whenever it occurs, Q.E.D.

It is important to note that the connection is also anti-symmetric in the curvature tensor. There is only one connection symmetry, which is generated directly by the commutator acting on a vector or tensor. The commutator of covariant derivatives arises by the well known fact [1] that the covariant derivative of any tensor in any space-time in a given direction measures how much the tensor changes relative to what it would have been if it had been parallel transported. The covariant derivative of a tensor in the direction along which it is parallel transported is zero [1]. The commutator of covariant derivatives measures the difference between parallel transporting the tensor clockwise and anti-clockwise. In a flat space-time with no connection such a commutator is zero, i.e. :

$$[\partial_\mu, \partial_\nu] = -[\partial_\nu, \partial_\mu] = 0. \tag{29.17}$$

For the arbitrary tensor in any space-time, the commutator operates on the tensor to produce the torsion and curvatures as follows:

$$\begin{aligned}[][D_\rho, D_\sigma]X^{\mu_1...\mu_k}_{\nu_1...\nu_l} = &-T^\lambda_{\rho\sigma}D_\lambda X^{\mu_1...\mu_k}_{\nu_1...\nu_l} + R^{\mu_1}{}_{\lambda\rho\sigma}X^{\lambda\mu_2...\mu_k}_{\nu_1...\nu_l} \\
&+ R^{\mu_2}{}_{\lambda\rho\sigma}X^{\mu_1\lambda...\mu_k}_{\nu_1...\nu_l} + ... - R^\lambda_{\nu_1\rho\sigma}X^{\mu_1...\mu_k}_{\lambda\nu_2...\nu_l} \\
&- R^\lambda_{\nu_2\rho\sigma}X^{\mu_1...\mu_k}_{\nu_1\lambda...\nu_l} - ... \end{aligned} \tag{29.18}$$

So it is always incorrect to use a null torsion, because that means a null commutator and a null curvature (i.e. removing the torsion removes all the information needed for relativity theory).

For example, let:

$$\mu = 0, \ \nu = 1, \ \rho = 2 \tag{29.19}$$

in Eq. (29.9). Then:

$$[D_0, D_1]V^2 = R^2{}_{\sigma 01}V^\sigma - T^\lambda{}_{01}D_\lambda V^2. \tag{29.20}$$

It follows that:

$$T^\lambda{}_{01} = \Gamma^\lambda{}_{01} - \Gamma^\lambda{}_{10} = -T^\lambda{}_{10} \tag{29.21}$$

i.e.
$$\Gamma^\lambda{}_{01} = -\Gamma^\lambda{}_{10}. \quad (29.22)$$

The curvature tensor is:
$$R^2{}_{\sigma 01} = -R^2{}_{\sigma 10} \quad (29.23)$$

For example, when:
$$\sigma = 0 \quad (29.24)$$

the curvature tensor is:
$$R^2{}_{001} = \partial_0 \Gamma^2_{10} - \partial_1 \Gamma^2_{00} + \Gamma^2_{0\lambda}\Gamma^\lambda_{10} - \Gamma^2_{1\lambda}\Gamma^\lambda_{00} \quad (29.25)$$

where there is summation over repeated λ indices. By anti-symmetry in μ and ν:
$$\partial_0 \Gamma^2_{10} = -\partial_1 \Gamma^2_{00} = -\partial_0 \Gamma^2_{01} = \partial_1 \Gamma^2_{00} \quad (29.26)$$

i.e.
$$\Gamma^2_{00} = -\Gamma^2_{00} = 0, \quad (29.27)$$
$$\Gamma^2_{01} = -\Gamma^2_{10} \neq 0. \quad (29.28)$$

The symmetric connections are zero, and the other connections are anti-symmetric Q.E.D.

Similarly we may systematically consider all other μ and ν indices:
$$\left.\begin{array}{l}\mu = 0, \nu = 2; \mu = 1, \nu = 3; \\ \mu = 0, \nu = 3; \mu = 0, \nu = 2;\end{array}\right\} \quad (29.29)$$

to find that:
$$\Gamma^\lambda_{00} = \Gamma^\lambda_{11} = \Gamma^\lambda_{22} = \Gamma^\lambda_{33} = 0 \quad (29.30)$$

for all λ. It is seen immediately that any metric of the Einsteinian era that does not obey the symmetry of Eq. (29.30) is incorrect geometrically, and cannot produce any meaningful results in physics. It is important to realize that all the commonly taught metrics of the Einsteinian era are incorrectly based on a symmetric connection, for example the mis-named [2–10] Schwarzschild metric, The Robertson Walker metric of big bang, all black hole metrics, and so on. So the Einsteinian era must be discarded as a matter of urgency and replaced by the ECE era.

Acknowledgments

The British Government is thanked for a Civil List Pension (2005) and Armorial Bearings (2008). The staffs of AIAS and TGA and many others are thanked for interesting discussions.

References

[1] S. P. Carroll, "Space-time and Geometry: an Introduction to General Relativity" (Addison Wesley, New York, 2004), chapters 1 to 3.

[2] M. W. Evans, "Generally Covariant Unified Field Theory: the Geometrization of Physics" (Abramis 2005 onwards), multi volume monograph (see www.aias.us).

[3] L. Felker, "The Evans Equations of Unified Field Theory" (Abramis 2007).

[4] K. Pendergast, "Crystal Spheres" (www.aias.us, Abramis to be published).

[5] K. Pendergast, "The Life of Myron Evans" (see www.aias.us).

[6] F. Fucilla (Director), "The Universe of Myron Evans" (scientific film, 2008).

[7] M. W. Evans, source ECE papers on www.aias.us.

[8] H. Eckardt, S. Crothers, L. Felker and others, ECE educational articles on www.aias.us.

[9] M. W. Evans and J.-P. Vigier, "The Enigmatic Photon" (Kluwer 1994 to 2002, softback and hardbamc), in five volumes.

[10] M. W. Evans and S. Kielich, "Modern Non-Linear Optics" (first and second editions in six volumes, (Wiley, New York, 1992, 1993, 1997 and 2001).

[11] C. Alley, address to the Czech Assembly 2006, Wheeler Fest 2006.

www.ingramcontent.com/pod-product-compliance
Ingram Content Group UK Ltd.
Pitfield, Milton Keynes, MK11 3LW, UK
UKHW051248180426
11947UKWH00020B/1599